MICRO TOTAL ANALYSIS SYSTEMS 2002

Micro Total Analysis Systems 2002

Proceedings of the μTAS 2002 Symposium,
held in Nara, Japan,
3–7 November 2002

Volume 2

edited by

Yoshinobu Baba
The University of Tokushima, Tokushima, Japan

Shuichi Shoji
Waseda University, Tokyo, Japan

and

Albert van den Berg
University of Twente, Enschede, The Netherlands

SPRINGER SCIENCE+BUSINESS MEDIA, B.V.

Library of Congress Cataloging-in-Publication Data.

ISBN 978-94-010-3953-6 ISBN 978-94-010-0504-3 (eBook)
DOI 10.1007/978-94-010-0504-3

Printed on acid-free paper

PROGRAM COMMITTEE
µTAS2002

PREFACE

The Sixth International Conference on Miniaturized Chemical and Biochemical Analysis Systems, known as μTAS2002, will be fully dedicated to the latest scientific and technological developments in the field of miniaturized devices and systems for realizing not only chemical and biochemical analysis but also synthesis. The first μTAS meeting was held in Enschede in 1994 with approximately 160 participants, bringing together the scientists with background in analytical and biochemistry with those with Micro Electro Mechanical Systems (MEMS) in one workshop. We are grateful to Piet Bergveld and Albert van den Berg of MESA Research Institute of the University of Twente for their great efforts to arrange this exciting first meeting. The policy of the meeting was succeeded by late Prof. Dr. Michael Widmer in the second meeting, μTAS'96 held in Basel with 275 participants. The first two meetings were held as informal workshops. From the third workshop, μTAS'98 (420 participants) held in Banff, the workshop had become a worldwide conference. Participants continued to increase in μTAS2000 (about 500 participants) held in Enschede and μTAS2001 (about 700 participants) held in Monterey. The number of submitted papers also dramatically increased in this period from 130 in 1998, 230 in 2000 to nearly 400 in 2001. From 2001, μTAS became an annual symposium. The steering committee meeting held in Monterey, confirmed the policy of former μTAS that quality rather than quantity would be the key-point and that the parallel-session format throughout the 3.5 days would be maintained. The upcoming μTAS2002 is the first meeting specialized on this field that is held in Asia. Again there is a very large increase in submitted abstracts (over 450), especially from the Asian Countries such as Japan, Korea, Taiwan, and China. The program committee had a very hard work to do and had to select 66 abstracts as oral presentations and more than 250 abstracts as poster presentations, an acceptance rate slightly over 70 %. In addition, we have excellent invited speakers from North America, Europe and Asia. The co-chairs of μTAS2002 wish everyone will understand the consistent μTAS policy and agree that the program committee did a very excellent job in selecting abstracts and arranging the program. This is the first μTAS meeting in Asia, we do hope the meeting will contribute to spread the latest knowledge of μTAS to Asia as well as North America and Europe.

Recent trend that appeared in the previous μTAS2001 is that the contributions are much more focused on the actual applications than on the basic studies. Microfluidc components, devices and integrated systems are still a main stream in the MEMS field. A droplet base fluidics, nanofluidics and a gas-liquid two phase flow are the new topics. Integration of the flow devices for a total system is obvious. Plastic micro fabrication, using PDMS are currently used and miniaturization technologies proceeds to realize the structure of nanometer scale. The interest in materials for biochemical and biomedical applications and nanotechnologies is showing remarkable growth. The genomics and proteomics analysis and separation science remain high from the previous meeting. The interests of the clinical diagnosis and cellular analysis have considerably increased. Devices and systems developments that address highly sensitive measurements of chemical and biochemical molecules continued in the micro-optics, mass spectroscopy and electrochemical detection.

We would like to express our great gratitude to program committee members of the 12 steering committee members and Takehiko Kitamori for their really tough work. We would like to greatly thank D. Jed Harrison, Albert van den Berg and J. Mike Ramsey for sharing their experiences with previous meetings and providing many useful suggestions. We also thank Hiroyuki Fujita, Takehiko Kitamori and other local organizing committee members. The excellent arrangement could not be done without the efforts by June Echizen of the secretariat of μTAS2002 and her colleagues. Finally we wish to thank the most important people of all, the invited speakers, the technical contributors and participants! We hope all participants will enjoy μTAS2002.

Yoshinobu Baba
Shuichi Shoji
μTAS2002 Co-Chairman
August 10, 2002

CONTENTS

VOLUME 1

Day 1

Keynote 1

Cell Growth and Monitoring

Separation, General

Gels for Biochemical Analysis

Micro Analysis Systems

Keynote 2

DNA Separation

Droplet Base Fluidics

Day 1 Posters

Fluid Mechanics & Design Tools

Micropumps & Microvalves

Clinical Diagnosis

Micro-Optical Systems

Day 2

Plenary 1

DNA Assay

Nano Fluidics

Magnetic

Materials

Plenary 2

Materials

VOLUME 2

Day 3

Plenary 3

Proteomics

Microfluidics II

Sample Preparation

Detection Systems

Plenary 4

Cell Manipulation in Flows

Microfluidic Components

Drug Discovery

Biochemical Applications

Novel Quantification Strategies

Day 4

Plenary 5

Single Cell Analysis

Nano-microfabrication

DNA System

Microfluidics III

Beads in Biochemical Microfluidics

Patrick Griss, Helene Andersson, Wouter van der Wijngaart, Göran Stemme
Royal Institute of Technology, Dept. Signals, Sensors and Systems, 100 44 Stockholm, Sweden

Abstract
The handling of beads, or microspheres, for specific applications in microfluidics is discussed. Beads add functionality to microfluidic devices which is difficult to achieve otherwise. They can be used as passive components (e.g. as the solid phase for chemical reactions) or as active components (e.g. as an actuator for liquid pumping).

Keywords: beads, microspheres, expandable, microfluidics

1 Introduction

The use of beads is relatively recent in microfluidics and offers novel features for μTAS.

Because of high temperatures and aggressive chemicals used during the fabrication process of microfluidic chips, chemical surface modifications are complicated. By using beads as a mobile solid-phase carrier for biochemical reactions in sealed micromachined devices these difficulties are circumvented. Beads provide a high degree of flexibility in terms of different surface coatings, materials and sizes.

The control of reagents and products in space and time requires actuators and valves that allow the handling of liquids on-chip. These components can be external or integrated on the chip, depending on the specific needs of the application. Expandable beads (Expancel® microspheres, Expancel, Sundsvall, Sweden) were shown to be suited for on-chip liquid actuation. They have a gas tight thermoplastic shell that encloses a liquid hydrocarbon. When heated, the thermoplastic shell softens, the hydrocarbon undergoes a phase change to gas, and the volume of the microspheres increases considerably. The volume expansion is irreversible, i.e. the beads do not shrink after cooling down.

2 Chemical Immobilization of Beads

A novel technique enabling selective trapping of micro- and nanometer-sized particles in microfluidic devices without the use of physical barriers has been presented [1, 2]. It is a fast, convenient and simple method, involving microcontact printing and self-assembly, that can be applied to structured and unstructured silicon, quartz or plastic substrates.

Different chemistries can be used to immobilize the beads. Figure 1 shows immobilized streptavidin-coated beads on a silicon substrate with a printed pattern of biotin-labeled BSA. Figure 1 (b) shows that beads are arranged into an almost perfect match of the printed pattern and demonstrates that features with as few as two or three beads can be formed reproducibly. The beads can also be immobilized on structured surfaces, e.g.

Y. Baba et al. (eds.), Micro Total Analysis Systems 2002, Volume 2, 605–607.

microchannels. They withstand liquid flow, hence allowing further functionalization or biochemical screenings.

3 Mechanical Immobilization of Beads

Micromachined chambers with physical barriers have been integrated into microfluidic devices for bead-based DNA analysis [3], chromatography [4], and immunoassays [5]. These devices mechanically trap beads submerged in a liquid, while the liquid itself can flow through the device unhindered.

In contrast to the design of a dam in a liquid channel [4, 5, 6], grid-like trapping component, as depicted in Figure 2, were presented. The grid consists of tiny deep reactive ion etched (DRIE) pillars [7, 8]. The filter pillars are 3 μm wide and 50 μm high with a spacing of 2 μm. The required reaction chamber volume and the shape of the reaction detection sensor define the shape of the filter device. It must contain a large number of pillars to guarantee clogging insensitivity.

Single nucleotide incorporations were successfully performed in such devices.

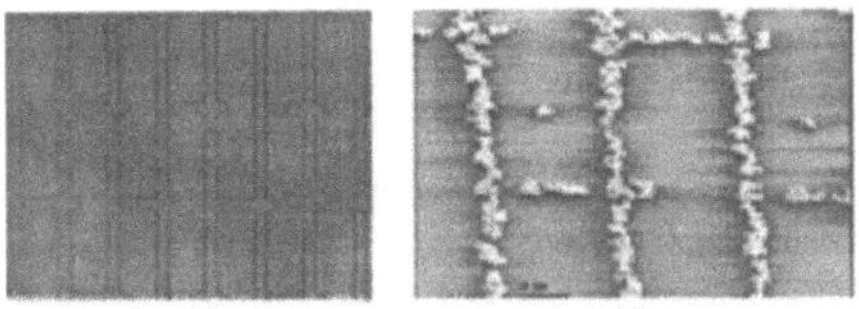

a) b)

Figure 1: Streptavidin-coated beads immobilized in a high-resolution pattern. In (a) a photo of the stamp is shown. In (b) it is shown that the beads are arranged into an almost perfect match of the printed pattern.

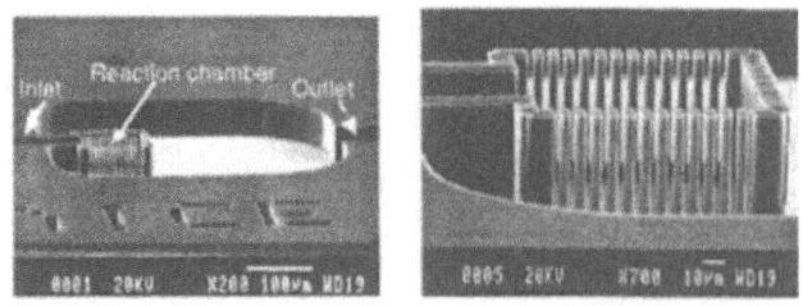

a) b)

Figure 2: SEM pictures of a filter chamber before bonding the top lid. The beads are trapped in the reaction chamber while a bead/liquid mix flows from the inlet to the outlet.

4 Expandable Beads for Microfluidic Liquid Handling

Expandable microspheres are available with expansion temperatures in the range of 70 – 200 °C, whereby the volume increases with the temperature. If the temperature is increased above their highest expansion temperature, they collapse.

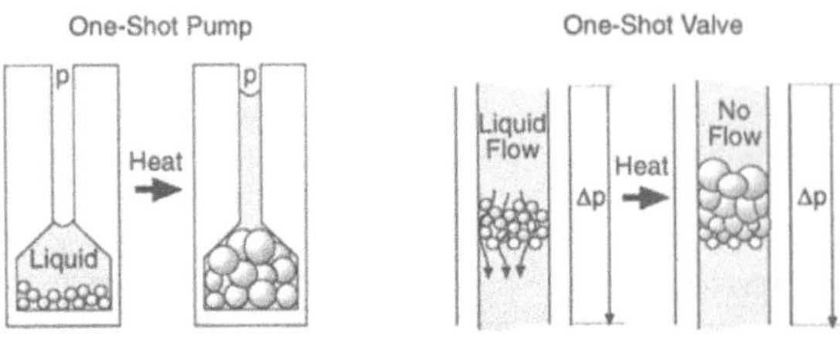

a) b)

Figure 3: Schematic representation of a one-shot micropump (a) and a one-shot microvalve (b) based on expandable microshperes.

Liquid handling in microfluidic channels was demonstrated using expandable microspheres that were mechanically trapped [9], as schematically shown in Figure 3. Water in the nanoliter range was displaced in a microchannel against a counter pressure of 100 kPa (*i.e.* the microspheres act as a one-shot pump) and the flow of water in a microchannel was completely blocked (*i.e.* the microspheres act as a one-shot normally open valve) after heating the beads to 70°C.

Because of their high chemical tolerance, liquid actuation and valving with expandable microspheres is largely independent of the liquid properties.

Expancel beads can also be localized and immobilized via surface chemistry and photoresist patterning [10], without inhibiting their expansion properties, as shown in Figure 4.

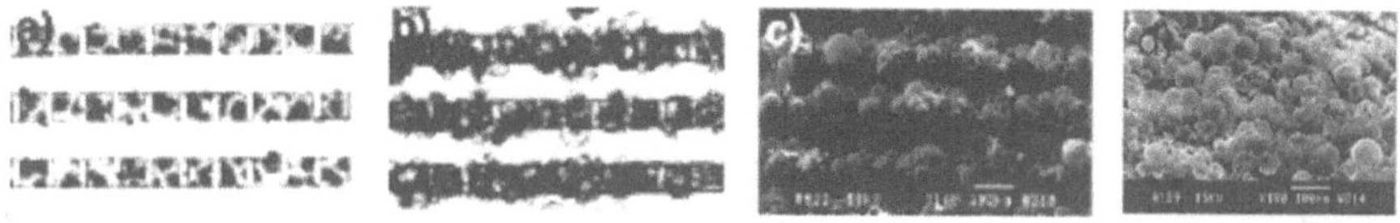

Figure 4: Photos of selectively deposited expandable microspheres before (a) and after (b,c) expansion, whereby the resist method is used. The horizontal lines are 100 μm wide, the vertical shorter lines 15 μm wide. (d) shows a SEM image of expanded microspheres immobilized by surface chemistry.

5 Conclusions

By decoupling chemical functionality from microfabrication technology, physically and chemically immobilized beads are powerful tools for solid phase chemistry in μTAS. In addition, expandable beads are active components in microfluidic networks.

6 References

[1] H. Andersson, C. Jönsson, C. Moberg, G. Stemme, Self-Assembled and Self-Sorted Array of Chemically Active Beads for Analytical and Biochemical Screening, Talanta, 2002, 56, 301.

[2] H. Andersson, C. Jönsson, C. Moberg, G. Stemme, Patterned Self-Assembled Beads in Silicon Channels, Electrophoresis, 2001, 22, 3876.

[3] Ahmadian, A., Russom, A., Andersson, H., van der Wijngaart, W., Uhlen, M., Stemme, G., Nilsson, P., SNP Analysis by Allele-Specific Extension in a Micromachined Filter-Chamber, *BioTechniques, In press.*

[4] Oleschuk, R., Shultz-Lockyear, L., Ning, Y., Harrison, J., *Analytical Chemistry,* 2000, 72, 585-590.

[5] Sato, K., Tokeshi, M., Odake, T., Kimura, H., Ooi, T., Nakao, M., Kitamori, T., *Analytical Chemistry,* 2000, 72, 1144-1147.

[6] Eckersten, A., Edman Örlefors, A., Ellström, C., Erickson, K., Löfman, E., Eriksson, A., Eriksson, S., Jorsback, A., Tooke, N., Derand, H., Ekstrand, G., Engström, J., Honerud, A., Aksberg, A., Hedsten, H., Rosengren, L., Stjernström, M., Hultman, T., Andersson, P., Micro Total Analysis Systems 2000, Proceedings of the μTAS 2000, 14-18 May 2000, Enschede, The Netherlands, 521-523.

[7] Andersson, H., van der Wijngaart, W., Enoksson, P., Stemme, G., *Sensors and Actuators B: Chemical,* 2000, 67, 203-208.

[8] Andersson, H., van der Wijngaart, W., Stemme, G., Micromachined Filter-Chamber Array with Passive Valves for Biochemical Assays on Beads, *Electrophoresis,* 2001, 22, 249-257.

[9] Griss, P., Andersson, H., Stemme, G., Expandable Beads for the Handling of liquids, Lab-on-Chip, 2002, Vol. 1, Issue 2

[10] Andersson, H., Griss, P., Stemme, G., Expandable microspheres-surface immobilization techniques, Sensors and Actuators B: Chemical, Volume 84, Issues 2-3, 15 May 2002, Pages 290-295

HIGH PERFORMANCE TWO DIMENSIONAL SEPARATIONS OF TRYPTIC DIGESTS ON MICROFLUIDIC DEVICES

Stephen C. Jacobson, Jeremy D. Ramsey, Christopher T. Culbertson, and J. Michael Ramsey
Oak Ridge National Laboratory, Oak Ridge, TN 37831-6142 USA

Abstract

High-efficiency, two-dimensional separations of tryptic digests were performed with a microfluidic device. The first dimension separation was micellar electrokinetic chromatography (MEKC) and the second dimension separation was capillary electrophoresis (CE). To maintain a small footprint, the microchip was designed with a 20-cm long serpentine channel for the first dimension and a 1.3-cm long channel for second dimension. The turns in the serpentine channel were asymmetrically tapered to minimize geometrical dispersion. The separation efficiency was 200,000 and 25,000 plates for MEKC and CE separations, respectively. In analysis times under 15 min, peak capacities up to 4500 were generated.

Keywords: separation, chromatography, electrophoresis, peptides

1. Introduction

Several electrokinetically driven separation techniques including capillary electrophoresis, micellar electrokinetic chromatography, open channel electrochromatography, and gel electrophoresis have been successfully demonstrated on microfluidic devices. The dexterity with which materials can be manipulated and the ability to fabricate structures with interconnecting channels that have minimal dead volume contribute to the high performance of these devices and enable efficient coupling of multiple separation techniques. If the separation mechanisms for a multi-dimensional separation are orthogonal, this system offers higher performance than a one dimensional separation because the peak capacity for the multi-dimensional system is the product of the peak capacities of the individual separation techniques [1]. One approach for performing multi-dimensional separations is to serially couple the columns and to rapidly inject the effluent from the first dimension into the second dimension at a fixed frequency. This comprehensive approach

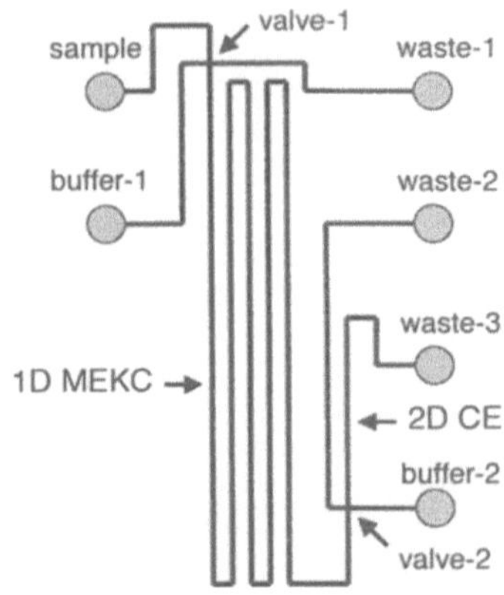

Figure 1. Schematic of a microfluidic device for two dimensional separations.

Y. Baba et al. (eds.), Micro Total Analysis Systems 2002, Volume 2, 608–610.

works most effectively when each peak eluting from the first dimension is sampled at least four times by the second dimension [2]. Serial coupling has been demonstrated for a variety of systems using standard instrumentation including liquid chromatography and capillary electrophoresis [3,4]. Two dimensional separations have also been extended to microfluidic devices where micellar electrokinetic chromatography [5] and open-channel electrochromatography [6] have been coupled with capillary electrophoresis for separating peptides and tryptic digests. We report here increased performance of microchip devices for two-dimensional separations by incorporating low dispersion turns [7,8], sampling the first dimension effluent more rapidly, and decreasing the injection plug length in the second dimension.

2. Experimental

The microchip was fabricated in white crown glass using standard photolithography, wet chemical etching, and bonding procedures. For the microchip schematic in Figure 1, the first dimension separation channel was 19.6 cm long, 70 μm wide, and 10 μm deep, and the second dimension separation channel was 1.3 cm long, 20 μm wide, and 10 μm deep. The low dispersion turns of the serpentine channel (see Figure 2) had a radius of curvature of 0.3 mm and asymmetrically tapered from 70 to 20 μm wide. The microchip was operated by making a single injection with valve-1 into the MEKC channel and multiple injections with valve-2 into the CE channel at a fixed frequency, e.g., 1 Hz. The separations were detected using laser induced fluorescence in the CE channel approximately 1 cm downstream from valve-2. Two-dimensional plots were generated by plotting the individual electropherograms at the injection time into the second dimension. The buffer-1 reservoir contained 20 mM sodium dodecylsulfate, 10 mM boric acid, and 10% (v/v) 2-propanol for the MEKC, and the buffer-2 reservoir contained 100 mM boric acid for the CE. High voltages were applied to the reservoirs of the device using six independently controlled and programmable power supplies. The electric field strengths in the first and second dimension channels were 200 and 2400 V/cm, respectively. The sample, bovine hemoglobin, was digested with trypsin and labeled with 5-TAMRA (5-carboxytetramethylrhodamine succinimidyl ester).

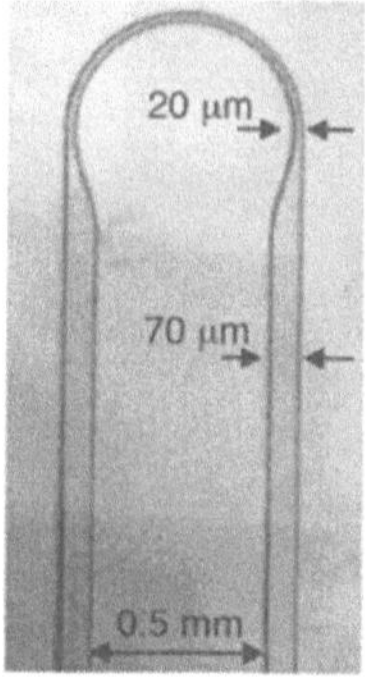

Figure 2. Image of a low dispersion turn used in the microchip depicted in Figure 1.

3. Results and Discussion

Figure 3 shows the 2D separation of a tryptic digest of bovine hemoglobin as an image plot. The plot depicts the time windows from 250 to 750 s for the MEKC and from 0.7 to 1.25 s for the CE which contained 95% of the separation information. In addition, the line plots in Figure 3 are the projections of the 2D data into each separation axis and mimics the one dimensional equivalent of the individual separation techniques. Peak capacities up to 4500 were estimated for this and other separations. These peak capacities were higher than peak capacities reported using conventional column-based analyses and a 2 to 4-fold increase over previously published 2D microchip separations. The increased performance can be attributed primarily to an increase in the second dimension sampling rate (~ 1 Hz) and a decrease in the injection time for the second dimension (20 ms). Another major advantage over conventional analyses was the reduced analysis time required for these separations, e.g., less than 15 min. The microchip analysis time was up to four times faster than previously reported column-based 2D separation techniques and at least 25 times faster than 2D slab gel electrophoresis. In addition, the footprint of the microchip was two times smaller than previously reported by using the serpentine channel design, and the separation performance of the serpentine channel was maintained by minimizing the geometric dispersion with low dispersion turns.

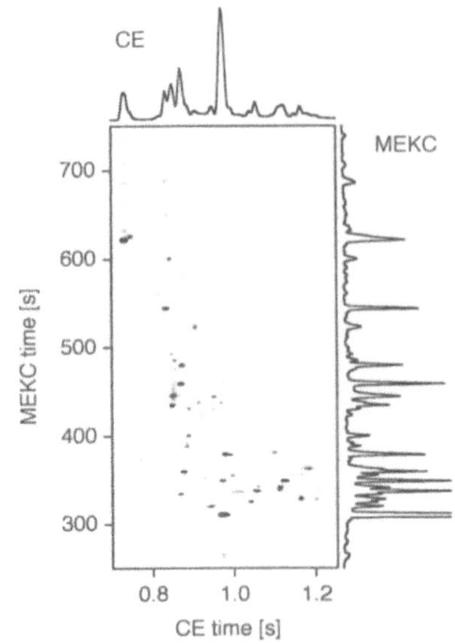

Figure 3. Two dimensional separation of a tryptic digest of bovine hemoglobin.

Acknowledgements

This research was supported by the National Cancer Institute under Grant R33CA83238. Oak Ridge National Laboratory is managed and operated by UT-Battelle, LLC under contract DE-AC05-00OR22725 with the U.S. Department of Energy. J.D.R. was supported by an appointment to the ORNL Postdoctoral Research Associates Program, administered by ORISE and ORNL. The authors thank Christopher D. Thomas for preparation of the microchip devices.

References

1. Giddings, J. C. *Unified Separation Science*; Wiley: New York, 1991.
2. Murphy, R. E.; Schure, M. R.; Foley, J. P. *Anal. Chem.* 1998, *70*, 1585-1594.
3. Bushey, M. M.; Jorgenson, J. W. *Anal. Chem.* 1990, *62*, 978-984.
4. Moore, A. W., Jr.; Jorgenson, J. W. *Anal. Chem.* 1995, *67*, 3448-3455.
5. Rocklin, R. D.; Ramsey, R. S.; Ramsey, J. M. *Anal. Chem.* 2000, *72*, 5244-5249.
6. Gottschlich, N.; Jacobson, S. C.; Culbertson, C. T.; Ramsey, J. M. *Anal. Chem.* 2001, *73*, 2669-2674.
7. Molho, J. I.; Herr, A. E.; Mosier, B. P.; Santiago, J. G.; Kenny, T. W.; Brennen, R. A.; Gordon, G. B.; Mohammadi, B. *Anal. Chem.* 2001, *73*, 1350-1360.
8. Griffiths, S. K.; Nilson, R. H. *Anal. Chem.* 2001, *73*, 272-278.

2-D PROTEIN SEPARATION WITH CHROMATOGRAPHY AND ISOELECTRIC FOCUSING

Masumi Yamada and Minoru Seki
Department of Chemistry and Biotechnology, School of Engineering,
The University of Tokyo, 7-3-1 Hongo, Bunkyo-ku, Tokyo 113-8656, Japan
(e-mail: mseki@chembio.t.u-tokyo.ac.jp)

Abstract

A novel two-dimensional separation system of proteins on a PDMS microdevice is presented. In the first dimension, proteins are separated chromatographically, and in the second dimension, separated perpendicular to the direction of the flow by isoelectric focusing (IEF). With this system, we successfully separated proteins and opened up a new possibility for proteomics.

Keywords: proteomics, two-dimensional separation, chromatography, isoelectric focusing

1. Introduction

In recent years, a lot of research has been done on 2D separation of proteins. However, conventional methods of 2D gel electrophoresis usually require a long separation time and complicated operations. Some researchers have reported the miniaturization of 2D separation of proteins for the direct connection to a mass spectrometer. Here, we present a novel system of 2D separation of proteins, which becomes feasible only in microscale devices, and the potential of this system.

2. Theory

Principle of the 2D separation used in this study is shown in Fig.1. In the first dimension, proteins are chromatographically separated by changing the ion concentrations of elusion buffer [1]. Next, proteins are introduced into the IEF column, which has a pH gradient perpendicular to the flow direction [2]. In the second dimension, the proteins are moved perpendicular to the flow direction by the effect of applied voltage and focused on the certain position according to each isoelectric point. This separation method could become feasible only in microscale devices, because the flow in microdevices is perfectly laminar and the required migration length is very short.

Advantages of this system are as follows: First, it is theoretically possible to separate proteins two-dimensionally for total injected amount in a microdevice. Second, it would

Y. Baba et al. (eds.), Micro Total Analysis Systems 2002, Volume 2, 611—613.

be easy to connect this microdevice directly to a mass spectrometer. Thirdly, various types of chromatography are available for the first dimension in this separation system.

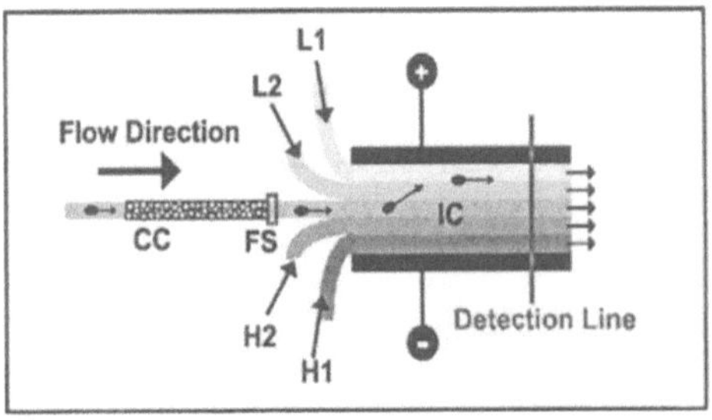

Figure 1. Schematic diagram of the two-dimensional separation of proteins: The arrows show the flow directions. CC is chromatographic separation column, FS is frit structure, IC is IEF column. Buffers with lower pH are introduced from L1 and L2, and buffers with higher pH are introduced from H1 and H2.

3. Experimental

An example of a fabricated microdevice is shown in Fig.2. The material of this microdevice is PDMS. An micromixer, a frit structure, Pd electrodes and a IEF column were integrated on this device. Average channel depth was 150 μm, and channel width was 100 μm. The width of IEF column was 1 mm, and the length was 14 mm. Ion-exchange beads (d = 30 μm) were packed into the chromatographic separation column.

TRITC conjugated BSA was used as a sample. In the chromatographic separation column, NaCl concentration were changed from 0.05~0.45 M. Different pH buffers (10 mM MES, pH ranging from 5.5 to 7.5) were introduced into the IEF column, and the retention time in the column was about 47 sec. The introduced position of BSA was between 600~800 mm from anode. Applied voltage between an anode and a cathode was 3V.

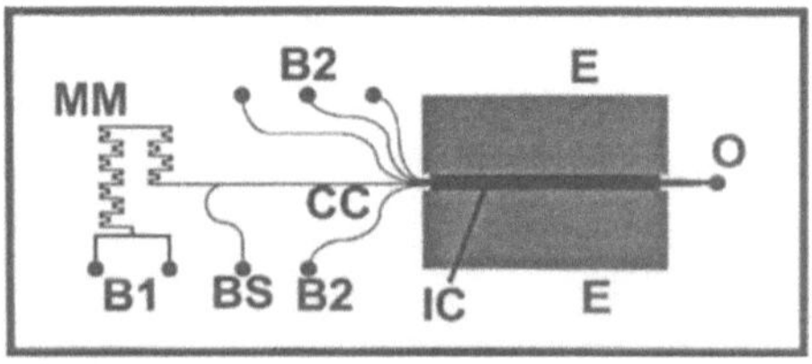

Figure 2. Microdevice for two-dimensional separation of proteins: B1 are inlets of buffers for NaCl gradient. B2 are inlets of buffers for pH gradient. BS is inlet for beads and sample. MM is a micromixer, E are palladium electrodes, and O is an outlet. The size of this device is 45×18 mm.

4. Results and discussion

First, we confirmed that the pH gradient in the IEF column was kept under the influence of applied voltage. The result is shown in Fig.3. It was seen that pH gradient was kept sharp enough to perform IEF under the applied voltage of 3V.

Next, we separated proteins two-dimensionally, and the result is shown in Fig.4. When applying no voltage, the peak didn't migrate to the direction of pH gradient

(Fig.4-A). However, by applying 3V between an anode and a cathode, it was observed that proteins were focused and the peak position was shifted perpendicular to the flow direction (Fig.4-B). The length and the width of IEF column, the kind of buffer and the applied voltage should be optimized for the higher resolution.

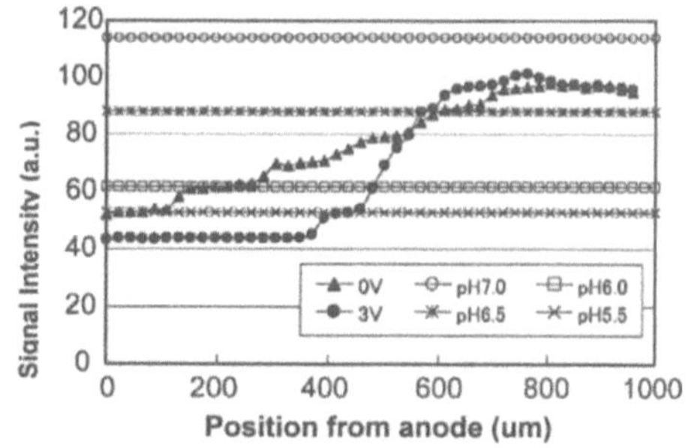

Figure 3. pH gradient at the end of IEF column under the applied voltage of 0 or 3 V: Fluorescein was used as a pH indicator.

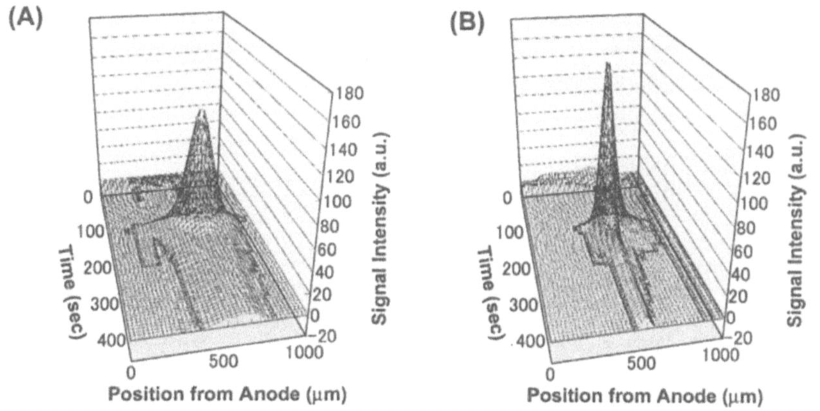

Figure 4. Two-dimensional separation of TRITC-BSA. (A) No voltage applied. (B) 3 V applied between an anode and a cathode.

Conclusions

We could successfully show the possibility of the new method for 2D separation of proteins. This method would become one of the important tools for proteomics.

Acknowledgement

This research was supported in part by grants for Scientific Research on Priority Areas [A] (No. 13025216) and Millennium Project (No. 12319) from the Ministry of Education, Culture, Sports, Science and Technology in Japan.

References

1. M. Seki, M. Yamada, R. Ezaki, R. Aoyama and J.W. Hong, *Proceedings of μTAS 2001 Symposium*, pp. 48-50, (2001)
2. K. Macounova, C. R. Cabrera and P. Yager, *Anal. Chem.* **73**, 1627-1633, (2001)

A 15-SECOND PROTEIN SEPARATION EMPLOYING HYDRODYNAMIC FORCE ON A MICROCHIP

Mari Tabuchi[1], Yasuhiro Kuramitsu[2], Kazuyuki Nakamura[2,3] and Yoshinobu Baba[1]

[1]Dept of Medicinal Chemistry, Faculty of Pharmaceutical Sciences, The University of Tokushima, Japan, CREST, Japan Science and Technology Corporation, JST, Japan,
[2]Department of Biochemistry & Biomolecular Recognition,
Yamaguchi University School of Medicine, Ube, Japan,
[3]Central Laboratory for Biomedical Research and Education,
Yamaguchi University School of Medicine, Japan
(tabuchi@ph.tokushima-u.ac.jp)

Abstract

We report a novel pressurization technique for microchip electrophoresis (μ-CE) that enables the separation of many biological samples (12-96 samples) involving complex mixtures of proteins (>20 proteins) in a single run only taking 15 seconds (s). Although μ-CE relies almost exclusively on electrokinetic forces and some loss of resolution is unavoidable, the method reported here employing hydrodynamic force has enabled a dramatic reduction in separation time without compromising resolution.

keywords: microchip, proteome analysis, Jurkat cells, high through-put screening

1. Introduction

The μ-CE is expected for high-through put screening systems of proteome analysis, but problem of reduced resolution by high-voltage in a short separation channel has been the major shortcoming over ten years. Usual μ-CE relies almost exclusively on electrokinetic manipulations [1], which have been believed to be unique technique for μ-CE. However, we speculated that pressurization applied to the separation channel before electrophoresis would give more unique separation beyond usual electrokinetic conditions. Herein, we focused on hydrodynamic force on a chip and developed two pressurized sample-loading methods, the VP method and the PP method.

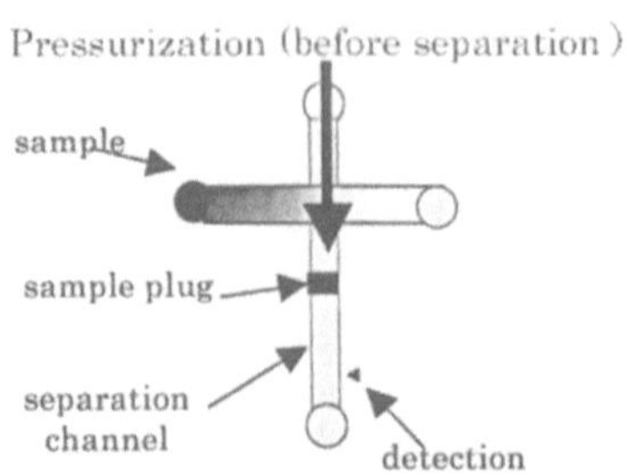

Figure 1 Illustration of a micro-channel we used and pressurization method we developed for μ-CE.

2. Experimental

A μ-CE system (cosmo-*i*; SV1100 or SV1210, Hitachi Electronics Engineering, Hitachi, Japan), and a microchip (*i*-chip 3 or *i*-chip12; Hitachi Chemical, Hitachi, Japan), having channels 100 μm wide and 30 μm deep, was used for the separations. After the sample was injected by voltage or pressure, sample plug was pushed forward by

Y. Baba et al. (eds.), Micro Total Analysis Systems 2002, Volume 2, 614–616.

pressurization after loading the sample but before electrophoresis separation, in the absence of an electric potential (PV method and PP method) (Fig. 1). Pressures of P_L: 1.0 - 2.0 kPa; P_M: 2.5 - 5.0 kPa; and P_H: 5.5 - 8.5 kPa, were applied. Standard proteins or protein mixture extracted from human Jurkat cells, which are T lymphoblastic cells, were used. Jurkat cells were cultured at 37 °C or 51 °C for 30 min.

3. Results and Discussion

When we used our methods to separate a sample protein mixture, Fig 2 clearly demonstrated that pressurization is effective, since higher pressurization produced higher speed separation (PP method); furthermore, medium pressurization produced resolution improvement (VP method), whereas some peaks were missing in this time course when it was not.

Usual pressure-driven flow provides inadequate separation because of laminar flow [2]. However, the result of computer simulation of Fig. 3 represents that our pressurization methods clearly produced flat flow, since pressure was applied before electrophoresis separation, not simultaneously, thereby enabling production of sharp peaks derived only from EOF flow. Moreover, the pressurization technique can produce higher density sample plug even free solution buffer.

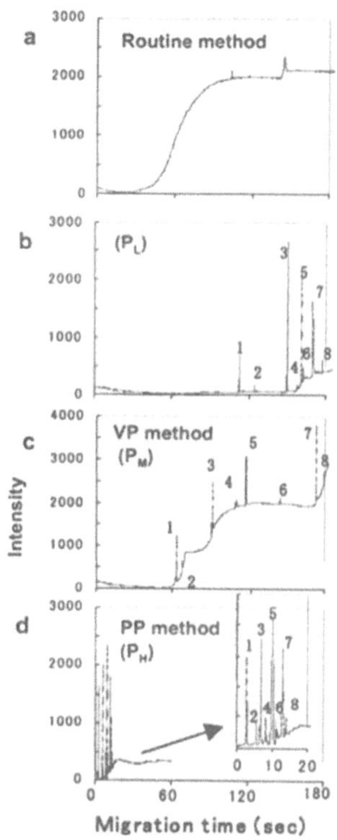

Figure 2 Electropherograms of standard proteins(14.4 - 200kDa) by routine or new method.

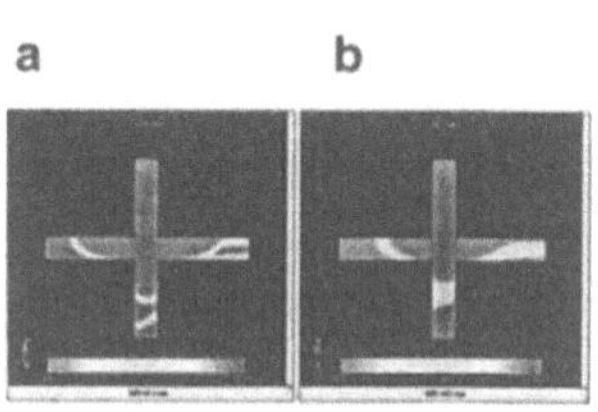

Figure 3 Simulated images of the sample mass 0.3 s after separation for new (a) and conventional (b) methods.

Utilizing this technique, 12 samples of complex protein mixture extracted from Jurkat cells were separated within 15 s at a single run (Fig. 4A) using a 12-microchannel array, with high reproducibility. The target proteins of Jurkat cells (I; 7 kDa: thymosinβ_4 and II; 18 kDa: phosphorylated stathmin), which increase after heat-treatment at 51°C for 30 min, were detected within only 15 s by the new method, while several hours were required for conventional 2-DE analysis. In addition, we found some other target proteins, which could not be detected as clear spots by 2-DE analysis (Fig. 4B).

We propose that our novel technique enables small amounts of marker proteins, those which are idiotypic for a specific physiologic or pathologic state of cells or tissues, can be quickly identified using this technique instead of commonly used 2-DE.

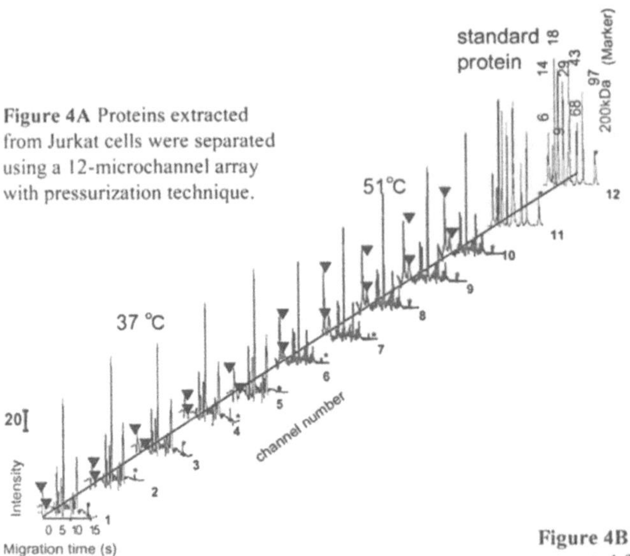

Figure 4A Proteins extracted from Jurkat cells were separated using a 12-microchannel array with pressurization technique.

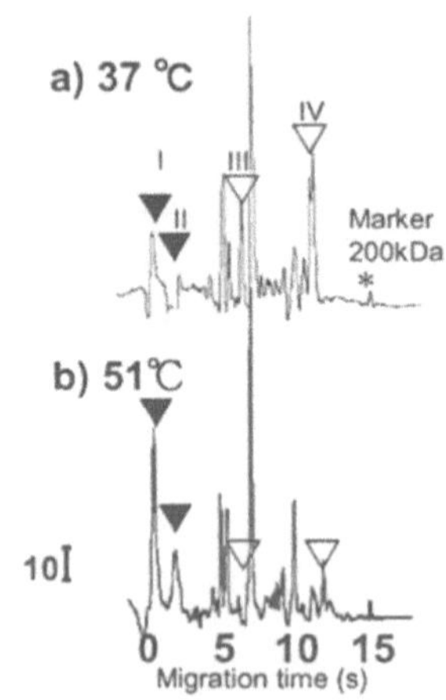

Figure 4B The μ-CE analysis of protein extracted from Jurcat cells in the culture condition of a) 37 °C and b)51 °C.
▼ target proteins ▽ other target proteins

4. Conclusions

In conclusion, hydrodynamic force seems to be more effective than electric field supply alone as a means of achieving high-speed, high-resolution, and high-intensity protein separation, by employing pressure prior to electrophoresis separation. This novel pressurization technique on a microchip will offer the enormous advantages for high through-put first screening systems of human proteome analysis over commonly used 2-DE.

Acknowledgments

The present work was partially supported by a Grant of CREST from the JST, a Grant from the New Energy and Industrial Technology Development Organization (NEDO) of the Ministry of Economy, Trade and Industry, Japan, a Grant in-Aid for Scientific Research from the Ministry of Health, Labour, and Welfare, Japan, and a Grant-in-Aid for Scientific Research from the Ministry of Education, Science and Technology, Japan.

References

1. Harrison, D. J. *et al. Science*, **261**, 895-897 (1993).
2. Ross, D., Johnson, T. J., Locascio, L. E. *Anal. Chem.* **73**, 2509-2515 (2001).

INTEGRATED TEMPERATURE CONTROL SYSTEM FOR MICROFLUIDIC DEVICES

A. Dodge[¥,1], R.M. Guijt[¥,2], G.W.K. van Dedem[2], N.F. de Rooij[1], E.Verpoorte[1]
[¥]Both authors contributed equally to this work
[1]Institute of Microtechnology, University of Neuchâtel, Neuchâtel, Switzerland
[2]Kluyver Institute for Biotechnology, TU Delft, Delft, The Netherlands

Abstract

For localized temperature control in microfluidic devices, integrated cooling and heating systems are required. To date, several heating systems for microfluidic devices have been presented based on resistive heating. However, no integrated cooling systems have been reported yet. Here, a microfluidic cooling and heating system based on endothermic and exothermic reactions is presented.

Keywords: Integrated cooling, temperature-dependent fluorescence

1. Introduction

Temperature control in microchannels is of great interest for microfluidics, especially for applications involving (bio)chemical reactions. For instance, there are many reports of PCR on chip, most of which use external components to achieve heating and cooling for fast thermocycling (see, for example, Ref. [1]). Micromachined heating elements have been incorporated in monolithic chip-based devices for various applications, including PCR [2] and reaction chemistry in the gas phase [3]. To date, however, no integrated cooling systems for microfluidic devices have been reported. In this contribution, a simple method for direct cooling and heating on chip, based on endothermic and exothermic reactions inside microchannels, is presented.

2. Concept

Chemical reactions can occur in an endothermic or an exothermic way. Endothermic reactions require energy from their environment, whereas exothermic reactions release energy, i.e. heat is respectively taken from or released to the environment. Heat exchange induced by chemical reactions was studied on a microfluidic device, where heat transfer is generally faster, due to the low thermal mass. Temperature control inside a microchannel through exploitation of endothermic and exothermic reactions is presented here. The chip layout used for the experiments (Figure 1) allows two reagents, coming from channel 1 and 2 (RC_1 and RC_2), to react in the heating/cooling channel (HCC) when a vacuum is applied. In microfluidic devices, flows are generally laminar, resulting in well-defined reagent flows. The flow resistances of RC_1 and RC_2 can be modified to change the flowrate ratio. The well-defined interface between the two reagent flows allows good control of the endothermic and exothermic reactions, and thus of the thermal effect.

3. Results and discussion

The evaporation of acetone and the dissolution of sulfuric acid were chosen as the endothermic and exothermic reaction, respectively. For cooling the central channel, localized evaporation of acetone in HCC was achieved by drawing acetone from RC_1 against an airflow coming from RC_2.

Y. Baba et al. (eds.), Micro Total Analysis Systems 2002, Volume 2, 617–619.

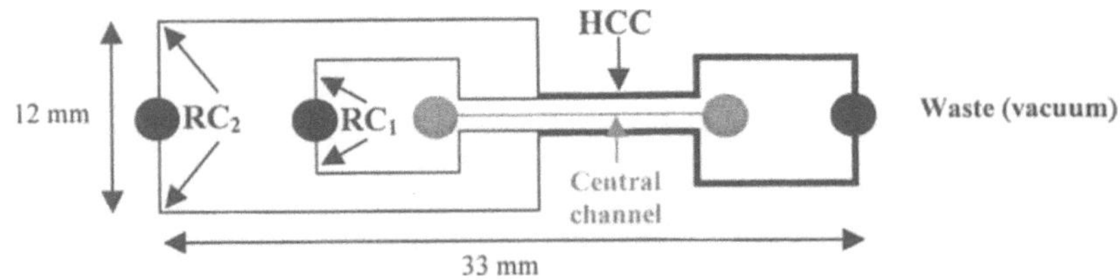

Figure 1: *Chip layout of the glass microfluidic system. Distance between the central channel and the HCC is 30 μm, all channel depths are 20 μm. The channel length of RC_1 is 30 mm, the channel lengths of RC_2 were 30 and 766 mm respectively.*

In the same fashion, heating was achieved by dissolution of sulfuric acid, drawing 97% H_2SO_4 from RC_1 against a water flow coming from RC_2. Temperature effects were visualized by exploiting the temperature-dependent fluorescence of rhodamine B [4]. A signal increase is expected if temperature decreases in the dye-filled central channel. Inversely, a signal decrease is expected for a temperature increase. These effects were indeed observed, as can be seen in Figures 2 (cooling) and 3 (heating). The temperature changes occurred almost instantaneously upon mixing of the reagents.

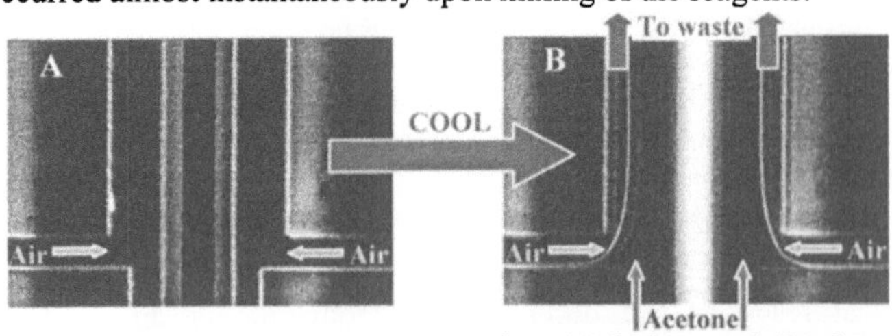

Figure 2: *Fluorescence microscope images taken (A) before and (B) during cooling of a 1-μM rhodamine B solution. Signal increases as temperature decreases. A glass-PDMS-hybrid was used with a glass channel depth of 20 μm. Acetone-to-air flow rate ratio was 1:7.*

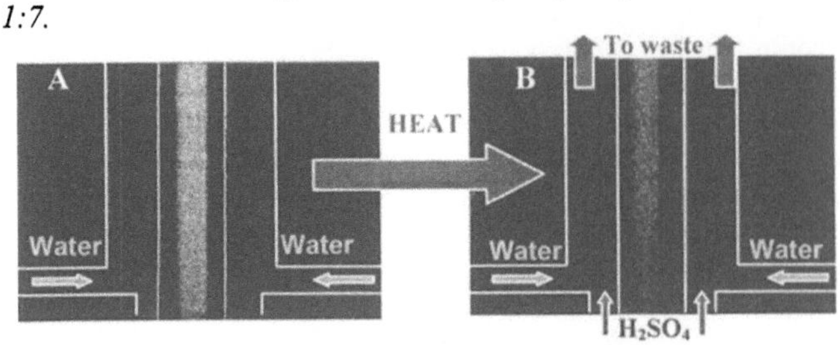

Figure 3: *Fluorescence microscope images taken (A) before and (B) during heating of a 1-μM rhodamine B solution. Signal decreases as temperature increases. A glass device with a channel depth of 20 μm was used. The H_2SO_4-to-water flow rate ratio was 1:11.*

To quantify the thermal effect inside the microchannel, fluorescence was measured as a function of temperature by heating an aqueous solution of 1 μM rhodamine B in a glass chip. A 500-nm-thick, transparent layer of indium tin oxide, deposited on one side of the device, served as a resistive heater. The resulting signal was collected with a CCD camera mounted on a microscope, and pixel-processed to obtain the quantitative data presented in Fig. 4.

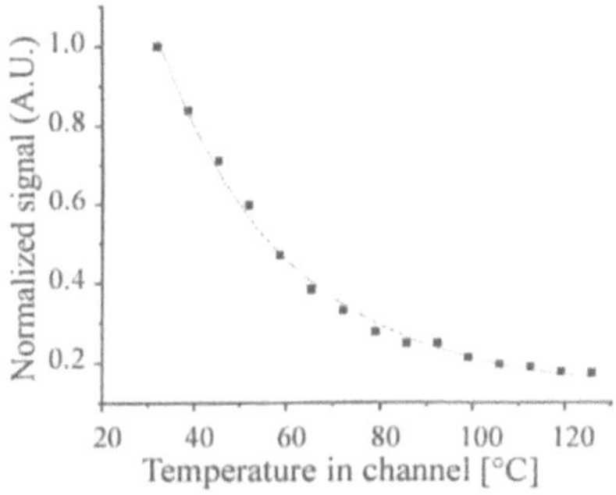

Figure 4 : *Fluorescence intensity of a 1-μM rhodamine B solution as a function of temperature .*

reagent$_1$-reagent$_2$	L_{RC1}: L_{RC2} (mm)	Flow rate ratio	T (°C)
acetone - air (COOL)	12 : 766 **12 : 30**	1 : 0.3 **1 :7**	15 **9**
H_2SO_4 - water (HEAT)	12 : 766 **12 : 30**	1 : 0.4 **1 : 11**	37 **74**

Table 1 : *Thermal effect in central channel for cooling / heating using different flow rate ratios.*

Two different designs were tested to see if the temperature effects could be regulated by varying the RC_1/RC_2 flow ratio. The flow resistance of the RC_2 channel was reduced by decreasing its length from 766 to 30 mm. The temperatures obtained using two flow rate ratios for both the endothermic and exothermic reaction are summarized in Table 1. Cooling was calculated by extrapolating the curve in Figure 4 to lower temperatures.

4. Conclusions

The temperature control methods described here find application in studying the influence of temperature on (bio)chemical reactions. Heating can be used to achieve optimal enzymatic activity in chip-based assays, while cooling could be used to rapidly quench chemical reactions. In microchip CE, on-chip cooling could be employed to decrease Joule heating. [5]. The usable temperature range of the system presented could be extended by selecting from a variety of exothermic and endothermic reactions.

Acknowledgments

R.G. was supported by the Technology foundation STW in the framework of the Biomas project DST664351. A.D. acknowledges funding from the Swiss Commission for Technology and Innovation (Project "Nanocontainer" No. 5493.2 TNS).

References

[1] J. Khandurina, T.E. McKnight, S.C. Jacobson, L.C. Waters, R.S. Foote and J.M. Ramsey, *Anal. Chem.*, 72, pp. 2995-3000, (2000).

[2] E. T. Lagally, P. C. Simpson, R. A. Mathies, *Sensors and Actuators B-Chemical* **63**, 138-146 (2000).

[3] I.M. Hsing, R. Srinivasan, M.P. Harold, K.F. Jensen, M.A. Schmidt, *Chem. Eng. Sci.* 55 (1), pp. 3-13, (2000).

[4] D. Ross, M. Gaitan and L.E. Locascio, *Anal. Chem.*, 73, pp. 4117-4123, (2001).

[5] T. Greibrokk, *Anal. Chem.*, 73, pp. 374A-378A, (2002).

PATTERNING FLOWS USING GROOVED SURFACES: APPLICATION TO MICROFLUIDICS

Armand Ajdari [1], Abraham D. Stroock [2], Stephan K. Dertinger [2], George M. Whitesides [2]
[1] Laboratoire de Physico-Chimie Théorique, UMR CNRS-ESPCI 7083, 10 rue Vauquelin, F-75231 Paris Cedex 05, France
[2] Department of Chemistry and Chemical Biology, Harvard University, Cambridge MA 02138, U.S.A.

Abstract
Pressure-driven flows over grooved surfaces are shown to develop transverse components. This induces helical recirculation in channels or capillaries with grooved walls, and more generally permits the design of 3D flows with simple patterns of grooved regions. This strategy is easy to implement with standard methods of microfabrication, and permits for example the realization of a low Reynolds number chaotic mixer.

Keywords : flow design, anisotropic surfaces, chaotic mixer, microfluidics

1. Introduction

In microfluidic systems, it is a challenge to create the three-dimensional flows that are required for manipulating streams within flows e.g. for mixing or for the spatially resolved delivery of reagents. A practical constraint is indeed that the common lithographic methods used in fabrication lead to planar, layered structures. as a consequence the flow structure follows locally a simple pattern, that is Poiseuille-like for pressure-driven flows, plug-like for electro-osmotic flows (EOF).

A solution is proposed that consists in using surfaces with patterns of grooves. This design is easily fabricated using standard lithography techniques. We demonstrate here the effect in a simple geometry before pointing out further implications.

2. Theory

We have estimated the flow in a channel of rectangular cross-section (thickness H, width w), the "floor" of which bears sinusoidal grooves of amplitude αH and wave-vector q, at an angle θ with the channel axis [1]. An expansion in the adimensional parameter α (groove amplitude relative to the channel height) shows that the flow in the vicinity of the bottom plate tends to orient along the groove valleys. Taking into account the (Poiseuille) recirculation imposed by the lateral walls, the overall flow pattern is helical (see Figure 1 for a schematic representation for the case of square shaped grooves).

Y. Baba et al. (eds.), Micro Total Analysis Systems 2002, Volume 2, 620–622.

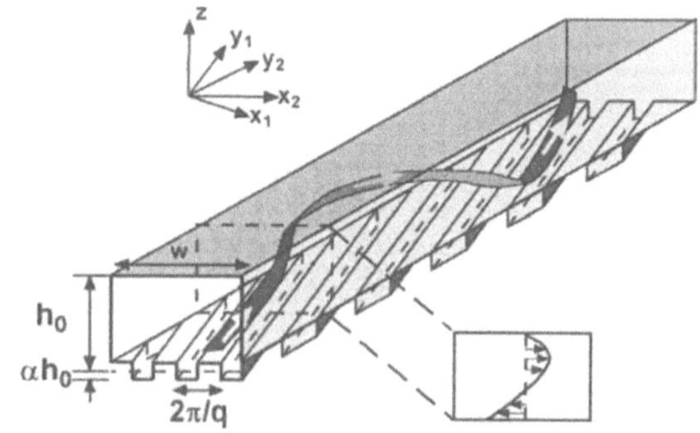

Figure 1: Geometry of the PDMS channels analyzed here, together with a schematic representation of the helical flow induced by the groove pattern on the channel floor

Characterizing the pitch by the angle Ω that the flow makes with the channel axis right below the top surface, we estimate:

$$\tan(\Omega) = \frac{\alpha^2(K_{||}(qH) - K_{\perp}(qH))\cos\theta\sin\theta}{1 - \alpha^2(3/2 - \tilde{K}(qH))} \tag{1}$$

with $K_{||}$, $K_{\perp}$ and $\tilde{K}$ functions of the adimensional number qH [1]. An estimate of the maximal angle corresponds to $\alpha qH \sim 1$, with $\tan(\Omega)|_{\max} \simeq \alpha \cos\theta \sin\theta$.

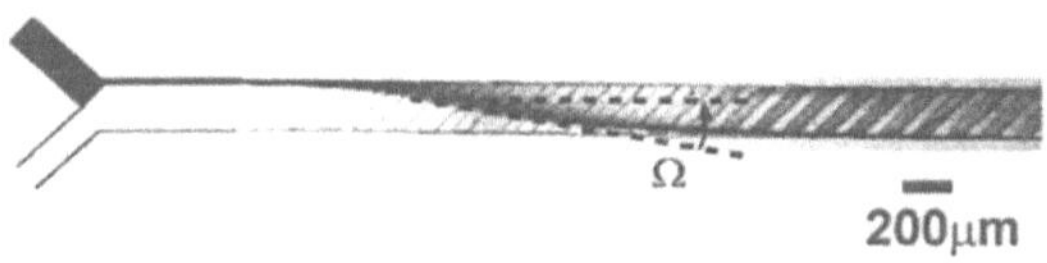

Figure 2: Optical micrograph for the determination of Ω. $w = 200\mu m$, $H = 79\mu m$, $q = 2\pi/100\mu m$, and $\alpha = 0.34$.

3. Experimental

To check the relevance of this estimate we performed experiments on "rifled" microchannels fabricated in PDMS (fabrication along the methods described in [3]).

In order to evaluate the pitch of trajectories, we injected a narrow stream of solution of dye along one side of the channel as shown in Figure 2; a broader stream of clear solution was injected alongside the dyed stream. We evaluated the values of $\tan(\Omega)$ on optical micrographs like the one in Figure 2.

	predicted : tan(Ω)(×100)	measured: tan(Ω) (×100)
$H = 79$; $qH = 5$; $\alpha = 0.134$	4.1	6.0
$H = 103$; $qH = 6.5$; $\alpha = 0.17$	7.7	6.4
$H = 67$; $qH = 4.2$; $\alpha = 0.20$	7.5	7.5
$H = 74$; $qH = 4.6$; $\alpha = 0.24$	11.1	8.8
$H = 79$; $qH = 5.0$; $\alpha = 0.34$	18.0	16.8
$H = 75$; $qH = 2.4$; $\alpha = 0.10$	1.5	2.4
$H = 69$; $qH = 2.2$; $\alpha = 0.20$	5.8	6.6
$H = 82$; $qH = 2.6$; $\alpha = 0.26$	9.6	10.0
$H = 87$; $qH = 2.7$; $\alpha = 0.37$	17.8	16.1

Figure 3: Predicted values from formula (1) for *sinusoidal* grooves, and experimental values evaluated as in Figure 2. Channel characteristics: $w = 200\mu m$, H given in microns, grooves oriented at $\theta = \pi/4$ with respect to the axis of the channel.

The satisfactory comparison in Figure 3 suggests that is possible to theoretically anticipate the amplitude of the effects.

4. Results and discussion

Although the simple helical flow analyzed here may have some interest on its own, a much richer set of flows can be generated by using more sophisticated patterns. For example a pattern with alternating chevrons allows to mix streamlines in a way that permits chaotic passive mixing [3]. Hydrodynamic dispersion for pressure-driven flows in such patterned channels can also be significantly reduced which could prove useful for separation micro-devices. Analysis along the lines of [1] should prove useful in apprehending the geometries achievable, possibly combined with some CFD calculation.

Acknowledgments

We are grateful to H.A. Stone and I. Mezic for their collaboration.

References

1. A.D. Stroock, S.K. Dertinger, G.M. Whitesides, A. Ajdari, *Patterning flows using grooved surfaces*, submitted to Anal. Chem. in April 2002.
2. J. C. McDonald et al., *Fabrication of microfluidic systems in poly(dimethylsiloxane)*, Electrophoresis, **1**, 27-40 (2000).
3. A.D. Stroock, S.K. Dertinger, A. Ajdari, I. Mezic, H.A. Stone, G.M. Whitesides, *Chaotic Mixer for Microchannels*, Science, **295**, 647-651 (2002).

MICRO-EVAPORATOR FOR PRETREATMENT OF ENVIRONMENTAL MONITORING SAMPLES

Akira Koide, Tadashi Sano, Takeshi Harada and Ryou Miyake
Mechanical Engineering Research Laboratory, Hitachi, Ltd.

Abstract

We have developed a new microfluidic device for concentration of solvent. To increase the evaporation rate of solvent, the micro-evaporator flows the solvent as a thin sheet. The prototype of the device was tested, and it was found that 0.11 [mL/min] of water could be evaporated if it was heated with a 10 [W] heater tip.

Keywords: environmental monitoring, pretreatment, evaporator, micromesh

1. Introduction

Toxic substances contained in the environment become detectable after a series of pretreatment, such as extraction, refining, and concentration (1). Micro fluidic devices for pretreatment of samples can reduce the time of this environmental monitoring. In this paper, we describe a micro-evaporator for the concentration of solvent.

2. Micro-evaporator with gas-liquid separation mesh

The conventional concentration process involves putting the sample together with a solvent in a flask that is rotating in hot water (Fig. 1). The evaporated solvent is sucked out of the flask in order to shorten the processing time.

Figure 2 conceptually illustrates the micro-evaporator. The evaporator has a three-layer structure of glass / silicon / glass. It has a serpentine channel for liquids, a gas-liquid separation mesh, thin channel for gas. The key to achieving a fast concentration capability is to ensure that the solvent flow has a depth of less than 100 μm in order to yield a relatively large contact surface area between the solvent and air. The gas-liquid separation mesh makes this possible. The gas-liquid separation mesh has about 50000 holes(ϕ 0.05[mm]) on a 10 x 13 x 0.02 [mm] diaphragm. It is covered with water-repellent film. It can reduce the adhesion of the solvent when it flows under the mesh. Holes make up about 70% of the diaphragm's area. Flow processing is performed to increase the amount of solvent processing. Furthermore, simultaneous processing by many

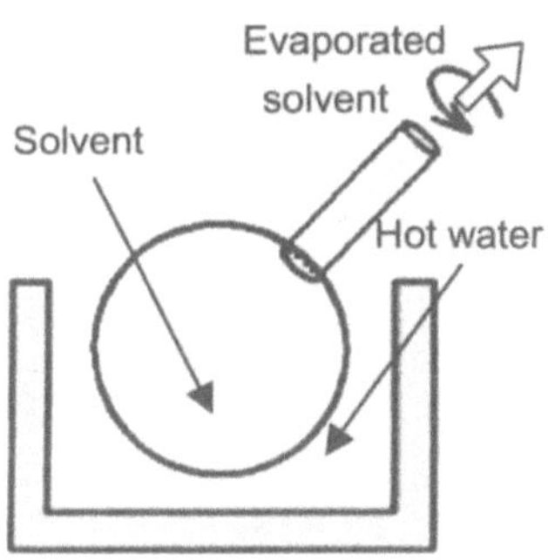

Fig. 1 Conventional evaporator.

Y. Baba et al. (eds.), Micro Total Analysis Systems 2002, Volume 2, 623–624.

devices can increase the amount of processing (Fig.3).

3. Performance evaluation

Figure 4 shows the manufactured micro-evaporator with the gas-liquid separation mesh. When water was condensed using the device's tip (heater power: 10[w]), we found 0.11[mL/min] could be evaporated. This suggests that the micro-evaporator with a gas-liquid separation mesh represents a superior way of condensing solvent.

Acknowledgments

This work was performed by Hitachi, Ltd. under the management of the Micromachine Center as a part of the Research and Development of Micromachine Technology supported by the New Energy and Industrial Technology Development Organization. The authors wish to thank these agencies for their support of this research.

References

(1) Andreas Manz, Fiona Bessoth and Martin U. Kopp, CONTINUOUS FLOW VERSUS BATCH PROCESS , Proc. of μTAS'98, pp. 235-240.

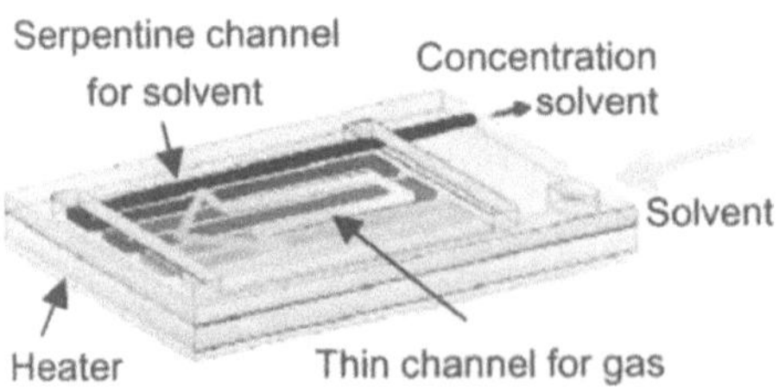

Fig. 2 Micro-evaporator illustration.

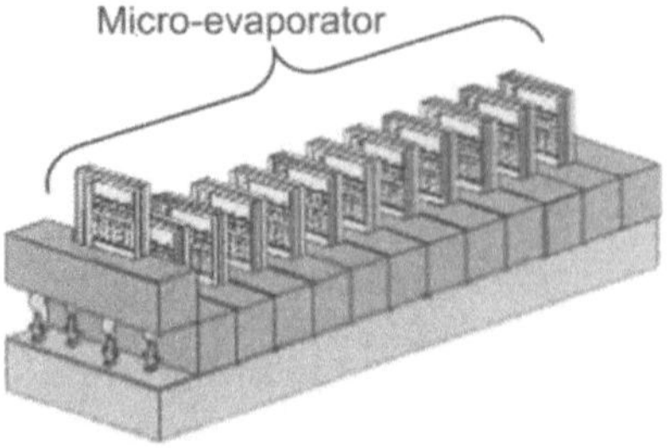

Fig. 3 Parallel processing module.

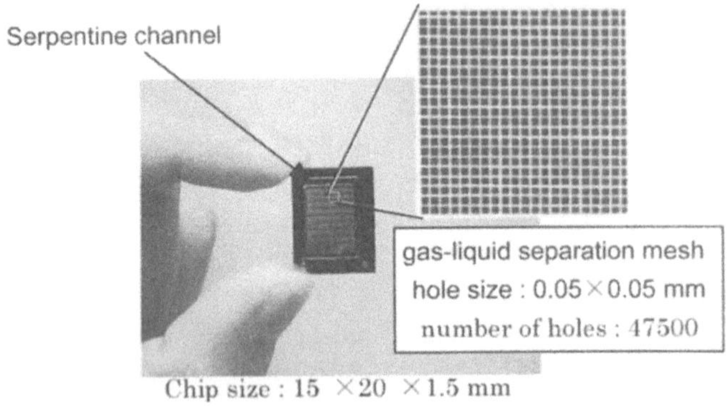

Fig. 4 Prototype micro-evaporator with gas-liquid separation mesh.

AUTOLOGOUS BLOOD RECOVERY AND WASH IN MICROFLUIDIC CHANNEL ARRAYS UTILIZING ULTRASONIC STANDING WAVES

Andreas Nilsson[1], Filip Petersson[1], Hans W. Persson[1], Henrik Jönsson[2] and Thomas Laurell[1]

[1]Department of Electrical Measurements, Lund Institute of Technology, P.O. Box 118, S-222 10 Lund, Sweden

[2]Department of Cardiothoracic Surgery, Lund University Hospital, Lund, Sweden

Keywords: radiation force, ultrasound, particle separation, erythrocytes, blood wash

Introduction

It is known that suspended particles are affected by a radiation force when exposed to ultrasound. The particles in a fluidic microchannel of a rectangular crossection are gathered in the nodes when exposed to an ultrasonic standing wave. By introducing a flow splitter at the end of the microchannel, the particles can be separated from the rest of the fluid. If the fluid used in the separation is whole blood, it is possible to separate the erythrocytes from the plasma (Fig. 1). Another aspect of this separation is that fat embolies released and mixed with the blood, e.g. during cardiac surgery, also can be separated from the erythrocytes. Recovered contaminated blood is usually given back (autologous recovery) to the patient uncleaned which results in micro occlusions in the brain capillaries, e.g. fat embolies. The fat embolies are the cause of SCADS (Small Capillary Arteriolar DilatationS) in the brain, i.e. small brain damages which gives rise to a cognitive decay for the patient. The need to wash the blood clean from the fat embolies prior to reinfusion are therefore important. The difference in acoustic impedance between the red blood cells and the fat causes the red blood cells to gather in the nodes and the fat in the antinodes of the acoustic standing wave. It is therefore possible to separate the red blood cells from the fat (Fig.2). This paper describes the development of a microfabricated multichannel array for erythrocyte separation and fat emboli exclusion.

Materials and methods

A multiseparator was constructed in <100>-silicon by double sided photolithography and anisotropic wet etching. A glass lid covered the channel and was attached by anodic bonding. To the inlets and outlets on the rear side, rubber silicon tubings were glued. The piezoelectric element was attached to the rear side with ultrasonic gel. For the pure blood tests, porcine blood was diluted with saline solution to a 10% concentration of red blood cells. For the fat tests a shed blood phantom that consisted of porcine blood mixed in saline solution (approximately 10% red blood cell concentration) and a sonicated emulsion of tritium-labelled triolein (1% concentration) was used. The phantom was sucked through the channels using two syringe pumps controlling the flow through the outlets. As an ultrasonic standing wave was applied, the erythrocytes was accumulated in the nodal planes and the plasma and the fat in the antinodes.

Results and discussion

Results showed that 90% (range 85-97%) of the tritium labelled triglycerides was removed and 95% of the red blood cells were recovered. While the red blood cells flow easily through with the fluid, even when aggregated in the nodal bands, it was noted that the fat emulsion "tumbled" along the channel walls. The fat seemed to be more easily affected by the radiation

Y. Baba et al. (eds.), Micro Total Analysis Systems 2002, Volume 2, 625–626.

force than the red blood cells, and if the radiation force was too high they would be locked up over the ultrasonic transducer and disrupt the flow, and in worst case occlude the channel.

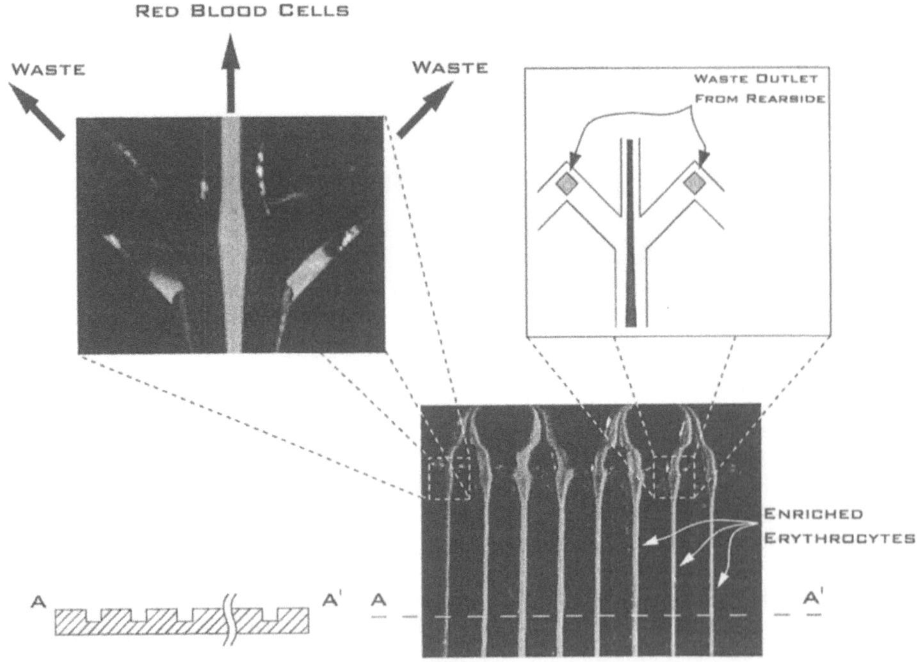

Figure 1. Red blood cells are separated from the rest of the fluid. The central flow channels collecting the enriched erythrocytes are connected to a single outlet at the top. The waste outlets collecting the rest of the fluid connects two waste channels each, and are positioned at the side channel connection.

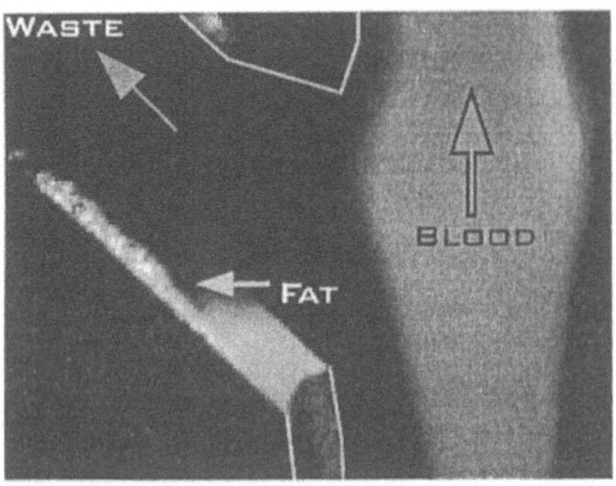

Figure 2. Close-up of the three way junction where the fat is separated from the red blood cells. Red blood cells are gathered in the single acoustic node in the ultrasonic standing wave field and the fat cells into the antinodes because of the difference in acoustic impedance between them.

PLANAR ULTRA-FILTRATION CHIP FOR RAPID PLASMA SEPARATION BY DIFFUSION

Kazuhiro Iida, Hisao Kawaura, Noriyuki Iguchi, Tohru Sano, and Masakazu Baba
Fundamental Research Laboratories, NEC Corporation,
34 Miyuki-ga-oka, Tsukuba, Ibaraki, 305-8501 Japan

Abstract

Plasma can be extracted in two seconds with a planar ultra-filtration chip without using centrifugation. The planar structure of the chip makes it suitable for mass production by indenting on plastic, and, the chip requires no extra equipment for separation. The chip is thus applicable to disposable devices for 'point-of-care' blood testing.

Keywords: ultra-filtration, plasma separation, capillary force, point-of-care

1. Introduction

A μTAS chip is small and costs little, but typically has to be used with a control box that is large and expensive. The chips generally used for clinical blood tests are such a chip. To obtain a cell-free plasma sample, the chip is spun for a few minutes in a control box equipped with a rotor, microcomputers, and a power supply [1, 2]. We have developed a planar ultra-filtration chip that can be used to rapidly separate plasma by diffusion.

2. Experimental

Figure 1 shows the structure of the chip (A: plane view, B: sectional view). The chip consists of two parallel micro-fluidic channels separated by a wall. The wall contains a row of funnel-shaped micro-openings, each with a diameter of 3.6 μm at their narrowest point (C: SEM image of an opening). The channels and wall are engraved on a (110) silicon wafer using an anisotropic etchant, 25%TMAH. The surface of the chip is oxidized with nitric acid to make it hydrophilic.

In our tests, ten-times-diluted mouse blood was applied to one channel, and saline was applied to the other. The migration of blood cells was monitored using an optical microscope. The eluted saline was mixed with reagents for detecting glucose and proteins.

3. Results and Discussion

Both liquids automatically entered the channel, and reached the terminals within two seconds. In the eluted saline, we detected glucose and proteins but no blood cells, nor collapsed ghosts of cells. These results suggest that plasma can be rapidly separated through diffusion.

Y. Baba et al. (eds.), Micro Total Analysis Systems 2002, Volume 2, 627–629.

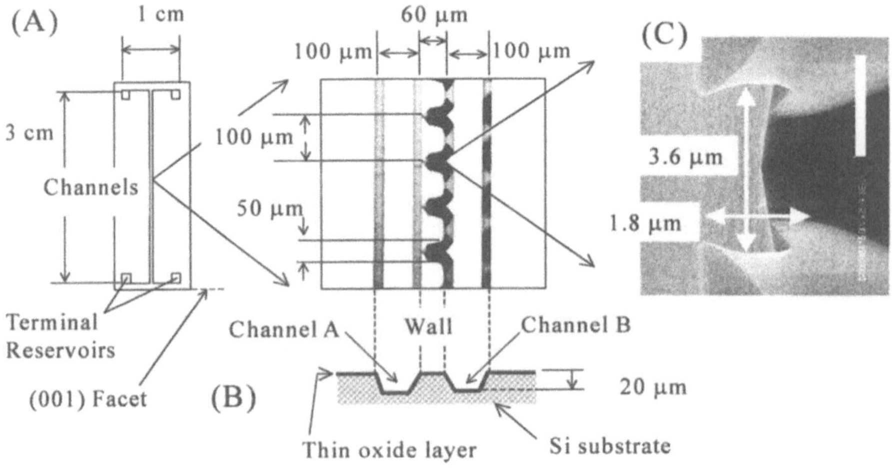

Figure 1. Structure of chip

We also observed a 'one-way effect' that allowed soft materials, such as blood cells, to pass through the wall in only one direction (the right panel in Figure 2). When the mouse blood entered the funnel-shaped openings from the wider end, red blood cells 7.5 μm in diameter could squeeze through the narrow opening, which was 6 μm in diameter, by changing their shape (the left panel in Figure 2). However, the red blood cells could not pass through the opening from the opposite direction. This effect suggests that we can separate materials based on their softness, and also that the one-way effect increases the separation efficiency because the effect should suppress the redistribution of separated materials.

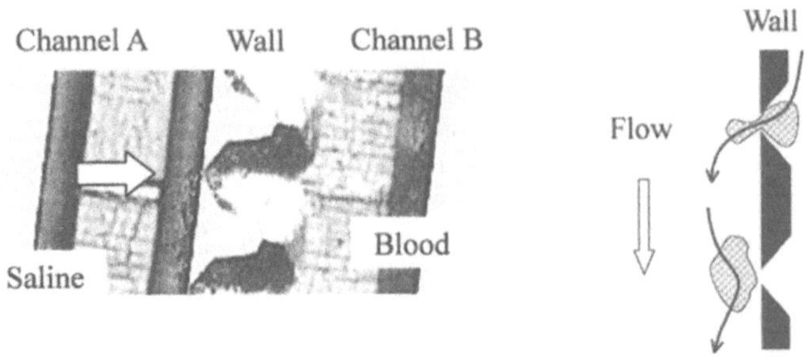

Figure 2. One-way effect

The separated plasma may then be easily introduced into subsequent micro-fluidic channels for analysis. If the openings are made with nano-scale diameters, we should be able to analyze the plasma composition by sorting molecules according to size, using the ultra-filtration device as a molecule sorter (Figure 3). Since the diameters of the openings are progressively smaller, from d1 to dN, each channel delivers molecules with diameters within a particular range.

We can thus add new functions to the device by designing openings with a particular shape. Since this device can be easily produced on a plastic surface by indention, or by injection molding, it is suitable for use as a disposable chip for 'point-of-care' clinical testing and daily laboratory analysis.

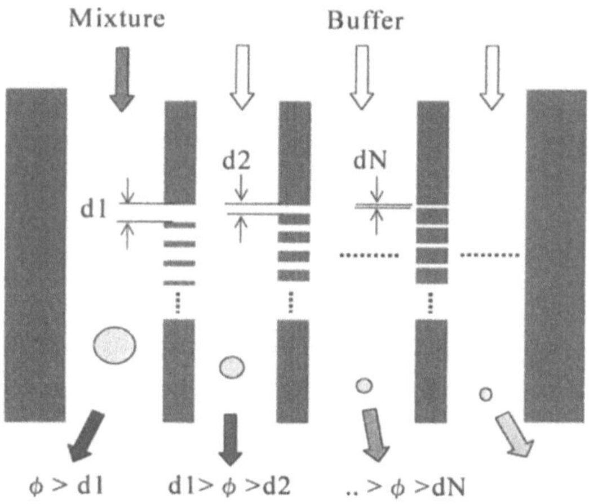

Figure 3. Molecular sorter

4. Conclusion

An ultra-filtration device was fabricated on a silicon substrate. The device separated plasma from ten-times-diluted mouse blood in two seconds. Glucose and proteins were detected in the separated fluid, but no blood cells, nor collapsed ghosts of cells.

Acknowledgements

We thank Dr. Kunihiko Naito, Laboratory of Applied Genetics, Department of Animal Resource Sciences, Graduate School of Agricultural and Life Sciences, the University of Tokyo, for the mouse blood samples.

References

1. US Patent Nos.5061381 (1991) – 5998031 (1999)
   (ABAXIS Inc., http://www.abaxis.com/default.htm)
2. Y. Horiike, et al. (2001) Biomedical Engineering Vol. 39 No. 2, 2p3 (in Japanese).

MICROFLUIDIC DEVICE FOR BIOANALYSIS USING FREELY MOVING BEADS TRAPPED IN A RECIRCULATING FLOW

G.-L. Lettieri, N. F. de Rooij, E. Verpoorte
Sensors, Actuators, & Microsystems Laboratory, Institute of Microtechnology, University of Neuchâtel, Rue Jaquet-Droz 1, CH-2007 Neuchâtel, Switzerland, Gian-Luca.Lettieri@unine.ch

Abstract
The dynamic binding of biotin-4-fluorescein (B4F) (Molecular Probes Europe BV) with 2-um streptavidin-coated (SC) beads (Polysciences) by using a continuous flow method is presented. A recirculating flow is created between micro-etched diffuser/nozzle elements. Polystyrene beads (2-μm diameter) are concentrated and held in place in a particular flow pattern while being sequentially perfused by different solutions.

Keywords: bead-based assay, recirculating flow, trapping channel

1. Introduction

In this paper, we present recent results obtained with a microfluidic device, in which beads may be trapped and controllably manipulated using fluid flows only. It was shown previous work [1] that by opposing an electro-osmotic flow (EOF) against a pressure-driven flow (PF), controlled vortex motion can be generated at diffuser and nozzle elements. The chip features a trapping channel which consists of planar diffuser and nozzle elements, connected by a narrow microchannel. Moreover, a recirculating flow is generated in the trapping channel. Beads close to and in the trapping channel are caught in this particular flow pattern and are preconcentrated by using fluid flow only. It was also shown that while the beads were recirculating, different solutions could be flushed sequentially over them [2]. This approach for bead-based assay makes it unnecessary to use physical barriers to trap the beads, does not require any channel surface modification, and ensures excellent bead handling control.

2. Experimental

Figure 1 shows a schematic of the different steps of the experiment, which resembles that described in [2] except for the bead loading step (1st step). Before use, the original SC bead buffer is replaced by 10 mM sodium tetraborate buffer (pH 9, $\mu_{EOF} = 6.2x10^4$ cm^2/Vs) by centrifugation (2x 1min at 5000 rpm). For each measurement, beads ($c_{beads} = 5x10^7$ beads/mL) are loaded as a plug into the trapping channel by EOF. This ensures that the newly bound SC bead/B4F complex concentration is constant during all experiments ($c_{beads} \cong 10^8$ beads / mL).

Y. Baba et al. (eds.), Micro Total Analysis Systems 2002, Volume 2, 630–632.

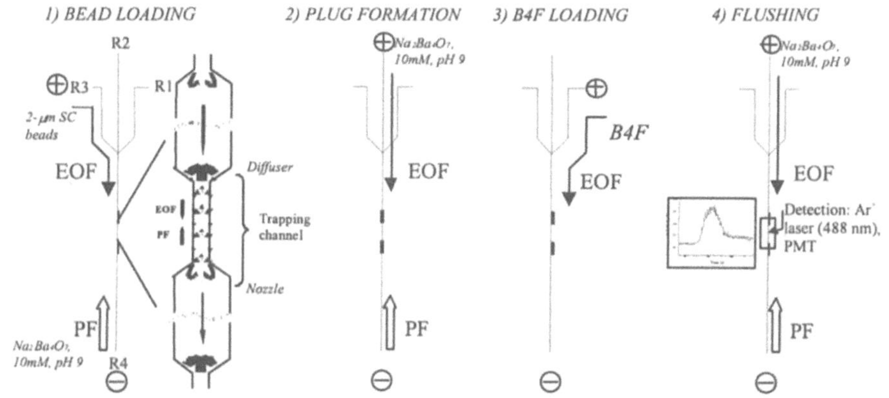

Figure 1. 1) A potential is applied between R3 and R4 (V_{34} = 2 kV, E = 290 V/cm), t_1 = 60 s. Beads are loaded electrokinetically while a counter pressure (P = 4 – 9 mbar) is applied to the waste reservoir (R4). 2) V_{24} = 2 kV, t_2 = 10 s. To define a plug of beads, a potential is applied between R2 and R4. 3) V_{14} = 0.5 kV, t_3 = incubation time. B4F is driven by EOF to the trapping channel to be mixed with the recirculating SC beads. 4) V_{24} = 0.5 kV, t_4 = t_3 + 480 s. Buffer flushes away excess unbound biotin. Once detection is done, beads are flushed away towards R4 (V_{24} = 3 kV).

3. Results and Discussion

Reproducible dispensing of beads is important, since different concentrations of trapped SC beads can result in different values of detected fluorescence for a given concentration of B4F or incubation time. This is illustrated in Fig. 2, which shows the fluorescent signal as a function of time for Protein A-Fluoresbrite YG-carboxylate beads (Polysciences, Inc.) recirculating at a diffuser element. By increasing E, the size of the bead cluster is reduced, which leads to an increased bead concentration in the detection window, and consequently, increased fluorescence. The trapping channel geometry also plays an important role in keeping the bead concentration constant. For diffusers with an opening angle, $\theta_{diffuser}$, ≥ 60°, and a cross-sectional area ratio, A_{wide}/A_{narrow}, of 5, the beads can be entirely confined at the diffuser. We did not reach this efficiency with the nozzle, but for $\theta_{nozzle} \geq 90°$ and $A_{wide}/A_{narrow} = 5$, only a few beads (~ 2 beads/s) were observed to escape. Figure 3 shows the signal as a function of incubation time for 1 μM B4F. The binding between B4F and SC beads increases up to a time of 350 s. At this point the signal starts to saturate, indicating that all the available streptavidin sites present on the beads are occupied by the biotin. For an incubation time of 150 s, the detection limit obtained was 50 nM for B4F (see inset of Fig. 3).

4. Conclusions

In particular, Figure 3 demonstrates the potential of this technology for bead-based bioanalysis. Efforts are now being focused on investigation of the kinetics of the dynamic bonding of B4F to SC beads, and on the realization of a multi-step immunoassay.

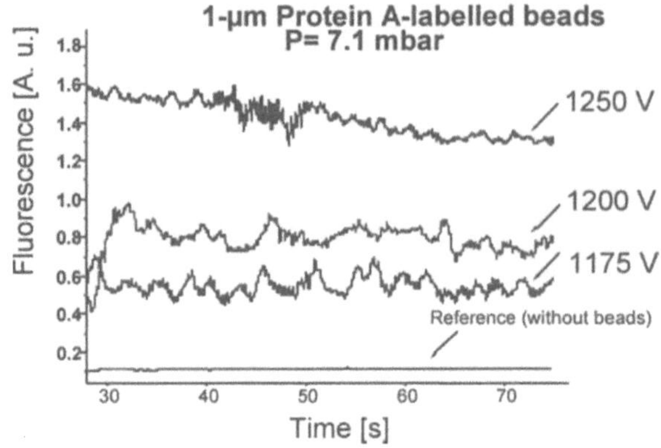

Figure 2
Fluorescence signal recorded in a cluster of freely moving, 1-μm Protein A-coated, fluorescent beads at a diffuser element for potential ranging from 1.175 kV to 1.250 kV. The higher the potentials, the more concentrated are the beads, leading to a larger fluorescent signal.

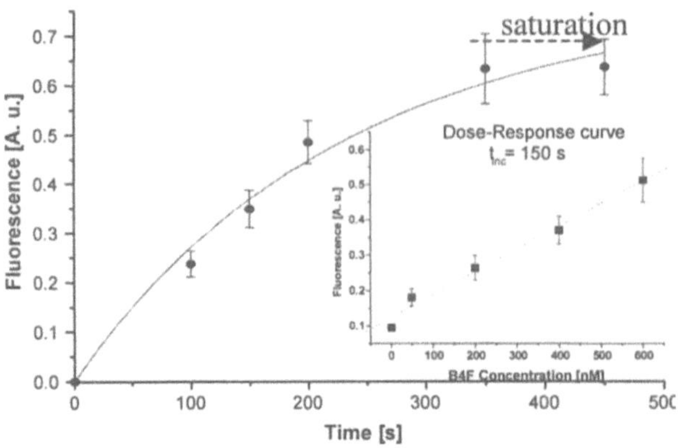

Figure 3. From the binding curves, a plot of fluorescence signal as a function of incubation time was constructed, by averaging the signal over 20 s. Inset: dose-response curve for an incubation time of 150 s.

Acknowledgement

This work was supported by the Centre Suisse d'Electronique et de Microtechnique (CSEM), located in Neuchâtel, Switzerland.

References

[1] G. Lettieri et al., Proc. Micro Total Analysis Systems 2000, Kluwer Academic Publishers, Dordrecht, The Netherlands, pp. 351-354 (2000).

[2] G.-L. Lettieri et al., Proc. Micro Total Analysis Systems 2001, Kluwer Academic Publishers, pp. 503-504, (2001).

SPR IMAGING MEASUREMENTS OF PEPTIDE MICROARRAYS IN PDMS MICROFLUIDIC CHANNELS ON GOLD THIN FILMS

Greta J. Wegner, Hye Jin Lee, and Robert M. Corn
University of Wisconsin-Madison, Department of Chemistry, 1101 University Avenue, Madison, WI 53706, USA

Abstract

The application of microfluidic channels created in polydimethylsiloxane (PDMS) to surface plasmon resonance (SPR) imaging is described. The microfluidic channels are used to both fabricate peptide line arrays and deliver small volumes (5 μL) of solution to the peptide array for the detection of sequence specific antibody adsorption onto chemically modified gold surfaces.

Key words: **SPR imaging, peptide microarrays, PDMS microfluidic channels.**

1. Introduction

Surface plasmon resonance is a label-free surface sensitive optical technique used to monitor binding interactions between biomolecules in solution and attached to a gold surface. We have recently described a fabrication methodology for creating DNA microarrays on gold surfaces for SPR imaging measurements of DNA and RNA hybridization adsorption [1]. These efforts have been extended to the study of peptide-protein interactions using SPR imaging. In this paper, we describe the use of microfluidic channels constructed in PDMS to fabricate peptide arrays on gold surfaces as well as to deliver small volumes of target molecules (e.g. antibody) for quantitative surface plasmon resonance (SPR) imaging measurements.

2. Fabrication of peptide line arrays using microfluidics

Figure 1 outlines the two-step process used to prepare the peptide array: (1) A set of parallel microchannels was used to deliver chemical reagents to covalently attach the peptide probes to the surface. The peptide array was created on a gold thin film that was modified with a self-assembled monolayer of 11-mercaptoundecylamine (MUAM). Chemical modification of this monolayer was used to attach thiol-containing peptides to the surface by a thiol-disulfide reaction that was previously developed for the attachment of thiol modified DNA [2]. After peptide immobilization was complete, the parallel PDMS was removed from the gold thin film and the MUAM background was exposed to a solution of PEG-NHS, which prevents non-specific adsorption of the antibodies onto the surface. (2) Next, a "worm" microchannel with a wraparound design was used as a small

Y. Baba et al. (eds.), Micro Total Analysis Systems 2002, Volume 2, 633–635.

volume flow cell to introduce antibody solutions (5 uL) to the peptide "line array" during SPR imaging measurements.

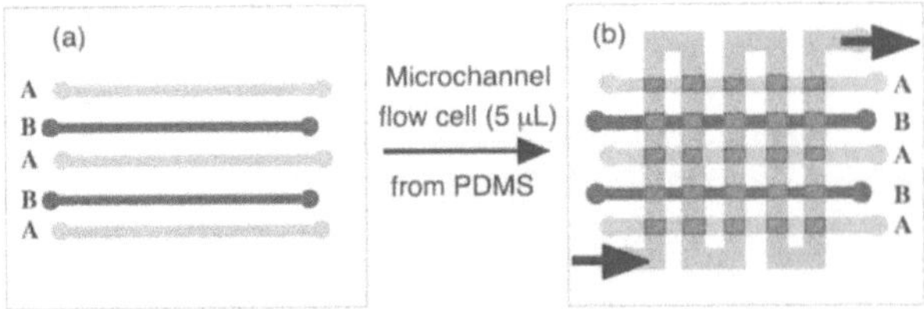

Figure 1. Schematic diagram showing peptide array fabrication using microfluidic channels from PDMS.

3. SPR imaging measurements of peptide line arrays

An array was prepared with two peptides based on the sequence of the FLAG peptide tag, an epitope tag frequently used in protein purification [3]. The introduction of a 50 nM solution of anti-FLAG M2 into the worm microchannel resulted in differential binding to the two sequences (see Figure 2 showing SPR difference imaging obtained before and after the introduction of anti-FLAG M2). As can be seen in the line profile taken across the peptide channel in Figure 2, more SPR signal is obtained for peptide A than to peptide B. This is due to the fact that these peptides differ by a single amino acid important to the binding motif of FLAG peptide to anti-FLAG M2. This data further demonstrates the utility of SPR to distinguish peptides differing by a single amino acid. The detection limit of this peptide array was about 5 femtomoles, making this technique amenable to protein studies, where sample is limited and costly.

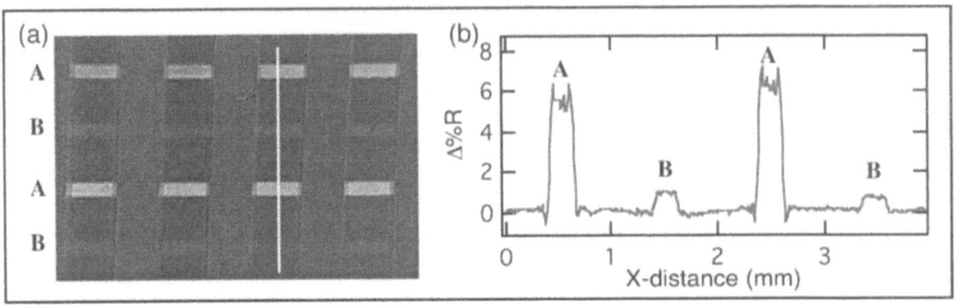

Figure 2. (a) SPR difference image of antibody adsorption onto a peptide array and (b) line profiles showing binding of anti-FLAG M2 onto a microarray of surface bound probe peptide with different epitopes listed A and B. A is CSGDYKDDDDK and B is CSGDYKDADDK.

Quantitative analysis of the antibody-binding interaction can also be determined using peptide arrays to determine the binding constant of a peptide-antibody interaction. An isotherm for peptide A was constructed by using the worm flow cell to deliver 5 μL aliquots of antibody in increasing concentration over the peptide array (Figure 3). The percent reflectivity was measured for each sample and plotted against the solution concentration. This data follows a Langmuir isotherm (solid line). The insert shows the linear relationship between the inverse surface coverage and the inverse solution concentration and the slope was used to determine the binding constant to be $1.5 \times 10^{8}\ M^{-1}$.

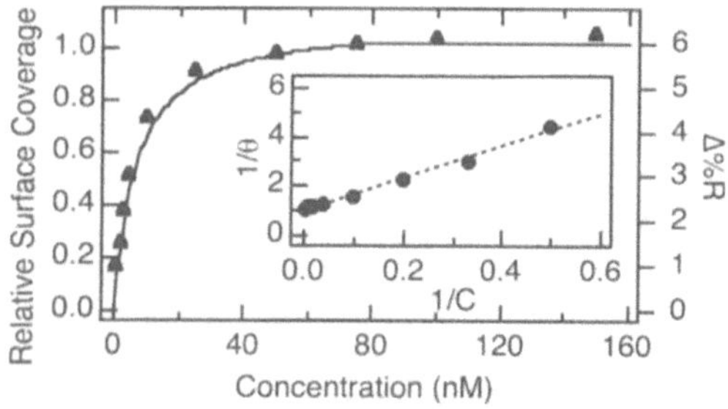

Figure 3. Langmuir isotherm of peptide-antibody interaction. The inset shows the ratio between inverse surface coverage and inverse concentration. The linear relationship demonstrates that this data follows the Langmuir isotherm theory.

4. Conclusions

The creation of peptide arrays on gold surfaces in this paper demonstrates that SPR imaging measurements can be used to study peptide-protein interactions in an array format. Peptide arrays can be made with up to 100 different peptides to permit the simultaneous quantification of peptide-protein binding interactions. This technology can further be employed to study DNA-protein and carbohydrate-protein interactions.

Acknowledgements

This research is funded by the National Institute of Health (Grant R01-GM59622-03) and the National Science Foundation (Grant CHE-0133151).

References

1. H.J. Lee, T.T. Goodrich, and R.M. Corn, *Anal. Chem*, 73, pp. 1-7, (2001).
2. E.A. Smith, M.J. Wanat, Y. Cheng, S.V. Barreira, A.G. Frutos, R.M. Corn, *Langmuir*, 17, pp. 2502-2507, (2001).
3. T.P. Hopp, K.S. Prickett, V.L. Price, R.T. Libby Raymond, C.J. March, D.P. Cerretti, D.L. Urdarl, and P.J. Conlon, *Biotechnology*, 6, pp. 204-210, (1988).

SHAH CONVOLUTION FOURIER TRANSFORM DETECTION OF PARTICLE VELOCITIES BY USING AN INTEGRATED 1x128 PLANAR WAVEGUIDE BEAMSPLITTER

Klaus B. Mogensen[1], Yien C. Kwok[2], Jan C. T. Eijkel[3], Nickolaj J. Petersen[1], Andreas Manz[3] and Jörg P. Kutter[1]
[1]MIC,Technical University of Denmark, 2800 Lyngby, DK
[2]Department of Chemistry, University of Virginia, Virginia, USA
[3]Department of Chemistry, Imperial College, London, UK

Abstract
A biochemical microdevice for particle velocity measurements is presented. The device consists of an electrically insulated channel network, a 1x128 planar waveguide beamsplitter for multiple point detection and stray light rejection structures that are monolithically integrated on a silicon substrate. Frequency information (and hence bead velocities) were obtained by Fourier transform detection while additional time information was obtained by wavelet transformation.

Keywords: SCOFT, planar waveguides, multiple-point detection, particle velocity measurements

1. Introduction

Particle velocity measurements using multiple-point detection schemes have previously been performed in a microsystem [1] where illumination of the channel segments was achieved by extensive use of free-space optical elements, such as lenses, mirrors and beam expanders. Such a set-up is difficult to miniaturize further and very sensitive to shock and vibrations. The motivation of this study was to overcome these restrictions by monolithic integration of a planar waveguide beam splitter for fluorescence excitation, which furthermore allowed alignment-free operation, because the chip could be directly connected to a light source by glued-on optical fibers.

2. Particle velocity measurements

The design of the device (4.5x5.0 cm^2) is shown in Fig. 1a. It was fabricated with silica-on-silicon technology with the use of two photolithographic mask steps. Excitation light from an argon-ion laser (488 nm) was coupled into one of the beamsplitters (the right side of the device, Fig. 1a) through a permanently attached optical fiber. Fig. 1b shows a microscope picture of a segment of the detection channel that has been filled with a 170 μM fluorescein solution in order to visualize the excitation regions. The waveguides had a width of 24 μm and were separated by 100 μm. Stray light rejection structures were furthermore integrated to reduce background excitation light between the detection regions.

Y. Baba et al. (eds.), Micro Total Analysis Systems 2002, Volume 2, 636–638.

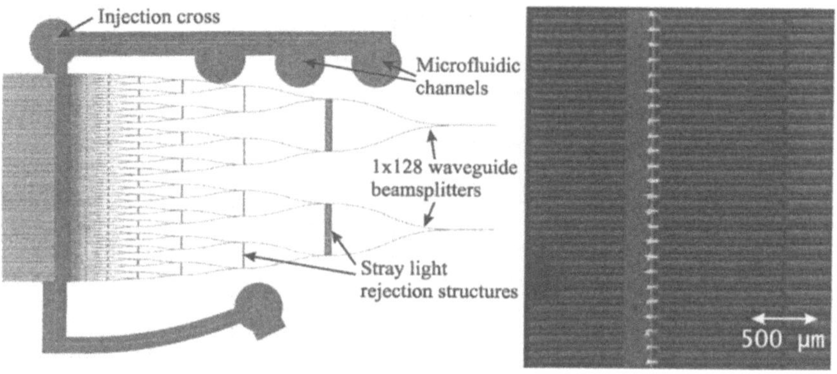

Figure 1. a) Mask design of a device for particle velocity measurements. Two 1x128 planar waveguide beamsplitters were monolithically integrated with the channel network in order to illuminate the channel at 128 regularly spaced points. b) Microscope picture of a 50 µm wide channel segment being illuminated at multiple points.

A photomultiplier tube (Hamamatsu Photonics R550) equipped with a fluorescein emission bandpass filter (520DF15 Omega Optical) and a high pass filter (LWP-25-515 Delta Light & Optics) was positioned above the channel for collection of the emitted signal. Fluorescing beads (PolyScience RG, 6 µm) were pumped down the detection channel using electroosmotic pumping with a field strength varying from 30 to 270 V/cm. The concentration of the particles was chosen to correspond to about 5-20 beads in the detection region (d=1.28 cm) at any given time. The generated emission pattern is shown in Fig. 2a.

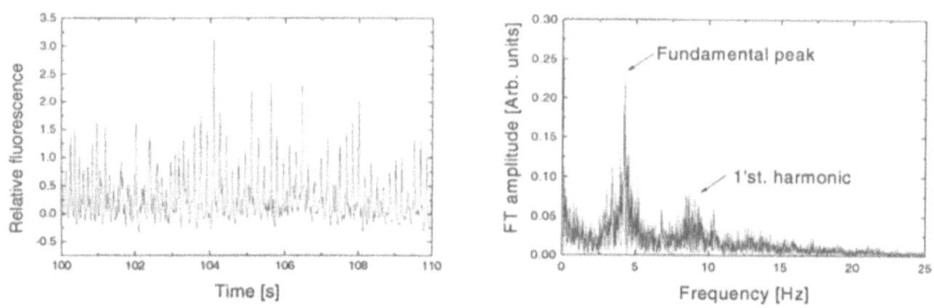

Figure 2. a) A part of an electropherogram obtained with 6 µm fluorescing beads passing the detection region (E=93 V/cm). b) Fourier transform of a 120 s. time-domain signal with similar experimental condition as in a). The position of the fundamental peak corresponded to a velocity of 430 µm/s.

The emission pattern was analysed by Fourier [1] and wavelet [2] transformation. By making a Fourier transformation of the time-domain signal the frequency components (which are proportional to the velocities) were extracted (Fig. 2b). An advantage of this approach compared to two-point measurements is that velocity information of complex samples can be obtained, because there is no restriction to having only one particle in the detection zone. The position of the fundamental peak in Fig. 2b at 4.3 Hz corresponded to a velocity of 430 μm/s. Changing the field strength resulted in a change in the position of the peaks, as expected. This is shown in Fig. 3a.

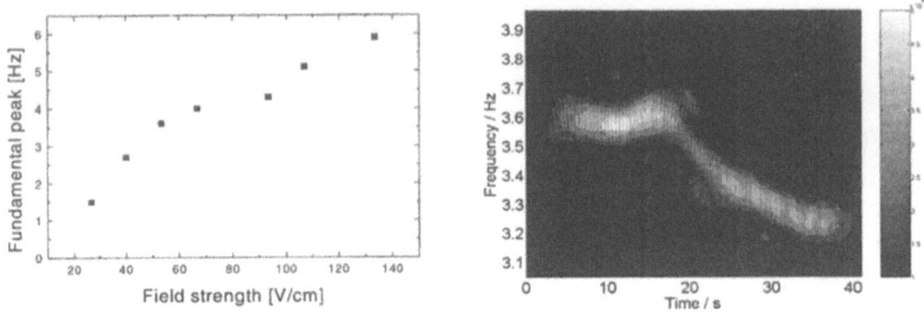

Figure 3. a) Position of the fundamental peak as a function of field strength. b) Wavelet transform detection of a single bead passing the detection region.

Fig. 3b shows a wavelet transformation of a signal generated from a single bead, which was hydrodynamically pumped past the detection region. This demonstrates the strength of wavelet analysis compared to Fourier transformations, because the frequency (and thus velocity) can be obtained as a function of time (and thus position along the channel). The decrease in frequency seen here is believed to be caused by the bead entering a different region of the parabolic flow profile due to a shift in its vertical position in the channel.

3. Conclusions

In conclusion, it has been shown that a planar waveguide beamsplitter can be used for Fourier and wavelet transform detection of particle velocities in a microsystem.

Acknowledgements

This work was funded by the Danish Government's Scientific Research Council, STVF (contract #9900683) and EPSRC UK, the UK Department of Trade and Industry (DTI) and the Lab-on-a-Chip consortium (LOC).

References

1. Y. C. Kwok, N. T. Jeffery and A. Manz. *Anal. Chem.*, 73, p. 1748-1753 (2001).
2. J. T. C. Eijkel, Y. C. Kwok, A. Manz. *Lab on a Chip*, 1, p. 122-1261 (2001).

A MICRO VISCOSITY DETECTOR FOR USE IN MINIATURIZED CHEMICAL SEPARATION SYSTEMS

M.T. Blom[1]*, E. Chmela[2], F.H.J. van der Heyden[1], R.E. Oosterbroek[1], R. Tijssen[2], M. Elwenspoek[1], A. van den Berg[1]
[1] MESA+ Research Institute, University of Twente, The Netherlands,
[2] Department of Chemical Engineering, University of Amsterdam, The Netherlands
* email: m.t.blom@el.utwente.nl

Abstract
A novel micromachined differential viscosity detector is presented that is suitable for integration with an on-chip hydrodynamic chromatography system. Viscosity detection is demonstrated using a prototype that shows a resolution in the specific viscosity of $3.0*10^{-3}$.

keywords: hydrodynamic chromatography, polymer separation, viscosity detection

1. Introduction

In liquid chromatography of polymers, viscosity sensing is one of the possible detection techniques. It is based on sensitive measurement of the specific viscosity (η_{sp}) change, caused by the dissolved polymers:

$$\eta = \eta_0(1+\eta_{sp}) \quad (1)$$

in which η is the viscosity of the solution and η_0 that of the solvent. Here, the development of a viscosity detector suitable for integration into a planar hydrodynamic chromatography (HDC) system is presented. The HDC system [1] is based on size separation of particles or polymers in a flow-field gradient in an extremely narrow channel (fig. 1). However, the viscosity detector is generally applicable as a differential viscosity sensor and can thus be applied as a detector for all types of polymer chromatography.

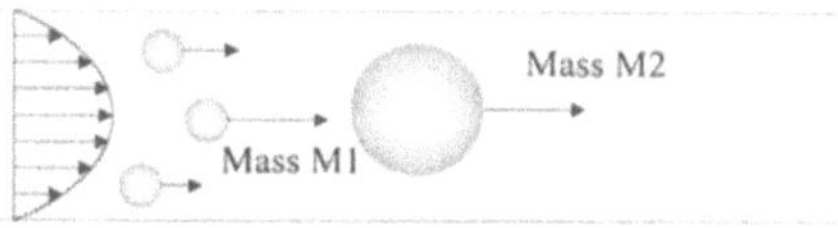

Figure 1: HDC principle. The average velocity of large molecules is higher than that of small ones, since larger molecules are restricted from the wall of the channel. In a small channel this effect is strong enough to allow analytical separation.

2. Design

In order to realize sensitive detection ($\Delta\eta_{sp} \approx 10^{-4}$), a differential measurement of the viscosity is required, using a fluidic Wheatstone bridge [2] as shown in figure 2. Initially all four fluidic resistances are filled with solvent and the bridge is balanced. If a polymer peak from a separation channel enters the bridge, it will be absorbed by the delay volume in one branch and will create an increase in pressure drop in the other branch, thus unbalancing the bridge.

Y. Baba et al. (eds.), *Micro Total Analysis Systems 2002, Volume 2*, 639–641.

The specific viscosity can be detected by measuring the total bridge pressure P_1 and the differential pressure P_2:

$$\eta_{sp} = \frac{4P_2}{P_1} \tag{2}$$

As the volume flows in HDC systems are in the nl/s range, a low hydraulic capacitance of the pressure sensor P_2 in figure 2 is essential to keep response times short. Therefore, a differential membrane pressure sensor, having an extremely low hydraulic capacitance of $C \approx 4*10^{-19}$ m^3/Pa, was developed [3].

The pressure sensor consists of a silicon nitride membrane supported by a silicon frame. The fluidic channels defined in the silicon are closed using a (transparent) Pyrex wafer. This is essential as the readout of the membrane deflection is performed optically using an AFM type system in which the angular deflection of a laser beam, reflected by the membrane, is detected with a split photodiode. The pressure sensor is anodically bonded to a Pyrex-silicon structure into which the Wheatstone structure is defined. A part of the Wheatstone bridge and the injection region used for definition of a predefined viscosity plug, is shown in figure 3. The injection structure is identical to the one used for definition of a sample plug in the hydrodynamic chromatography chip that is described in [1].

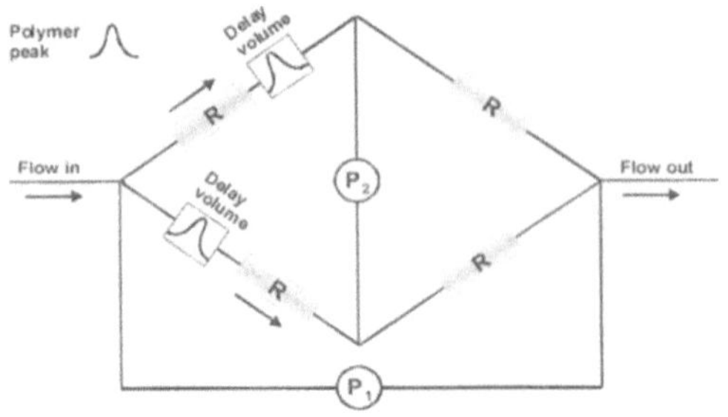

Figure 2: Differential viscosity detection by using a Wheatstone bridge of fluidic resistances. A polymer plug entering a balanced bridge will be delayed in one branch creating a detectable differential pressure P_2.

3. Results and discussion

The characterization of the stand-alone differential pressure sensor P_2 resulted in a pressure resolution of 2 Pa for a 125 μm wide and 200 nm thick membrane. For the first integrated viscosity detector a larger membrane was used. This 206 μm wide and 470 nm thick membrane theoretically is less sensitive than is required for a functional HDC detector. The integrated differential viscosity detector is shown in figure 4. It incorporates the differential pressure sensor and a commercial Honeywell sensor for measurement of the absolute bridge pressure. Fused silica capillaries are used for fluidic connection to an external valve system.

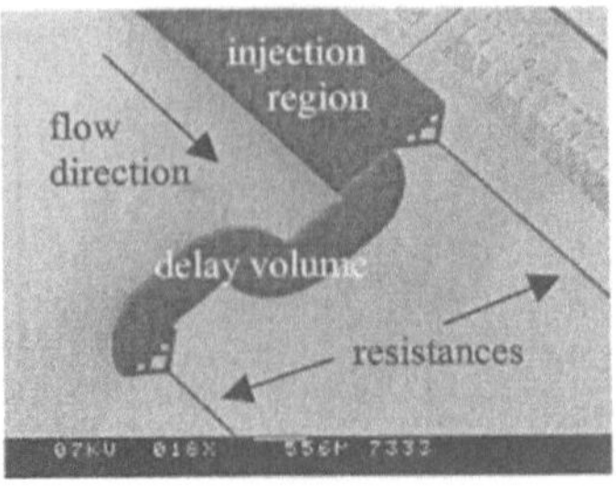

Figure 3: SEM picture of part of the Wheatstone bridge showing the deeper delay volume and two bridge resistances, preceded by a nozzle.

The prototype detector shown in figure 4 was characterized using the injection of a known plug, thus creating a predefined viscosity signal. Both the bridge pressure P_1 and the differential pressure P_2 were measured. The result for a plug of ethanol in DI water injected at 0.25 bar is shown in figure 5. Injection of ethanol plugs at different pressures allows an estimate of the resolution of $\Delta\eta_{sp} = 3*10^{-3}$ for this non-optimized micro viscosity detector. Besides resizing of the differential pressure membrane, several enhancements in the differential pressure sensing are possible that can be applied for optimization of the viscosity detector. Upscaling of the complete system is beneficial as well, possibly enabling detection in micro liquid chromatography (LC) applications.

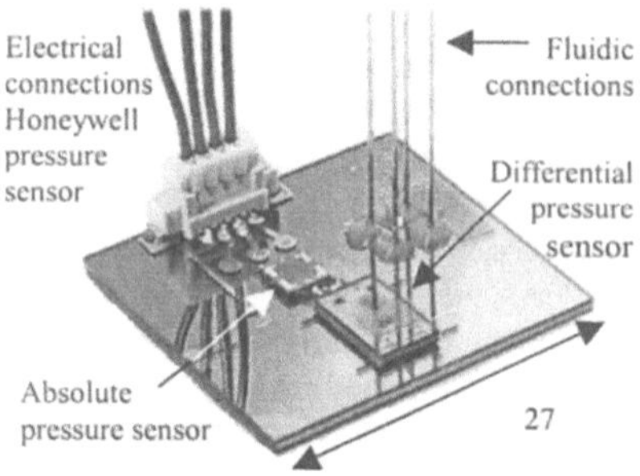

Figure 4: Close-up of a ready-to-use differential micro viscosity detector.

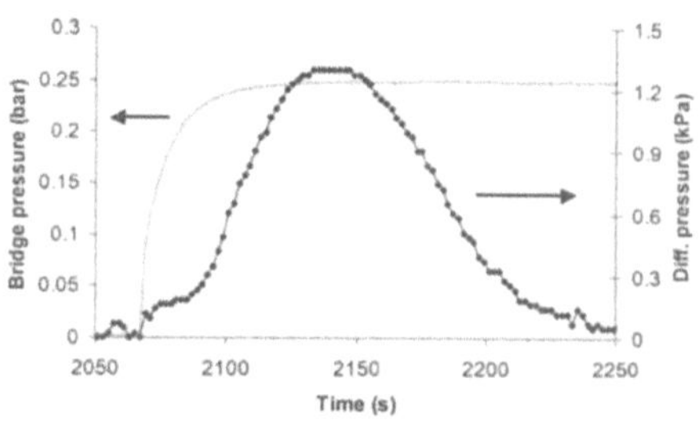

Figure 5: Pressure response of an ethanol plug that was injected at 0.25 bar. A viscosity peak is clearly visible.

4. Conclusions

A functional differential viscosity detector was designed and manufactured that is suitable for integration with an on-chip hydrodynamic chromatography system. Using a non-optimized system, viscosity sensing was demonstrated with a resolution in specific viscosity of $3.0*10^{-3}$.

Acknowledgements

This research was supported by the Dutch Technology Foundation ("Stichting Technische Wetenschappen"), project number AAC4556.

References

1. E. Chmela, M.T. Blom, J.G.E. Gardeniers, A. van den Berg, R. Tijssen, A Chip System for Size Separation of Macromolecules and Particles by Hydrodynamic Chromatography, Anal. Chem. 74, no. 14, p. 3470-3475, 2002.
2. Max A. Haney, The differential viscometer I. A new approach to the measurement of specific viscosities of polymer solutions, J. Appl. Polymer Science 30, p. 3023-3036, 1985.
3. F.H.J. van der Heyden, M.T. Blom, J.G.E. Gardeniers, E. Chmela, M. Elwenspoek, R. Tijssen, A. van den Berg, A micro viscosity detector for a planar hydrodynamic chromatography (HDC) system, Proceedings μTAS Enschede, pp. 595-598, 2000.

TOWARDS INTEGRATED MICROSYSTEMS FOR CHEMICAL SYNTHESIS

Klavs F. Jensen
Chemical Engineering and Materials Science & Engineering
Massachusetts Institute of Technology, Cambridge, MA, USA
(kfjensen@mit.edu)

Abstract

Steps towards the realization of integrated chemical systems are presented with particular emphasis on multiphase reactions, chemical separation, process monitoring, and the integration of these components. Flow visualization techniques for multiphase flows and microfabrication enabled methods for characterizing mass transfer are described. Techniques that exploit surface tension effects are introduced for efficient separation of gas-liquid flows. Microsystems capable of creating and breaking high interfacial area emulsions are proposed for extraction of products. Integration of analytical tools is achieved in terms of on-chip integration of optical spectroscopy as well as in-line use of gas-chromatography and mass spectroscopy to microfluidic devices. Challenges in the integration of microfluidic, reaction, separation, and analytical units into a microchemical system are considered starting from experiences with a multireactor system for testing heterogeneous catalytic reactions.

Keywords: microreactors, separations, process monitoring, integration

1. Introduction

Integration of advances in microTotal Analysis Systems (μTAS) [1] with chemical synthesis promises to yield a wide range of efficient devices for high throughput screening, reaction kinetic studies, and process optimization. Rapid screening of reaction pathways, catalysts, and materials synthesis procedures could provide faster routes to new products and optimal operating conditions. Moreover, microchemical systems would clearly require less space, be easier to vent, use fewer utilities, produce less waste, and offer safety advantages over conventional synthesis set-ups in chemical fume hoods.

Microreactors with sub millimeter dimensions have been demonstrated for a wide range of chemical reactions [2-4]. Reduction in size from conventional synthesis platforms has been accomplished for both homogeneous and heterogeneous chemical reactions and in many cases produced improved performance relative to macroscopic systems. In particular, the high heat and mass transfer rates possible in microfluidic systems allow reactions to be performed under more aggressive conditions with higher yields than can be achieved with conventional reactors. More importantly, new reaction pathways deemed too difficult to control in conventional macroscopic equipment can be conducted safely because of the high heat transfer and ease of confining small volume.

Y. Baba et al. (eds.), Micro Total Analysis Systems 2002, Volume 2, 642–645.

These inherent safety characteristics of microreactors also imply that systems of multiple microreactors could be deployed in distributed point-of-use synthesis of small volume chemicals with storage and shipping limitations, such as highly reactive and toxic intermediates.

Development of integrated microchemical systems requires advances in the individual components (microreaction technology, separation methods at μm scales, and chemical analysis) as well as integration techniques (e.g., microfluidic elements, control, and packaging). Examples of advances in components are given along with a discussion of integration challenges.

2. Multiphase microreactors

Both single phase and multiphase reactions have been demonstrated in microchemical reactors [2-4]. This contribution focuses on gas-liquid-solid reactions, including direct fluorination and oxidation. Flow visualization techniques provide basic understanding of the underlying complex flow phenomena needed to design multichannel and multiphase microreactors (cf. Figure 1) with optimal contact between reactants as well as uniform flow distribution. Microfabrication techniques are used to build sensing ports into reactor channels allowing local sampling of gas-liquid reactions and thus, direct measurement of local mass transfer rates necessary to optimize and predict reaction rates. Multiphase reactor design strategies are formulated on the basis of reactor geometry and combined data on pressure drop, flow regimes, and mass transfer rates

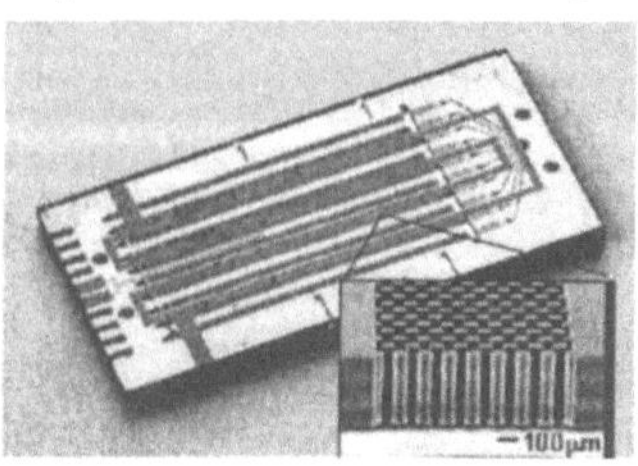

Figure 1. Multichannel gas-liquid microreactor with microfabricated catalyst

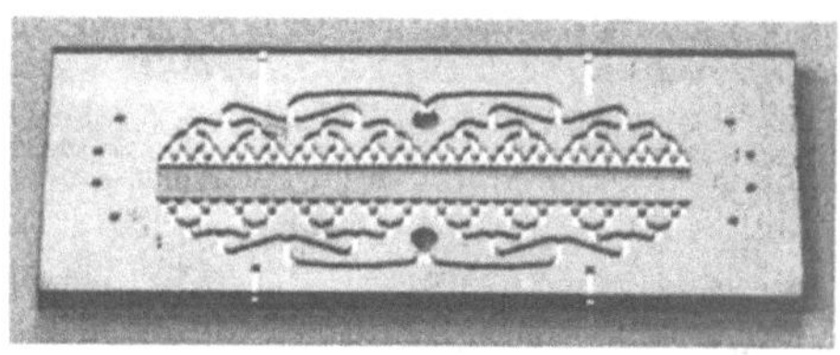

Figure 2. Cross-flow microreactor for evaluating catalytic reactions

systems.

3. Separation in microchemical systems

Application of multiphase reactions in microchemical systems requires integration of efficient microfabricated gas-liquid separation units. Surface tension effects in multiple capillary columns are shown to enable high liquid separation rates without gas entrainment. In conventional chemical synthesis, the product is typically separated from the reactor effluent by methods such as crystallization, extraction, or distillation. Solid formation is typically not desired in microsystems due to the potential for plugging. The exception is materials synthesis where microfluidics can provide control of particle size distribution, surface chemistry, and materials characteristics. This use of microchemical systems is illustrated with specific examples, including the synthesis of colloidal particles.

Extraction in microsystems has been done previously by using the laminar nature of microfluidic flows to co-flow the extracting fluid and reactor effluent so that the product is able to diffuse into the product stream. However, the liquid-liquid interface can be difficult to stabilize for immiscible liquids, e.g., water – organic solvent. Moreover, the interfacial area for mass transfer between two co-flowing stream is small – even for a microsystem – when compared to reactor effluent broken up into small droplets in the extracting solvent. To address theses issues, extraction techniques are presented that are based on first forming an emulsion and then using electrical fields to initiate coalescence and segregate the two phases.

4. Monitoring reactions in microchemical systems

The integration of optical spectroscopy, specifically Fourier transform infrared and ultra violet–visible spectroscopy, by transparent windows, wave-guides, and fibers provide opportunities for monitoring reaction conversion and selectivity. This ability, when combined with flow and temperature sensing provides opportunity for on-line optimization of reaction products as well as obtaining reaction kinetic data. Challenges in applying these optical techniques to multiphase reaction systems are discussed. Gas chromatography and mass spectrometry can also be used to monitor reaction as long as the sample size and its collection are properly matched to the microfluidic volume.

Characterization of catalytic systems presents special challenges in terms of measuring the catalytic activity as well as measuring properties of small amounts of catalyst. Applications of microchemical systems in catalyst development are reviewed with particular emphasis on reactors capable of handling catalyst in powdered form (see Figure 2) from common catalysts synthesis procedures and allowing for post-reaction catalyst characterization. Methods for characterizing reactions without confounding mass and heat transfer are described along with techniques for measuring properties of the small amounts of catalysts (~mg). Compared to previously reported values from other catalysts, the agreement of kinetic parameters and mechanistic information obtained in the microreactor demonstrates the potential of using microfabricated reactors as laboratory tools for heterogeneous catalyst testing.

5. Integration of microchemical systems

The necessity of synthesizing sufficient quantities for subsequently evaluation usually dictates that microchemical systems be operated as continuous systems. These systems will need fluid controls for adjusting reagent volumes and isolating defective units. Early implementation of integrated microchemical systems is likely to involve modular systems with microreactors, separation units, and analytic components mounted on electrical, fluidic, and optical "circuit boards" (see Figure 3). Critical challenges in realization such systems are introduced starting with experiences from a multireactor system for investigating heterogeneous gas phase reactions (see Figure 3) [5].

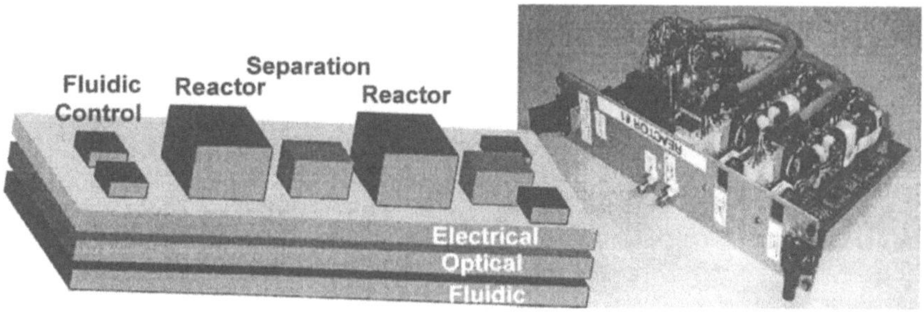

Figure 3. Schematic of modular, integrated microchemical system (left). Example of microreactor electrical-fluid board from mutireactor catalyst test station (right).

6. Conclusions

The integration of μTAS with chemical synthesis into microchemical systems could ultimately revolutionize chemical research by providing flexible tools for rapid screening of reaction pathways, catalysts, and materials synthesis procedures as well faster routes to new products and optimal operating conditions. Moreover, such microchemical systems would clearly require less space, use fewer resources, produce less waste, and offer safety advantages. Progress towards integrated systems will require continued development and characterization of microreactors, separation units, and integrated analytical methods as well as new, innovative approaches for connecting modular microfluidic components into flexible fluidic networks allowing active control and with the potential for matching components to a particular chemistry.

Acknowledgements

The author thanks Professor Martin A. Schmidt and members of the research group for collaborations forming the basis for this contribution, as well as DARPA and the members of the MicroChemical Systems Technology Center for funding.

References

1. J.M. Ramsey and A. van den Berg, (Eds.), *MicroTotal Analysis Systems (μTAS) 2001*, Kluwer Academic Publisher, Dordrecht, The Netherlands, (2001).
2. W Ehrfeld, V. Hessel, and H. Löwe, Microreactors: New Technology for Modern Chemistry, Wiley-VCH, Weinheim, Germany (2000).
3. P.D.I. Fletcher S.J. Haswell, E.Pombo-Villar, B.H.Warrington, P. Watts, S.Y.F. Wong, and X.L. Zhang "Micro reactors: principles and applications in organic synthesis," *Tetrahedron*, **58** 4735-4757 (2002)
4. K.F. Jensen, "Microreaction engineering—is small better?" Chem. Eng. Sci., **56,** 293-303 (2001).
5. D.J. Quiram," *Characterization and Systems Integration of Microreactors*," Ph.D. Thesis, Massachusetts Institute of Technology (2001)

Enabling Technologies for a Personal Flow Cytometer (Part I)

C. Cabuz, A. Padmanabhan, B. Fritz, E. Cabuz, J. Schwichtenberg, P. Reutiman, B. Demers, T. Rezachek, E. Satren, F. Battrell[†]
Honeywell Intl., 12001 State Highway 55, Plymouth, MN 55441, USA
[†]Micronics Inc., 8463 154th Ave NE, Redmond, WA 98052, USA

Abstract

The paper reports on three key technologies that have enabled the development of a wearable flow cytometer: flow sensing and control based on low-power MEMS-based sensors/actuators; microfluidic processing based on miniaturized disposable plastic cartridges; and optical detection based on arrays of solid-state lasers/detectors and integrated micro-optics. This instrument can provide early detection of viral and bacterial infections based on quick (30 seconds) measurement of lymphocyte and neutrophil counts [1]. Personnel with little or no medical training can use this low-cost, low-maintenance instrument in doctor's offices, homes, and public/private-sector facilities. Multiparameter scattering-based or fluorescence-based cytometers can be developed based on this technology for clinical, industrial and military applications.

Keywords: Cytometry, Flow controller, VCSEL array, Light scattering, Flow sensor, Microvalve, Cartridge

1. Introduction

The counts of different types of white blood cells is the commonly used indicator for the diagnosis of infections, for monitoring cancer patients during chemotherapy, for monitoring AIDS patients, or for monitoring patients with blood disorders. Such blood analysis is currently being performed in central labs on expensive, high-maintenance equipment and can be time consuming. There is therefore a need for a miniaturized and portable cytometer system that could be used in each doctor's office, as a point-of-care instrument at home for patients with chronic diseases, at all admission sites in clinics or in places with high risk of exposure to biowarfare agents. In this paper we report on a highly integrated, low-cost, low-maintenance wearable/personal flow cytometer that is capable of counting and characterization of white blood cells. The cytometer uses disposable analysis cartridges pre-loaded with reagents to support the measurement. A drop of blood is drawn into the cartridge by the user. The cartridge is then loaded into the instrument which automatically conducts the optical scatter measurements to provide the counts of different types of white blood cells. The following sections describe the features and operation of the various subystems and the fully integrated personal cytometer.

2. Fluid Driving Subsystem: The Micro flow controller

Most of the existing precision fluid delivery systems (electroosmosis/electrophoresis-based, peristaltic pumps, syringe pumps) have pulsatility, power consumption and/or size/weight issues that prevent them from being used in a wearable cytometer application. We have developed a three-channel micro flow controller which is highly miniaturized (about 12 in^3

Y. Baba et al. (eds.), Micro Total Analysis Systems 2002, Volume 2, 646–648.

size), low-power/low-cost and capable of producing pressure-driven, pulse-free liquid flows in the nL/sec to mL/sec range [2]. Each channel of the controller operates in closed-loop and relies on two high-speed (~ 1 ms response time), low-power (< 1 mW) microvalves and a highly sensitive liquid flow sensor. The flow sensors can either be mounted in the instrument body, as shown in the brass board cytometer prototype (Fig. 1b) or can be embedded directly in the disposable cartridge (Fig. 1a). The overall response time of the flow controller is about 10 milliseconds and the system can be manually pressurized for this application. A figure of merit defined as: *size x cost x pulsatility* for this flow controller is more than *two orders of magnitude better* than that for currently available, state-of-the-art open-loop fluid pumping systems.

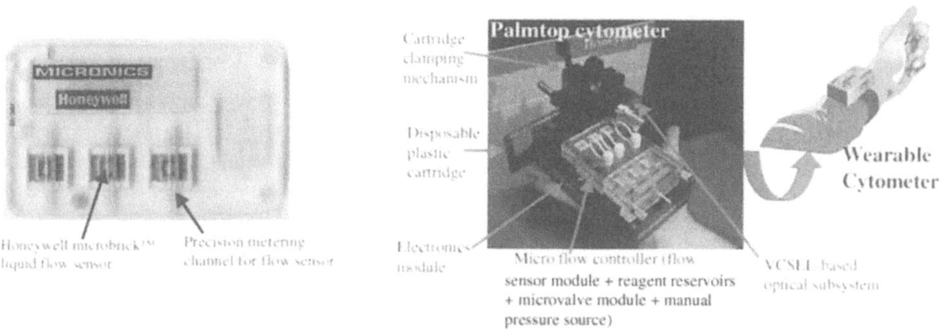

(a) (b)

Figure 1. (a) Photograph showing three MEMS liquid flow sensors embedded into a 8 cm x 5 cm size disposable fluidic cartridge; (b) Photograph of the 3 lb/144 in^3 palmtop cytometer prototype shown along with a schematic picture of the next generation 1 lb/16 in^3 wearable cytometer.

3. Disposable microfluidic analysis cartridges

The cartridge/microfluidic process technologies are presented in detail in a separate but related abstract (Enabling Technologies for a Personal Flow Cytometer - Part II).

4. VCSEL-based optical subsystem

In flow cytometry the optical alignment of the focussed laser spot with the flow path of the particle to be analyzed is critical and most instruments require periodic realignment and maintenance, typically performed by qualified personnel. The alignment problem is exacerbated by the variability in the core location in a low-cost, disposable plastic cartridge. We are proposing a novel solution to this alignment problem, based on Honeywell's red (670 nm) Vertical Cavity Surface Emitting Laser (VCSEL) arrays. The single laser used in conventional cytometers is replaced with a linear array of VCSELs (as shown in Fig. 2a) that provides: a) A means of determining the exact path of a particular cell (in the core flow); b) A means of actively locating the cytometer flow channel; c) Multiple measurements of size for each cell; and d) Time-of-flight measurement of velocity of each cell.

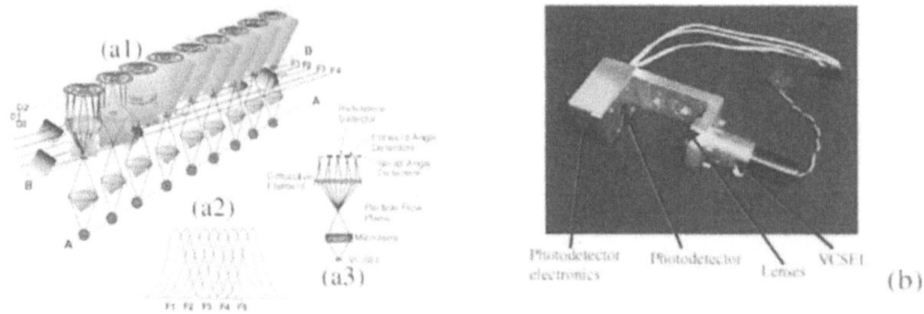

Figure 2. (a) (a1) The VCSEL array optical subsystem concept showing a linear VCSEL array AA' and focussing microlenses form a line BB' of focussed spots in the particle flow plane. (b) Photograph of the compact (5" x 2") integrated optical module

5. System integration

A modular palmtop brassboard prototype of the microcytometer with an overall footprint of 6" x 6"x 4", and weight of 3 lbs was built (Fig. 1b), that represents a *100x reduction in weight x power relative to existing* cytometers. The next generation wearable version of this cytometer will have specifications of: 4" x 4'x 1"/1 lb/0.5 W.

6. Experimental results

The brassboard palmtop cytometer has been extensively characterized using bead and blood samples. Repeatable counting and separation was demonstrated with 5 and 6 μm bead mixtures, as shown in Fig. 3a. In addition, counting and classification of three types (lymphocytes, neutrophils and monocytes) of white blood cells has been demonstrated, as shown in Fig 3b.

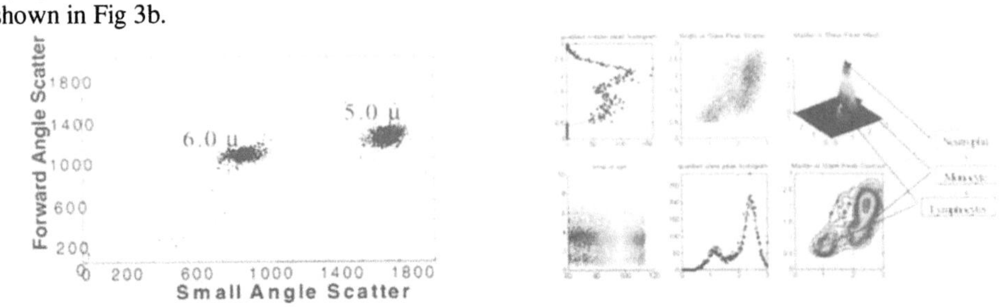

Figure 3. (a) Scatterogram showing the separation of 5 and 6 um beads; (b) Data showing the separation of 3 types of white blood cells – lymphocytes, neutrophils, and monocytes, obtained from a whole blood sample after lysing the red blood cells.

Acknowledgements

This work, done in collaboration with Micronics Inc., was supported by DARPA Contract MDA972-00-C-0029. Micronics Inc. designed/fabricated the microfluidic cartridges tha was used for this work.

References

1. [http://www.darpa.mil/MTO/bioflips/presentations/2001-1/index.html [DARPA BioFlips Program]

2. Cabuz, E., et al., "MEMS-based flow controller for flow cytometery," *Proceedings Hilton Head 2002: Solid-State Sensor, Actuator and Microsystems Workshop*, June 2 - 6, 2002, Hilton Head Island, South Carolina, pp. 110-111.

LEUKOCYTES DISCRIMINATION BY IMPEDANCE SPECTROSCOPY FLOW CYTOMETRY

Shady Gawad[1], Pascal Batard[2], Urban Seger[1], Stefan Metz[1], Philippe Renaud[1]
[1]STI-LMIS, Swiss Federal Institute of Technology, Lausanne, Switzerland.
[2] Ludwig Institute for Cancer Research, Lausanne, Switzerland.
shady.gawad@epfl.ch

Abstract

A device is presented which allows precise measurement and control of single cells using electrical means in a flow-through microchannel. The manipulation is based on the negative dielectrophoresis principle and the measurement is done by impedance spectroscopy. New results show the improvements obtained from cell pre-focussing on the measurement reproducibility as well as a fast sorting geometry. An application of the device to the discrimination of leukocytes sub-populations has been demonstrated.

Keywords: Cytometry, Impedance, Cell sorting, Leukocytes, Dielectrophoresis.

1. Introduction

Several chips for automated in-flow cell characterization and separation have been presented in the recent years. Some of theses devices use dielectrophoretic forces to separate cells from a heterogeneous population based upon differences in their dielectric properties [1,2]. Another approach which combines impedance spectroscopy flow cytometry (ISFloC) for cell analysis and dielectrophoresis for cell sorting has been presented last year [3,4]. Early measurements have shown that particles of different sizes as well as ghost-erythrocytes and normal erythrocytes can be differentiated by this methode.

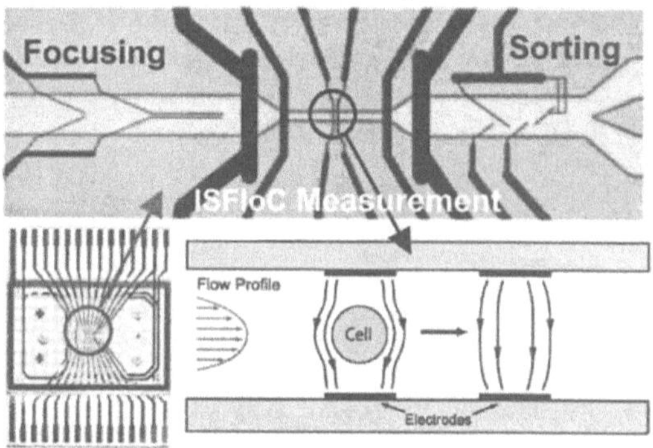

Figure 1. The new version of the chip includes a focusing electrode arrangement upstream of the measurement area. The cross section schematics shows the top and bottom electrodes used for the differential impedance measurement. The cell impedance is compared to the impedance of the media. Finally the sorting is done with the arrangement of electrodes on the right.

Here, we present a new version of this device with some critical improvements such as: A. changes in the detection electrode geometry, B. particle focusing and C. fast sorting functionality (Fig. 1). New measurement results are also reported.

Y. Baba et al. (eds.), Micro Total Analysis Systems 2002, Volume 2, 649–651.

2. Experimental

A. The use of overlapped electrodes on top and bottom of the channel produces a more uniform electric field and a smaller detection area. Noise removal and accurate cell speed measurement is achieved with a differential detection area.

B. A major limitation of the first version was fluctuations in the measurement due to the positioning of the cell inside the detection channel. This was caused by off-centered cells which were close to the electrodes thus giving higher signals. Our experiments demonstrate that dielectrophoretic focusing is a very efficient way to achieve a reproducible positioning in miniaturized flowcytometry as demonstrated by less scattered data (Fig. 2). Interestingly, it also regulates the cell speed through the measurement and sorting areas, thus making the sorting mechanism control more simple and efficient.

C. The geometry used in the sorting dielectrophoretic electrodes is optimized to extract the desired cells in the smallest time window (Fig. 3).

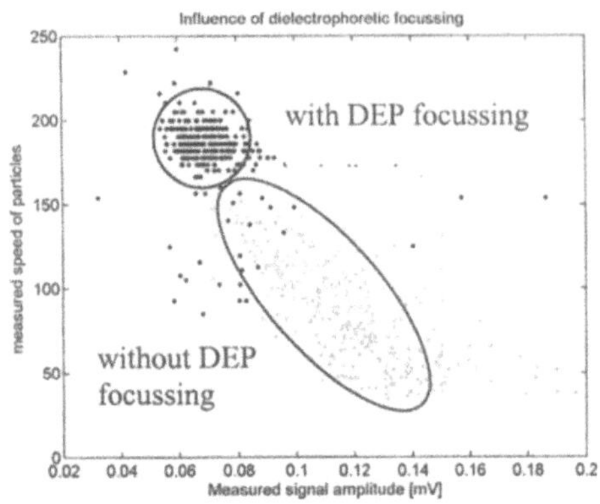

Figure 2. With the use of DEP the particles are positioned in the center of the channel, their speed is increased and the impedance signal dispersion is considerably decreased. This measurement has been done with 5 micrometers calibrated latex particles (serva).

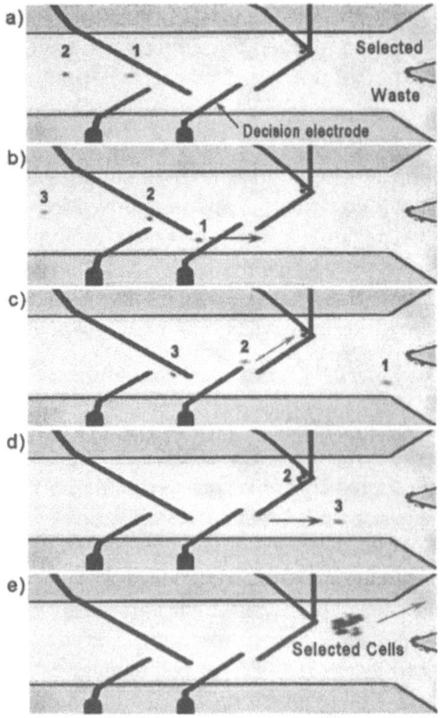

Figure 3. Video sequence showing the sorting of cell n°2 with a DEP arrangement of electrodes. **a)** Cells 1 and 2 arrive from the measurement channel centered. **b)** They are deflected from the first upper electrode. The decision electrode is inactive thus cell 1 goes to the waste. **c)** The decision electrode is activated and cell 2 goes to the collection area. **d)** The decision electrode is inactived again and cell 3 goes to waste. **e)** When enough cells have been collected they are released to the upper outlet.

3. Results and discussion

Some new biological experiments have been performed using FACS presorted leukocytes (BD, FACSVantage SE®) from donors. The results show that mixed populations of lymphocytes and granulocytes are discrimated by the ISFloC chip (Fig. 4). Another measurement run has been performed for validation with a mixed (lymphocytes+granulocytes) cell population which shows two distinct clouds at the same locations.

Also, paramagnetic microbeads (MACS®) were used as an alternative to FACS sorting for cell purification prior the impedance measurement, this methode being much faster than sequential FACS sorting. No modification in the obtained results could be noticed at the chosen excitation frequency.

As a conclusion, this measurement shows that it is possible to achieve discrimination of cell populations which is similar to the standard FACS forward and side size scatter technic using lower cost on-chip electrical detection.

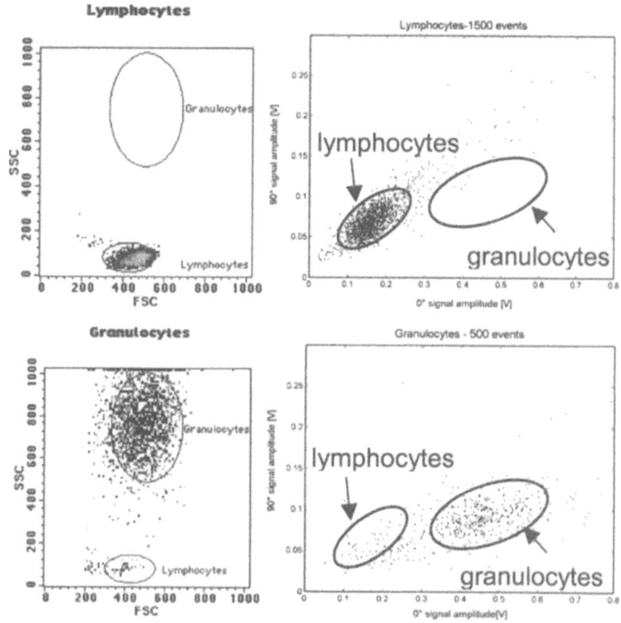

a) FACS data b) ISFloC data

Figure 4. Cloud plots of pre-sorted lymphocytes (top row) and granulocytes (bottom row).

a) Forward and side-scatter values as measured by FACS. The population of granulocytes is relatively smaller and slightly contaminated with lymphocytes.

b) The same populations as measured by impedance spectroscopy in the ISFloC chip (1.7 MHz sine excitation, in-phase vs. out-of-phase signal).

REFERENCES:

1. T. Schnelle et al., Journal of Electrostatics vol.47 (3), 1999,pp 121-132.
2. D. Holmes, H. Morgan, MicroTAS 2001, pp. 111-112.
3. S. Gawad et al., Lab on a Chip, vol.1, 2001, pp. 76-82
4. S. Gawad et al., MicroTAS 2001, pp. 253-255.

DESIGN AND FABRICATION OF MICRO/NANO-FLUIDIC CHIP PERFORMING SINGLE-CELL POSITIONING AND NANOLITER DRUG INJECTION FOR SINGLE-CELL ANALYSIS

Kwang-Seok Yun[1], Sang-Il Lee[2], Gyun Min Lee[2], and Euisik Yoon[1]
[1]Department of Electrical Engineering and Computer Science,
Korea Advanced Institute of Science and Technology (KAIST),
373-1, Guseong-dong, Yuseong-gu, Daejeon 305-701, Korea
[2]Department of Biological Sciences, KAIST, Daejeon, Korea
ksyun@iml.kaist.ac.kr

Abstract

We have developed a microfluidic chip for single-cell analysis and drug screening. The proposed device consists of surface-modified silicon channels capped with a grooved PDMS cover layer. Some region of the channel surface is treated to be hydrophobic for air leak during drug injection. We have demonstrated that the proposeded device can provide self-controlled single-cell isolation/positioning and independent drug injection onto the captured single-cell.

Keywords: cell isolation, cell capture, single-cell analysis, sample injection, surface modification

1. Introduction

Cells are usually influenced by a mixture of hormones, ions, and neurotransmitters released from neighboring cells. Dynamic monitoring of single cell is important for individual cell analysis without the influence of neighboring cells in independently controlled environments. Previously, Thielecke *et al.* reported the single-cell positioning chip which places single-cells on ring electrodes [1]. However, it was difficult to independently control the environment of each cell located in individual ring electrode structure. In this work, we report a microfluidic chip in which single-cell is autonomously captured in a cell-positioning site by a pre-defined fluid stream and a specific liquid flow including drugs or reagents can be consecutively supplied to each single-cell through a flow injection channel.

2. Design

The schematic view of the proposed structure is shown in Figure 1. The cell suspension media is introduced into the inlet by external pressure and single cell is positioned on each cell-positioning site. The drug or other chemicals can be injected on each captured cell through the corresponding injection channels and the resulting cell secretions may be monitored. The hydrophobic region is formed to prevent the suspension media from flowing out or into the isolated region and allows the air to leak out during drug introduction.

Y. Baba et al. (eds.), Micro Total Analysis Systems 2002, Volume 2, 652–654.

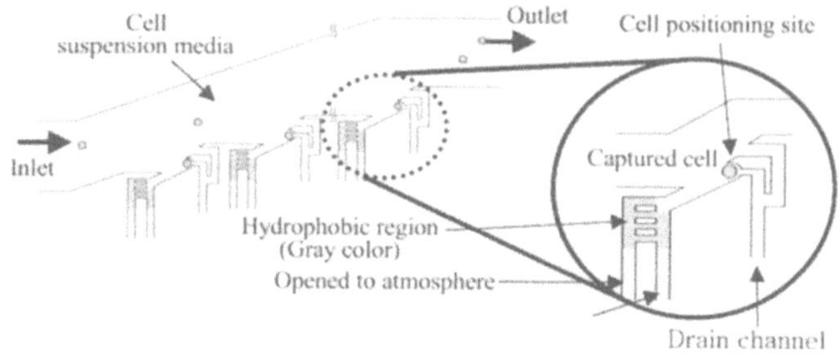

Figure 1. Schematic view of the proposed structure.

3. Fabrication

The device consists of a grooved PDMS and silicon substrate bonded together. Figure 2 shows the fabrication processes of the device. The PDMS structure is formed by replica molding method using 20 μm-thick SU-8 molds. For the silicon structure, the silicon substrate is etched using RIE and the PECVD silicon nitride is deposited. Next, for the preparation of hydrophobic region, fluorocarbon (FC) film is spin coated [2] and patterned using lift-off process. Finally the PDMS surface is modified to hydrophilic surface using O_2 plasma treatment and bonded with silicon substrate. Figure 3 shows the photograph of the fabricated device. The initial design of the devices has three cell-positioning sites.

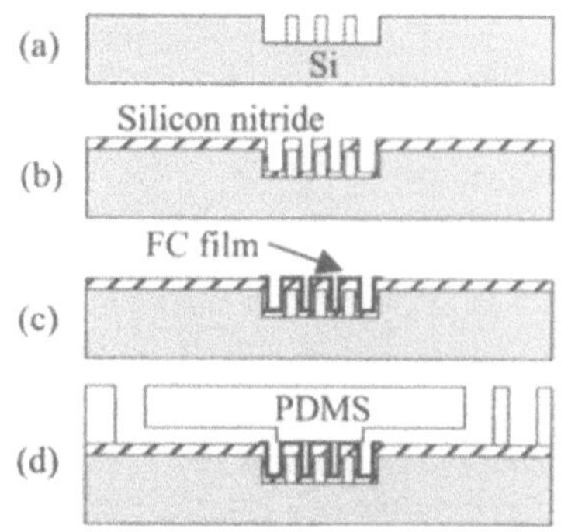

Figure 2. Fabrication processes: (a) Silicon etch. (b) Silicon nitride deposition. (c) FC film coating. (d) Bonding with PDMS structure.

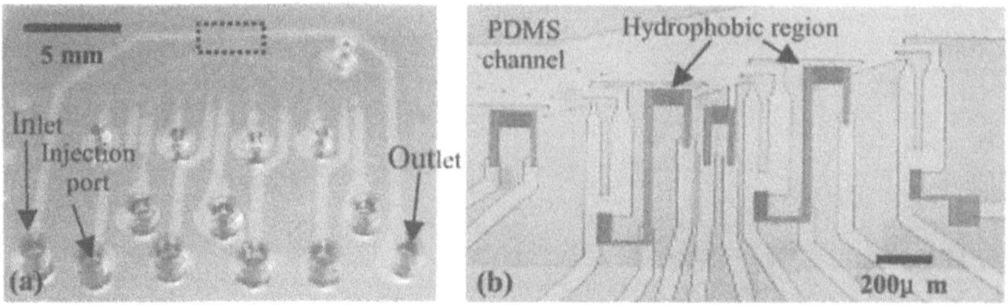

Figure 3. (a) Photograph of fabricated device. (b) Detailed view of dotted region in (a).

4. Experiments and discussion

Figure 4 shows the cell-positioning region after the introduction of sample liquid. This figure distinctively shows that the sample media is blocked in the hydrophobic regions maintaining the air in the dotted channel region.

In our experiment we tested the feasibility of cell positioning using polystyrene beads with a diameter of 15 μm in distilled water. The bead can be exactly placed on a cell-positioning site from the pre-determined flow stream as shown in figure 5(a). After the bead positioning, blue ink is injected from the injection channel (Figure 5(b)). In this experiment, the flow rates of the mainstream (water) and the injection stream (blue ink) are 170 nL/min and 17 nL/min, respectively. The excess ink stream is drained out through the buffer drain channel located above the main drain channel, demonstrating it does not flow into other neighboring captured cells. The experiment using living cells is shown in Figure 6. This photograph shows the actual living cell, CHO DG44, captured in a cell-positioning site. The cell suspension media is phosphate buffered saline (PBS) solution.

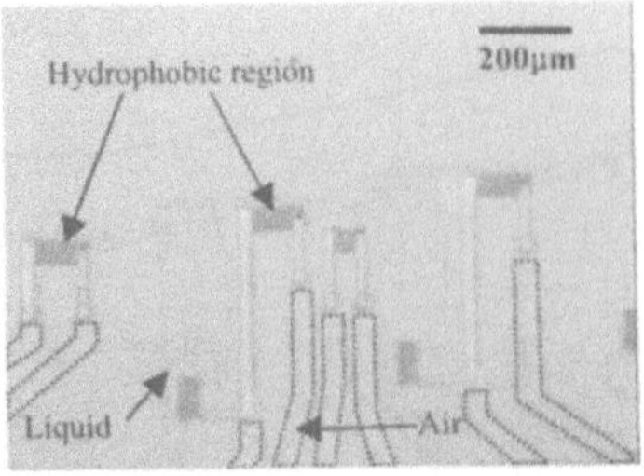

Figure 4. Photograph of the cell-positioning region without electrodes after the sample injection. Liquid stopped at hydrophobic region maintaining air in the dotted regions.

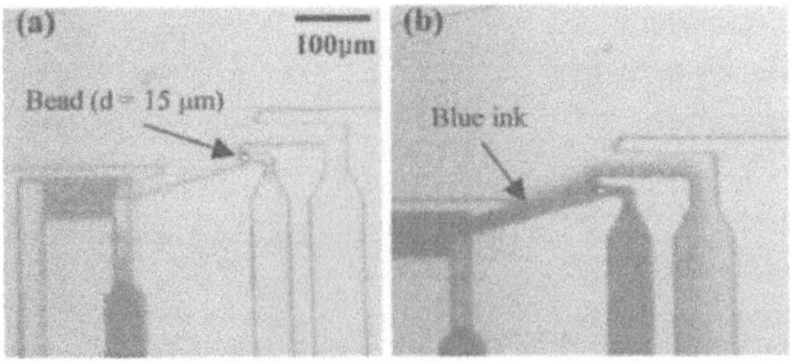

Figure 5. Photograph of the cell-positioning region with captured bead (a) before ink injection (b) after ink injection.

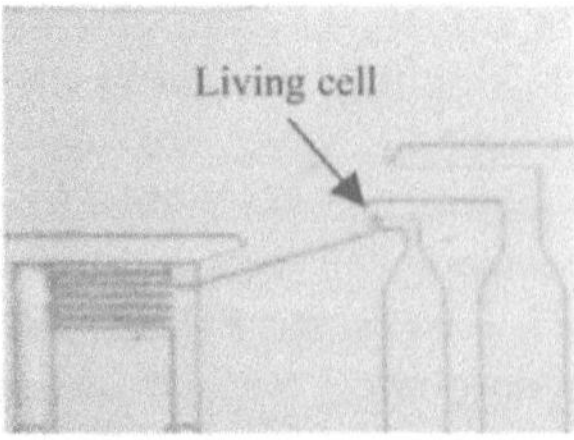

Figure 6. Placement of CHO DG44 cells on cell-positioning sites.

5. Conclusions

In this paper, we have demonstrated that the fabricated microfluidic chip can exactly position single-cell in a dedicated location and nanoliter drug injection can be independently introduced for single-cell analysis and high throughput drug screening.

Acknowledgements

This work has been partially supported by the KOSEF through the MICROS center at KAIST, the NRL program from the Ministry of Science and Technology and the project of "Development of micro optical and thermofluidic devices with high functionality" from the Ministry of Commerce, Industry and Energy of Korea.

References

1. H. Thielecke, T. Stieglitz, H. Beutel, T. Matthies, H. H. Ruf, and J.-U. Meyer, IEEE Eng. Med. And Biol., vol. 18, no. 6, pp. 48-52, 1999.
2. H. V. Jansen, J.G.E. Gardeniers, J. Elders, H.A.C. Tilmans and M. Elwenspoek, Sensors and Actuators A, vol. 41-42, pp. 136-140, 1994.

A CIRCULAR AC ELECTROOSMOTIC MICROPUMP FOR CHROMATOGRAPHIC APPLICATIONS

S.Debesset[1*], C.J.Hayden[2], C.Dalton[2], J.C.T.Eijkel[1] and A.Manz[1]
[1]Department of Chemistry, Imperial College London, London SW7 2AY, UK
[2]Institute of Bioelectronics and Molecular Microsystems,University of Wales,Bangor,UK
*on leave from: Ecole Nationale Supérieure d'Arts et Métiers, Paris, France
a.manz@ic.ac.uk

Abstract

Flow rates of up to 50 μm s^{-1} have been successfully achieved in a circular AC electro-osmotic pump. The flow rate was found to increase linearly with the applied voltage and to decrease linearly with the applied frequency. The pump is expected to be suitable for circular chromatography.

Keywords: circular micropump, chromatography, AC electroosmotic flow

1. Introduction and theory

Circular chromatography promises to be uniquely different from conventional liquid chromatography because analysis time and plate number (separation efficiency) can be varied at will, simply by increasing the number of sample passages around the circle. High plate numbers could be generated for difficult separations, whilst a short analysis time would be sufficient for simple separations [1]. It can be calculated from basic chromatographic theory that the time-dependent resolution, R_s, that is obtained in a rectangular channel of height h [m] will be

$$R_s \propto \frac{u_{opt}}{2\sqrt{D}}\sqrt{t} = \frac{5\sqrt{D}}{2h}\sqrt{t}, \tag{1}$$

where u_{opt} [m s^{-1}] is the flow velocity optimal to minimize the plate height at the channel height used (and therefore maximize resolution), and D [m^2 s^{-1}] is the analyte diffusion coefficient. It can be seen from Equation 1 that it is favorable to minimize the height h. However, since the optimal flow velocity is inversely proportional to h, a lower limit to the height will be posed by the pumping method. The maximum attainable resolution will be determined by the moment that the front and back of the concentration band of the fastest diffusing analyte meet in the circular channel, and will therefore be proportional to the diameter of the channel circle.

As a pumping method, AC electroosmotic flow (AC EOF) was chosen. Recently this method was theoretically described and experimentally demonstrated in linear channels [2,3]. Such flow is generated in channels with an interdigitated array of asymmetric electrodes at the channel floor when an AC voltage is applied. The method is suitable for circular channels and generates flow rates that are sufficiently high to perform open

Y. Baba et al. (eds.), Micro Total Analysis Systems 2002, Volume 2, 655–657.

channel chromatography (a velocity of up to 80 μm s^{-1} has been reported [3]). In addition, the instrumental requirements (an AC voltage of typically 1 V_{RMS} and 5 kHz) are modest. In this paper an AC EOF pump for circular chromatography is presented.

2. Experimental

The electrode design, manufactured on a chromium/gold covered glass plate, consists of a circular array of asymmetric electrodes (Figure 1). Each electrode pair has the following dimensions: electrode (4 μm wide), gap (5 μm wide), electrode (26 μm wide), gap (16 μm wide). This 50 μm wide unit is repeated 564 times along a circle with a 4.49 mm radius, a section of which is shown in Figure 1. The bottom glass plate with the electrode array was sealed either to a PDMS layer patterned with a 100 μm high and 100 μm wide channel structure (Figure 2) or to a laminate/glass sandwich. In the latter case, the channels were defined in the 35 μm thick laminate by excimer laser ablation and access holes for electrodes and solution were micro-milled.

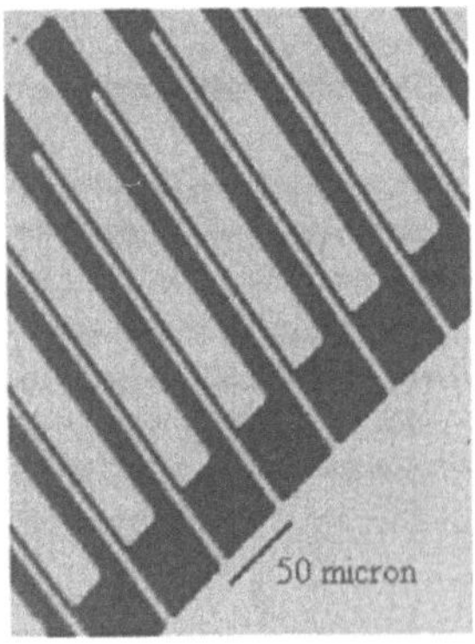

Figure 1. A section of the electrode structure (in gray).

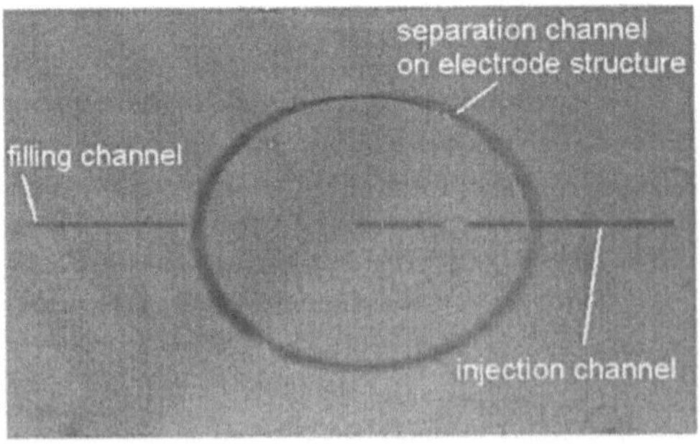

Figure 2. A device consisting of a channel structure in PDMS on top of the circular electrode array. The circle diameter is 9 mm.

3. Results and discussion

The pumping characteristics of both types of device were investigated for a solution of 10^{-4} M KNO_3 in water, containing polystyrene beads to measure the velocity. Solution flow was observed in the direction from the large electrode to the large gap. Using a glass/PDMS device, a linear relationship between applied voltage and measured velocity was found (Figure 3) with a maximal velocity of about 50 μm s^{-1} at an applied voltage of 1.8 V_{RMS}. The application of higher voltages led to electrolysis and bubble formation. Using a glass/laminate/glass device, the pumping velocity was found to be approximately inversely proportional to applied frequency between 5 and 20 kHz (Figure 4). Lower frequencies were not applied because of bubble formation.

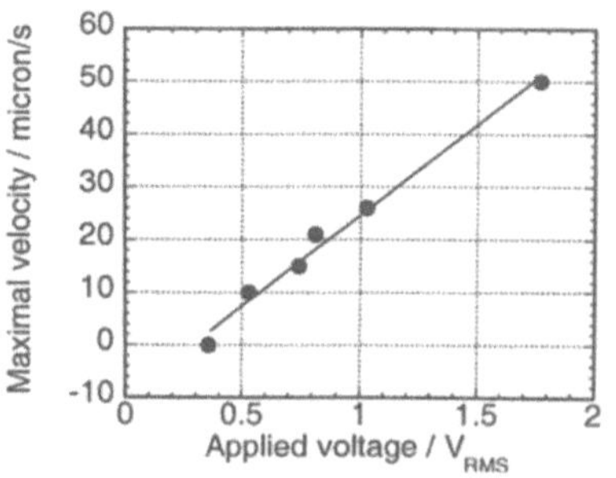

Figure 3. Circular pumping velocity as a function of applied RMS voltage at 5 kHz in a 100 μm high channel (glass/PDMS device).

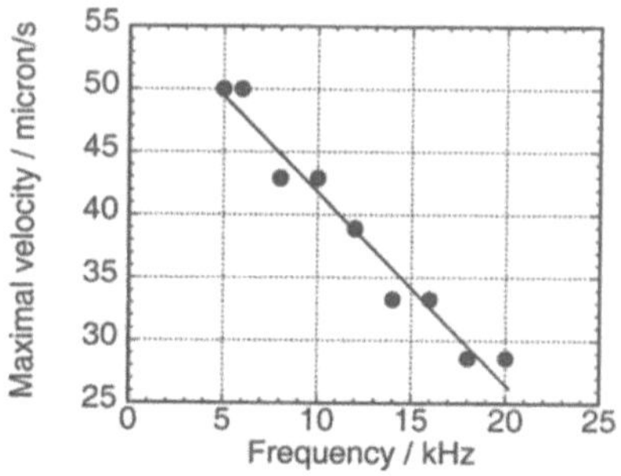

Figure 4. Circular pumping velocity as a function of frequency (V_{RMS} = 1 V) in a 35 μm high channel (glass/laminate/glass device).

The suitability of the devices for open channel chromatography was investigated by plotting the theoretical curve for the plate height as a function of average velocity (half the observed maximal velocity). Figure 5 shows the curves for the device with the 100 μm high channel at three values of the capacity factor *k'* (the ratio of analyte time spent on stationary and mobile phase). The average velocity obtained (25 μm s^{-1}) is slightly lower than the optimal velocity for *k'*=2. Plate heights of around 100 μm are expected, implying a plate generation rate of 0.25 plates s^{-1}. The maximum plate number theoretically obtainable with this device is therefore 5,000 plates after 20,000 s.

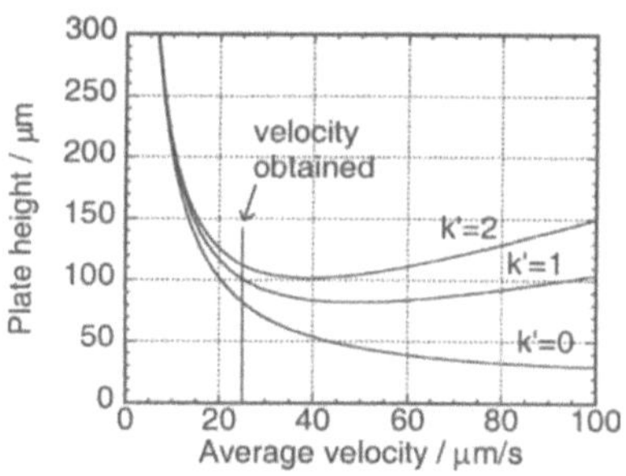

Figure 5. Plate height as a function of velocity for a rectangular channel of infinite width and 100 μm depth.

Acknowledgements

The work presented here was partly sponsored by the EPSRC UK, the Department of Trade and Industry (DTI) and the Lab-on-a-Chip Consortium (LOC).

References

1. J.C.T.Eijkel, C.Dalton, C.J.Hayden, J.A.Drysdale, Y.C.Kwok and A.Manz, Micro Total Analysis Systems 2001, Kluwer, Dordrecht, the Netherlands, pp.541-542.
2. A.Ajdari, Phys.Rev.E, **2000**, 61, R45-R48.
3. A.B.D.Brown, C.G.Smith, A.R.Rennie, Phys.Rev.E, **2000**, 63, 016305.

PULSED FIELD ELECTROPHORESIS: A NEW APPROACH FOR HIGH THROUGHPUT EXPERIMENTATION FOR NON-FLUOROGENIC ASSAYS

Andrea Chow, Anubhav Tripathi, Yue Suo, Roger Dettloff, John Owicki and Wally Parce
Caliper Technologies Corp., 605 Fairchild Dr., Mountain View, CA 94043, USA

Abstract
A new method of detecting changes in the electrophoretic mobility of fluorescent molecules in a microfluidic device suitable for high throughput experimentation is presented. This approach is based on applying a pulsed electric field to modulate the species velocity and the apparent species concentration in a continuous flow device.

Keywords: Pulsed field, electrophoresis, high throughput, non-fluorogenic assays

1. Introduction

Unlike conventional capillary electrophoresis in which a steady electric field is applied to separate mixtures in a sample band into distinct components spatially due to their differences in electrophoretic velocity, this new approach uses a pulsed electric field which induces a temporal change in the electrophoretic velocity of the species in a sample as it flows through a microchannel under an applied pressure gradient. When the electric field is turned off, the species move at the pressure driven velocity. When the electric field is on, the species move at a velocity equal to the vector sum of the electrokinetic and the pressure driven velocity. A detector located in the electric field region measures this time-dependent perturbation of the velocity in a form of a concentration perturbation of the species through the mass flux conservation. This temporal change in concentration of species can then be related back to the electrophoretic mobilities of the species.

2. Theory

The mass flux (J) of a given species is defined as the product of its concentration (C) and velocity (V).

$$J = CV \tag{1}$$

In the absence of an electric field, the velocity of a species flowing in the microchannel is that due to pressure driven flow (V_p) only. When the field is on, the velocity has two additional components due to electrophoresis (V_{ep}) and electroosmosis (V_{eo}). So, in the time scale when there is insignificant chemical reaction rate to create or destroy the species in question, the mass flux across the channel cross section in the separation region when the electric field is off (J_{off}) and on (J_{on}) must be conserved, leading to the equation:

Y. Baba et al. (eds.), Micro Total Analysis Systems 2002, Volume 2, 658–660.

$$J_{off} = C_{off}V_{off} = J_{on} = C_{on}V_{on} \tag{2}$$

$$\frac{C_{off}}{C_{on}} = \frac{V_{on}}{V_{off}} = \frac{(V_p + V_{eo} + V_{ep})}{V_p} \tag{3}$$

C_{on} is the species concentration when the field is on, and C_{off} is the concentration when the field is off. If the net electrokinetic velocity is in same direction as pressure driven flow, then C_{on} is always lower than C_{off}. Both electrokinetic velocities, V_{ep} and V_{eo}, are directly proportional to the applied electric field (E) by the mobility μ, i.e. $V = \mu E$. Thus, Eq. (3) can be rewritten as

$$\mu = \frac{V_p}{E}\left(\frac{C_{off}}{C_{on}} - 1\right) \tag{4}$$

In the absence of electroosmosis (by coating the channel wall, for example), μ in Eq. (4) is the electrophoretic mobility of the species.

In a reaction where fluorescently labeled reactant and product have different electrophoretic mobility, the detected fluorescence intensity can be used to examine the extent of the reaction. This pulsed field method is amenable to high throughput, continuous flow experimentation because it does not require an additional step of sample injection. Moreover, the problem associated with dispersion of the sample band in the separation channel is eliminated.

The sensitivity of the pulsed field electrophoresis method to mobility change can be optimized with respect to experimental parameters (V_p and E) using the mass flux balance equations. According to Eq. (3), the measured concentration of any single species, C, flowing under action of applied electric field and pressure gradient, is related to its measured concentration, C^0, at zero applied field as

$$\frac{C}{C^0} = \frac{V_p}{V_p + \mu^{app} E} \tag{5}$$

where μ^{app} is the apparent mobility of the species. The apparent mobility is $(\mu_{ep} - \mu_{eo})$ when the electro-osmotic mobility is smaller than the electrocrophoretic mobility, whereas it is $(\mu_{eo} - \mu_{ep})$ when the electro-osmotic mobility is larger than the electrophoretic mobility of the species. In detecting the extent of reaction in which the concentration on the reactant (C_r) and that of the product (C_p) changes, we are primarily interested in the difference in the concentrations of the two species. Hence, the measured signal using pulsed field electrophoresis is given by:

$$\frac{dC}{C} \equiv \frac{C_p}{C_p^0} - \frac{C_r}{C_r^0} = \frac{V_p(\mu_r^{app} - \mu_p^{app})E}{(V_p + \mu_r^{app} E)(V_p + \mu_p^{app} E)} \tag{6}$$

Again, the subscripts r and p denote the properties of the two labeled species, i.e., reactant and product. To optimize the measured signal, we determine the conditions when the first derivative of Eq. (6) with respect to the pressure driven velocity is at zero. This operation yields the relation

$$V_p^{opt} = \sqrt{\mu_r^{app} \mu_r^{app}} E \tag{7}$$

Hence, the optimal pressure-driven velocity at a given magnitude of the pulsed electric field is given by the geometric mean of the apparent mobilities of two species.

3. Experimental

Experiments were performed using a sipper microchip and a prototype microfluidic instrument to validate this approach and the results were compared with the theoretical predictions. The sipper microchip consisted of a sipper capillary connection to provide automated external sample accession to the microchip, for pulsed field mobility shift assays. The electrophoretic mobilities of two fluorescent dyes were evaluated using the measured intensity in the pulsed field method. These mobilities were found to be in good agreement with the values measured independently using the conventional steady field electrophoresis method. The predicted criterion for maximizing sensitivity of the mobility shift measurement was also confirmed by experimental data as illustrated in Figure 1, in which the measured signal is plotted as a function of the applied field strength at a given value of V_p for both theoretical prediction and experimental measurements.

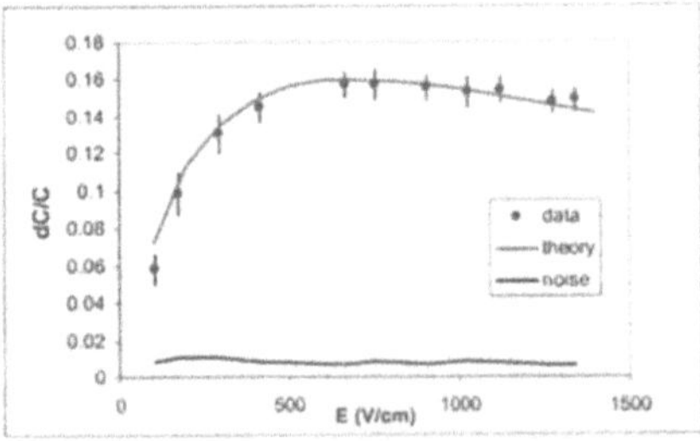

Figure 1. Pulsed field data (Eq.6) on fluorescent dyes, comparison of data and calculations

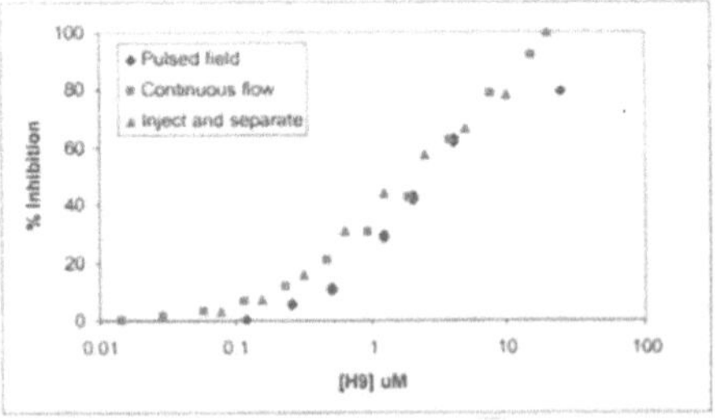

Figure 2. Comparison of dose response measurements of a known PKA inhibitor H9 using three microfluidic based electrophoresis methods.

This pulsed field mobility shift method has been used for screening for Protein Kinase A (PKA) inhibitors. Figure 2 is a comparison of the dose response of a known inhibitor, H9, measured by the pulsed field method compared with two other microfluidic electrophoresis-based approaches—inject-and-separate and continuous flow with a steady electric field—for high throughput experiments. All three methods give compared Ki values.

4. Conclusions

We demonstrated that the pulsed field electrophoresis technique could be used to perform high throughput mobility shift assays. Both theory and experiments show the feasibility of the concept.

Temperature Gradient Focusing

David Ross, and Laurie E. Locascio

National Institute of Standards & Technology, Gaithersburg, MD

david.ross@nist.gov

Abstract

A new gradient focusing method, temperature gradient focusing (TGF), is described. In TGF, a buffer is used with a temperature-dependent ionic strength, so that when a temperature gradient is applied to the system it results in a corresponding electrophoretic velocity gradient. If an opposing bulk buffer flow is also applied to the system, very high degrees of focusing can be achieved.

1. Introduction

Temperature gradient focusing (TGF) is a recently described new technique for the focusing and separation of charged species in solution [1]. In TGF, the electrophoretic motion of an analyte is balanced against the bulk motion of the solvent containing the analyte as is illustrated schematically in Fig. 1. Using a suitable buffer, the application of a temperature gradient gives rise to a corresponding gradient in the electrophoretic velocity. The bulk velocity is then adjusted so that the total analyte velocity is equal to zero at a unique point in the system, and all of the analyte in the system is focused at that point.

Used purely for sample concentration, TGF has been shown capable of focusing analyte to a peak concentration more than 10000 times greater than the initial input concentration. As far as we are aware, this factor of 10000 is the best that has been reported for any single method of sample concentration.

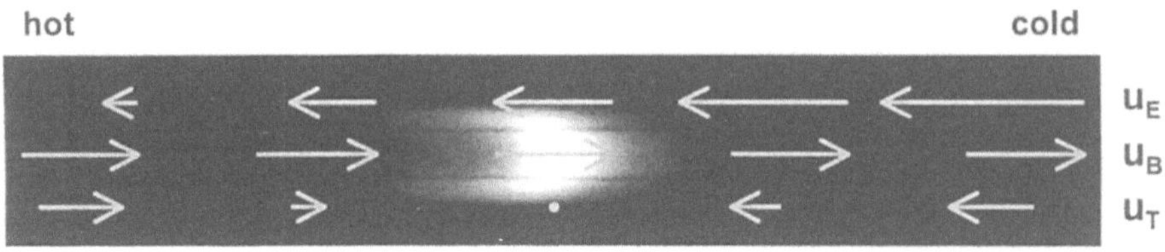

Fig. 1. Schematic Representation of TGF. Using an appropriate buffer, the electrophoretic velocity of an analyte, u_E, can be temperature dependent. If it is balanced by a bulk fluid velocity, u_B, the total analyte velocity, $u_T = u_E + u_B$, will sum to zero at a unique point near the middle of the channel, and the analyte will be focused at that point.

Y. Baba et al. (eds.), Micro Total Analysis Systems 2002, Volume 2, 661–663.

2. Experimental

Chemicals and Materials. All aqueous solutions were prepared using water from a Millipore Milli-Q [2] system (Bedford, MA). Tris(hydroxymethyl)aminomethane base (Tris) was obtained from Sigma [2] (St. Louis, MO). Boric acid was obtained from Mallinckrodt [2] (Phillipsburg, NJ).). Recombinant green fluorescence protein was obtained from Clontech [2] (Palo Alto, CA). All reagents were used as received.

Microchannel Fabrication. For the focusing and separation of green fluorescence protein, a 30 μm ID silica capillary (Polymicro Technologies Inc. [2], Phoenix, AR) was embedded between two 125 μm thick polycarbonate sheets. The two polycarbonate sheets were pressed together around the silica capillary and bonded at 13000 N, 160 °C for 10 min.

Copper heating/cooling blocks. For the examples of TGF described here, zones of different temperatures and the temperature gradients between them were produced by thermally anchoring the thin polycarbonate microchannel chips to alternately heated or cooled copper blocks. Three temperature zones were used: two cold zones covering much of the ends of the channels and the buffer reservoirs, and a narrow hot zone near the middle of the channel. The middle, heated copper block was heated using a small high-power resistor embedded into the copper and its temperature was regulated using a PID temperature controller (Omega Engineering Inc [2], Stamford, CT). To regulate the temperature of the cold zones, ¼ inch diameter holes were drilled through the cooled copper blocks, and cold water from a thermostatted bath (Neslab [2], Portsmouth, NH) was passed through them. The heated and cooled copper blocks were arranged so that there was a 1 mm gap between the hot and cold blocks on one side and a 2 mm gap on the other.

The capillary imbedded in polycarbonate (see above) was attached to the copper blocks using thermally conductive adhesive transfer tape (3M [2]). The chips were pressed against the copper blocks from above with 3 mm thick PDMS (Sylgard 184 [2], Dow Corning, Midland, MI) gaskets and a 2 mm thick acrylic (Acrylite OP-4 [2], Cyro Industries, Mt. Arlington, NJ)) top plate, which was secured to the outer copper blocks using nylon screws.

Focusing measurements. Samples to be focused were diluted into high-concentration Tris/boric acid buffer (900 mM Tris, 900 mM boric acid). The microchannel and reservoirs at each end of the channel were then filled with the diluted sample solutions, so that before application of the electric field, the entire channel/reservoir system contained a sample/buffer solution of uniform concentration. High voltage was then applied to the channel, and the samples were focused. Fluorescence imaging of the focused samples was performed using a research fluorescence microscope equipped with a long working distance 10x objective, a mercury arc lamp, appropriate filter sets, and a 3 chip color CCD camera (Dage-MTI[2],

Michigan City, IN). Digital images were acquired using Scion Image [2] software and a Scion CG-7 [2] frame grabber (Scion, Inc.[2], Frederick, MD).

3. Results and discussion

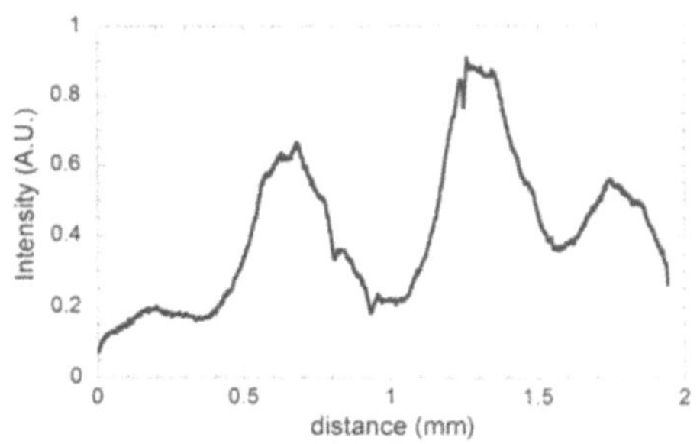

Fig. 2. Focusing of green fluorescence protein. The plot of intensity vs. distance along the microchannel shows 4 resolvable peaks. This result is to be compared with that shown in Ref[3].

In addition to very high degrees of sample concentration, TGF can also provide separations comparable to those obtained with isoelectric focusing (IEF). This is demonstrated in Fig. 2 for the focusing and separation of green fluorescence protein into four distinct bands. The intensity profile was obtained after 40 min. of focusing under an average applied field of 240 V/cm. A 30 μm i.d. fused silica capillary was used for the focusing with a linear temperature gradient spanning the range from 10 °C to 60 °C over a length of 2 mm. This protein separation is to be compared to that obtained with the same protein, for example in Ref[3].

TGF offers a number of potential advantages over traditional IEF. First, since it is much easier to generate a stable temperature gradient than a stable pH gradient, TGF is a much easier technique than IEF, and is consequently more suitable for miniaturization and integration for lab-on-chip applications. Second, analytes are not focused at their isoelectric points (pIs), so that TGF is not restricted to use only with analytes having an accessible pI. In fact, TGF has been demonstrated for a wide range of charged analytes including small molecules, amino acids, proteins, DNA, particles, and cells. In addition, since TGF allows for focusing of analytes at a pH other than their pI, analytes can be focused to much higher concentrations using TGF than IEF.

References

1. D. Ross and L. E. Locascio, Analytical Chemistry **74**, 2556-2564 (2002).
2. Certain commercial equipment, instruments or materials are identified in this report to specify adequately the experimental procedure. Such identification does not imply recommendation or endorsement by the National Institute of Standards and Technology nor does it imply that the materials or equipment identified are necessarily the best available for the purpose.
3. D. P. Richards *et al.*, Journal of Chromatography a **853**, 21 (1999).

DEVELOPMENT OF ELECTROPHORETIC ANALYSIS SYSTEM USING MULTIPLE CHANNEL MICROCHIPS

Hisashi Hagiwara[1,5] Yoshinori Mishina[2,5] Mika Inoue[1] Masumi Yamada[3] Toshiyuki Sakurai[2] Teruo Fujii[4] Minoru Seki[3] and Yoshiyuki Sakaki[5]
[1]Arbiotec,LTD. 3F CCR The University of Tokyo, 153-8904 Japan (e-mail: hagiwara@arbiotec.com), [2]Hitachi Electronics Engineering Co.,Ltd. Japan, [3]Dept. of Chemistry and Biotechnology, Graduate School of Engineering, The University of Tokyo,Japan, [4]Institute of Industrial Science, the University of Tokyo, Japan, [5]Human Genome Center , Institute of Medical Science, University of Tokyo, Japan

Abstract
We have developed a high-speed and high-throughput electrophoretic analysis system by use of microfabrication-technology. This analytic system can carry out the DNA fragments analysis of 120 samples in one hour.

Keywords: microfabrication-technology, chip-based CE system, 12-channel CE chip, automated preparation

1.Introduction

To realize the benefit of high-speed, high-throughput separations on capillary electrophoresis (CE) microchips, automated microchip preparation and sample loading, will be required. In this work, we report DNA fragmentation on a 12-channel CE microchip with automated sample loading using a robotic instrument.

2.Experimental

This analytic system is composed of two instrument and CE microchip. The instrument of the figure 1 has the function which filling the gel and samples in a CE microchip. Figure 2 shows the device has function of the electrophoretic analysis. The CE microchip for electrophoresis is made of polymethyl metacrylate (PMMA) We designed the independent channel which consist of twelve lanes for CE microchip shown in figure 3.

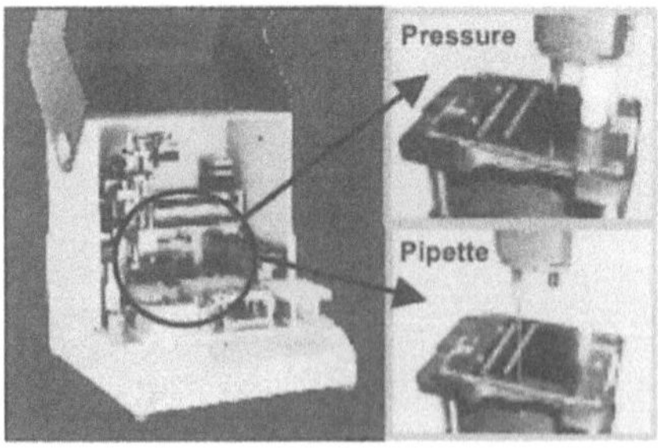

Figure 1. Robotic instrument
FEATURES
1)Processing Time: 5 min / 12 samples
(Includes time of gel filling: 40 sec)
2) Automatic sample injection and Gel filling. Accuracy of pipetting : ±0.5 μl.
3) The function to adjust parameter.
4) It corresponds to two types chip. (3 or 12 samples chip)(see:Figure.3).
5) Space: 412(W) × 512(D) × 485(H)mm

Y. Baba et al. (eds.), Micro Total Analysis Systems 2002, Volume 2, 664–666.

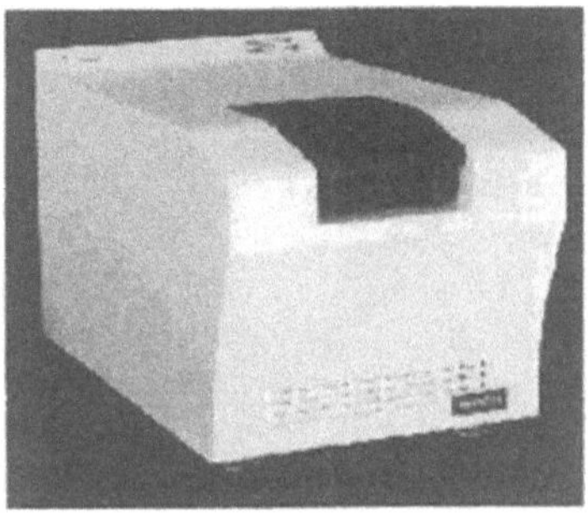

Figure 2. Electrophoretic device
FEATURES
1) Size range: 100-800 bp
2) Sizing resolution: ±5%
3) Sizing accuracy: ±10 bp
4) Ran time: 6 min / 12 samples
(simultaneous analysis)
5) Sensitivity: 0.1ng / ml
6)Space:300(W) × 460(D) × 330(H)mm

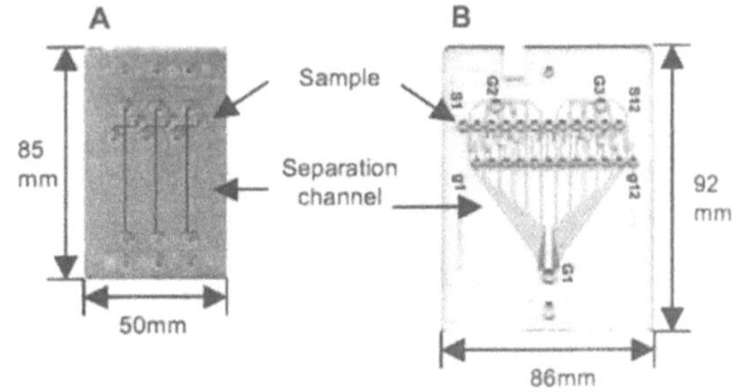

Figure 3. Design of microchips
FEATURES
1) Material: Polymethyl metacrylate (PMMA)
2) Channel size:100(W) × 30(D) μm
3) Separation channel length:
A) 3 channels microchip 33 mm
B) 12 channels microchip 42 mm

The first process that is filling up a gel to individual channels and loading the samples to a CE microchip takes five minutes. The second process, electrophoresis and a data analysis take six minutes in all 12 channels. The separated patterns are converted to digital form and automatically kept to PC. A software program was used to convert the two-dimensional images into a series of electropherograms corresponding to each of the CE separation channels. The raw data and the processing data were saved in either a text (.csv), bit map (.bmp) or a binary format compatible with common use software.

3.Results and discussion

This system can be used for separate fluorescent dyes, fluorescently labeled amino acids, DNA restriction fragments (RFLP etc. Figure 4), PCR product, oligonucleotides, short tandem repeats (STR), single strand (SSCP etc).

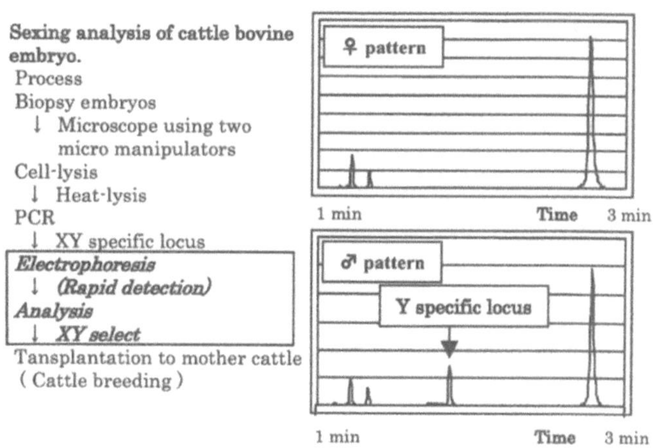

Figure 4. Application of rapid detection system to Cattle breeding

Furthermore, the production technology of this chip is being developed into not only the correspondence to electrophoresis analytic technology but also the chemical analysis technology [1]. We begin a microchip production for the micro-fluidics by using another material with PMMA at present (Figure 5)[2].

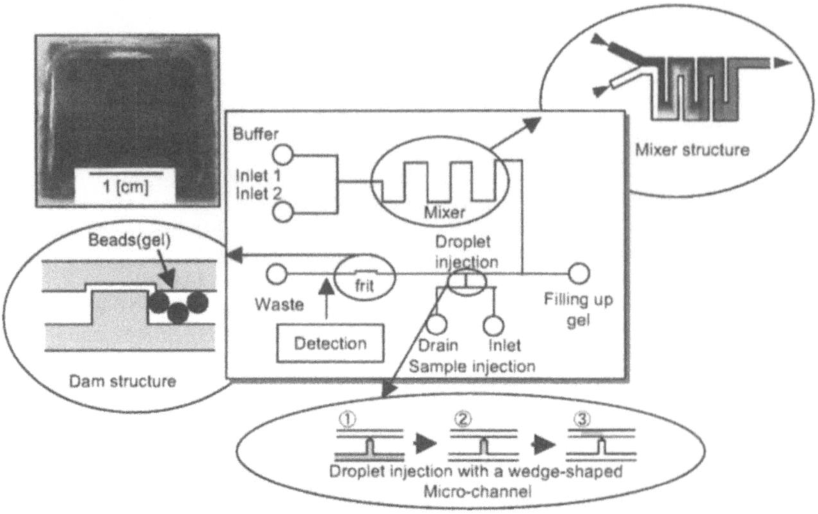

Figure 5. Schematic illustrations and photograph of liquid micro-chromatography chip

[This works was performed as a part of the research and development project of industrial science and technology program supported by NEDO (New Energy and Industrial Technology Development Organization).]

References

1. J.W.Hong, T.Fujii, M.Seki, T.Yamamoto and I.End , Integration of gene amplification and capillary gel electrophoresis on a polydimethylsiloxane-glass hybrid microchip, *Electrophoresis*, 22(2), pp.328-333, (2001).
2. M.Seki, M.Yamada, R.Ezaki, R.Aoyama, J.W.Hong , Chromatographic separation of proteins on a PDMS-polymer chip by pressure flow , Proceedings of the μTAS 2001 Symposium, pp48-50, (2001)

IMPROVED SAMPLE INJECTION METHOD ADAPTING HYDROPHOBIC PASSIVE VALVE SYSTEMS FOR MICROFLUIDIC DEVICES

Nae Yoon Lee, Masumi Yamada, and Minoru Seki
Department of Chemistry and Biotechnology, School of Engineering,
The University of Tokyo, Japan
(e-mail: mseki@chembio.t.u-tokyo.ac.jp)

Abstract

The needs for microscale sample injection are becoming more and more important with the development of proteomics and its application to the human disease diagnosis. In this research, we developed an improved microstructure for quantitative sample injection that does not necessitate the delicate control of the pneumatic pressure. Through this method, high reproducibility in sample injection was realized with less than 0.5% of error range.

Keywords: Sample injection method, hydrophobic passive valve, microfluidic device

1. Introduction

Exact-volume sample injection is very important and is one of the determining factors for the success of microscale reaction and separation. In our previous research [1], we proposed a channel structure to which a hydrophobic passive valve was introduced. With this method, the sample injection of certain exact volume was realized with high reproducibility. However, the system requires a relatively delicate control of the pneumatic pressure for avoiding the incorporation of air bubble which disturbs the reproducible separation. Therefore, in this study, we revised the microchannel structure for the simple sample injection.

2. Theory

Figure 1 shows the comparison between the previously reported injection method and the improved injection method. In the improved system, the sample droplet is incorporated into the separation channel just after the front of it reaches the solution in the separation channel, because an improved hydrophobic passive valve has the shape

Y. Baba et al. (eds.), Micro Total Analysis Systems 2002, Volume 2, 667–669.

of such that the sample droplet touches the separation-channel solution before it reaches the inlet of air vent microchannel. This structure eliminates the air vent closing, therefore, enables almost complete and easy injection of the sample every time.

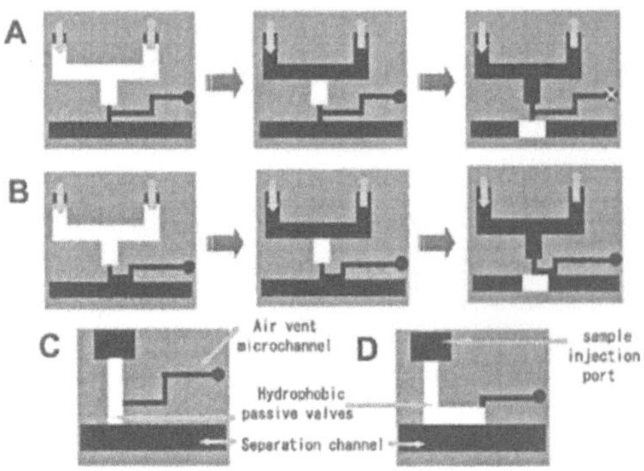

Figure 1 . The comparison between the previously reported method and the improved method of liquid injection. (A) Previously reported method: Without the air vent closing, the sample flows into the air vent microchannel. (B) Improved method: Sample is easily and almost completely injected into the separation channel without the air vent control. (C) Enlarged picture of previously reported hydrophobic passive valve. (D) Enlarged picture of the improved hydrophobic passive valve.

3. Experimental

Figure 2 represents the schematic diagram of the microfabrication of the microchip. This microchip is realized by combining two different PDMS polymer plates together, each having different channel depths.

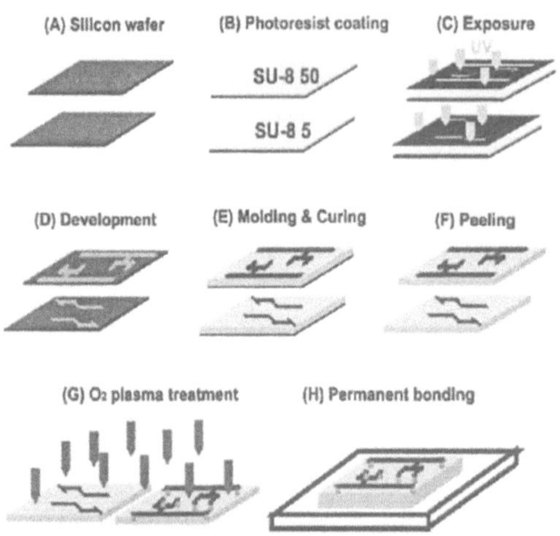

Figure 2 . Schematic diagram of the microfabrication (A) Two pieces of 64 mm x 21 mm silicon wafer were cut. (B) Two different kinds of photoresists were coated: SU-8 50 for the thicker microchannel, and SU-8 5 for the thinner microchannel. (C) UV was exposed for the formation of the microchannel. (D) Microchip was developed with 1-Methoxy-2-propyl -acetate. (E) PDMS prepolymer solution was poured and cured. (F) After baking, PDMS was peeled off the master. (G) O_2 plasma was exposed for the bonding of two PDMS polymer plates. (H) PDMS microchip was bonded to the glass plate.

4. Results and discussion

We fabricated four different structures of hydrophobic passive valve and tested which type of the valve completely injects the sample, and the results are shown in Figure 3. It is proven that the smaller the width of the hydrophobic passive valve and the shorter the distance between the sample injection port and the separation channel, lesser sample remained in the hydrophobic passive valve as residual, and almost perfect sample injection was realized within the 1% error range (Figure 4). When the length of hydrophobic passive valve was halved, the error range decreased further to less than 0.5% (Figure 4C).

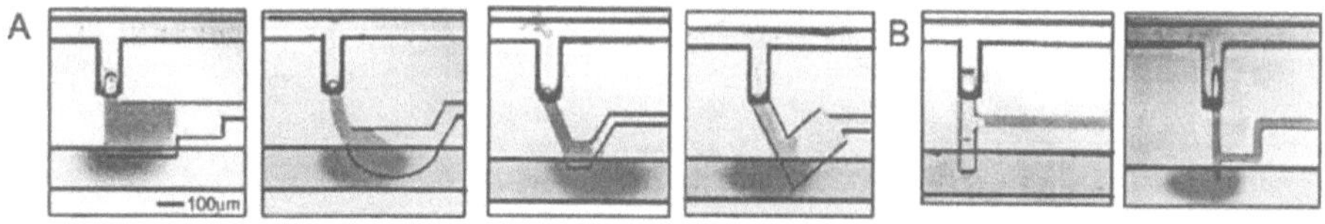

Figure 3. Many different structures of hydrophobic passive valves: (A) Four different types of improved hydrophobic passive valves. (B) Two different types of previously reported hydrophobic passive valves.

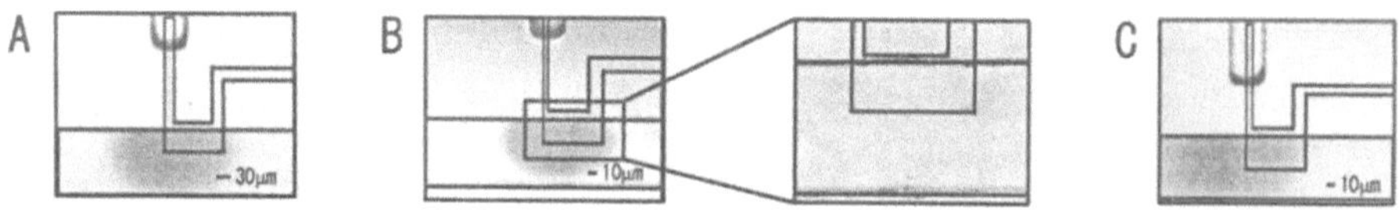

Figure 4. Further improved hydrophobic passive valves: (A) The width of the hydrophobic passive valve is 30 µm. (B) The width of the hydrophobic passive valve is reduced to 10 µm. (C) The length of the hydrophobic passive valve is reduced to half

5. Conclusion

Improved sample injection method is simple in structure and was proven to be easy to use. Exact-volume liquid injection can further be applied effectively in the field of chromatographic and electrophoretic separation, especially for pico-liter scale sample analysis.

Acknowledgement

This research was supported in part by Grants for Scientific Research on Priority Areas [A] (No. 13025216) and Millennium Project (No. 12319) from the Ministry of Education, Culture, Sports, Science and Technology in Japan.

Reference

1. N.Y. Lee *et al.*, 15th International Symposium on Microscale Separations and Analysis HPCE, p.237, (2002)

A CLOSED-LOOP CALIBRATION SYSTEM FOR A MICRODIALYSIS-BASED LAB-ON-A-CHIP

A.J.Sprenkels[1], D.G.Pijanowska[1,2], J.G.Bomer[1], S.Herber[1], K.M.van Delft[1], W.Olthuis[1] and P.Bergveld[1]
[1] MESA+ Research Institute, University of Twente, P.O.Box 217, 7500 AE Enschede, The Netherlands
[2] Institute of Biocybernetics and Biomedical Engineering, Polish Academy Sciences ul. Trojdena 4, 02-109 Warsaw, Poland
(E-mail: a.j.sprenkels@el.utwente.nl)

Abstract

In this contribution a closed loop calibration facility for a microdialysis based lab-on-a-chip is presented. Each facility consists of a channel structure etched in silicon, closed by a Pyrex® cover comprising two small platinum electrodes. In the silicon structure the calibration liquid is stored in the meander shaped channel that starts in an electrolyte filled reservoir. By applying an electric current via both electrodes through the electrolyte in the reservoir, gas bubbles are generated by electrolysis, forcing the calibration liquid into the main carrier channel leading to the sensor area of the lab-on-a-chip. Above the electrolyte reservoir two additional interdigitated electrodes are present that are used to measure the gas bubble volume by measuring the impedance of the gas/liquid mixture in the reservoir. Since the volume of the gas bubbles is equal to the dispensed calibration liquid, its total volume can be controlled by a closed loop impedance measuring system. In this way precise plugs of 60nl calibration liquid have been dispensed into the main carrier channel.

Keywords: Electrochemical actuation, sensor calibration, lab-on-a-chip, μTAS

1. Introduction

Microdialysis has become a standard sampling technique in clinical applications nowadays. In this technique a double-lumen microdialysis probe samples chemical substances such as glucose or ions through the semi permeable dialysis membrane into a perfusion liquid. This dialysate is then pumped from the probe to appropriate analytical equipment to measure its composition. Major drawbacks of such a system are the bulky equipment and the time lag between the sampling and the actual measurement. By integrating the microdialysis probe with the actual sensors on a single chip the size of the system is drastically reduced, resulting in for instance smaller sample volumes and shorter lag times. However for a reliable operation of this lab-on-a-chip, that is discussed elsewhere in this proceedings [1], an integrated calibration system for the sensors is necessary. First a brief description of the theory of operation of the electrochemical actuated calibration facility will be given, followed by an experimental section and results and discussion. Finally some conclusions are given.

Y. Baba et al. (eds.), Micro Total Analysis Systems 2002, Volume 2, 670–672.

2. Theory

Figure 1 shows the complete dual calibration facility. Each system comprises two reservoirs with small electrodes for the generation of hydrogen and oxygen bubbles by electrolysis of water in the reservoirs. Both reservoirs are connected to each other through a chevron-shaped barrier to limit the catalytic back-reaction of both gasses, but maintaining the necessary ohmic contact. Both reservoirs are in liquid contact with the meander containing the calibration solution, which ends in the main carrier channel, containing an additional conductivity detector that is part of the complete lab-on-a-chip.

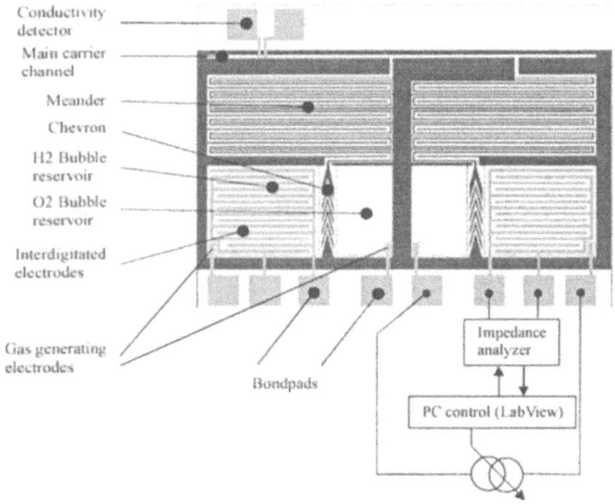

Figure 1. Dual calibration system, layout and electrical connections.

Starting from the basic electrochemical reactions that take place at the anode and cathode during electrolysis of water, and making a number of assumptions [2] it can be shown that the total volume of generated hydrogen gas and thus the amount of dispensed calibration liquid from the meander into the main carrier channel can be determined from the measured impedance between the two interdigitated electrodes above the hydrogen reservoir. This value is used in a feedback system to control the amount of dispensed calibration liquid.

3. Experimental

Fabrication

The silicon part, containing all channels, reservoirs and the chevron structure, was manufactured by dry reactive ion etching to a depth of 180µm. The filling ports were etched from the backside through the wafer in a second step. After sputtering a 25nm titanium adhesion layer and a 500nm thick platinum layer on a Pyrex® wafer the electrodes were patterned by ion beam etching. Finally both wafers were diced into separate chips after standard anodic bonding (420°C, 650V, 60 min.).

Filling

Although in the final application the meanders have to be filled with sensor-specific calibration solutions, in this experiment all systems were completely filled by immersing in a 500mM KNO_3 electrolyte solution and subsequent evacuation in a desiccator. Finally all filling ports were closed with small glass covers.

Measurement setup and protocol

As indicated in figure 1 the system is connected to an impedance measuring circuit, which is controlling the gas generating current using a feedback. This feedback system is implemented on a PC using an appropriate LabView application and a suitable I/O card (National Instruments PC1200AI). The actual dosed volumes are determined by connecting a fused silica capillary to the outlet of the main carrier channel monitored by a video microscope and subsequent analysing the video.

4. Results and discussion

Figure 2 shows the results of four subsequent dosing cycles of 60nl each. The lower graph shows the gas generating actuation current with time, while the upper graph represents the dosed volume calculated from the impedance measurement. The actual measured dosed volumes from the capillary video, deviate no more than 5% from the desired values.

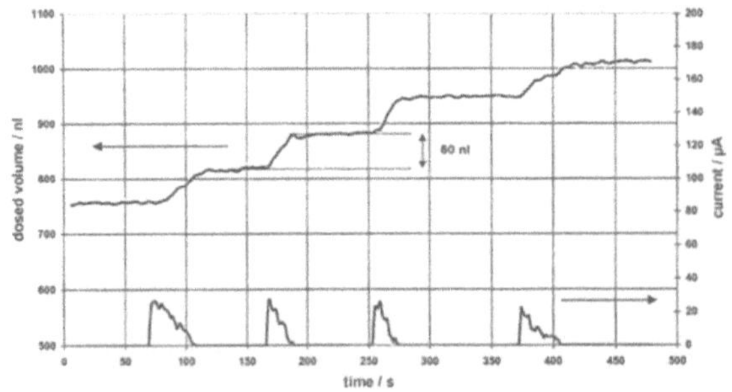

Figure 2. Dosed volumes calculated from the measured reservoir impedance (upper); Electrolysis current for each dosing (lower).

5. Conclusions

It was demonstrated that accurate control of dosing a calibration solution into the main carrier channel is possible with a closed loop feedback by measuring the impedance of the gas/liquid mixture in the hydrogen reservoirs. An overal accuracy of 5% has been obtained.

Acknowledgements

This project is financially supported by the Dutch foundation for Fundamental Research on Matter (FOM) and the Royal Dutch Academy of Science (KNAW).

References

1. D.G. Pijanowska et al., *An integrated microdialysis-based system*, Proc. Micro TAS 2002, Kluwer Academic Publisher, Dordrecht, The Netherlands, (2002).
2. S. Böhm et.al., *A closed-loop controlled electrochemically actuated micro-dosing system*, J. Micromech. Microeng., 10, pp. 498-504, (2000).

FACTORS INFLUENCING RESPONSE TIME OF DEP MICROFLUIDIC ACTUATION

T. B. Jones[1], T. Ito[2], H. Yoo[1], and R. Ahmed[1]

[1] University of Rochester, Rochester, NY (USA)

[2] Toba National College of Maritime Technology, Toba (Japan)

Abstract

Dielectrophoretic (DEP) microfluidic actuation features rapid on-chip manipulation, picoliter droplet dispensing, and mixing of aqueous liquids. In droplet dispensing applications, two critical time scales are identified: (i) the time required to fill the flow structure when voltage is applied and (ii) the time required for the capillary instability to form droplets after voltage is removed. For ~10 μm flow structures, filling takes much longer than droplet formation.

Keywords: microfluidics, laboratory-on-a-chip, dielectrophoresis, electrowetting

1. Introduction

DEP microfluidic actuation utilizes dielectric-coated, co-planar electrodes excited by AC voltage to pattern non-uniform electric fields for liquid manipulation on open substrates. It exploits the attractive tendency of dielectric liquids toward regions of stronger electric field. Fig. 1 shows a typical flow structure in cross-section. DEP microactuation is related to electrowetting [1], differing primarily in that DEP uses AC voltage at sufficient frequency so that the electric field penetrates the liquid and controls its shape [2]. In contrast, electrowetting uses DC voltage (or low frequency AC), meaning that free electric charge builds up on the surface and alters the contact angle but not the liquid profile.

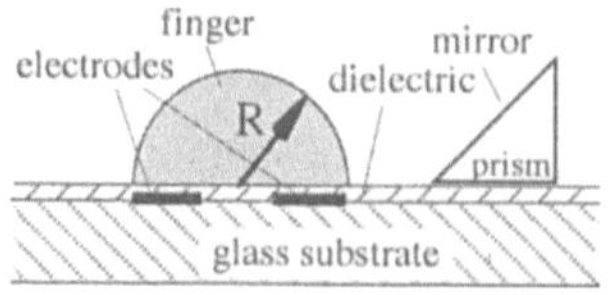

Fig. 1. Cross-sectional view of the co-planar electrodes, showing profile of liquid contained by DEP force. Liquid flows into or out of plane of paper. Mirror placed beside structure permits side view of moving finger.

DEP is fast, an important attribute in certain μTAS applications; in some cases, the leading edge of a finger is observed to move at speeds of ~50 cm/s. However, the transient fluid mechanics that control droplet dispensing are not well-understood, so we are using a high-speed video camera to record and study the behavior. Fig. 2 shows selected frames from a video of a liquid finger on a co-planar electrode structure like Fig. 1. The microscope is trained downward

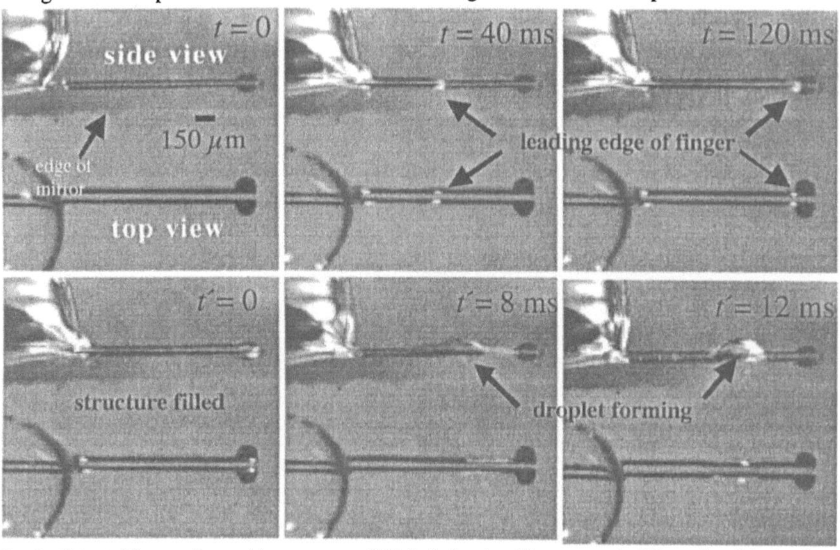

Fig. 2. Selected frames from video sequence (250 fps) showing (i) movement of finger in oil bath when voltage is applied at $t = 0$ and (ii) formation of a droplet when voltage is removed shortly after $t' = 0$.

Y. Baba et al. (eds.), Micro Total Analysis Systems 2002, Volume 2, 673–675.

on the horizontally oriented substrate. A mirror positioned at a 45° angle on the substrate alongside the electrodes allows simultaneous viewing from the side. When AC voltage is applied at $t = 0$ s, the finger emerges from the microliter-sized reservoir droplet of de-ionized water at the left, moves rapidly to the right, and stops at the end of the electrodes. When voltage is removed, two things happen: (i) the capillary force starts to draw the finger back into the large droplet and (ii) the capillary jet instability deforms the surface and pinches off a droplet. For the structure in Fig. 2, which has electrodes of width 50 μm, spacing 50 μm, and length 3 mm, typically only one or two droplets form. On the other hand, for structures with feature size down to ~10 μm, the capillary instability overwhelms capillary-driven finger retraction, making it possible to dispense multiple droplets in the ~50 picoliter volume range.

2. Dynamics of finger motion

Reviewed here is a simple model for the transient dynamics, wherein we assume that the cross-sectional profile of the advancing finger is semicircular (radius R), and constant along its length. Momentum may be ignored and so the force balance on a control volume containing the finger yields

$$f^e + f_\gamma + f_\mu = 0 \qquad (1)$$

The DEP and surface tension forces, $f^e = V^2 dC/dZ$ and $f_\gamma = -\pi R\gamma$, respectively, are constant, and the drag force, $f_\mu = -2RZ\tau_\mu$, is area-dependent. $C(Z)$ is the system capacitance, V is rms-applied voltage, and γ is the surface tension. Assume viscous shear is proportional to finger velocity, i.e., $\tau_\mu = K dZ/dt$, where K is a proportionality factor dependent on radius R and liquid viscosity μ. The solution to Eq. (1) takes the form: $Z(t) \propto \sqrt{t}$. Representative experimental data for a water finger in air, along with a fitted square-root law curve, are plotted in Fig. 3. For the curve, the so-called dielectric thickness d/κ_d (where d is the thickness of the dielectric coating and κ_d is the dielectric constant) was used as an adjustable parameter. A best fit is achieved for $d/\kappa_d = 2.17$, whereas, if we use $d = 10$ μm and $\kappa_d = 3$ (for polyimide), the correct value should be 3.33. Factors contributing to this discrepancy include electrical breakdown in the dielectric and uncertainties about the experimental parameters.

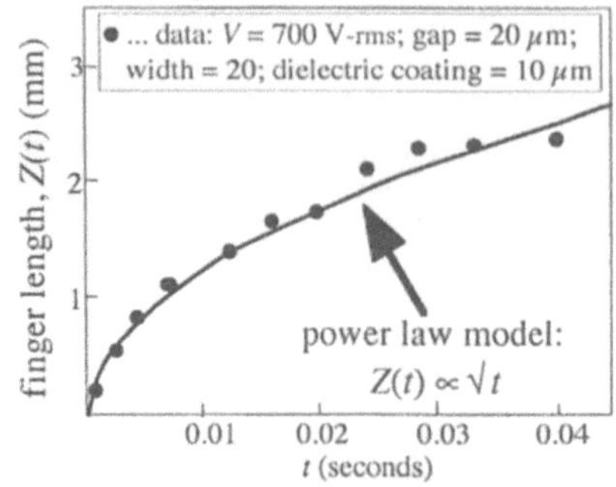

Fig 3. Finger length Z versus time t for a de-ionized water finger in air. Plotted curve is the best-fit curve for $Z(t) \propto \sqrt{t}$.

For experiments performed in oil, the transient motion of a water finger is smoother and good actuation is achievable at lower frequencies. This performance enhancement may result from the oil film that lubricates the finger. Heating effects are diminished because of heat transfer to the oil. It should be noted, however, that the cross-section of the moving finger is no longer uniform along its length, making application of the above model invalid for this case.

3. Droplet formation dynamics

Droplets form when the voltage is removed because, without the non-uniform AC electric field to stabilize the hydrostatic equilibrium, the capillary instability takes over. For structures with >50 μm cross-sections immersed in oil (cf., Fig. 2), this instability, which breaks up the finger into droplets, and surface tension, which tries to retract the liquid finger, compete after voltage removal. We estimate that, for larger structures, 50% or more of the liquid is returned to the reservoir droplet. By contrast, we believe thinner structures retain most of the liquid after voltage is removed, but because the finger is slower to fill, there is typically much less liquid distributed along the structure when the voltage is removed. This leads to formation of large numbers of small droplets. Fig. 4 shows 26 droplets, ranging in volume from ~20 to ~70 picoliters, distributed along a ~35 micron wide, ~6 mm long structure. In this experiment, the droplets formed in ~2 ms, while it took ~300 ms for the finger to reach the end of the structure.

One may test the correspondence of droplet spacing to the fastest growing wavelength as predicted by hydrodynamic theory. The mean spacing for the droplets in Fig. 4 is 200 μm (±58 μm), while the theory of rivulets predicts 143 μm [3]. This probably stems from the questionable assumption that the finger profile is semicircular, though the influence of oil viscosity may also be substantial. The speed and physical dimensions involved make determination of the time-dependent profile of the finger very difficult.

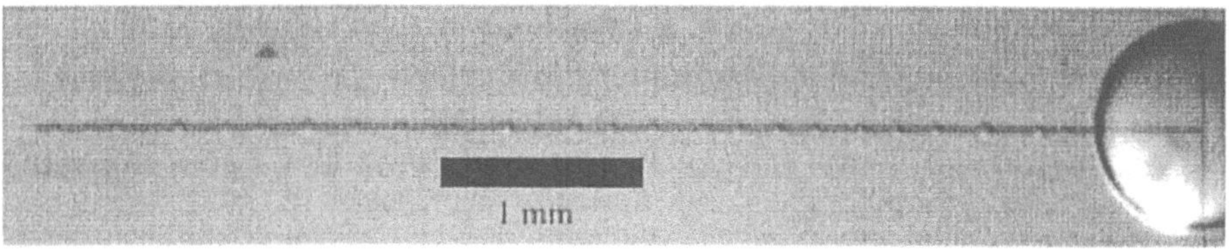

Fig. 4. Dispensing of 26 droplets with volume ranging from ~20 to ~70 picoliters (average ~ 37 picoliters) on co-planar strip electrodes (evaporated Al on glass): width $w \approx 10\ \mu m$, spacing $g \approx 16\ \mu m$. Experimental liquid: DI water in an oil bath. Dielectric coating: ~1.5 μm of spin-coated photoresist (Shipley #1805) applied in two layers with very thin, intervening layer of Teflon AF™. Volume of reservoir droplet at right is ~1 microliter.

4. Discussion

Substituting ethylene glycol for water in the oil bath experiment provides an instructive example of the interplay of the two time scales. In this case, the finger does not break up into droplets when the voltage is turned off, but instead remains intact and withdraws smoothly back into the reservoir droplet. Instability is almost entirely suppressed, presumably because the interfacial tension of ethylene glycol against oil (~0.02 Nt/m) is about one-fourth the water/oil value. Reduced interfacial tension increases the minimum unstable wavelength and decreases the unstable growth rate. The increased viscous drag due to the oil also inhibits the instability.

5. Conclusion

On-chip, DEP droplet dispensing speed is controlled by the time required to fill the electrode structure rather than by the hydrodynamic instability that actually forms the droplets. Thus, making DEP droplet dispensing accurate and reliable requires better knowledge of the dynamics of the finger. It is clear that many factors influence these dynamics, including the dielectric constant and thickness of the insulating coating, the interfacial tension and viscosity of the liquid, the wetting properties of the dielectric, and the structure dimensions. At least for actuation of water in air, the $Z(t) \propto \sqrt{t}$ law can be used to estimate scaling factors. For example, doubling the electrode length increased the filling time by a factor of four. Compared to most microfluidic schemes, DEP actuation is very fast; however, this scaling law imposes strict limits relating speed to the size of structures. In the case of DEP actuation in insulating oil, the fluid dynamics are far more complex. Numerical modeling will be required to establish the scaling law.

There are other important time scales that influence DEP actuation. For example, even DI water tends to heat up. We have seen instances where the leading edge of a finger starts to boil if it moves too slowly. The leading edge boils first because it is subjected to Joule heating for the longest time. Increasing the voltage to speed up the finger reduces exposure time of the water to the electric field and alleviates boiling. But increasing the voltage can lead to another problem: electrical breakdown. Practical μTAS applications of DEP involve optimization with respect to materials properties including dielectric constant, breakdown resistance, and wetting properties.

Acknowledgments

The National Science Foundation (USA), the National Institutes of Health (USA), and the Center for Future Health (University of Rochester) supported this work. T. Ito acknowledges a fellowship from the Ministry of Education, Culture, Sports, Science and Technology (Japan). M. Gunji, R. Mulgund, D. Hsu, and P. Osborne assisted with the experiments.

References

1. B. Berge, "Électrocapillarité et mouillage de films isolants par l'eau," *C. R. Acad. Sci. Paris*, Série II, 157-163 (1993); M. G. Pollack, R. B. Fair, A. D. Shenderov, "Electrowetting-based actuation of liquid droplets for microfluidic actuation," *Applied Physics. Lett.* **77**, 1725-1726 (2000); J. Lee, H. Moon, J. Fowler, T. Schoellhammer, and C-J. Kim, *Sensors and Actuators* **A 95**, 259-268 (2002).
2. T. B. Jones, M. Gunji, M. Washizu, and M. J. Feldman, "Dielectrophoretic liquid actuation and nanodroplet formation," *J. Applied Physics* **89**, 1441-1448 (2001).
3. S. Schiaffino and A. A. Sonin, "Formation and stability of liquid and molten beads on a solid surface," *J. Fluid Mech.* **343**, 95-110 (1997).

DROPLET FLIGHT STABILITY AND SAMPLE ENRICHMENT BY GAS-FLOW GUIDED DISPENSING

Simon Ekström, Jonas Bergkvist, Thomas Laurell and Johan Nilsson
Lund Institute of Technology, Lund University, P.O. Box 118, SE-221 00 Lund, Sweden
(e-mail: johan.nilsson@elmat.lth.se)

Abstract

It is demonstrated that the use of a gas-flow guided dispenser provides increased droplet stability and directionality. This facilitates dispensing of liquid gradients, were the surface tension of dispensed liquid varies.

Keywords: gas-flow guided dispensing, liquid chromatography, mass spectrometry

1. Introduction

In the recent proteome mapping efforts liquid chromatography (LC) has become an important tool, especially in combination with isotopic labelling and mass spectrometry [1]. Currently most of these chromatographic approaches to protein identification rely on on-line coupling to electrospray ionisation mass spectrometry for generation of tandem MS data, but the newly introduced Matrix-Assisted Laser Desorption/Ionisation (MALDI) instruments capable of tandem MS will provide an optional way of identifying proteins from single peptides [2,3]. These new MS techniques could benefit greatly from dispensing techniques that can facilitate the on-line fractionation of an entire chromatographic peptide separation (*i.e.* thousands of sample spots) onto the MALDI-target plate.

2. Theory

In order to facilitate on-line fractionation the dispensing technique must be both robust and generic. A typical chromatographic run takes between 30-60 minutes and during this time the liquid phase gradient may go from 0 to 50% organic solvent. Previous experience of coupling flow-through piezo-actuated dispensing to LC [4] have shown that the changes in surface tension of the dispensed liquid can lead to decreased droplet stability, directionality (formation of satellite droplets that may have a different trajectory) and in the worst case discontinued dispensing.

One way to compensate for the variation in surface tension could be to actively control the actuation of the dispenser, but the drawback is that any change in the gradient set-up will require a new calibration of the active control. A more convenient way to ensure that the dispensed droplets are deposited in precise positions is to stabilise the ejected droplets with a gentle stream of gas [5].

Y. Baba et al. (eds.), Micro Total Analysis Systems 2002, Volume 2, 676–678.

3. Experimental

The dispenser was fabricated by anisotropic etching of <100>-oriented, p-type silicon and consists of three wafers where the inlets, the outlet, the flow-through channel and the nozzles are defined. The three wafers were joined by silicon direct bonding and silicone rubber tubes were attached to the inlets and the outlet by silicone rubber adhesive. The inner diameter of the silicone rubber tubes was selected to fit standard 1/16" tubing but other dimensions can also be used. The piezo-electric element was mounted in a Plexiglas^{TM} backing which in turn was glued to the silicon parts using epoxy adhesive. Further details on the dispenser fabrication can be found in Laurell et al [6].

A CMA/100 pump (CMA/microdialysis, Solna, Sweden) was used to generate the liquid flow (mobile phase 0.1% TFA: 10-50 % ACN). For MALDI analysis samples were deposited onto a target precoated with a thin film of matrix/nitrocellulose [4].

4. Results and discussion

Figure 1 shows droplet ejection with satellite droplets from an ordinary flow-through dispenser (1A) and from a gas-flow guided dispenser (1B). The guiding gas-flow is achieved by adding a second nozzle with a single inlet for gas to the dispenser depicted in Figure 1A. The use of a guiding gas-flow ensures that all the droplets are deposited in the desired position. The applied gas flow leads to an increased droplet velocity (2A) and as the velocity increases the droplet volume decreases (2B), Figure 2. This is predominantly caused by changes in the droplet formation, arising from the increased pressure on the nozzle exit.

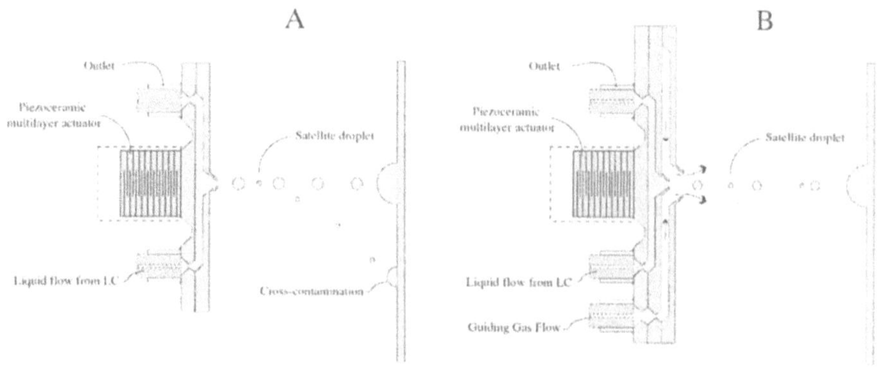

Figure 1. Droplet ejection with satellite droplets from an ordinary flow-through dispenser (1A), note the cross-contamination arising from the satellite droplets. (1B) Droplet ejection from a gas-flow guided dispenser where the satellite droplets are deposited on the same position as the main droplet.

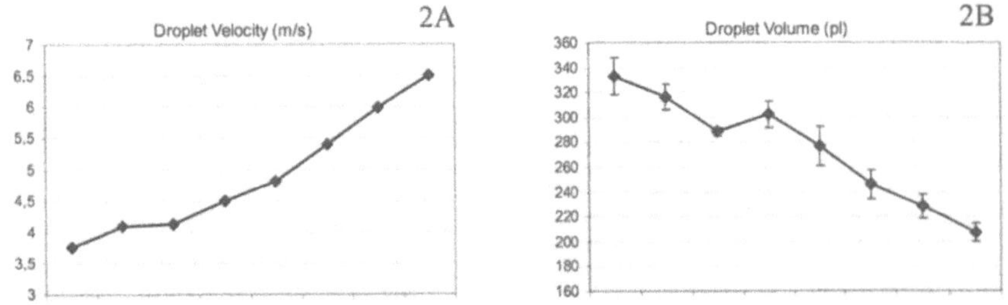

Figure 2. The applied gas flow leads to an increased droplet velocity (2A) and as the velocity increases the droplet volume decreases (2B).

5. Conclusions

The use of a gas-flow guided dispenser provides increased droplet stability and directionality during fractionation of liquid gradients. This in combination with the flow-through (for the liquid) format of the dispenser, which allows the user to determine the ratio of the liquid flow to be deposited simply by tuning the dispensing frequency, provides a good and generic tool for fractionation of LC separations.

References

1. Gygi, S.P., Rist, B., Gerber, S.A., Turecek, F., Gelb, M.H.Aebersold, R., Nat. Biotechnol. 1999, 17, 994-999.
2. Medzihradszky, K. F., Campbell, J. M., Baldwin, M. A., Falick, A. M., Juhasz, P., Vestal, M. L., Burlingame, A. L., Anal. Chem. 2000, 72, 552-558.
3. Griffin, T.J., Gygi, S.P., Rist, B., Aebersold, R., Loboda, A., Jilkine, A., Ens, W.Standing, K.G., Analytical Chemistry 2001, 73, 978-986.
4. Miliotis, T., Kjellstrom, S., Nilsson, J., Laurell, T., Edholm, L.E.Marko-Varga, G., J Mass Spectrometry 2000, 35, 369-377.
5. Miura M. U.S Patent 4106032, Aug 8, 1978.
6. Laurell, T., Wallman, L., Nisson. J., J. Micromech. Microeng. 1999, 9, 369–376.

PROTEIN MICROARRAY PATTERNED BY A SURFACE-TENSION-DRIVEN STAMPING SYSTEM WITH DISCRETE DISPENSING CHANNELS

Shih-Chang Lin, Yi-Chin. Tsai, *Fan-Gang Tseng, Haimei Huang, and Ching-Chang Chieng
National Tsing Hua University, 300 Hsinchu, Taiwan
*fangang@ess.nthu.edu.tw

Abstract

This study proposes a novel surface-tension-driven stamping system using a micro-stamp array with discrete dispensing channels to massively spot different kinds of bio samples with the same size and shape into micro arrays within a short period of time. Testing result demonstrated the feasibility of uniform and clean deposition of protein samples into an accurate array with less than 3% spot size and intensity variation. This system has the potential to spot hundreds of different bio-reagents into a dense array for the application of disease diagnosis or drug screening.

Keywords: microcontact printing, protein microarray, bio chip, micro channel, micro fluidics

1. Introduction

"Microcontact printing"[1-4] is a cost-effective and rapid means to pattern thousands of bio samples in parallel for use in biomedical applications, such as disease diagnosis and drug screening. However, most of the stamping systems can imprint only several different kinds of biosamples [1-3]. The stamping system presented in this study, as shown in Fig.1-2, can easily fill in 25 different kinds of biosamples and then imprint them into an array on a bioassay chip. This concept can be expanded to spot hundreds of different bio-samples. The micro-stamp chip was composed of three components: a micro-stamp array, an embedded micro-channel array, and a reservoir unit. Different bio-samples could be dispensed into different reservoirs and directed by surface tension to the tips of the micro stamps through the micro channels. The biosample-filled micro-stamp was then

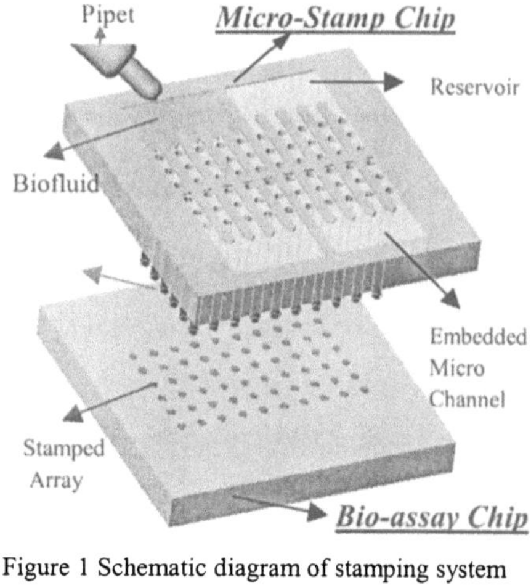

Figure 1 Schematic diagram of stamping system

Y. Baba et al. (eds.), Micro Total Analysis Systems 2002, Volume 2, 679–681.

brought into contact with the bio-assay chip to generate sample arrays for further bio processing (as shown in Fig. 2). The micro-stamp array was made of elastomeric polydimethylsiloxane (PDMS) to make conformal contact with non-planar surfaces.

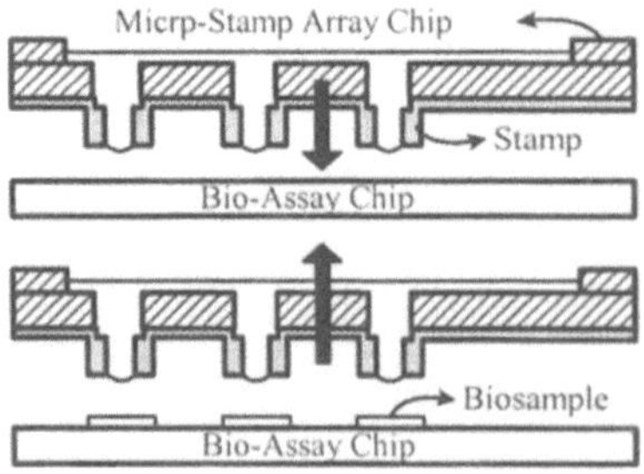

Figure 2 Design and operation of stamping system

2. Fabrication

This stamping system was fabricated by using Polymer-MEMS (micro-electro-mechanical system) technology. Fig. 3 illustrates the fabrication process of the micro-stamp chip. The fabrication involved thick photoresist lithography, PDMS molding, oxygen plasma oxidization, and etching of PDMS. The most crucial step was the oxidization of PDMS surface using oxygen plasma, which resulted in complete spreading-out of thick photoresist SU8 on PDMS surface and strong bonding between SU8 and PDMS. Fig. 4 shows the SEM pictures of the fabricated micro-stamp array and reservoir unit. The stamp size was 100 μm or 200 μm in outer diameter, 50 μm in inner diameter, and 100 μm in height. The bio-assay chip was manufactured by modifying the slide with APTS (aminopropyltrimethoxysilane) first and then activating the slide by BS3 (bis-sulfosuccinimidyl suberate)[5-6].

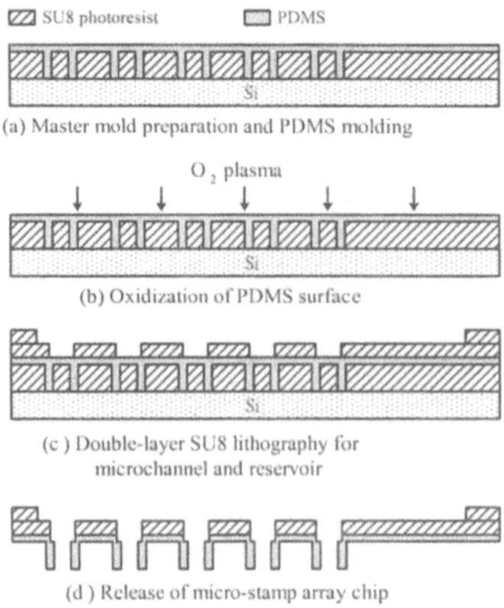

Figure 3 Fabrication procedure for the Micro Array Stamper Chip

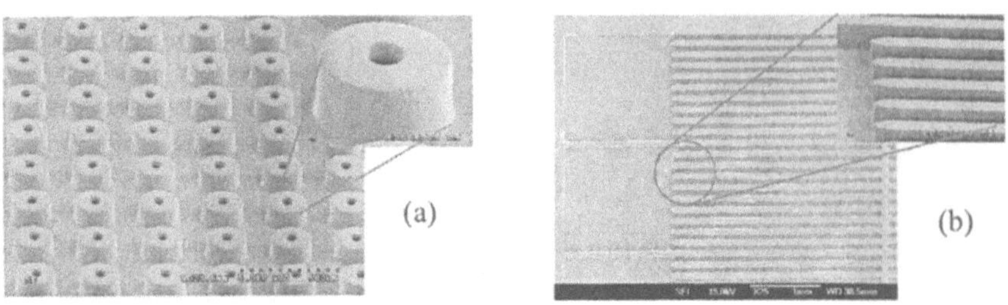

Figure 4 SEM pictures of fabricated (a) micro-stamp array and (b) reservoir unit

3. Results and discussion

To demonstrate the functionality of the stamping system, the micro-stamp was used to stamp anti-rabbit IgG-Cy3 conjugate onto the bio-assay chip, as shown in Fig. 5. Fig. 6 showed the stamped protein microarray after excitation of Cy3. The fluorescent bright spots in Fig. 6 were where the protein was bound and indicated that clean and separate printing could be achieved. The stamped spot size is 200μm in diameter. The fluorescence intensity and spot area of different spots on the same array were normalized to the average and the results were showed in Fig.7. The standard deviations of the normalized intensity and area were less than 3%, manifesting the uniformity of stamp printing.

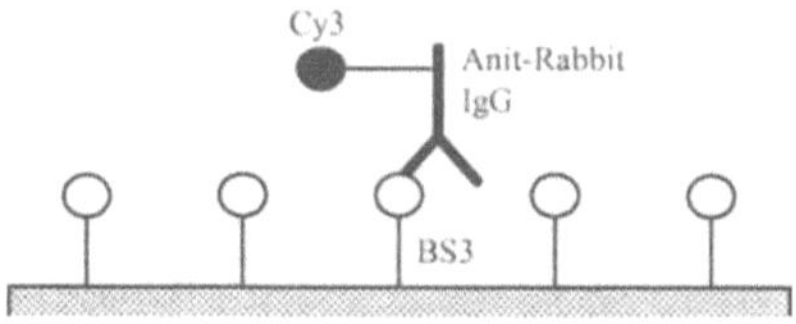

Figure 5 Schematic diagram of bio-assay configuration

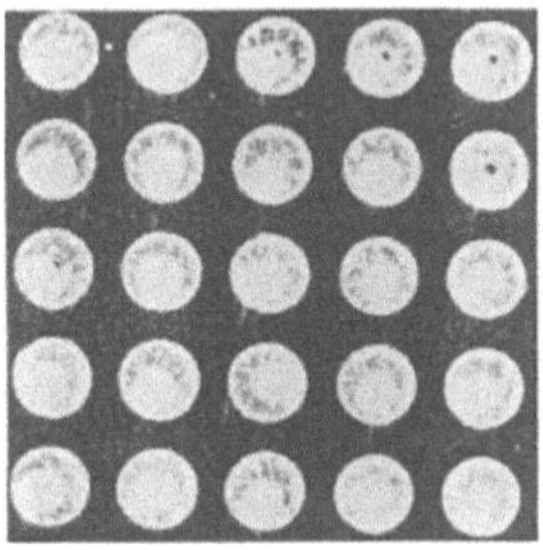

Figure 6 Protein spots on the APTS-BS3 slide

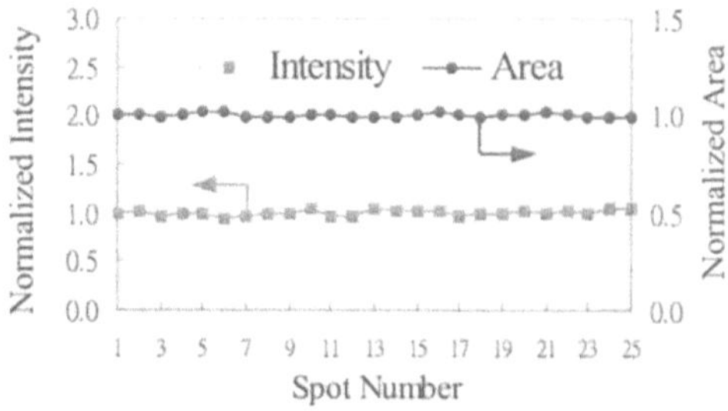

Figure 7 Intensity and area variation of different spots on the same protein array

Acknowledgements

The bio-assay chip prepared by Y.H. Cu is greatly appreciated. The work was supported by the National Science Council of Republic of China under the grant NSC 89-2323-B-007-005.

References

1. X.M. Zhao, Y. Xia, and G.M. Whitesides, "Soft Lithography for Nano-Fabrication" *J. Mater. Chem.*, Vol. 7, pp. 1069-1074, (1997).
2. A. Bernard, E. Delamarche, H. Schmid, B. Michel, H.R. Bosshard, and H. Biebuyck, "Printing Patterns of Proteins," *Langmuir*, Vol. 14, pp. 2225-2229, (1998).
3. R.S. Kane, S. Takayama, E. Ostuni, D.E. Ingber, and G. M. Whitesides, "Patterning Proteins and Cells Using Soft Lithography," *Biomaterials*, Vol. 20, pp. 2363-2376, (1999).
4. S.C. Lin, F.G. Tseng, H.M. Huang, C.Y. Huang, and C.C. Chieng, "Microsized 2D Protein Arrays Immobilized by Micro-Stamps and Micro-Wells for Disease Diagnosis And Drug Screening," *Fresenius J Anal Chem*, Vol. 371, pp. 202-208, (2001).
5. L.G. Mendoza, P. McQuary, A. Mongan, R. Gangadharan, S. Brignac, and M. Eggers, "High-Throughput Microarray-Based Enzyme-Linked Immunosorbent Assay (ELISA)," *Biotechniques*, Vol. 27, pp. 778-788, (1999).
6. F.G. Tseng, Y.H. Cu, H.M. Huang, S. C. Lin, and C. C. Chieng, "Protein Immobilization Properties of Various SAMs for the Application of Stamping Protein Micro Array," *Pacific Rim Workshop on Trans. and Micro/Nano Tech.* Xiamen University, China, July 22~24, pp. 753-756, (2002).

MICROMACHINED HOLLOW NEEDLE WITH INTEGRATED PRESSURE SENSOR FOR PRECISE, CALIBRATED INJECTION INTO CELLS AND EMBRYOS

S. Zappe [1], X.J. Zhang [1], I.W. Jung [1], R.W. Bernstein [1], E.E.M. Furlong [2], M. Fish [2], M. Scott [2] and O. Solgaard [1]

[1] Department of Electrical Engineering, Stanford University, USA
[2] Department of Developmental Biology, Stanford University, USA
(zappe@stanford.edu)

Abstract

We have developed surface micromachined hollow silicon nitride needles, specifically designed for injection of DNA and proteins into fruit fly embryos. Precisely controlled calibration of the injected volume is possible using an integrated membrane-based pressure sensor with interferometric read out. The membrane can withstand pressures in excess of 3 bar. The excellent mechanical stability of the hollow needle was verified in fruit-fly embryo injection experiments and the injection of water and a solution of GFP (green fluorescent protein) were successfully performed. Flow measurements of air through a needle with channel dimensions: 75 μm x 15 μm x 2.5 μm, resulted in a flow rate of r_{air} = 1.4 nl / (s bar).

Keywords: Surface micromachined injector, integrated pressure sensor, interferometric deflection read-out

1. Introduction

In the emerging field of bioengineering there is increasing demand for precise and controlled injection of drugs, DNA, structural proteins, enzymes, ions, antibodies or metabolites into living cells and embryos [1]. Investigations on the fruit fly, Drosophila melanogaster, are of special interest since the fruit fly serves as a model organism for studies of developmental and cellular processes common to higher eukaryotes, including humans. The complete fruit fly genome sequence was presented in March 2000, giving rise to an increased effort to identify the function of the approx. 13,600 genes [3]. A method to study the function of genes is to silence specific genes during fruit fly embryo development by injection of suitable dsRNA (double stranded RNA) [4].

Common injection tools like drawn glass capillaries are unsuitable for many high-precision experiments because their large and ill-defined diameters, mechanical instability, and high back pressure lead to poor repeatability and often unacceptable damage to injected cells. Significant improvements can be achieved by using micromachined injectors defined with high-resolution photolithography and micromachining technology.

Y. Baba et al. (eds.), Micro Total Analysis Systems 2002, Volume 2, 682–684.

2. Device fabrication

The schematic structure and SEMs of a microinjector are shown in Fig. 1. Silicon micromachining technology allows us to separate the penetrating needle tip from the liquid outlet to reduce the risk of clogging. The needle diameter can be scaled down to a few micrometer, and using e-beam lithography even sub-micrometer dimensions can be realized. The mechanical stability is maintained as very short needles can be created. Due to the compact needle design, the backpressure is low. The needle can be fabricated as an integrated part of a fluidic microsystem for automated injection into embryos and more functionality for example the integration of particle filters can be added.

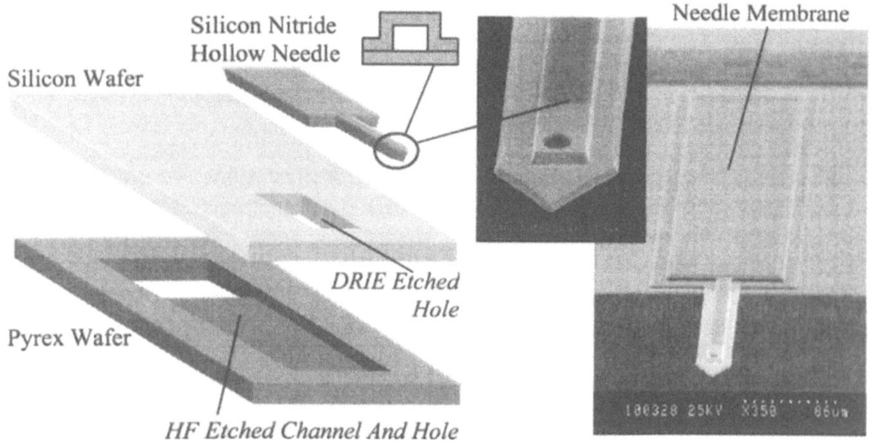

Figure 1: Left: Schematic structure of the microinjector. Right: SEMs of a fabricated device. The hollow needle is realized using LPCVD (Low Pressure Chemical Vapor Deposition) of silicon nitride and LTO (Low Temperature Oxide) and sacrificial etching of the patterned LTO layer in buffered HF. Interconnection to the needle is realized by Deep Reactive Ion Etching of a 500 µm x 100 µm wide hole through the silicon wafer. A structured Pyrex wafer with a 2.4 mm wide hole for the attachment of tubing is bonded to the silicon wafer.

3. Device characterization

The applied pressure close to the microinjector can be monitored using the silicon nitride membrane covering the supply channel that is etched through the silicon wafer. Figure 2 depicts the interferometric optical set-up that was used to monitor the deflection of an aluminum-coated membrane under pressure. Pressure changes can be detected by analysing the movement of the interference pattern. Pressure sensing very close to the needle allows us to study system dynamics, to calibrate the injectors and to ensure a highly precise dosage.

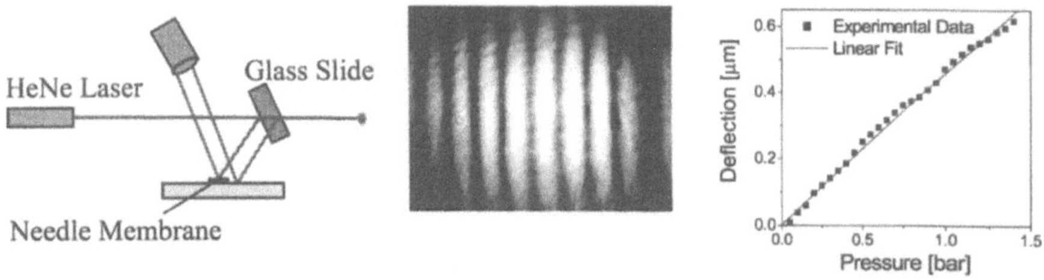

Figure 2: Left: Schematic set-up for interferometric measurement of the sensor displacement. A HeNe beam is focused to a spot size of 50 µm and split in two by reflections from the front and back of a glass slide. One beam is reflected from the aluminum-coated membrane, while the other is reflected from a reference on the silicon chip. The interference of the beams is recorded on a video camera. Center: Interference pattern of the two beams. Right: Membrane deflection, derived from changes of the interference pattern, as a function of applied pressure.

Flow measurements of air through a needle with dimensions: 75 µm x 15 µm x 2.5 µm showed a flow rate of r_{air} = 1.4 nl / (s bar). Taking into account the lower flow rate of water due to its 55 times higher kinematic viscosity, and the fact that it is possible to scale down the needle diameter, we expect to be able to control the volume down to the sub-picoliter regime using pressure-feedback controlled injection.

4. Discussion and conclusions

We have successfully demonstrated the operation of micromachined needles for injection in fruit-fly embryos, and showed that an integrated pressure sensor allows monitoring of the applied pressure close to the injector to improve calibration of the injected volumes. Currently, a microinjector package is developed to enable systematic screening of gene functions by automated injection of dsRNA into embryos in self-positioned arrays [4].

Acknowledgements

Support through a DARPA sponsored project (DARPA Project RA-0014) within the Bio:Info:Micro program is gratefully acknowledged.

References

1 Lacal JC, Perona R, Feramisco J (Eds.), *Microinjection*, Birkhaeuser, Basel, 1999
2 Adams MD et al., Rubin GM, Venter JC, *The Genome Sequence of Drosophila melanogaster*, Science, Volume 287, Number 5461, Issue of 24 Mar 2000, pp. 2185-2195
3 Carthew RW, *Gene silencing by double-stranded RNA*. Curr Opin Cell Biol. 2001 Apr;13(2):244-8
4 R.W.Bernstein et al, *Surface-tension mediated positioning and immobilization of Drosophila embryos in 2-D arrays for drug injection*, Proc. The Sixth Int. Symp. on Micro Total Analysis System (mTAS), Nara, Japan, November 3-7, 2002

FALLING-DROP INTERFACE FOR CONTINUOUS SAMPLE INTRODUCTION IN MICROFLUIDIC CHIPS

Qun Fang, Zhi-Jian Jia, Zhao-Lun Fang
Institute of Microanalytical Systems, Department of Chemistry, Zhejiang University, Hangzhou, China

Abstract

A continuous sample introduction system for chip-based capillary electrophoresis (CE) systems was developed using a falling-drop interface in order to achieve electrical isolation of the CE chip from the sample introduction system, and minimise sample flow into the sampling channel due to hydrostatic pressure. Effective sample change was achieved in the continuous CE separation of FITC-labelled amino acids by LIF detection with <5% carryover, 20-70 μL sample volume and 5-15 s sample introduction time.

Keywords: Microfluidic chip, falling-drop interface, sample introduction, world-to-chip interfacing, capillary electrophoresis

1. Introduction

Chip-based capillary electrophoresis (CE) systems, particularly with laser induced fluorescence (LIF) detection, have undergone fast development in recent years. Currently, particular interests are focused on achieving continuous introduction of a series of samples to meet the requirements for analysis of real-world samples [1-4]. A continuous sample introduction system for chip-based capillary electrophoresis (CE) systems was reported by our group, characterized by a flow-through sample reservoir with guided overflow [4]. In this work, modifications were made on the previous design to further improve the performance by employing a falling-drop interface with hydrophobic capillary outlet, wide conical inlet, PDMS-treated chip bottom, and guided waste flow using a filter-paper wick.

2. Experimental

The design of the sample introduction interface is shown in Fig. 1. A crossed-channel glass chip was used for the CE separations, with reservoirs fixed at the waste and working electrolyte buffer terminals. A sequential injection system was used for continuously presenting the samples to the flow-through reservoir through a 75 μm quartz sampling capillary, the tip of which was made hydrophobic to reduce the drop size to 2.5-6 μL. A 1.2 mm-hole was drilled through the chip at the sampling terminal, functioning as a flowthrough reservoir. A plastic pipette tip with 0.4-mm i.d. was inserted into the hole to the depth of the cover plate to reduce the dead volume, and an 1 mm thin plexiglas cell with a wide (7 mm) conical inlet was epoxyed over the inlet hole to facilitate collection of the falling drops without splashing losses. The surface of the chip bottom was covered with a layer of PDMS to avoid spreading of waste flow on the

Y. Baba et al. (eds.), Micro Total Analysis Systems 2002, Volume 2, 685–687.

chip. A filter paper tube positioned at the outlet functioned as a wick guide to remove the waste smoothly.

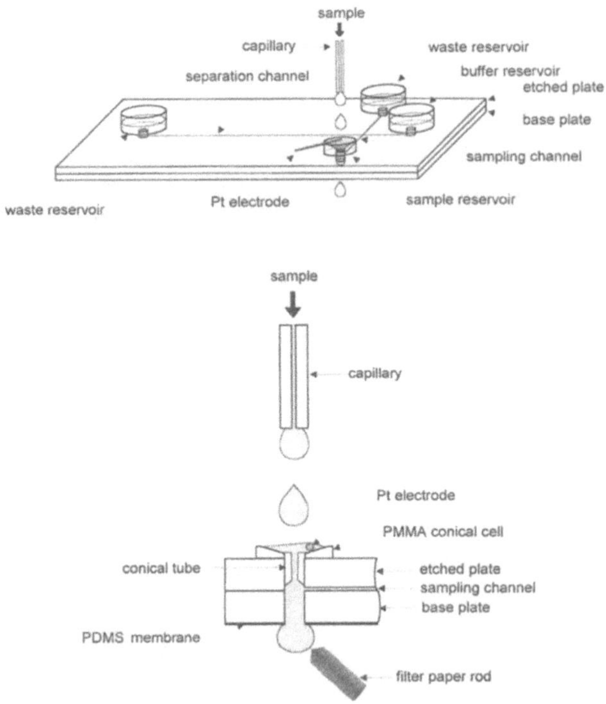

Figure 1. Structure of the microfluidic chip system with falling-drop flow-through sample introduction interface (not to scale).

3. Results and discussion

The falling-drop design allowed electrical isolation of the CE chip from the SI system, even when the sample reservoir was connected to the HV terminal. This feature is an improvement over the previous design [4], where grounding of the sample reservoir was required. The design also minimized sample flow into the sampling channel due to hydrostatic pressure, and reduced the sample volume.

The performance of the system was tested using FITC-labelled amino acids as model samples with LIF detection (as shown in Fig. 2). Employing a separation channel of 4 cm (2 cm effective length), 1000 V separation voltage, throughput of 30/h was achieved for separation of FITC-arginine and phenylalanine. A minimum sample volume of 20 μL was required for sample change, requiring 5-10 s for achieving a carryover of less than 5%.

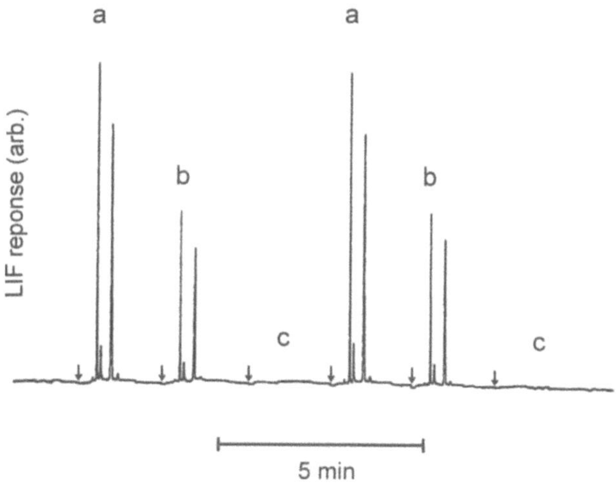

Figure 2. Electropherogram recorded by sequentially introducing (a) 200 μM, (b) 100 μM FITC-arginine and phenylalanine, and (c) a blank sample buffer solution, for two cycles to show carryover effects, with applied sampling and separation field strengths of 600 and 250 V/cm, respectively.

4. Conclusions

The proposed interface proved to be an efficient means for achieving sample change on microfluidic chip systems, allowing electrical isolation of the CE chip from the sample introduction system.

Acknowledgements

This work was supported by Natural Science Foundation of China under project No. 20299030 and National Ministry of Education and Personnel Research Foundation.

References

1. S. Attiya, A.B. Jemere, T. Tang, G. Fitzpartrick, K. Seiler, N. Chiem and D.J. Harrison, *Electrophoresis*, **22**, 318 (2001).
2. W.A. Ausserer, L. Bousse, S.J. Gallagher, C.B. Kennedy and H.L. Phan, *JALA*, **6**, 69 (2001).
3. Y.H. Lin, G.B. Lee, C.W. Li, G.R. Huang and S.H. Chen, *J. Chromatogr. A*, **937**, 115 (2001).
4. Q. Fang, G.M. Xu and Z.L. Fang, *Anal. Chem.*, **74**, 1223 (2002).

New Assays Based on Laminar Fluid Diffusion Interfaces - Results from Prototype and Product Testing

Christopher J Morris, Troy Daiber, Bernhard H Weigl, Natasa Kesler, and Ron L Bardell

Micronics, Inc. 8463 154th Ave. NE, Redmond, WA, 98052 USA

Abstract

Central to Micronics' microfluidics technology platforms is the generation and use of Laminar Fluid Diffusion Interfaces (LFDIs). One of the embodiments of LFDIs is in diffusion-based detection technology, which is based on the parallel flow of individual streams in a single microfluidic channel. Micronics' T-Sensor® based Chemical Analysis System (CAS) harnesses diffusion to identify unknown concentrations of chemical species, allowing researchers to perform multiple assays on a single version of a disposable microfluidic cartridge. Implementations of both a prototype portable device and a benchtop system are demonstrated, with CV's ranging from 2.6 to 10%.

Keywords: diffusion, chemical analysis, T-Sensor

1. Introduction

Under microfluidic conditions, fluids flow in a predictable, laminar fashion, which allows miscible fluids to flow next to each other without turbulent mixing, and without being physically separated by a membrane. Diffusion, however, also occurs under these conditions. This effect may be used to perform various chemical assays with the T-Sensor® lab card, in which the response of an indicator in the Laminar Fluid Diffusion Interfaces (LFDI) zone may be interrogated providing data on an analyte in real time in a continuous flow process.

2. Theory

Most of Micronics' microfluidic technologies may be implemented as either stand- alone lab cards or integrated, computer-controlled, pump-driven systems. Stand-alone lab cards function without external hardware; fluids are driven instead by such forces as hydrostatic pressure, capillary flow, or (in the case of the AHT-series card) fluid absorption. These passive lab cards offer the benefit of single use, low-cost disposables that can provide rapid qualitative and semi-quantitative detections. These are ideal devices for point-of-care applications, and lend themselves to a low-cost, portable T-Sensor system using a scanning LED for detection.

Pump-driven, disposable ActiveT™ Lab Cards are computer-controlled through a pneumatic microcircuit of on-chip valves and air conduits (Fig. 1) and an external bank of precision pumps (Fig. 2). These disposables are designed to automatically perform all fluidic functions, including initial wetout, fluid mixing, generation of

Y. Baba et al. (eds.), Micro Total Analysis Systems 2002, Volume 2, 688–690.

LFDIs, and separation of fluid streams at the end of the interaction—all by pressing a single button on an attached computer. The external fluid and electronic hardware is contained in the small benchtop Microfluidics Pump Station (MPS) shown in Fig. 2.

3. Experimental

In one series of experiments, a T-Sensor scanner was prototyped to use passive AHT-series cards. Reagents were loaded on the disposable card containing a 100 μm deep LFDI channel. As the light of a blue LED scanned across the channel, the filtered fluoresced signal was detected by a photomultiplier tube and sampled at 1000Hz. One scan was acquired every 5.5 seconds over about 10 minutes. Each scan was corrected for background variation, and the intensity profile was extracted as shown in Fig. 4. The peak-height ratio was the ratio of one peak (the unknown) to the other peak (the known, or reference). Fluorescein in a 4.0 pH buffer was used as an indicator to determine pH of Hepes solutions.

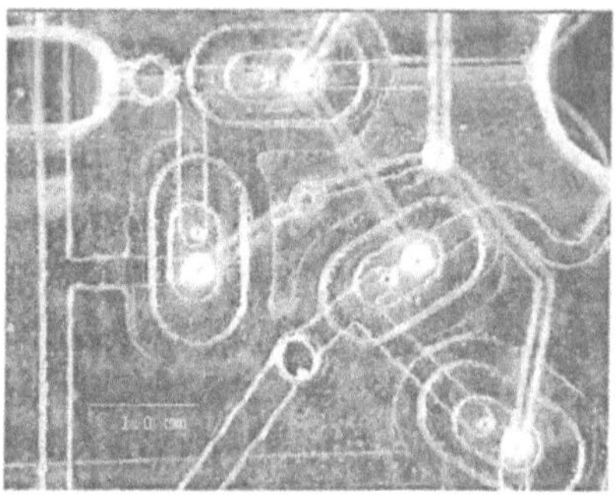

Figure 1. Close up of a Micronics disposable card showing the air-pressure driven valves that provide automated computer-controlled fluid operations.

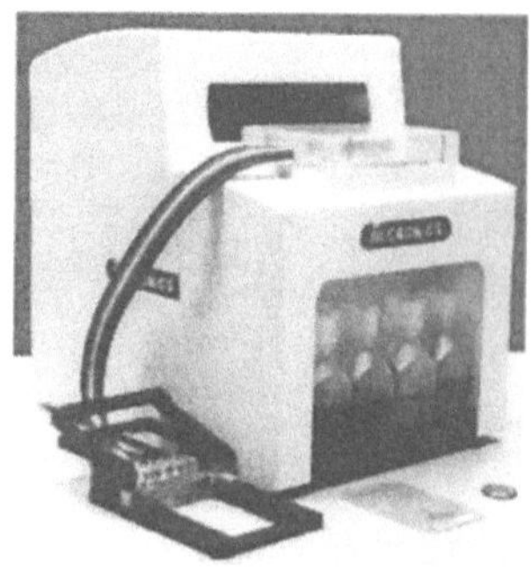

Figure 2. Microfluidics Pump Station with MicroHydro manifold, disposable MTS-series lab card, and a $0.25US coin.

A second series of experiments tested Micronics' Chemical Analysis System (CAS) consisting of an MPS and an ActiveT™ card mounted in a MicroHydro Manifold on an inverted microscope (Zeiss IM35) as shown in Fig. 3. A 50W Mercury Lamp was used for illumination, and images were captured by a 16-bit, cooled CCD camera (MX516, Starlight X-press). The images were taken at a distance that allowed diffusion in the LFDI zone for 0.6 seconds. Prepared calcium buffers and Magnesium-Green indicator at 100 μM (kits C-6775 and M-3733, Molecular Probes) were used.

4. Results and Discussion

The intial study of a scanning LED prototype instrument used to detect pH had a card-to-card CV of 2.6% in terms of peak height ratio for nine cards. The next scanning LED prototype could have increased sensitivity, be portable and battery-powered.

Figure 5 shows the concentration curve of ionized calcium measured by Micronics' CAS and the corresponding variation at each concentration in terms of peak height ratio. Figure 6 shows the predicted calcium levels and error bars in terms of calcium concentration over the course of a four day experiment involving 70 cards. Only one of these cards failed to function as expected and is shown as an outlier. The CV of the error bars shown in Fig. 6 is never higher than 4% in terms of peak height ratio, or 10% in terms of predicted calcium concentration.

5. Conclusions

Micronics' T-Sensor®–based CAS harnesses diffusion to identify the concentrations of unknown species using most detection chemistries. This allows researchers to perform multiple assays using multiple samples on a single disposable microfluidic cartridge version, while minimizing reagent use. We have demonstrated implementations in both a prototype portable system, and a benchtop system.

Figure 3. ActiveT™ card in microscope-mounted MicroHydro manifold.

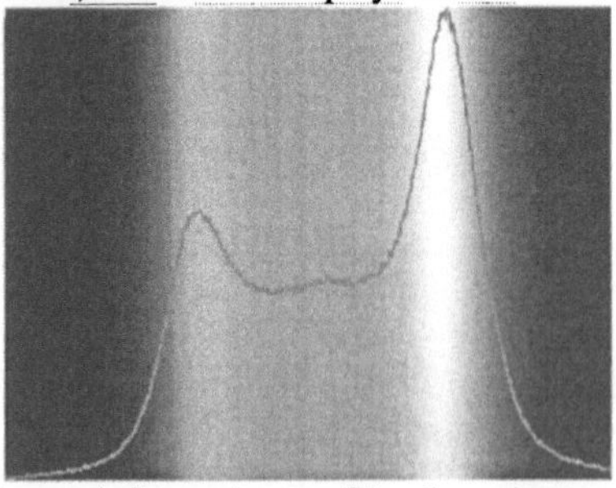

Figure 4. Intensity profile plot extracted from the underlying image.

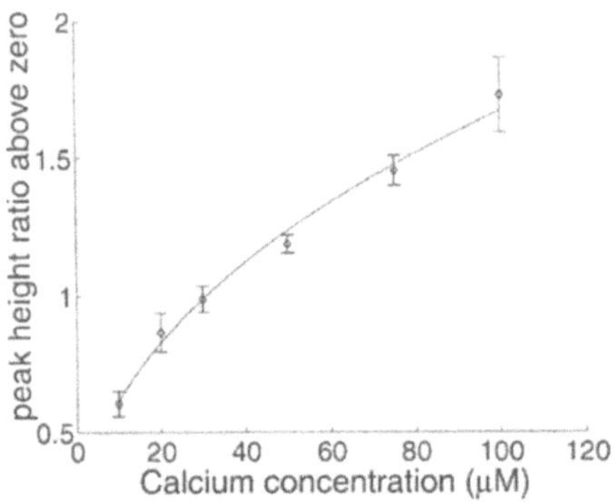

Figure 5. Calibration curve of peak height ratio vs. calcium concentration. Error bars are shown for 1-σ deviations.

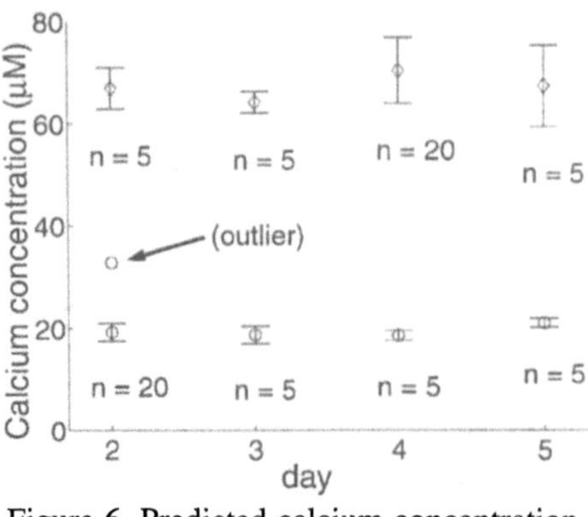

Figure 6. Predicted calcium concentration at 20 and 75 μM over 4 days. 1-σ Deviation error bars in terms of predicted calcium concentration over ***n*** cards.

A NEW APPROACH FOR FABRICATING A ZERO DEAD VOLUME ELECTROSPRAY TIP FOR NON-AQUEOUS MICROCHIP CE-MS

Daria Petersen[1], **Sami Varjo**[2], **Oliver Geschke** [1], **Marja-Liisa Riekkola**[2], **Jörg P. Kutter**[1]

[1] Mikroelektronik Centret (MIC), DTU, Building 345 east, 2800 Lyngby, Denmark; e-mail: da@mic.dtu.dk

[2] Department of Analytical Chemistry, Helsinki University, Finland

Abstract

We report a new approach for fabricating a fully integrated zero dead volume electrospray interface (ESI) on a glass CE microchip.

Keywords: non-aqueous microchip CE-ESI-MS, glass micromachining

1. Introduction

A low dead volume connection for ESI is essential for maintaining separation efficiency because of the small sample volumes typically injected on microchip devices (less than 1 nL) and the narrow band widths of the resolved compounds.

Borofloat glass was chosen as a substrate for our microsystems due to its chemical resistance to non-aqueous media and its compatibility with high voltages required for electrophoretic separations.

Many groups have previously demonstrated a chip-ESI-MS interface by a variety of fabrication and assembly methods, mostly on polymer substrates. A few groups who worked with glass microsystems (for example [1-2]), used either capillaries inserted directly into the channel at the edge of the device, or spraying from the flat edge of the chip. Our approach to fabricate a monolithically integrated ESI tip eliminates the dead volume often present in conventional needle junctions, resulting in reduced band broadening and thus increased separation efficiency.

2. Microfabrication of monolithically integrated tip

An integrated glass electrospray tip was fabricated using the following approach (see Fig. 1). First, a channel was etched into the bottom glass chip, while supporting structures for the needle were created by etching into both bottom and top chips. In a second step, the supporting beams were further etched from the top and the bottom of the bonded wafer sandwich to weaken those support structures. After this, most of the now very fragile support beams were mechanically cut to release the tip structure in the center containing the microchannel, leaving a few supporting beams intact around the tip. Finally the remaining area was heated and drawn to break the last supporting structures and to produce a very fine needle (see Fig. 2). After depositing gold on the tip, the chip was ready for use in CE-MS (see Fig. 3).

Y. Baba et al. (eds.), Micro Total Analysis Systems 2002, Volume 2, 691–693.

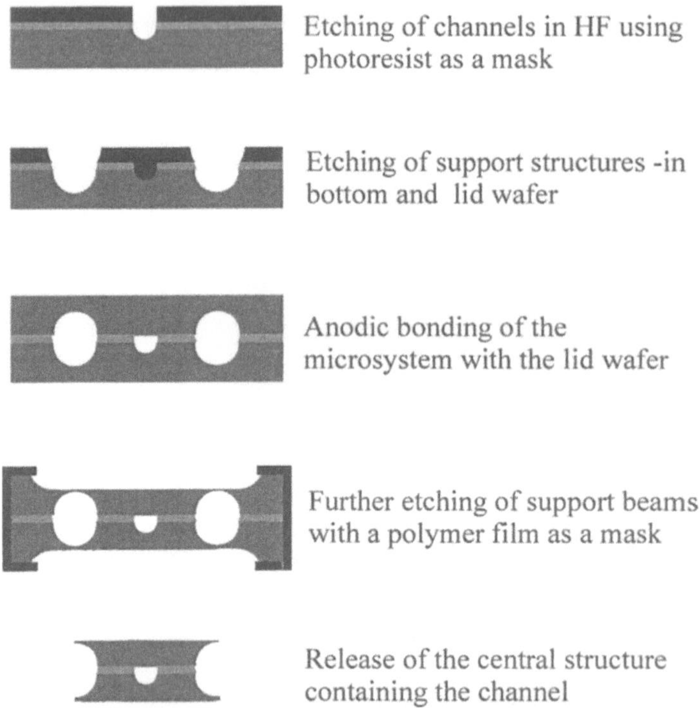

Figure 1. Fabrication sequence for the microsystem. Cross sectional drawing is shown for the area where the support structures are located.

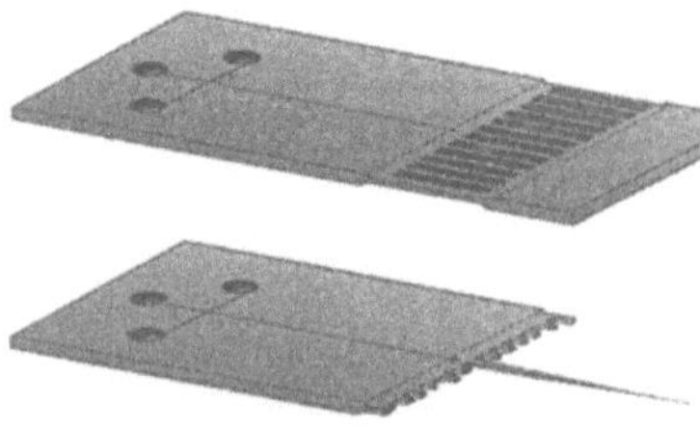

Figure 2 A. Sketch of the microchip: with the temporary support structures and after the release of the tip.

Figure 2 B. Photograph of an actual device with the released tip.

3. Results and discussion

First prototypes have been fabricated and tested, displaying excellent stability of the electrospray, as seen in Figure 3.

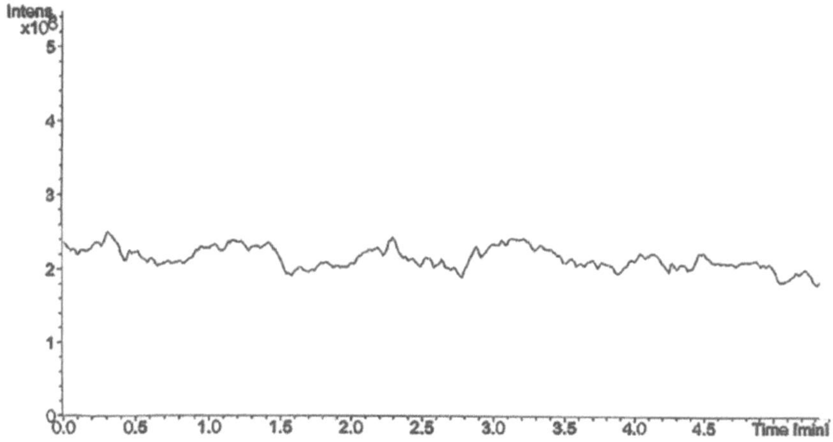

Figure 3. Signal of the electrospray marker versus time.
The data was obtained on a Bruker DataAnalysis Esquire-LC 1.5I MS. Levorphanol (M = 257 g/mol) at a concentration of 50 mg/mL in buffer (10 mM HAc/NH_4Ac in methanol) was used as a spray marker.

4. Conclusions

Stable electrosprays were generated using the new zero dead volume tip. More experiments are necessary to investigate the performance of the entire chip, and the chip coupled to a mass spectrometer, during non-aqueous CE separations.

References

1. Qifeng Xue, Frantisek Foret, Yurij M. Dunaevskiy, Paul M. Zaracky, Nicol E. McGruer, and Barry L. Karger, Multichannel Microchip Electrospray Mass Spectrometry, Anal. Chem., 69, pp. 426-430, (1997).
2. J. M. Lazar, Roswitha S. Ramsey, Stephen C. Jacobson, Robert S. Foote, J. Michael Ramsey, Novel fabricated device for electrokinetically induced pressure flow and electrospray ionization mass spectrometry, Journal of Chromatography A, 892 (1-2), pp. 195-201, (2000).

SILICONE RUBBER COUPLING FOR MICRO FLUIDIC DEVICES

-THEORETICAL ANALYSIS OF SEALING PROPERTIES FOR INTERCONNECTION-

Tadahiro Hasegawa and Koji Ikuta
Dept. of Micro System Engineering, School of Engineering, Nagoya University
Furo-cho, Chikusa-ku, Nagoya 464-8603 Japan
TEL: +81-52-789-5024, FAX: +81-52-789-5027
E-mail: ikuta@mech.nagoya-u.ac.jp

Abstract

We have developed the simple and effective method “Silicone Rubber Coupling” for holding and connecting of the modular micro fluidic device "Biochemical IC" proposed by Ikuta [1][2]. The experimental data of this method showed very unique characteristics. The maximum leakage pressure increases while the coupling part gets thin, differing from general sealing technology "O-ring". Thin coupling part is effective in practical use. Additionally, it is of importance in academic. Therefore, the sealing properties of the silicone rubber coupling method was analyzed theoretically, whose results proved the validity of the experiments. This method is low-cost and useful not only for the biochemical IC chips but also the micro fluidic devices in general

Keywords: Silicone rubber coupling, Sealing properties, Biochemical IC, μ-TAS

1. Introduction

We have been developing new interconnecting method “silicone rubber coupling”, which is one of key technology for the modular micro fluidic devices [3]. So far some fluidic couplers have addressed to connect external fluidic reservoirs to a micro fluidic device, by inserting capillary tubes [4][5]. In case of interconnection between micro fluidic devices, they have the disadvantages of dead volume, complicated fabrication, and not accurate positions between multiple channels. On the other hand, our silicone rubber coupling is simple and effective. The experiments showed a good seal under high pressure (>400[kPa]). Moreover, the maximum leakage pressure increased while the thickness of the silicone rubber got thin. In this paper, the validity of the experiments will be demonstrated by theoretical analysis of sealing properties.

2. Silicone Rubber Coupling Method

The silicone rubber coupling method utilizes thin silicone rubber films and the holder unit. Fig.1 shows the holder unit to stack up the biochemical IC chips. Thin silicone rubber films are sandwiched between the chips. All of the chips are held together by the spring force to prevent fluid leakage. Thin silicone rubber films in various shapes, formed

Y. Baba et al. (eds.), Micro Total Analysis Systems 2002, Volume 2, 694–696.

from a polymer mold made by micro stereo lithography, are fabricated easily [3]. Moreover, multiple channels between the IC chips are adjustable precisely, since thin silicone rubber films and the IC chips are of the same shape (14x14 mm).

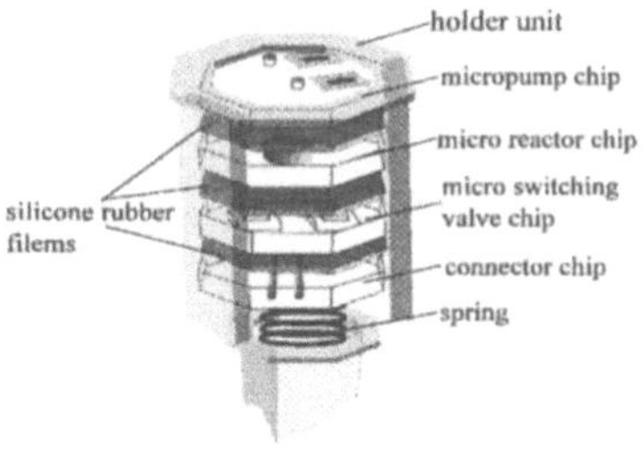

Figure 1. Holder unit with Biochemical IC chips

3. Theoretical Model of Sealing Property

It is impossible to derive a rigorously strain-energy of silicone rubber films, caused by its incompressible solid properties. Therefore, the theoretical model of the silicone rubber films was derived by introducing the ratio between contact area A_L and the lateral area A_F (the shape ratio S)(Fig.2), to analysis the relationship between practical design parameter and leakage pressures.

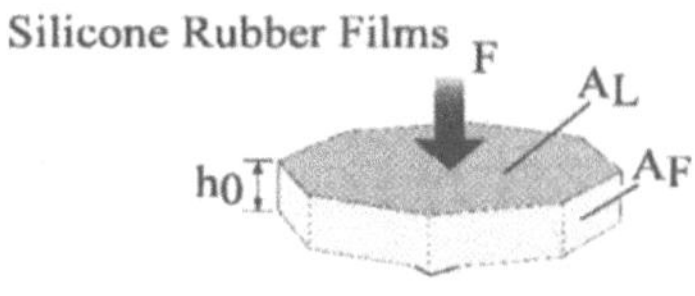

Figure 2. Calculation of shape ratio

$$S = \frac{A_L}{A_F} = \frac{A_L}{D_1 h_0} = \frac{D_2}{h_0} \quad (1)$$

Where D_1 and D_2 is the constant value. Then, we assume that the spring constant K is proportional to the displacement Δh and the squares of the shape ratio S, because the compressive force F changes dramatically depending on S and Δh.

$$K \propto S^2 \Delta h \quad (2)$$

Next, the displacement Δh is calculated by using the Hooke's law.

$$\Delta h = \frac{F}{K} = \frac{F}{C_1 S^2 \Delta h} = \frac{h_0^2 F}{C_1 D_2^2 \Delta h} = C_2 \frac{h_0^2 F}{\Delta h} \quad (3)$$

$$\Delta h^2 = C_2 h_0^2 F \quad (4)$$

Where C_1 and C_2 is the constant value. Furthermore, we assume that Young's modulus E, which increases drastically depending on the strain ε, is proportional to the squares of the strain ε.

$$E \propto \varepsilon^2 \quad (5)$$

Finally, the internal stress σ is calculated. The internal stress, which is equal to the surface pressure of the silicone rubber films, indicates the maximum leakage pressure.

$$\sigma = E\varepsilon = C_3 \varepsilon^3 = C_3 \frac{\Delta h^3}{h_0^3} = C_3 \frac{C_2 h_0^2 F}{h_0^3} \Delta h = C_4 \frac{F}{h_0} \Delta h \quad (6)$$

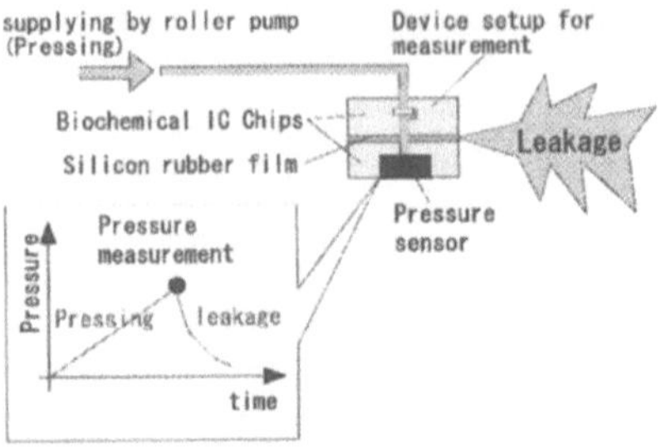

Figure 3. Experimental setup for leakage test

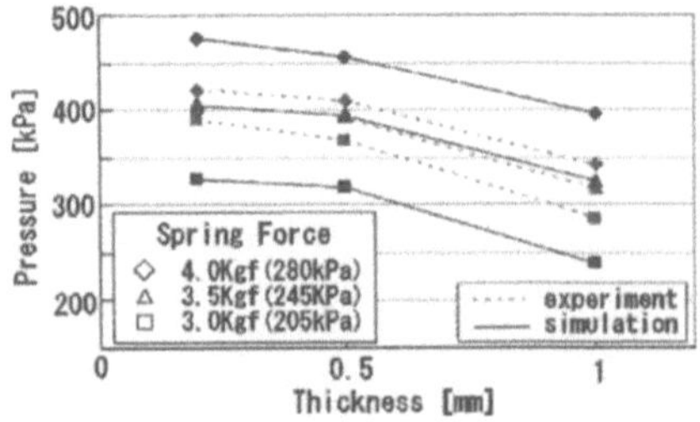

Figure 4. Thickness dependence of leakage pressure

Where C_3 and C_4 is the constant value. Equation (6) shows that the maximum leakage pressure is in proportion to the compressive force F and the displacement Δh, and in inverse proportion to the silicone thickness h_0. Its theoretical model showed the validity of the experiments satisfactorily.

4. Results and discussion

Fig.3 shows the experimental setup for leakage tests. The connector chip with one-way valve, thin silicone rubber film and the chip with a pressure sensor are stacked up into the holder unit and held by spring force. Thin silicone rubber films with different thicknesses (200, 500, 1000 μm) are made. The internal pressure of the IC chip was measured continuously until occurring leakage, while the device setup was pressurized by the roller pump. Fig.4 shows comparison of computed values with the experimental data. The constant values $C_1 \sim C_4$ were estimated by the experimental data in the leakage test. A unique sealing property of the silicone rubber coupling indicates qualitatively. Moreover, a quantitative value also indicates agreement, according to set up equation (2) and (5) strictly. And it becomes easy to design the silicone rubber films quantitatively.

5. Conclusions

In the results of the leakage test, the effective sealing properties in practical use, to miniaturize the coupling part, was appeared. Accordingly, the relationship between the practical design parameter of the silicone rubber coupling and leakage pressure was analyzed by introducing the theoretical model, so that the validity of the leakage test was proved. From these results, the silicone rubber coupling method is useful not only for the biochemical IC chips but also the general micro fluidic devices.

References

1. K.Ikuta, K.Hirowatari, and T.Ogata, MEMS'94, pp.1-6
2. K.Ikuta, Artificial Life, MIT Press. pp.17-23 (1996)
3. T. Hasegawa and K. Ikuta, Micro-TAS'01, pp.377-378
4. N. J. Mourlas, D. Jaeggi, N. I. Maluf, and G. T. A. Kovacs, Transducers'99, pp.1988-1989
5. Tze-Jung Yao, Sangwook Lee, Weileun Fang, Yu-Chong Tai, MEMS'00, pp.624-627
6. K.Ikuta, T.Hasegawa, T.Adachi, and S.Maruo, MEMS'00, pp.739-744

A Rapid, Flow-through, DNA Extraction Module for Integration into Microfluidic Systems

Phillip Belgrader, Kevin Ness[1], Farzad Pourahmadi, Kevin Brounstein, M. Allen Northrup, and Gary Long[2]
Microfluidic Systems Inc., 3918-B Valley Ave., Pleasanton, CA, 94566
[1]Lawrence Livermore National Lab, 7000 East Ave., Livermore, CA 94550
[2]Tetracore Inc., 11 Firstfield Rd., Gaithersburg, MD 20878
(email: belgrader@mfsi.biz)

Abstract

A reusable flow-through micropillar chip module for efficient, pathogen nucleic acid extraction, purification, and concentration was evaluated on large volume samples. The module was tested on aqueous samples of *Francisella tularensis* genomic DNA. Only 30 min were required to process a 75-ml sample, resulting in a 1000-fold concentration effect. The data demonstrates the potential of microfluidic approaches for flow-through processing, detection and genetic identification of pathogenic agents at low concentrations in real-world samples.

Keywords: sample preparation, DNA, PCR, microfluidics

1. Introduction

Rapid, autonomous monitoring of the environment for pathogenic agents is critical for military and civilian biodefense. Types of samples of interest include water, air, and soil, each of which present inherent challenges for sample processing. In addition, backgrounds for a particular sample type exhibit very diverse characteristics that are dependent on location. Typically, air samples are collected either into liquid via a wet-wall cyclone collector or on a filter and then resuspended into liquid. Soil is typically added to a liquid to release and separate microbes from the soil colloids. Thus, the ability to process large volume liquid samples is ubiquitous to many sample types and critical for sensitive detection and identification of pathogenic agents. We are building complete integrated microfluidic instruments to monitor samples for genetic signatures of pathogenic agents. Genetic testing offers high sensitivity to maximize the probability of detection and high specificity to minimize the probability of false positives. A key component to this technology is the high-surface area micropillar chip module for nucleic acid extraction, purification, and concentration. Here, we present work using the micropillar chip, which has been described previously, to demonstrate processing large volume liquid samples containing pathogenic microbial DNA.

2. Experimental

The silicon micopillar chips (Figure 1) were fabricated using deep reactive ion etching. Thin-film resistive heaters were deposited on the backside of the chips.

Y. Baba et al. (eds.), Micro Total Analysis Systems 2002, Volume 2, 697–699.

Solutions were driven though the chips with a syringe pump. A typical protocol consisted of passing through the chip: 1) a 2:1 mixture of sample and 1X Bind solution (6 M GuSCN, 10 mM Tris-HCl pH 6.5, 1 mM EDTA) at 5 ml/min, 2) 1 ml of 70 % ethanol at 1 ml/min, 3) air, and 4) 10 mM Tris-HCl pH 8.0 for elution. During the elution, heat was applied and fractions of approximately 25 ul were collected. Real-time PCR analysis was performed using assays for *F. tularensis*. 25-μl reactions consisted of 1X PCR Buffer (Life Technologies), 0.3 uM of each PCR primers, 0.15 uM of fluorogenic probe, 5 mM $MgCl_2$, 0.2 mM each dNTP, 0.05 U/μl of Platinum Taq polymerase (Life Technologies), 100 ug/ml BSA, and 10 μl of sample. The probe contained a FAM dye reporter at the 5 end and a TAMRA quencher dye at the 3 end. Thermal cycling was performed for 45 cycles of 95 °C for 3 second and 60 °C for 30 seconds on a Smart Cycler (Cepheid).

3. Results and discussion

The objective of this work was to demonstrate efficient processing of bacterial genomic DNA in a large sample volume using the micropillar chip module (Figure 1). To accomplish this, the system was challenged with a 75-ml sample of *Francisella tularensis* genomic DNA at a concentration of only 10^2 copies/ml. *F. tularensis* is a small, nonmotile, aerobic, gram-negative coccobacillus that can infect humans through the skin, mucous membranes, gastrointestinal tract, and lungs. It is one of the most infectious pathogenic bacteria known, requiring inoculation or inhalation of only 10 organisms to cause disease (1). It is considered to be a dangerous potential biological weapon because of its extreme infectivity, ease of dissemination, and substantial capacity to cause illness and death. Therefore, the ability to detect trace amounts of *F. tularensis* requires fast and efficient sample processing prior to sensitive PCR analysis.

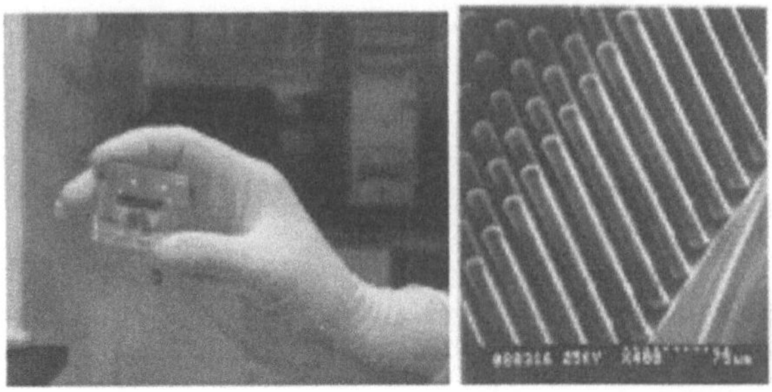

Figure 1. (left) The micropillar chip module for integration into a microfuidic system. (right) An electron micrograph of a micopillar chip. The pillars are 200 um in length and 7 um in diameter.

As shown if Figure 2A, a sample consisting of 10^2 copies/ml of *F. tularensis* genomic DNA analyzed directly by real-time PCR is well below the detection limit of 10^4 copies/ml. However, the tremendous capability of the chip module was expected to significantly improve the detection limit. The 75-ml sample of *F. tularensis* genomic DNA at 10^2 copies/ml was processed through the chip module, with the extracted DNA eluted into the first 25-ul fraction. PCR analysis of the eluted DNA resulted in a signal equivalent to unprocessed *F. tularensis* DNA at 10^5 copies/ml, representing nearly a 1000-fold concentration effect.

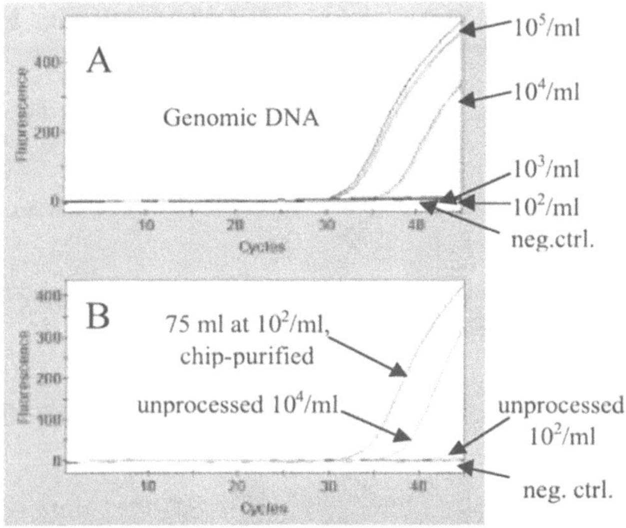

Figure 2. Real-time, quantititave PCR results. A) Serial 10-fold dilutions of unprocessed *F. tularensis* DNA to determine a detection limit of 10^4/ml. B) 1000-fold concentration effect using the chip module.

4. Conclusions

The micropillar chip module is a flow-through component that offers several advantages, including: system compatibility for microfluidic integration, the ability to accommodate large volume samples, and efficient extraction of nucleic acids at low concentrations. In addition, the high concentration effect permits the use of smaller PCR reaction volumes to achieve faster thermal cycling without losing sensitivity.

Acknowledgements

We wish to thank Mike Ascher and the California Department of Health Services for use of their facilities.

References

1. D.T. Dennis, T.V. Inglesby, D.A. Henderson, J.G. Bartlett, MD; M.S. Ascher, et. al. (2001) Tularemia as a biological weapon. JAMA 285: 2763-2773.

THE INFLUENCE OF FLOW CHANNEL GEOMETRY ON CAPTURE EFFICIENCY OF RARE CELLS USING PROTEIN A-ANTI HUMAN CD3 MAGNETIC BEADS

Vasile I. Furdui and D. Jed Harrison
Department of Chemistry, University of Alberta, Edmonton, AB, T6G 2G2, CANADA

Abstract
A new planar fluidic device for analyzing the influence of large channels on capture efficiency of rare cells was fabricated. The flow path for optimal capture of cells by a magnetic bead bed was determined.

Keyword: Magnetic beads, Immunomagnetic separation, Cells, Blood, DNA

Introduction

Separation of rare cells from whole blood is required for DNA sample preparation. In some cases only a few cells of interest will be present in one mL of whole blood, requiring large sample volumes be used. Using a large sample in a microfluidic device requires a large volume channel for cell separation in order to decrease the analysis time. The low stability of a magnetic bead bed, coated with specific antibodies and used for cell capture in a large open channel, means the fluid flow path must be optimized.

Results and Discussion

At μTAS 01 we presented a fluidic device for the immunocapture of rare cells on a magnetic bead bed [1]. For a 2 μl blood sample about 20 min was required, due to the low flow rates achieved in the small-channel device. A new device with much larger channels has been designed and fabricated to reduce this time to less than 2 min. A glass chip with 50 μm deep channels was prepared, with different fluid flow paths designed to evaluate the effect of the layout on the cell capture efficiency. The nine different channel designs are shown in Figure 1. Each design has the same cross-section, but the width of the individual channels making up the total cross-section was varied, as was the fluid distribution path. Magnetic beads were introduced into the channels and trapped with a magnet to form beds. Blood samples were then passed through the beads, as shown in Figure 2. The difficulty associated with forming high quality bead beds with magnetic capture, discussed previously [2], meant the flow path design was critical to good performance.

Devices were tested using reconstituted blood samples consisting of 5×10^5 Jurkat cells/ml and 5×10^9 horse red blood cells /ml in a 1:1 volume ratio of RPMI buffer to horse plasma. Anti-CD3 coated magnetic beads were introduced first and trapped with a magnet, then a 2 μl sample of the reconstituted blood was introduced in each of the flow

Y. Baba et al. (eds.), Micro Total Analysis Systems 2002, Volume 2, 700–702.

channels of the glass chip and transported downstream using suction generated by a syringe pump.

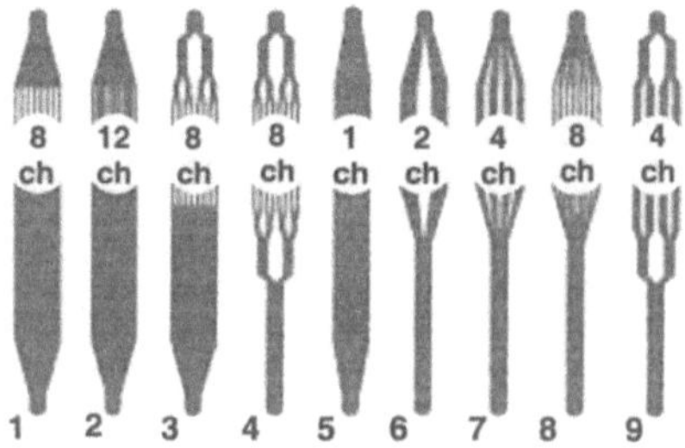

Figure 1: The nine different channel designs

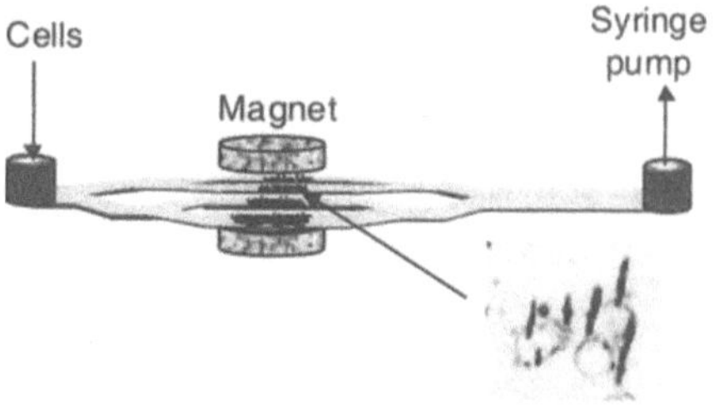

Figure 2: Immunomagnetic separation of T cells. Anti-human CD3 magnetic beads are captured by magnets and then cells are introduced from reservoir under negative pressure. Image shows Jurkats captured by magnetic bead strands.

After washing the chip with RPMI buffer, the magnets were removed and the beads and isolated Jurkat cells were collected in the outlet reservoir. The DNA was extracted from the samples recovered from the chip using DNA Direct Dynabeads, amplified by PCR, and analyzed by capillary gel electrophoresis (CGE) along with a set of standards. A standard calibration curve of signal versus number of cells present is shown in Figure 3. PCR was done for all samples and standards at the same time, to reduce differences in amplification from run to run.

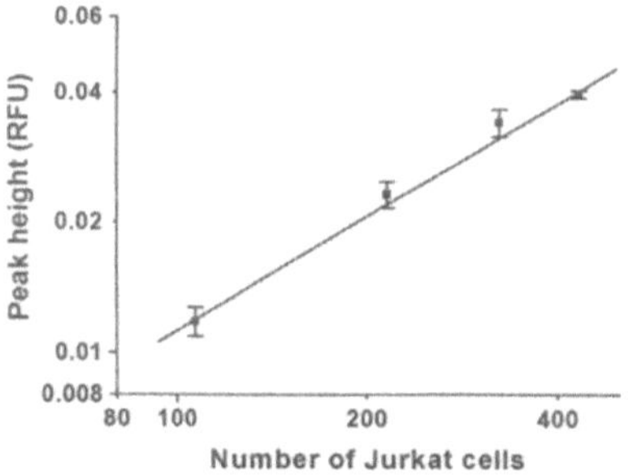

Figure 3 : Calibration curve of signal for PCR amplified gene versus the number of cells present. Jurkat specific PCR amplification was used with CGE analysis.

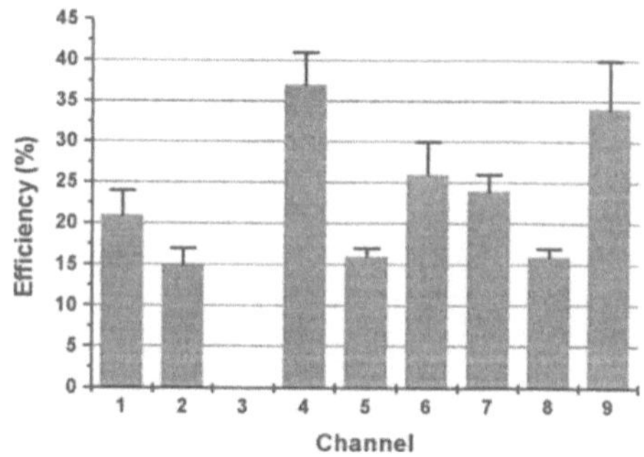

Figure 4 : Capture efficiency determined for channels 1-2 and 4-9.

All channels except 3 were tested in these experiments, since channel 3 was so similar to channel 4. The comparison of results shown in Figure 4 illustrates that devices 4 and 9 gave the highest capture efficiency. The results make it clear that the stepwise flow bifurcation design in devices 4 and 9 more effectively distributes the flow of cells and beads into each of the channels, than does the simple splitter design of layouts 1, 2, 7 and 8. The poor performance of the open layout 5 shows that narrower channels are much more useful in structuring dense enough magnetic bead beds to trap the cell streams. The same trend for efficiency was observed in different experiments as presented in Figure 5.

In large flow beds, long open leads, on flow channels appear in the magnetic bead bed, giving poorer interaction between the cell flow and the bed [2]. Our results show that many narrow channels are preferred, but that a well developed bifurcation scheme is needed to create uniform flow for the highest capture efficiency.

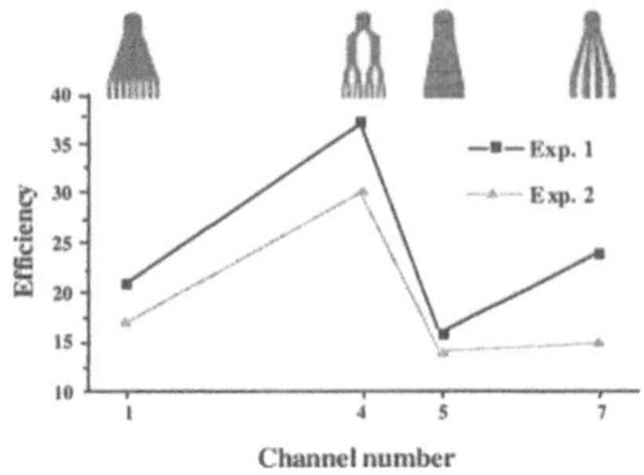

Figure 5: Capture efficiency determined in different experiments for devices 1, 4, 5 and 7.

Conclusions

The results demonstrate that efficient capture of rare blood cells can be performed at high speed using large flow channels on a planar fluidic device. The flow paths must be optimized for effective use of magnetic bead beds on these devices.

Acknowledgements

We thank the Defense Advanced Research Project Agency (BioFlips) for financial support of this project and the University of Alberta for support of the NanoFab Lab.

References

1. V. Furdui and D. J. Harrison, *Immunomagnetic separation of rare cells on chip for DNA Assay Sample Preparation*, Proceedings of the Micro Total Analysis Systems 2001, Kluwer Academic Publishers, Dordrecht, The Netherlands, pp.289-290.
2. D. Jed Harrison, E. Majid, S. Attiya, G. Jiang, *Enhancing the microfluidic toolbox for functional genomics and recombinant DNA methods*, Proceedings of the Micro Total Analysis Systems 2001, Kluwer Academic Publishers, Dordrecht, The Netherlands, pp.10-12.

ENGINEERED IMPROVEMENT OF THE GENERATION-2 μCHEMLAB™ BIOTOXIN DETECTOR

Ronald F. Renzi, James F. Stamps, Robert W. Crocker, Boyd J. Wiedenman, Scott M. Ferko, Brent A. Horn, Dan Yee, Victoria A. Vandernoot, Jason A. West, Julia A. Fruetel, Yolanda Fintschenko, Mary Clare Stoddard
Sandia National Laboratories, Livermore, CA USA

Abstract
The μChemLab™ program is developing hand-portable systems for detecting a broad range of chemical, biological, and viral agents in both gas and liquid samples. The μChem Lab liquid sample analyzer employs electrokinetic sample injection, chip-based electrophoretic microseparations and laser-induced florescence detection to analyze liquid samples. A second-generation liquid phase prototype is described. The device incorporates improvements from technological advances and applied research experience. New features include a modular design that readily accommodates on-chip preconcentration and additional separation techniques. The redesign reduces hardware failures, minimizes downtime during component replacement, improves usability, and provides increased sensitivity. Improvements have been made without compromising previous system performance.

Keywords: lab-on-a-chip, microanalytical system, electrophoresis, fluorescence detection, electrodes, power supply, optics, integrated

1. Introduction

The μChemLab™ liquid-phase microanalytical system prototype is being developed to detect a broad range of biological, and viral agents. A second-generation prototype has been developed as a platform for applied research and engineering evaluation. The system uses multiple parallel microfluidic separations with laser-induced fluorescence detection to measure protein or peptide characteristics. Agents are identified by comparing the sample characteristics to a database by an analytical algorithm running on an embedded microprocessor.

2. System Description

μChemLab™ is an automated, hand-portable, stand-alone, and modular prototypical instrument. With exterior dimensions of 7" x 4" x 5", it occupies approximately 100 in^3, weighs 5.5 pounds, and consumes 3.6 watts of power during operation. The instrument contains individual microseparations modules, as shown in figure 1. The current system operates two modules, with room for future expansion. Capillary zone (CZE) and capillary gel (CGE) electrophoresis modules have been demonstrated [3]. Chromatographic separations data are obtained and analyzed within several minutes after sample injection. The menu-driven user interface consists of an LCD display and a four-button keypad. An embedded microprocessor performs system control, diagnostics, data acquisition and analysis. A convectively-cooled enclosure protects all system

Y. Baba et al. (eds.), Micro Total Analysis Systems 2002, Volume 2, 703–705.

subcomponents, and an easily-accessed vibration-isolated platform carries the individual separation modules.

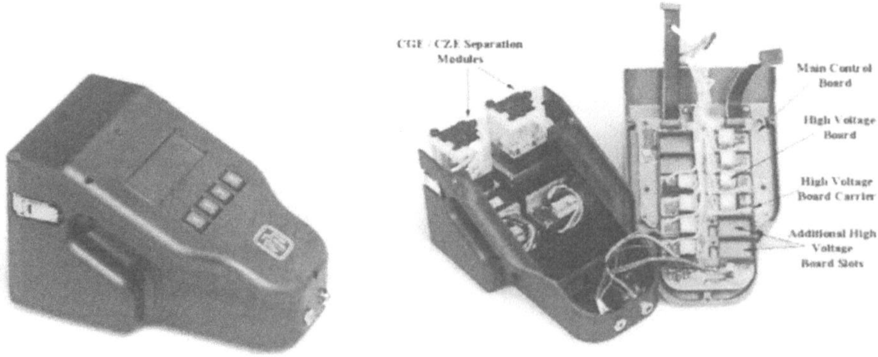

Figure 1: The uChemLab Generation 2 Biotoxin Analyser. Modular system components shown on right including two separations/detection modules and multichannel controller.

All subsystems are also now modular. Each microseparations module contains a fused silica microfluidic separations chip, microfluidics electrode/reservoir modules, and an optics module. Internal components are easy to access and replace, and all parts are interchangeable. Simplification of adjustment, conditioning, diagnosis, and component replacement has resulted. Modularity is also intended to simplify future component development and methods implementation.

For example, fluidics modules now use a port geometry generalized for future microchip designs. With better chip and excitation source registration, second-generation optics modules are no longer CZE or CGE specific. Simple module adjustment and alignment supports ready adaptation to any μchip-based method. The 12-channel high voltage boards are easily reconfigured to support any electrokinetic-driven experimentation.

New electronic features include a twelve-channel high voltage controller board designed to survive repeated arcs that might result from fluid leaks or device mishandling. Each channel is controlled by a modular daughter board with an electronically controlled potential float feature, bi-polar current capability, and real-time current monitoring, which have improved fluids control during injection and separation. The result has been improved separations quality and greater reproducibility.

In operation, a buffered and labeled protein sample is syringe-injected into the analyzer. The injected sample is transported through the liquid manifold portion of the fluidics module to microchannels in the fused-silica chip. The fluidic manifold interfaces the chip to a sample introduction port, flush port, and sample, sample waste, buffer, and buffer waste fluid reservoirs with electrodes.

New miniaturized chip designs have been developed to improve sample plug geometry, reduce injection loop volumes (to 200 nanoliters), provide filtration at channel inlets, simplify fabrication, and increase chip yields and quality. Reducing sample volumes from 1.5 microliters to 400 nanoliters has mitigated injection carryover. Separation column lengths of 10 cm have been maintained. A 20mm x 20mm design accommodates either CGE or CZE separation methods. A 25mm x 25 mm version provides two additional ports for various uses. These currently enable sample preconcentration. Dispersion minimized column winding [2] allows an overall 67% savings of chip area.

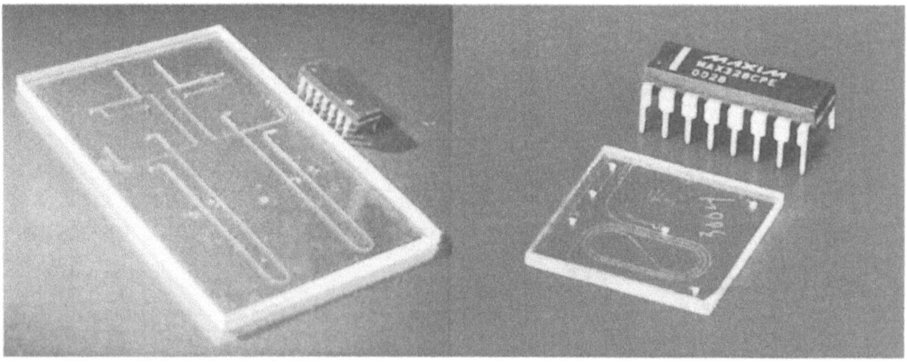

Figure 2. First (50 x 75 mm, left) and second (20 mm sq, right) generation microfluidic chip designs. The first generation was a dual channel chip; generation two, single.

Detection of analytes is accomplished by laser-induced fluorescence (LIF) at 392 nm using laser diodes. The orthogonal excitation/collection optics have been replaced by a miniature confocal epi-fluorescent optical design. The optical train redesign has resulted in a four-fold excitation energy gain. A higher numerical aperture lens and increased quantum efficiency detector have improved sensitivity [3]. Sub-nanomolar concentrations of various proteins are currently detectable. The elimination of alignment dyes and the incorporation a single-screw positional adjustment reduces channel alignment times from over an hour to seconds.

Acknowledgements

The authors gratefully acknowledge the assistance of Wen-Yee Choi and Evelyn Cruz, Susan Jamison and William Kleist in the development of the microChemLab liquid analysis system. This program is supported by the US Dept. of Energy Chemical and Biological National Security Program. Sandia is operated by Sandia Corp, a Lockheed-Martin company, for the US DOE under contract DE-AC04-94AL85000.

References

1) D Lindner, *The uChemLab Project: Micro Total Analysis System...*, LabChip, 1, (2001) p 15N-19N.

2) SK Griffiths, RH Nilson, *Modeling Electrokinetic Transport for the Design and Optimization of Microchannel Systems*, Proc. Micro Total Analysis Systems 2001, Kluwer Academic Publisher, Dordrecht, The Netherlands, pp. 456-458 (2001)

3) JA Fruetel, et al, *Results from uChemLab/CB...*, Proc. Micro Total Analysis Systems 2002, Kluwer Academic Publisher, Dordrecht, The Netherlands, (2002).

SPATIALLY LOCALIZED VOLTAGE CONTROL IN GLASS MICROCHANNELS FOR ENHANCED SAMPLE HANDLING FLEXIBILITY

R.M. Guijt[1, 2], G.W.K. van Dedem[2], N.F. de Rooij[1] and E. Verpoorte[1]

[1]SAMLAB, Institute of Microtechnology, University of Neuchâtel, Rue Jaquet-Droz 1, 2000 Neuchâtel, Switzerland

[2]Kluyver Institute for Biotechnology, Delft University of Technology, Julianalaan 67, 2628 BC Delft, the Netherlands

Abstract

Electro-osmotically induced pumping (EOIP) can be used for sample and fluid handling in field-free regions in microfluidic devices. Here, EOIP for fluid handling in field-free regions is combined with electrophoretic separation, resulting in enhanced flexibility in sample handling. The influence of the EOIP-driven sample zone formation on the resulting peak area and separation efficiency was investigated. Increased EOIP voltages resulted in the injection of larger sample zones, as indicated by increased peak areas and decreased separation efficiency.

Keywords: electro-osmotically induced pumping, localized voltage control, fluid handling

1. Introduction

Electro-osmotic pumping (EOP) is often used for sample transport within microfluidic devices. EOP commonly involves imposing an electric field (E) along the full length of the channel by immersing electrodes into open solution reservoirs situated at both ends of the channel. Electrolysis gases generated at the electrodes drift to the surface of the solution reservoirs and escape into the air. Increased liquid handling flexibility can be achieved by spatially localized voltage control in microchannel subsections. However, this requires integration of electrodes in a microchannel, which is challenging since electrolytic gas bubbles will effectively interrupt the electric field. Several approaches to prevent bubble formation in localized voltage control have been presented in the literature. Dissipation of gases through a gas-permeable coverplate has been reported [1]. Porous structures have also been used as electrically conducting barriers to isolate gas formation from the main channel [2,3].

2. Theory

For spatially localized voltage control in glass devices, an electrically conducting barrier is required for separating the main channel from the channels and electrodes where the field is applied and where electrolytic gases are formed. Electrically connected side channels for EOIP have been presented before by our research group [4]. Electrical contact between adjacent channels was established by electrical breakdown of a thin glass wall. Small pores were formed between the main and side channels, allowing electrical contact but preventing electrolytic gases from entering the main channel. Restriction of fluid flow through the pores allowed EOIP, driven by EOP generated in the main channel in the section located between two side channels.

Y. Baba et al. (eds.), Micro Total Analysis Systems 2002, Volume 2, 706–708.

3. Experimental

In figure 1a, the layout of a cross injection device is given. As illustrated in figure 1b, a potential difference applied between side channels 3 and 4 results in EOP in the electric field in channel S_w, which induces a fluid flow (EOIP) in channels S, B and B_w. Sample is dragged from reservoir S towards the intersection, where the sample flow is 'pinched' by buffer flows from B and B_w (figure 1c, left). EOIP can thus be used as an alternative for sample loading using electrodes in reservoirs S, S_w, B and B_w. After sample loading, voltages are switched, and sample present at the intersection is subjected to an electric field and injected into the separation channel, B_w (figure 1c, right).

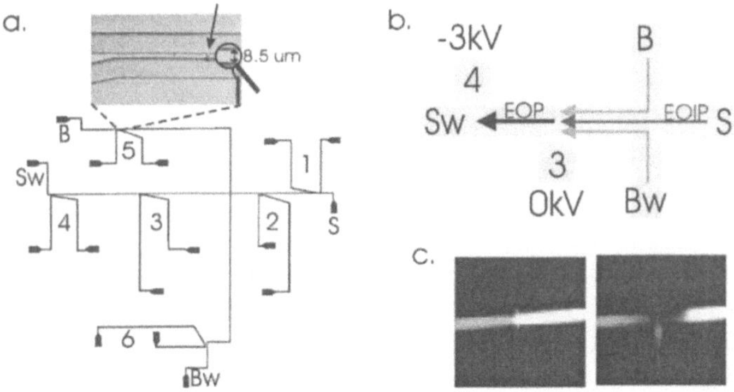

Figure 1. a. *Layout of the cross injection device. Distances S_w, B and B_w to the cross are 28 mm, S to the cross, 14 mm. The distance between main and side channels (glass wall thickness) is 8.5 μm. The photograph illustrates damage to the glass wall caused by the electrical breakdown process.* **b.** *Schematic diagram of sample loading using EOIP.* **c.** *Sample flow at the intersection is 'pinched' by the buffer flows in channel B and B_w, resulting in a well-defined sample plug. After switching the voltages, this plug can be injected into channel B_w.*

4. Results and discussion

A combination of EOIP for sample loading using side channels 3 and 4, and injection in an electric field applied between side channels 5 and 6, was used for the separation of a mixture of FITC-labeled amino acids (figure 2).

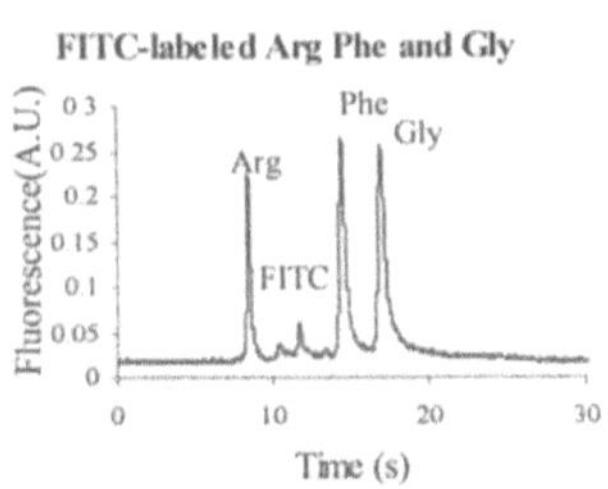

Figure 2. *Combination of EOIP and EOP for the separation of 10 μM FITC-labeled Arg, Phe and Gly. Sample loading took 30 seconds using EOIP, with a potential difference applied between channels 3 and 4 (4 kV). Injection was performed using a potential difference applied between reservoirs 5 and 6 (4 kV), and pushback voltages applied to side channels 1 and 3 (2.3 kV). Separation length (L_D) : 11 mm, buffer : 32 mM carbonate buffer* pH 9.5.

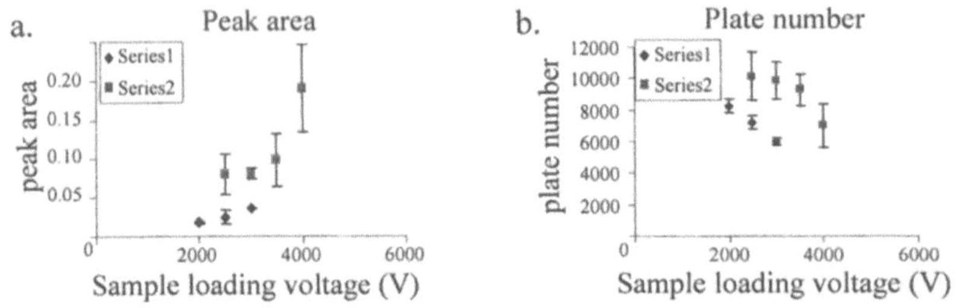

Figure 3 *Dependence of peak area (a) and plate number (b) on the EOIP sample loading voltage. Buffer : 32 mM carbonate pH 9.2, sample : 0.1 mM fluorescein in buffer. Injection conditions as described in figure 2. Series 1 and 2 originate from two series of measurements in the same device at different* L_D.

During sample loading, the sample plug is defined by the sample loading voltage and the hydraulic resistance of channels S, B and B_W. The influence of the sample loading voltage on the plate number and peak area was investigated by the injection of 0.1 mM fluorescein. Increased peak areas are observed at higher injection voltages, indicating that the injected sample plug size grows as EOIP rates increase (figure 3a). Longer sample plugs would also account for the decreasing plate numbers observed at higher sample loading voltages (figure 3b), since initial plug length can significantly influence detected peak width in separations done over short lengths on chip [5]. A possible explanation for increased plug sizes involves the time required for the induced fluid flow to stop after switching voltages to the injection mode. At higher EOIP rates, this may require more time, and leakage of sample from channel S could occur during injection.

5. Conclusions

Combination of EOIP and EOP for sample injection gives more flexibility for sample handling, since no electrodes in the sample environment are required. However, care must be taken to prevent secondary flow effects, which may result in undesired variation of the plug volume and/or re-injection of sample from the waste reservoir.

Acknowledgements

RMG was financially supported by the Dutch technology foundation STW, applied science division of NWO and the technology program of the Ministry of Economic Affairs, in the framework of the BIOMAS project (DST 66.5341).

References

[1]McKnight, T.E. *et al. Anal.Chem.* **2001**, *73*, 4045-4049.
[2]Guenat, O.T. *et al. Sensors and Actuators B,* **2001**, *72*, 273-282.
[3]Lazar, I.M. *et al. J.Chrom.A.* **2000**, *892*, 195-201.
[4]Guijt, R.M. *et al. Journal of the Association for Laboratory Automation* **2002**, *7*, 62-64.
[5]Effenhauser, C.S. *et al. Anal.Chem.* **1993**, *65*, 2637-2642.

INTEGRATED ELECTROSPRAY CHIP FOR MASS SPECTROMETRY

Jun Xie[*], Qing He[*], Yu-Chong Tai[*], Jun Liu[**] and Terry Lee[**]
*Department of Elec. Eng, California Inst. of Tech., Pasadena, CA, 91125 USA
**Beckman Research Inst., City of Hope Research Center, Duarte, CA, USA
(Email: junxie@mems.caltech.edu)

Abstract

We report here a novel micromachined fluid dispensing system for ESI-MS (Electrospray Ionization - Mass Spectrometry) that has integrated electrolysis-based micro pump, micro mixer and ESI nozzle on a single chip. The whole structure is built by multi-layer Parylene surface micromachining technology. Using this system, precise and multi-sample dispensing for ESI-MS can be achieved on a single chip.

Keywords: Microfluidics, Micropump, Electrolysis, ESI-MS

1. Introduction

Electrochemical microactuation based on electrolysis has been demonstrated as a successful technique for microfluidic application [1][2]. But usually the device involves relatively complicated packaging, such as wafer bonding. Here, we made an electrolysis-based micropump using the Parylene surface micromachining technology we developed [3]. With this technology, an electrolysis micropump, a passive micro mixer, and an ESI nozzle have all been integrated on a single chip. This stand-alone system doesn't need complicated packaging and only electrical wire (no external fluidic coupling) is connected to the chip. That simplifies the operation. Samples are stored and sealed in reservoir, then are pumped out by bubble pressure generated by electrolysis. After mixing at micro mixer, samples are fed to MS inlet through an integrated ESI nozzle. Operation principle and fabricated device are shown in Fig. 1.

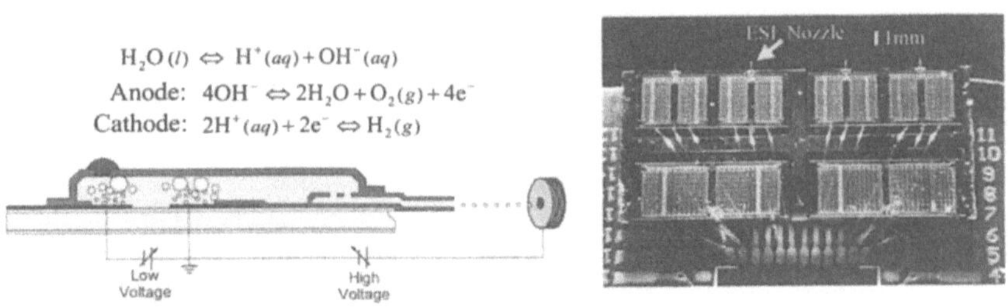

(a) (b)

Figure 1. (a) Operation principle of the integrated system. (b) Fabricated device with electrical connections.

Y. Baba et al. (eds.), Micro Total Analysis Systems 2002, Volume 2, 709–710.

2. Experimental

The device is fabricated by multi-layer Parylene surface micromachining technology. Electrodes for electrolysis are formed by Cr/Au (100Å /2000Å) layers on silicon substrate. The micro nozzle and mixer are constructed by one 4 μm photoresist sacrificial layer sandwiched by two 2 μm Parylene layers. A 2000Å Al is patterned as Parylene mask to create sharp ESI nozzle. The reservoir is formed by 15 μm photoresist layer and 4 μm Parylene. Freestanding nozzle is created by etching away the silicon underneath. Finally, sacrificial layer is released by acetone with ultrasonic stirring and followed by methanol and DI water cleaning. Fig. 2 shows the detailed process flow. To demonstrate the capability of this system, a multi-sample analysis has been done as shown in Fig. 4. Sample solution was dropped onto inlet and photoresist was used to seal the inlet after the surface tension filled the reservoir (~100nL) with the solution. An electrolysis voltage was applied along with a high voltage for ESI. Before the electrolysis voltage was applied there was no signal that means no electrospray. After electrolysis pumping happened, the expected isotope peak was observed in the mass spectrum. The flow rate is estimated to be around 80nL/min. Two different target

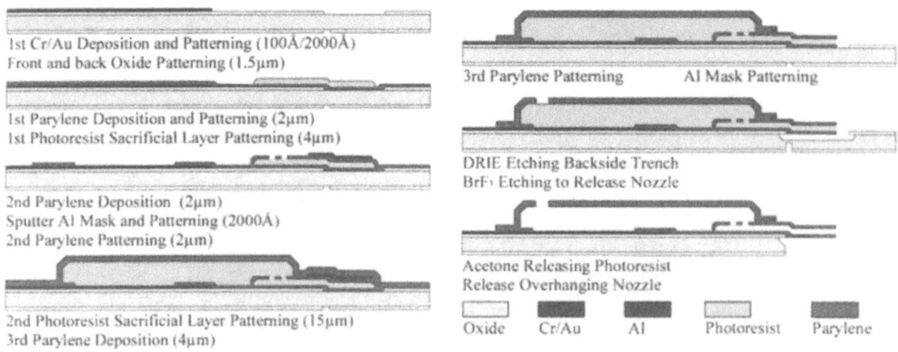

Figure 2. Fabrication process flow.

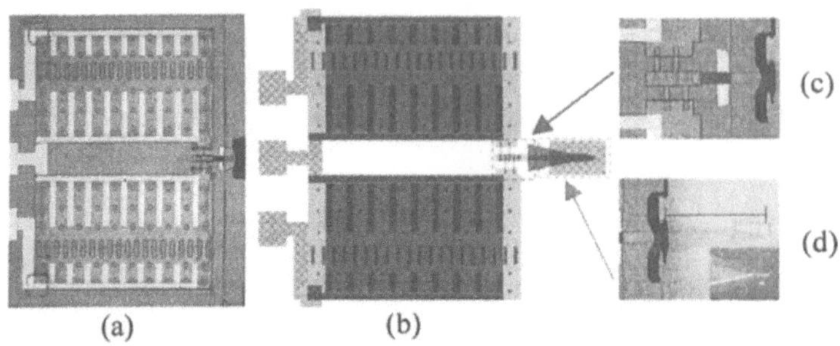

Figure 3. Detailed pictures of integrated system. (a) two pumping chambers, (b) design layout, (c) micro passive mixer and (d) ESI nozzle.

samples were stored in the two reservoirs that are connected by mixer. Then by controlling the electrolysis voltage applied to each reservoir, we could control which sample solutions was fed into the nozzle and got electrosprayed into MS. Two chambers were filled, respectively, with Tetrabutyl ammonium iodide (25 pmol/μl, marked as m/z 242) and a peptide called leuprolide (30-50 pmol/μl, marked as m/z 606). Both samples were dissolved in 5% methanol and 5% acetic acid. Fig. 4 shows ion chromatograms for each sample and representative mass spectra at the times indicated.

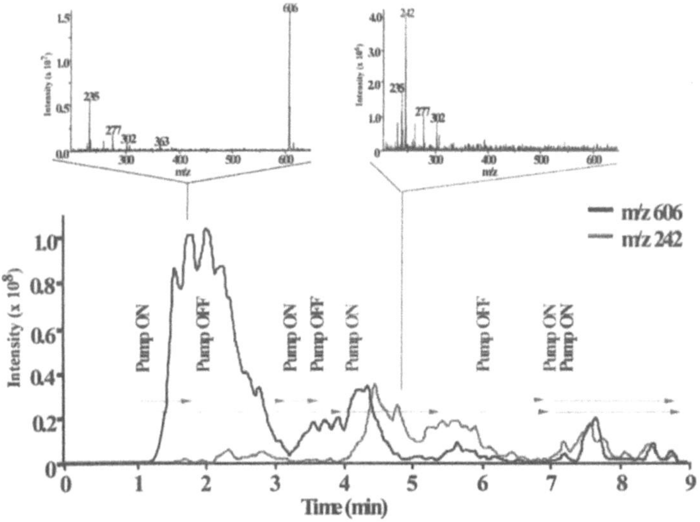

Figure 4. Chromatograph and Mass Spectrum of the multi-sample analysis.

3. Conclusions

A novel integrated microfluidic system for ESI-MS has been developed using multi-layer Parylene surface micromachining technology. The experiments demonstrated the effectiveness of electrolysis micropump, performance of the whole system and the essential idea of on-chip gradient elution.

Acknowledgements

This work is supported in part by NSF(Grant No. EEC-9402726) and NIH (Grant No. 5R01 RR06217-10).

References

1. C.R. Neagu, J.G.E. Gardeniers, etc., *An electrochemical microactuator: principle and first results*, Journal of MEMS, Vol. 5, No. 1, March 1996, pp.2-9
2. S. Böhm, B. Timmer, etc., *A closed-loop controlled electrochemically actuated micro-dosing system*, J. Micromech. Microeng., 10(2000), pp.498-504
3. J. Xie, X. Yang, X.Q. Wang and Y.C. Tai, *Surface micromachined leakage proof Parylene check valve*, MEMS 2001, Interlaken, Switzerland, pp.539-542

Dissolvable and Asymmetric Hydrogels as Components for Microfluidic Systems

Qing Yu*, Jeffrey S. Moore* and David J. Beebe^

*Beckman Institute for Advanced Science and Technology
University of Illinois at Urbana-Champaign, Champaign, IL 61801 USA
^Department of Biomedical Engineering, University of Wisconsin at Madison
Madison, WI 53706 USA
(dbeebe@engr.wisc.edu)

Abstract

We demonstrate the development of novel hydrogel materials by controlling the crosslink density. A pH-responsive hydrogel with a crosslink density gradient, created by laminar flow and diffusion, showed asymmetric swelling during volume transition. Hydrogels with cleavable crosslinkers have also been designed and fabricated inside microfluidic channels. Upon exposure to reagents that can chemically cleave such crosslinkers, the de-crosslinked hydrogels dissolve completely, making them promising as sacrificial structures and sensors for microfluidics.

Keywords: Hydrogel, Microfluidics, Sensor

1. Introduction

The swelling characteristic of a hydrogel is extremely sensitive to the crosslinking density. Hydrogels having an uneven crosslink density could exhibit asymmetric swelling as a result of the crosslink density gradient, leading to the development of shape-changing hydrogels. When hydrogels are composed of specially designed crosslinkers that can be cleaved by chemical reagents, the hydrogels can be dissolved completely upon de-crosslinking.[1] Such dissolvable and asymmetric hydrogels may thus function as components (e.g. sensors or sacrificial structures) for microfluidic systems.

2. Experimental

Hydrogels were polymerized inside microfluidic channels using the liquid-phase photopolymerization technique that we have previously developed.[2,3] In the fabrication of hydrogels having a crosslink density gradient, a laminar flow of two pre-hydrogel mixtures of different crosslinker concentrations was established. The crosslink density gradient obtained due to the diffusion at the interface between the two streams was then "locked in" via photopolymerization (Figure 1).

In order to demonstrate sacrificial hydrogel structures, pre-hydrogel mixtures with cleavable crosslinkers such as N,N'-cystamine bisacrylamide were photopolymerized into hydrogels. A solution of cleaving reagent such as dithioerythritol (DTT) was then flowed through the microchannel to initiate de-crosslinking and subsequent dissolution of the hydrogel structures.

Y. Baba et al. (eds.), Micro Total Analysis Systems 2002, Volume 2, 712–714.

3. Results and Discussion

Hydrogels having asymmetric crosslink density exhibit asymmetric swelling during volume transition, leading to the demonstration of a shape-changing hydrogel (Figure 1). This observation confirms that the crosslink density plays a significant role in determining the maximum degree of swelling within hydrogels.

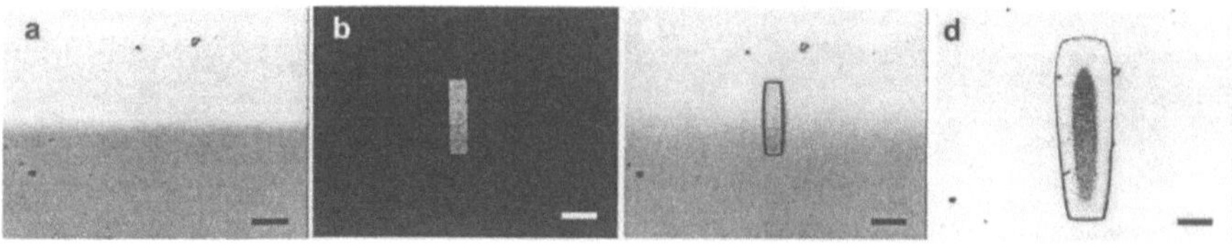

Figure 1: Fabrication of a shape-changing pH-responsive hydrogel with asymmetric crosslink density gradient using laminar flow. (a) A two-stream laminar flow of two dyed pre-hydrogel mixtures is formed; the mixture in the top stream contains 0.5 % crosslinker while the mixture in the bottom stream contains 1.5 % crosslinker. (b) A 250 x 1000 μm rectangle-shaped photomask is placed over the stream interface, and the flow is stopped to allow diffusion at the interface. (c) Photopolymerization results in the formation of a hydrogel with crosslink gradient. (d) In basic buffer, the hydrogel swells asymmetrically, with higher degree of swelling at the end where the crosslink density is lower. Bars, 500 μm.

The dissolution of hydrogels with cleavable crosslinkers via de-crosslinking has also been demonstrated. Two types of cleavable crosslinkers were used. One contains a disulfide bond that can be cleaved by disulfide reductants such as dithioerythritol, while the other contains a 1,2-diol functionality that can be cleaved by oxidants such as $NaIO_4$. Because of the chemical specificity of the de-crosslinking reaction, hydrogels containing a cleavable crosslinker have potential as sensors with facile detection readout. Figure 2 shows an array of nine hydrogel posts, some with cleavable disulfide crosslinker, the others with non-cleavable crosslinker. In the presence of a disulfide reductant, the hydrogel posts function as a sensor array providing a visual readout in the form of a pattern change.

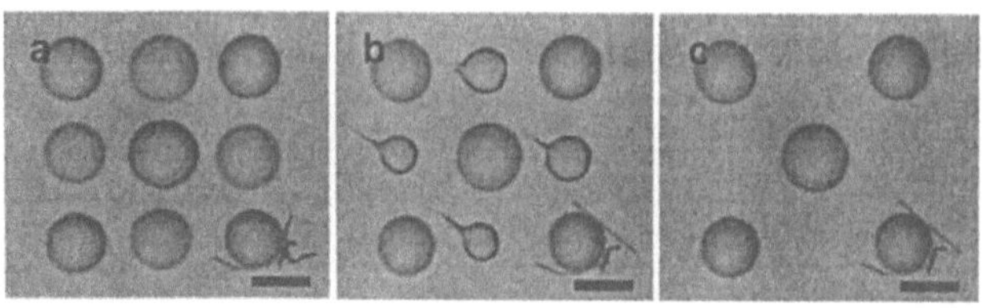

Figure 2: Demonstration of dissolvable hydrogels as a sensor readout system. (a) A 3 x 3 array of hydrogel posts was polymerized sequentially from two different pre-hydrogel mixtures, one with cleavable crosslinker, the other one with non-cleavable crosslinker. (b) When a de-crosslinking reagent was flowed through the microchannel at 0.1 mL/min, the hydrogel posts with the cleavable crosslinker dissolved away, revealing a different pattern of posts as a form of read out (c). Bars, 500 μm.

Dissolvable hydrogels can also be used in microfluidics as selective sacrificial structures. With proper chemical and engineering design, dissolvable hydrogels can be used for autonomous flow control. Figure 3 shows an example where three types of hydrogels were photopolymerized inside a microchannel using sequential polymerization. The hydrogel walls blocking the entrances to the top and bottom branches have different types of cleavable crosslinkers, while the hydrogel post in the center has non-cleavable crosslinker. The hydrogel walls respond to specific de-crosslinking reagent to open the branch selectively, and the hydrogel post in the center was not effected.

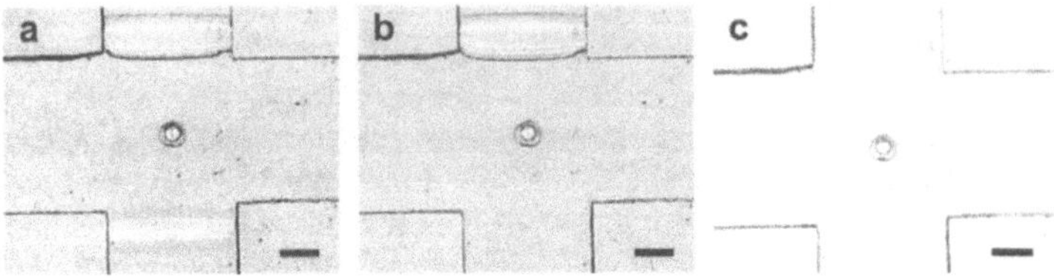

Figure 3: Dissolvable hydrogels as selective sacrificial structures. (a) Two hydrogel walls were polymerized with different cleavable crosslinkers, while a hydrogel post was polymerized with non-cleavable crosslinker. (b) One of the hydrogel walls dissolved in the presence of 5 % $NaIO_4$ solution, while the other hydrogels were not affected. (c) The other hydrogel wall was dissolved using 50 mM DTT solution. The hydrogel post in the center remains. Bars, 500 μm.

4. Conclusion

Dissolvable hydrogels, which contain cleavable crosslinkers, have been demonstrated. Hydrogels capable of asymmetrically swelling due to a crosslink density gradient in the network have also been achieved. These hydrogel materials are promising as components for microfluidic systems. Asymmetric volume changes opens up new possibilities in actuator design while the ability to create selective sacrificial structures or sensors has the potential to greatly simplify the design of micro total analytical systems.

Acknowledgements

This research was supported by a grant from DARPA (#F30602-00-2-0570).

Reference

[1] Lee, H.; Park, T. G. *Polymer Journal* **1998**, *30*, pp. 976-980.

[2] Beebe, D. J.; Moore, J. S.; Bauer, J. M.; Yu, Q.; Liu, R. H.; Devadoss, C.; Jo, B.-H. *Nature* **2000**, *404*, 588-590.

[3] Beebe, D. J.; Moore, J. S.; Yu, Q.; Liu, R. H.; Kraft, M. L.; Jo, B.-H.; Devadoss, C. *Proc. Nat. Acad. Sci.* **2000**, *97*, pp. 13488-13493.

The Development of Quite Low Cost Disposable, Rapid, Individually Controlled Matrix (10 x 10) Microreactor System

Masahiro Kuwata, Hajime Sudo, Koichiro Kawano, Kaname Miyazaki
Mechanical Systems Laboratory, Corporate Research & Development Center,
Toshiba Corporation, Japan
(masahiro.kuwata@toshiba.co.jp)

Abstract

We have developed a unique new matrix (10 x 10 matrix, 500 μm diameter test tube, 2 mm pitch, within 20x20 mm area) microreactor system which is not only fast but also individually controllable. This reactor system can perform 100 different kinds of processes in parallel more quickly than is possible with any conventional microreactor with quite lower cost due to the use of disposable sheet.

Keywords: microreactor, micromachining, PCR, process control

1. Introduction

Important advances have been made in biochemistry and biotechnology in recent years. Nevertheless, to develop these advances for practical use, analysis systems with much higher throughput are required. Thus, many approaches to achieve faster analysis has been studied [1]. However we consider that the speed may have reached a limit (Parhaps, the limit will be 1 sec for heating and cooling). So to achieve much higher throughput analysis, we proposed a not only miniaturized, faster but also "individually controllable in parallel" reactor system with quite low cost disposables.

2. Structure of the reactor system

Figure 1 shows the structure of the reactor system. The reactions are carried out in quite low cost disposable micro test tubes (10 x 10 matrix, 500 μm diameter, 2 mm pitch) fabricated from polypropylene sheet (30 μm thickness) by the thermal press technique. Each test tube is inserted into a microchamber equipped with a dedicated microheater. The temperature of the sample in each test tube is measured and controlled individually by feedback loop and PWM heating. The hat shape of the microchamber ensure efficient heat conduction. The size, distance, and materials of the reactor system are finely tuned by numerical thermal analysis (Figure 2). The structure is obtained by micromachining technologies.

3. Results

Figure 3 shows the temperature response of one reactor, the maximum speed and time required to reach the target temperature. Due to the high degree of miniaturization (500

Y. Baba et al. (eds.), Micro Total Analysis Systems 2002, Volume 2, 715–717.

µm diameter) and optimal design for heating and cooling, the maximum heating speed reaches 96.5 K/s and even the maximum cooling speed reaches 68.5 K/s. The rapidity of heating and cooling enables each reactor to reach the target temperature within 1 sec. Normally, the use of a flexible material such as polypropylene for reactor such as a quite low cost disposable micro test tube results in the instability in temperature control. However, this control system has a self-tuning function to aboid unstable oscillations and achieve control with repeatable accuracy of within ±0.05 K.

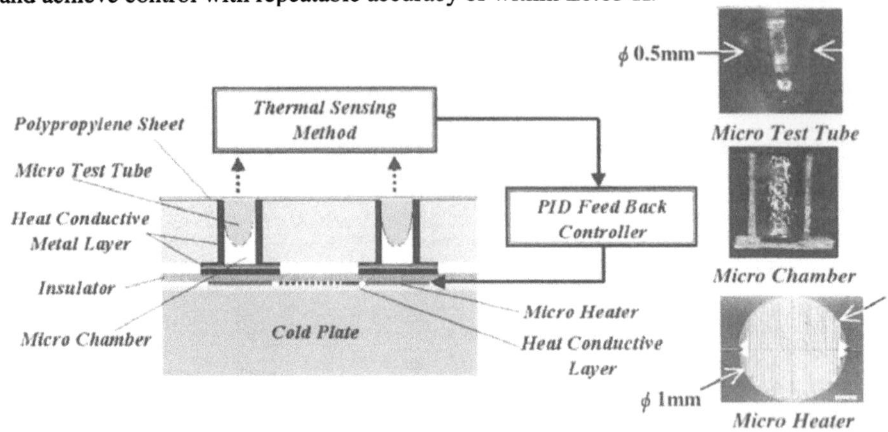

Figure 1. Structure of the reactor.

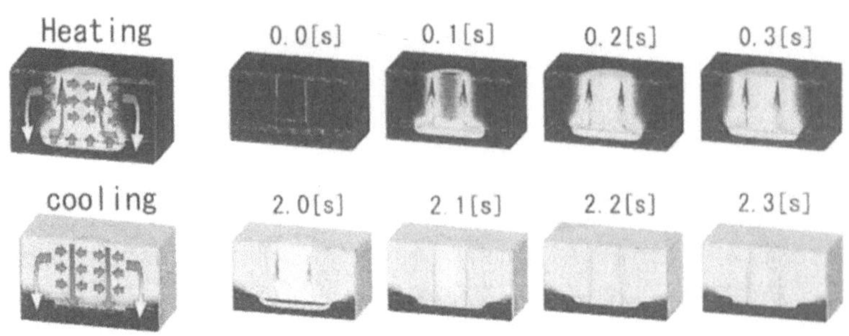

Figure 2. Numerical thermal analysis.

In addition to their rapid response, the reactors (10 x 10 matrix) can be controlled individually in parallel. Figure 4 shows the qualitative verification of parallel and individual control by the color changes of a thermochromic liquid crystal (TLC). The TLC is designed to change color from black to green (white in this paper) at approximately 60°C. The controller system controls the temperature of each reactor temperature to plot out letters "TOSHIBA" continuously, and thus only the reactors set to 60°C appear green. In this reactor design (with a pitch of only 2mm), heat diffusion effects are minimized by placing insulation between the reactors and carefully selecting of the positions of the heaters, radiators and so on. Therefore, reactors that should be black are not turned green by the diffusions of high temperature from nearby green reactors. This individually controlled reactor system can be easily applied to parallel multiprocesses reactions or to combinatorial methods to determine the optimal reaction conditions.

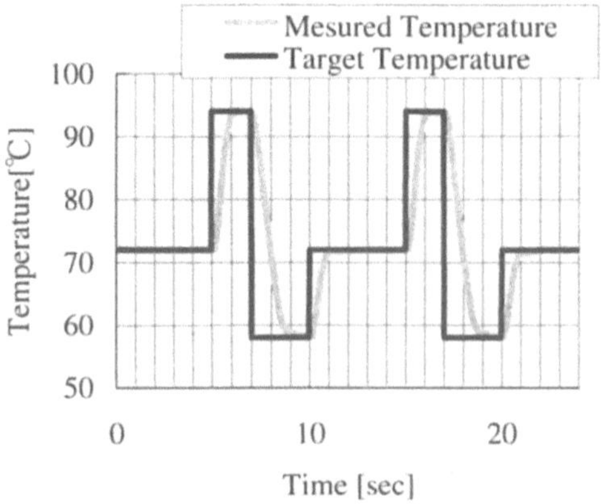

Maximum speed
Heating: 96.5 K/s
Cooling: 68.5 K/s

Time required for 5% control
Heating: 0.96 sec
Cooling: 0.96 sec

Time required for 2% control
Heating: 1.36 sec
Cooling: 1.24 sec

Figure 3. Results for temperature response, maximum speed, and time required.

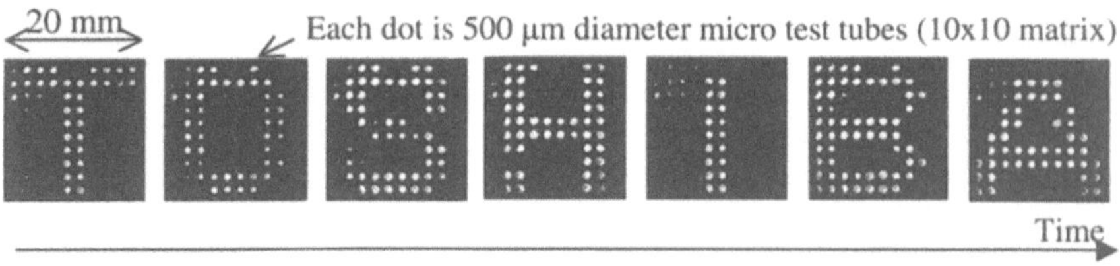

Figure 4. Temperature distributions described by TLC.

4. Conclusions

We have developed a unique new microreactor system and have demonstrated its outstanding performance of rapidity and parallel control. These features are extended to be useful various applications in the fields of biochemistry, point-of-care-testing, on-site environment monitoring, and so on.

References

1. (for example) E.Shinohara, A.Kondo, K.Akahori, K.Tashiro, S.Shoji, Development of Biochemical Micro Reactor Array System, IEICE Trans. Electron., vol.E84-C, No.12, pp.1807-1812, (2001)

DEMONSTRATION OF HYDROGEL VOLUME CONTROL USING PULSE WIDTH MODULATION

Michael J. Bassetti and David J. Beebe
Dept. of Biomedical Engineering, University of Wisconsin-Madison, Madison, WI 53703, USA (mjbasset@mmb.bme.wisc.edu, dbeebe@engr.wisc.edu)

Abstract

Hydrogels are fabricated inside microfluidic networks using liquid phase photopolymerization techniques. Square voltage waveforms of differing amplitudes and duty cycles are applied to electrodes in the channel, causing the hydrogels to change volume.

Keywords: microfluidics, hydrogel, electrode.

1. Introduction

Hydrogels that respond to such environmental factors as pH, molecular concentrations, light, temperature, and more have begun to show themselves useful as sensors and actuators in microfluidic systems [1-3]. Hydrogels that change volume under the influence of an electric field have also been demonstrated previously [4], but few, if any experiments have been geared towards developing them for use in microfluidic systems. Initial testing on hydrogels in electric fields showed that the time response of volume change can be very fast compared with hydrogels that respond to pH, temperature, light, and other stimuli [5]. Here we present further results showing that applying non-DC voltage waveforms with varying pulse widths to the hydrogel allows precise control of hydrogel volume.

2. Experimental

The current test device consists of a quartz substrate (onto which Pt electrodes have been patterned) with microfluidic channels assembled on top by following the μFT cartridge construction procedure [6]. Briefly, a polycarbonate cartridge is placed on the substrate and filled with a photopolymerizable construction material. This liquid is exposed through a photomask that defines the channel geometry. Once the unpolymerized liquid is removed and the channels flushed, active hydrogel structures can be

Fig. 1 - Current test device. Channels 2000 μm wide are fabricated atop the quartz substrate. The top channel has been highlighted for clarity.

Y. Baba et al. (eds.), Micro Total Analysis Systems 2002, Volume 2, 718–720.

fabricated inside. This is done by flowing in the hydrogel pre-polymer mixture (acrylic acid and 2-hydroxyethyl methacrylate in a 1:4 molar ratio, ethylene glycol dimethacrylate [1.0 wt %], and 2,2-dimethoxy-2-phenyl-acetophenone [3.0 wt %]) and polymerizing through a photomask (30 mW/cm^2). Then, the channel is flushed with methanol and fluidic contacts are added. Fig. 2 shows the geometry of the test device. The fluid medium used is a 10:1 water and potassium buffer (SB116, Fisher Scientific, Fair Lawn, NJ) solution flowing at a constant rate of 0.1 ml/min (approximately 0.3 cm/s).

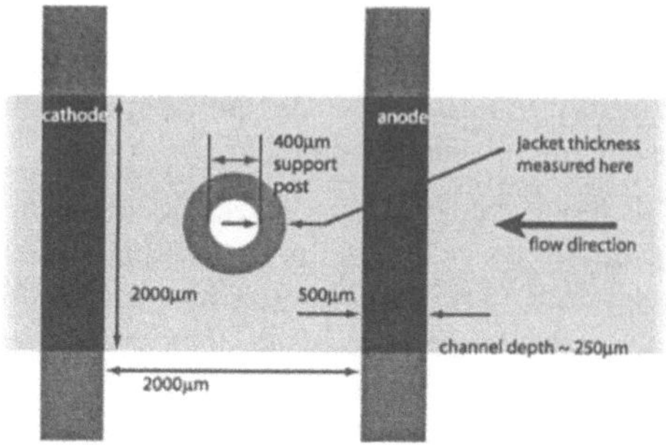

Fig. 2 – Geometry of the test device.

3. Results and discussion

Fig. 3 shows hydrogel jacket thickness as a function of time for three different square wave voltage signals. At time = 0 seconds, a 50 Hz signal with duty cycle of 20% is applied to the electrodes. At 10 seconds, it is raised to 50%, at 20 seconds, 80%, and at 30 seconds the voltage is turned off. These experiments show that, for constant amplitude signals, the equilibrium hydrogel volume is proportional to the duty cycle. As amplitude is increased, the speed of volume change (in transition between different duty cycles) is increased, as is the final equilibrium volume at higher duty cycles. The hydrogel response is non-isotropic (fig. 4). Volume change is first observed nearest the anode (where jacket thickness is reported in fig. 3) and then in areas further away from the anode.

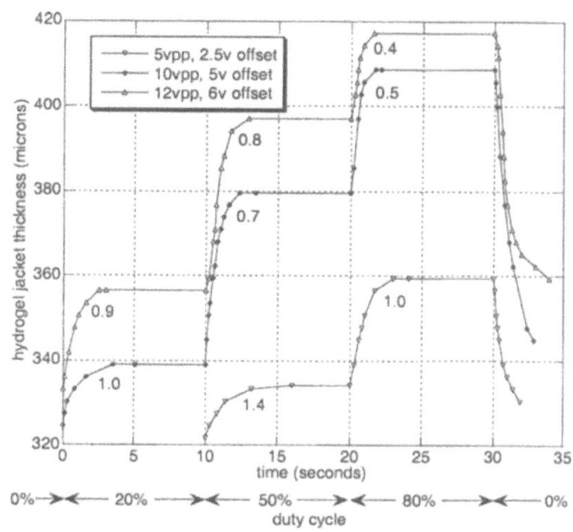

Fig. 3 – Hydrogel response curves. Measured time constants are expressed in seconds next to each response.

Using square voltage waveforms with varying pulse widths to control hydrogel volume has several advantages over previously demonstrated DC control. First, with DC control it was observed that hydrogels would continue expansion, whereas in the current experiment it is seen that the hydrogels are quick to reach and hold a steady jacket thickness. Second, the amount of gas generated by electrolysis tends to be less than with DC voltages, keeping the channel from becoming blocked by bubbles.

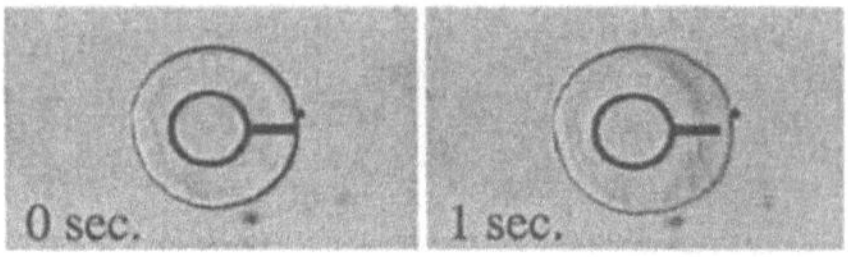

Fig. 4 – Typical hydrogel response. The response begins towards the anode and spreads across the remainder of the hydrogel. A scale bar has been added to highlight the difference in jacket thickness before and after turning the voltage on.

It should be noted that the results presented here are for a single set of experimental conditions. The opportunities for refining the parameter space are broad, and a desired response could be tailored by altering such things as hydrogel composition, fluid medium, flow rate and direction, geometry and placement of electrodes, shape of the electrical signal, and several other parameters.

4. Conclusion

Precise volume control in hydrogels has been demonstrated by means of pulse width modulation. With modifications, this technique could be used to develop components in microfluidic systems such as valves, pumps, and other types of actuators.

Acknowledgements

This work has been supported under a grant from DARPA (#F30602-00-1-0570).

References

1. D. Beebe, J. Moore, J. Bauer, Q. Yu, R. Liu, C. Devadoss, and B.-H. Jo, "Functional hydrogel structures for autonomous flow control inside microfluidic channels," *Nature*, v. 404, pp. 588-590, 2000.
2. Y. Gu, A. Baldi, B. Ziaie, and R. Siegel, "Modulation of drug delivery rate by hydrogel-incorporating MEMS devices," *Proceedings of the 2nd annual international IEEE-EMBS special topic conference on microtechnologies in medicine and biology*, pp. 406-409, 2002.
3. D. Eddington, R. Liu, J. Moore, D. Beebe, "An organic self-regulating microfluidic system," *Lab on a Chip*, v. 1, pp. 96-99, 2001.
4. T. Tanaka, I. Nishio, S.-T. Sun, S. Ueno-Nishio, "Collapse of gels in an electric field," *Science*, v. 218, pp. 467-469, 1982.
5. M. Bassetti, J. Moore, D. Beebe, "Development of electrically triggered hydrogels for microfluidic applications," *Proceedings of the 2nd annual international IEEE-EMBS special topic conference on microtechnologies in medicine and biology*, pp. 410-413, 2002.
6. C. Khoury, G. Mensing, D. Beebe, "Ultra rapid prototyping of microfluidic systems using liquid phase photopolymerization," *Lab on a Chip*, v. 2, pp. 50-55, 2002.

DEP-driven simultaneous 2 x 2 droplet mixer array

Masahide Gunji[1], Thomas B.Jones[2] and Masao Washizu[1]
[1]The University of Tokyo, 7-3-1 Hongo Bunkyo-ku, Tokyo, 113-8656, Japan
[2]University of Rochester, Rochester, NY, 14627-0126, U.S.A.

Abstract

This paper describes the experimental demonstration of an on-chip mixer array for nanoliter-sized liquid droplets for combinatorial chemistry. The device employs dielectrophoretic (DEP) liquid microactuation to divide microliter-sized volumes of two different samples or reagents into nanoliter droplets, distribute them in a 2 x 2 array, and then mix adjacent droplet pairs simultaneously using electrostatic coalescence.
Keywords: Dielectrophoresis, Microfluidics, Combinatorial chemistry

1. Introduction

Dielectrophoresis (DEP) is the motion of particle or liquid masses resulting from the interaction of the applied electrostatic field and the induced polarization. It was originally developed for the actuation of particles in liquid, such as biological cells, but the authors have recently shown that it is also effective for the shaping and actuation of liquid itself[1]-[3]. The DEP force attracts dielectric liquids to regions of strong electric field. Fig. 1 depicts field-induced, transient flow of a water droplet on a simple co-planer electrode structure, which consists of two parallel, co-planer strips imbedded in insulating layer. When rf voltage is applied, a water droplet is pulled along the gap of the electrodes by the DEP force. Using this microfluidic system, we demonstrate on-chip micro liquid mixing device for combinatorial chemistry, in which m kinds of samples and n kinds of reagents are first sub-divided into nano-liter volumes using liquid DEP, and then electrically fused to obtain m x n product simultaneously.

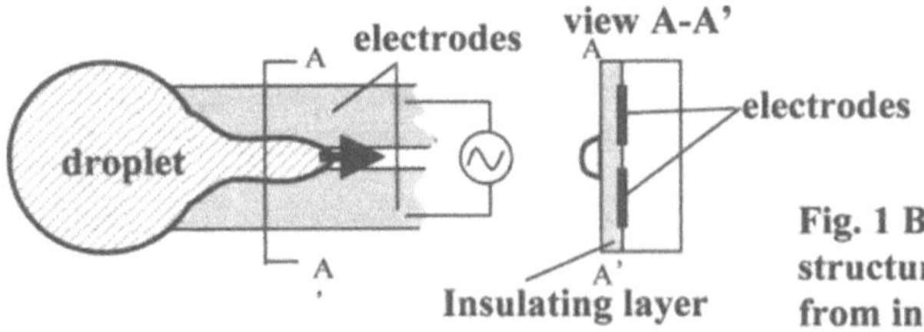

Fig. 1 Basic co-planar electrode structure with finger projecting from initial droplet

2. DEP droplet formation

Fig. 2 depicts the basic DEP droplet generation mechanism. Parallel, co-planar strip electrodes with a number of regularly spaced, semi-circular regions (hereafter called stations) are patterned on a planar surface and covered with a 2 micron insulating layer. After liquid is first dispensed atop one end of this structure (Fig. 1-a), voltage is applied. In response to the DEP force, the liquid is pulled along the electrodes until it is distributed along the entire length of the electrodes (Fig. 1-b). The voltage is then

Y. Baba et al. (eds.), Micro Total Analysis Systems 2002, Volume 2, 721–723.

removed and the liquid pinches off into droplets, one at each circular station, by capillary instability (Fig. 1-c). This entire process, effected remotely by application of AC voltage, takes place in less than 0.1s.

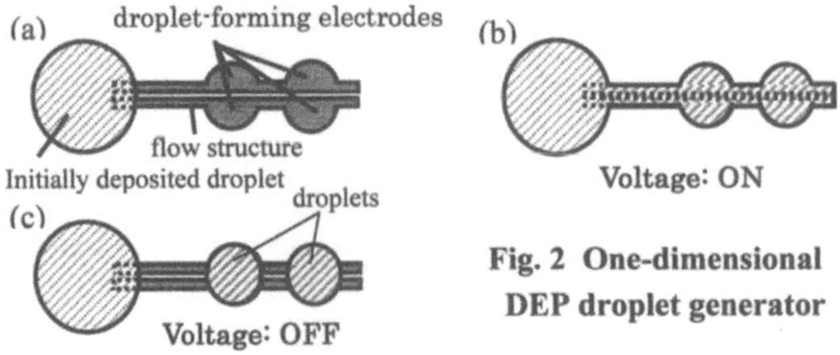

Fig. 2 One-dimensional DEP droplet generator

3. 2 x 2 droplet mixer array

By arranging these electrode structures so that they intersect at right angles, a droplet-based liquid mixer can be created. Fig. 3, featuring 2-each pairs of x-directed and y-directed electrode structures patterned on the substrate, shows the device tested in our experiments. Pairs of stations are located adjacent to each other near crossings, as shown in Fig. 3-b. Between neighbouring stations are special fusion electrodes used to induce droplet coalescence. In order to apply voltage independently to the structures, the electrode strips are segmented at the crossings. In a typical experiment, a micropipette is used to deposit samples (say, A and B) and reagents (a and b) at the ends of the y-directed and x-directed electrode structures, respectively. Voltage is first applied to the y-directed electrode pairs to dispense sample liquid droplets at stations A1, A2, B1, and B2. Next, voltage is applied to the x-oriented electrodes to form reagent droplets at stations a1, a2, b1, and b2. Finally, a voltage pulse is applied to the four fusion electrodes, which causes the droplets at the adjacent stations to be attracted together and to coalesce simultaneously.

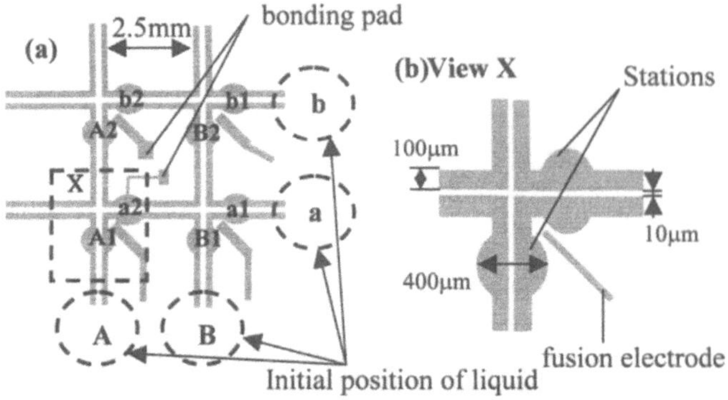

Fig. 3 The structure of n x m liquid mixer

4.Experimental observation

The video frames in Fig. 4 show the process steps for the 2 x 2 mixer array. DI water is used for this experiment. Nanoliter-sized water droplets are formed at each station and then the neighbouring droplet pairs are coalesced by sequences of voltage application (500Vrms, 100kHz) and appropriate changes of electrode connections.

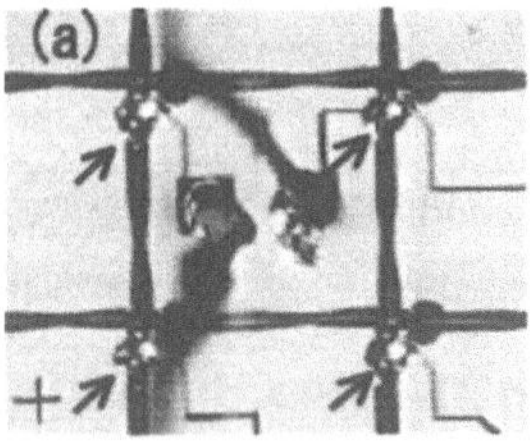

(a) droplet formation 1

(b) droplet formation 2

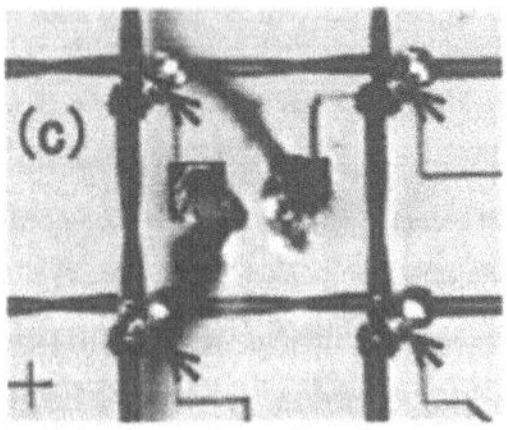

(c) droplet fusion

Fig. 4 Mixing based on droplets fusion
(a) Droplets are formed on A1, A2, B1, B2.
(b) Droplets are formed on a1, a2, b1, b2.
(c) Voltage applied to the fusion electrodes and the neighbouring droplets are fused and mixed.

5.Conclusion

For droplet volumes ranging from microliters down to nanoliters, dielectrophoresis provides a unique, controllable force for the manipulation of liquid. Using the DEP force, we have demnstrated an on-chip mixer array for nanoliter-sized liquid droplets. Scaling this simple nanoliter droplet mixing scheme up to much larger array sizes might make it possible to achieve massively parallel wet chemical processing for applications in combinatorial chemistry.

References

1. T. B. Jones, M. Gunji, M. Washizu and M. J. Feldman, *Dielectrophoretic liquid actuation and nanodroplet formation*, Journal of Applied Physics, Vol. 89, No. 2, pp. 1441-1448, 15 Jan. 2001
2. M. Gunji, T. B. Jones and M. Washizu, *DEP microactuation of liquids*, Proc. MEMS2001, Interlaken, Switzerland, (2001)
3. M. Gunji, T. B. Jones and M. Washizu, *Dielectrophoretic Microfluidic Devices*, Proc. IEJ-ESA Joint Symposium on Electrostatics, Kyoto, Japan, pp.78-87, (2000).

INTEGRATED THIN FILM TEMPERATURE SENSORS FOR POLYCARBONATE MICROFLUIDICS

Chris Kimball and Don L. DeVoe
Maryland MEMS Laboratory
Department of Mechanical Engineering
University of Maryland, College Park, MD 20742

Abstract

A technique for lithographic patterning of extremely thin metal films on polycarbonate substrates has been developed. This is used to fabricate integrated arrays of thin film temperature sensors and heaters in a polycarbonate microfluidic device.

Keywords: Temperature sensors, polycarbonate, integrated temperature control, lithography, temperature gradient

1. Introduction

One critical function for a number of biological and biochemical microsystems is temperature control. While several plastic-based microfluidic systems have been fabricated with temperature control [1], these systems typically employ macro-scale heaters and temperature sensors attached to the outside of the device. More accurate spatial temperature sensing, with more localized control and faster cycling times can be achieved by integrating thin film heaters and sensors into microfluidic systems using traditional deposition and lithographic techniques directly on the substrate. This has been successfully demonstrated using silicon and glass substrates [2]. While lithographic patterning of metal layers on polyimide or polyolefin has been demonstrated [3,4,5], this has not yet been feasible using plastics which are suitable for injection molding or imprinting. The surface of such plastics, i.g. acrylic and polycarbonate, are affected by the solvents in all commercially available resists. To avoid this problem, other methods such as metal shadow masking [6] and screen printing with conductive ink [5] have been employed. However, these methods cannot provide the resolution and thickness control of traditional physical vapor deposition combined with photolithography.

2. Results

We have developed a method of patterning extremely thin evaporated metal layers (200 Angstroms) directly on polycarbonate using standard photolithography with a modified photoresist formulated in our group. The modified photoresist has been designed specifically to prevent swelling of the polycarbonate surface. Using this technique, integrated heaters and resistive temperature devices (RTDs) have been developed. A thin sensor allows for higher sensitivity with less Joule heating. We have demonstrated integrated sensors and Wheatstone bridges with a sensitivity of 120 μV/°C, which provides a resolution of 0.009 °C when measured with a data acquisition

Y. Baba et al. (eds.), Micro Total Analysis Systems 2002, Volume 2, 724–726.

unit capable of one microvolt resolution. These noise in the sensor output was below 0.014°C. The results were repeatable to within 0.4°C (Figure 1). An annealing effect was observed, causing the resistivity of the sensor to decrease slightly at elevated temperatures. More complete annealing of the sensors should improve the repeatability.

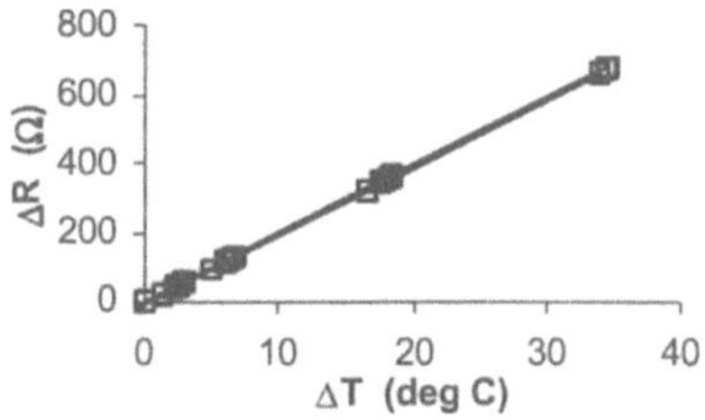

Figure 1. Repeatability of a 20 nm thick Cr RTD. The sensor was cycled four times demonstrating the measurements to be repeatable to within 0.4°C.

An array of such temperature sensors is integrated onto a polycarbonate substrate, and bonded with spin on PDMS to a second polycarbonate wafer containing a microfluidic channel. The primary channel is 500 microns wide and 4 cm long, with 5 sensors fabricated across the width of the channel, with 4 rows of sensors down the length of the channel for a total of 20 temperature sensor elements in the array. A photograph of the device is shown in Figure 3, and an array of sensors across the width of the channel is shown in Figure 2. The resistance changes of the sensors are measured with integrated Wheatstone bridge circuits, which are also patterned directly onto the polycarbonate substrate. The bridges are patterned at a sufficient distance from the channel to ensure no temperature change occurs at the bridges. Fluid is passed through the channel over a lithographically patterned heater, and then over the array of temperature sensors (Figure 4). The heater is not thermally isolated from the substrate, causing some heating of the substrate even with no flow (Figure 5). The spacing between the heater and start of the sensor array is 3.5mm. The change in the temperature distibution caused by fluid flow has been measured (Figure 6). A smaller device will be made to measure the thermal diffusion as a hot and cold fluid are brought together in a microchannel. These results will be compared with a theoretical model of the system.

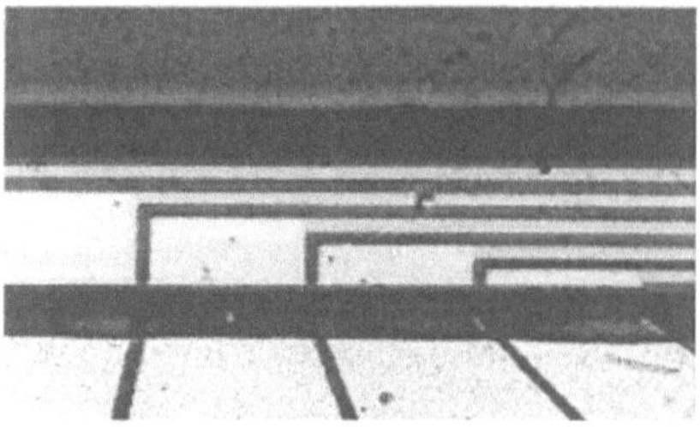

Figure 2. Photograph of microchannel with array of 20 nm Cr sensors across the width of the channel.

Figure 3. Photograph of a polycarbonate microfluidic platform with integrated temperature sensing and control elements collocated with the microfluidic elements.

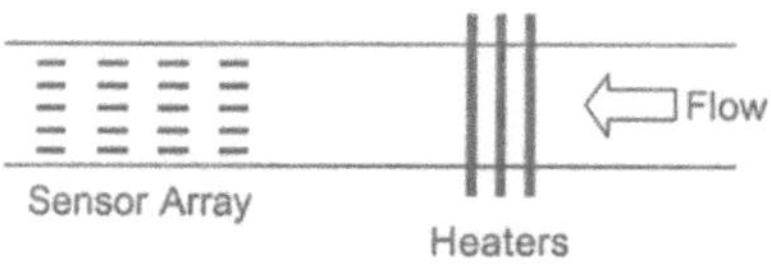

Figure 4. Schematic diagram of current device. Fluid is passed over heaters and then over an array of temperature sensors.

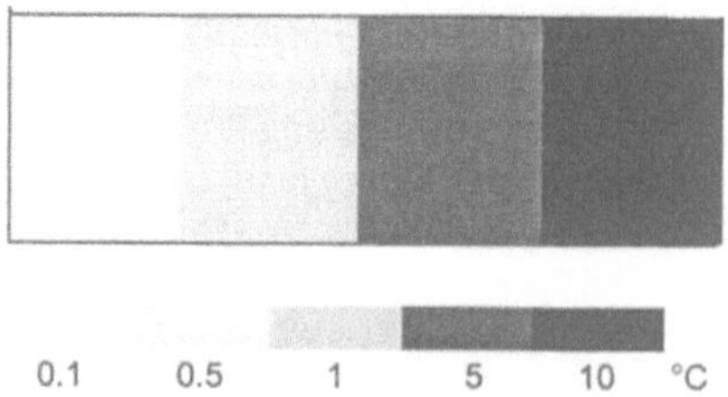

Figure 5. Temperature distribution in channel (with no fluid flow) due to heat conduction through the PC substrate. Temperature shown is change from room temperature. The heater was located 3.7 mm from the beginning of the sensor array, and was raised to a temperature was 55°C.

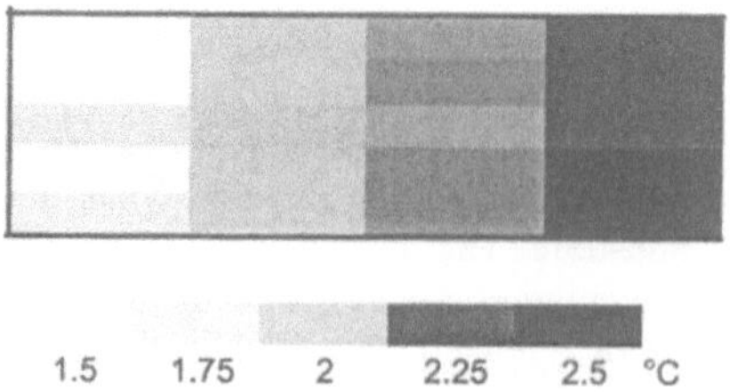

Figure 6. Temperature change in microchannel due to a fluid flow rate of 3.7 mm/sec.

3. Conclusion

This simple method of patterning many microscale thermal sensors and heaters in plastic microfluidic devices can ultimately be used to control both temperature and temperature gradients very accurately on the micro scale.

References

1. Yu, H. et. al., *A Miniaturized and Integrated Plastic Thermal Chemical Reactor for Genetic Analysis*, Micro Total Analysis Systems 2000, pp. 545-548, (2000).
2. Lagally, E. T., and Mathies, R. A., *Integrated PCR-CE System for DNA Analysis to the Single Molecule Limit*, Micro Total Analysis Systems 2001, pp. 117-118, (2001).
3. Meng, et. al., *A MEMS Body Fluid Flow Sensor*, Micro Total Analysis Systems 2001, pp. 167-168, (2001).
4. Metz, et. al., *Flexible Polyimide-Based Microchannels with Embedded Microelectrodes*, Micro Total Analysis Systems 2001, pp. 171-172, (2001).
5. Zhao, et. al., *Functional and Efficient Electrode-Integrated Microfluidic Plastic Devices*, Micro Total Analysis Systems 2001, pp. 193-194, (2001).
6. Ropke, et. al., *Manufacturing Issues of Polymer Microfluidic Devices with Integrated Electrodes for Electrochemical Detection of Heavy Metals in Environmental Samples*, Micro Total Analysis Systems 2001, pp. 193-194, (2001).

POLYIMIDE-BASED MICROFLUIDIC DEVICES WITH NANOPOROUS MEMBRANES FOR FILTRATION AND SEPARATION OF PARTICLES AND MOLECULES

S. Metz[1], S. Gawad[1], Ch. Trautmann[2], A. Bertsch[1], and Ph. Renaud[1]
[1]Swiss Federal Institute of Technology (EPFL), 1015 Lausanne, Switzerland
[2]Gesellschaft für Schwerionenforschung (GSI), 64291 Darmstadt, Germany

Abstract

The development of polyimide-based, microfluidic devices with nanoporous membranes is reported. Microchannels are fabricated from spin-on polyimide by a lamination technique. High aspect ratio nanopores are created in the polyimide channel walls by ion irradiation and subsequent ion-track etching. The devices can be used for on-chip filtration and separation of particles or as microstructures for drug delivery.

Keywords: Nanoporous, Membrane, Polyimide, Microchannel, Filtration

1. Introduction

Porous materials for BioMEMS structures, such as porous silicon, have gained interest in the field of bio-separation and drug delivery [1]. However, devices with porous polymers were mostly obtained by the assembly of pre-fabricated membranes and micromachined structures [2]. Here, we present a different approach where submicron pores with high aspect ratios are generated in-situ by ion irradiation and ion-track etching.

2. Experimental: Microchannel Fabrication and Nanopore Generation

A modified lamination technique is used to fabricate polyimide-based microchannels [3]. Layers of chrome and sacrificial aluminum are evaporated on silicon substrates. On the aluminum surface two 5 to 20 µm thick layers of photosensitive polyimide (PI-2732, DuPont) are applied, photostructured and cured to define the microchannels (Fig. 1a). On a separate carrier substrate a thin Mylar™ sheet is made to adhere temporarily by inserting water between the foil and the wafer. The foil is spin coated with a 5 to 20 µm thick layer of photosensitive polyimide. The second substrate is flipped upside down after a softbake step and bonded to the first wafer (Fig. 1b). After removing the carrier wafer and the Mylar™, the top layer of polyimide can be photostructured and the devices are cured. The structures are released from the fabrication support by anodic metal dissolution of the sacrificial aluminum layer (Fig. 1c).
Nanopores are then created in the polyimide channel walls by ion irradiation and ion-track etching. The devices are first irradiated perpendicular to the channels with Xenon ions at the Linear Accelerator Unilac at the GSI, Darmstadt. The ion fluence ranges typically from 10^4 to 10^8 ions/cm^2. The ion energy is varied between 190 and 1550 MeV by placing thin (10 to 80 µm) aluminum foils in front of the polymer samples. Depending on the foil thickness the ion energy is reduced in such a way, that

Y. Baba et al. (eds.), Micro Total Analysis Systems 2002, Volume 2, 727–729.

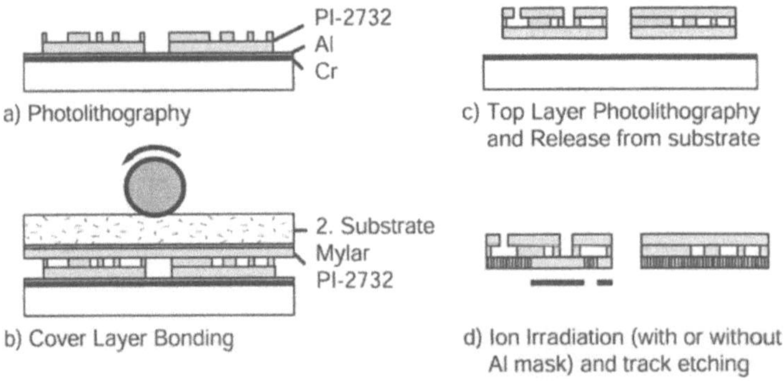

Figure 1. Fabrication process for polyimide-based microfluidic devices with nanoporous membranes by lamination and ion-track etching.

the ions penetrate either only the top layer (Fig. 2a, Fig. 3 upside down), or both, the top and the bottom layer of the device (Fig. 2b, Fig. 4 upside down). Ion irradiation can also be completely masked with a thick aluminum foil (200 μm) defining selected zones to be irradiated (Fig. 2c). In a second step, the latent ion-tracks are etched to the desired pore diameter in sodium hypochlorite solutions at different concentrations, temperatures and pH values to obtain the nanopores within the channel walls (Fig. 1d).

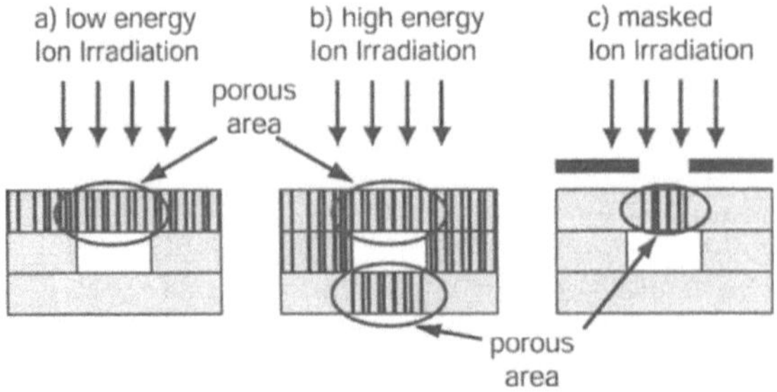

Figure 2. Illustration of ion irradiation and resulting porous areas after track etching: a) low energy ions only penetrate top layer and create etchable tracks, b) ions with higher energy cross two layers and yield etchable tracks in channel top and bottom layer, c) Ion irradiation can be masked to create pores in a predefined area.

3. Results and Discussion

The micromachined polyimide channels had widths from 30 to 500 μm, heights from 5 to 20 μm and lengths up to 6 cm. The layer transfer and lamination technique yields a high bond strength interface between the polyimide layers [3]. For the nanoporous channel walls the pore density can be controlled by the ion flux during the irradiation (e.g. 10^4 to 10^8 ions/cm^2), but the pores remain stochastically distributed. The pore depth can be adjusted by the ion energy resulting in different filter areas. Figure 3 shows a microchannel with pores only in the bottom wall, which results from irradiation with low energy ions. Ions with higher energies penetrate deeper into the polyimide and lead to pores in the channel top and bottom walls (Fig. 4). The pore diameter is only determined by the etching conditions (temperature, pH-value, etch time) and can be controlled from 20 nm to several μm. The porous, microfluidic channels were characterized quantitatively by cross-flow filtration of fluorescent microspheres from an aqueous solution (Fig. 5).

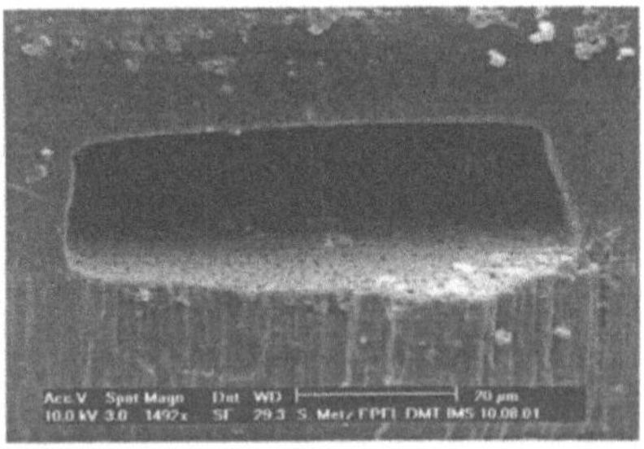

Figure 3. SEM image of cleaved microchannel (20 μm height, 50 μm width) with porous area at bottom (pore diameter 200 nm).

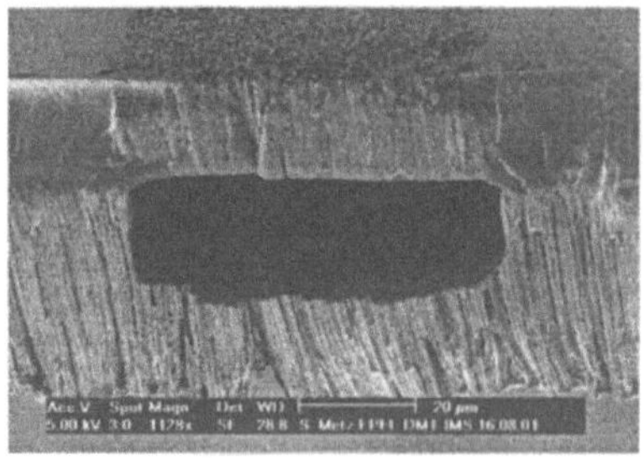

Figure 4. SEM image of cleaved microchannel (20 μm height, 50 μm width) with porous area at top and bottom (pore diameter 200 nm).

4. Conclusions

Polyimide-based microchannels with integrated nanoporous membranes were fabricated by lamination and ion-track etching. The properties of the porous channel walls, such as pore density, pore depth and pore diameter can be independently controlled by the irradiation and the ion-track etching conditions. The pores in the channel top and bottom layer allow for cross-flow filtration of particles and molecules in fluids.

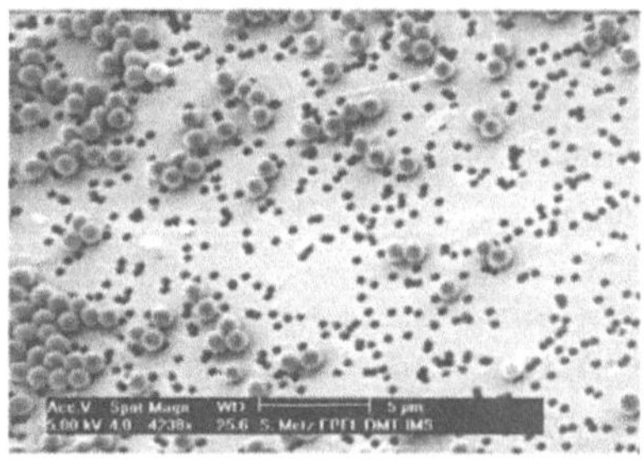

Figure 5: Polyimide membrane with pores (300 nm diameter) after cross-flow filtration of polystyrene beads (1 μm diameter).

References

[1] T. A. Desai, D. J. Hansford, and M. Ferrari, "Micromachined interfaces: new approaches in cell immunoisolation and biomolecular separation," *Biomolecular Engineering*, vol. 17, pp. 23-36, 2000.

[2] P. C. Wang, D. L. DeVoe, and C. S. Lee, "Integration of polymeric membranes with microfluidic networks for bioanalytical applications," *Electrophoresis*, vol. 22, pp. 3857-3867, 2001.

[3] S. Metz, R. Holzer, and P. Renaud, "Polyimide-based microfluidic devices," *Lab on a Chip*, vol. 1, pp. 29-34, 2001.

Micro Electrophoresis Chips with On-chip Optical Waveguides Utilizing Novel Buried SU-8/SOG Double Layers

Che-Hsin Lin[1], Gwo-Bin Lee[2], Chun-Che Lin[3], Shu-Hui Chen[3]
[1]Department of Electrical Engineering
Southern Taiwan University of Technology, Tainan, 710 Taiwan, ROC
[2]Department of Engineering Science, [3]Department of Chemistry
National Cheng Kung University, Tainan, 701 Taiwan, ROC
E-mail: Lin_chehsin@hotmail.com

Abstract

This paper reports an innovative micro electrophoresis chip, which is integrated with buried optical waveguides on glass substrates for on-line detection of biomedical samples. A novel buried optical waveguide structure using SU-8/SOG (spin-on-glass) double layers is demonstrated, which can increase light guiding efficiency due to large difference of refractive index between SU-8 ($n = 1.8$, after hard-baking) and organic-based SOG ($n = 1.36$). With this novel approach, delicate optical system and even microscopy are not required for biomedical sample detection, resulting in a more compact or even a portable detection system.

Keywords: electrophoresis, optical waveguides, SU-8, spin-on-glass, on-line detection

1. Introduction

Integration of optical detection mechanism in a chip has great attractions to researchers trying to miniaturize micro capillary electrophoresis (CE) system. Webster *et al.* reported a microfabricated photodiode with integration of micro CE system using surface micromachining techniques [1]. Chabinyc *et al.* placed a highly sensitive photo-detector under a sample flow channel to detect fluorescence signals [2]. However, expensive fabrication process and high cost for the APD (avalanche photo-diode) sensors made their devices not disposable. Another approach using optic fibers for coupling the detection light and collecting the signals was reported by Liang *et al.* [3]. Nevertheless, time-consuming fabrication process and complicated light coupling procedure made their design not suitable for batch fabrication and practical applications. An alternative approach is to fabricate integrated optical waveguides for fluorescence detection where detection light is simply focused around the detection point. Figure 1 shows a schematic of the proposed micro CE device for on-line detection. An integrated optical waveguide is place at the downstream of a separation channel and used to collect the fluorescence emission from the sample flow.

2. Fabrication

Y. Baba et al. (eds.), Micro Total Analysis Systems 2002, Volume 2, 730–732.

Figure 2 shows a simplified fabrication process of the microfluidic device. A fast and reliable fabrication process for formation of sealed microchannels using wet chemical etching and fusion bonding of soda-lime glass was first developed [4]. After microchannel was formed, a layer of SOG was coated on the wall, followed by filling of SU-8. Not only was organic-based SOG layer used to improve the surface smoothness of the etched channel, but it also increased the difference of reflection index between core/wall interfaces, resulting in better light guiding efficiency. Etched single-mode and multi-mode optical fibers were then inserted into the waveguide channel to connect the optical waveguide and a photo detector. The length of the separation channel is 40 mm and the optical waveguide is placed 30 mm away form the intersection of the CE channels. Cross-section view and the light guiding effect of the buried optical waveguides are shown in Fig. 3. The optical loss of the SU-8/SOG optical waveguide was measured to be 3.5 dB/cm, which is good enough for most optical detection in microdevices.

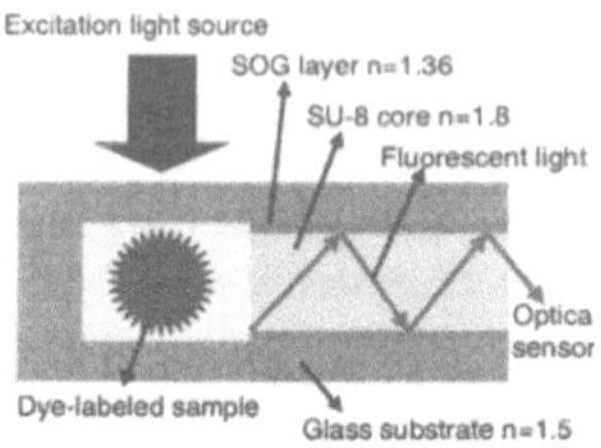

Fig. 1 Working principle of the proposed micro capillary electrophoresis chip integrated with buried optical waveguide structures for on-line separation and detection.

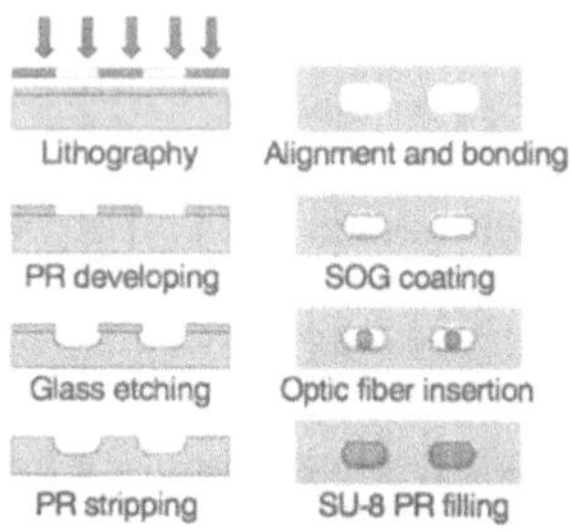

Fig. 2 A simplified fabrication process for micro electrophoresis chips integrated with buried optical waveguides in soda-lime glass.

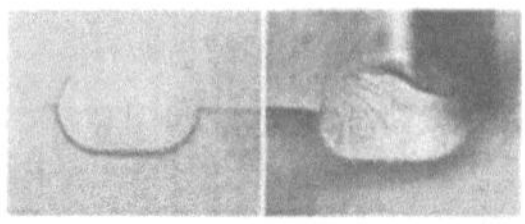

Fig. 3 Cross-section view of SU-8 optical waveguides. (A) glass channel was filled with SU-8/SOG double layers to form the optical waveguide structure. (prior to lighting), (B) 633-nm He-Ne laser was coupled into the SU-8/SOG optical waveguide. The channel was 80 μm in width and 50 μm in depth.

3. Experimental

The experimental setup for the device is shown in Fig. 4. With this approach, no microscopy and delicate optical alignment were used, resulting in a more simple and compact apparatus. The capillary electrophoresis was operated in "double L" mode using two high-voltage power supplies. Diluted Rhodamine B was first used to test the performance of the proposed device. Figure 5 shows the relation between relative signal amplitude and fluorescein concentration of Rhodamine B. The signal amplitude for the multi-mode optic fiber is about 39-fold bigger than one for the single-mode fiber. Results showed the detection limit could be as low as 10^{-7} M for the multi-mode fiber and 10^{-6} M for the single-mode fiber. Lower fluorescein concentration could be also detected if the PMT was set at a higher voltage level. Figure 6 shows electropherograms of a mixture of Rhodamine B (10^{-4} M) and Cy 3 ($6.5x10^{-3}$ M) fluorescent dye. Results showed the fabricated microchip could separate the mixture with highly efficiency. All three peaks of Cy 3 fluorescein after hydrolysis were separated and detected. The performance of the proposed device is comparable with the conventional equipment.

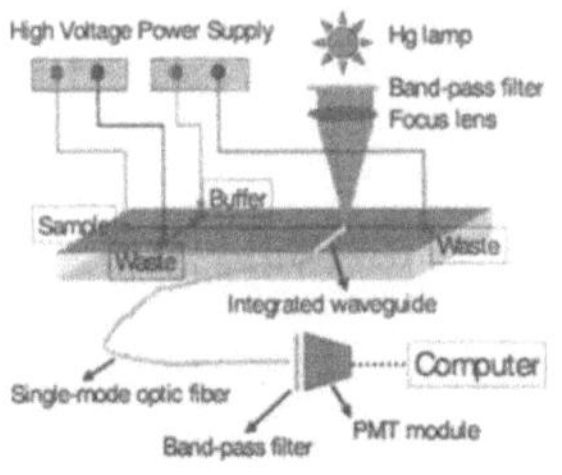

Fig. 4 Schematic representation of the experimental setup.

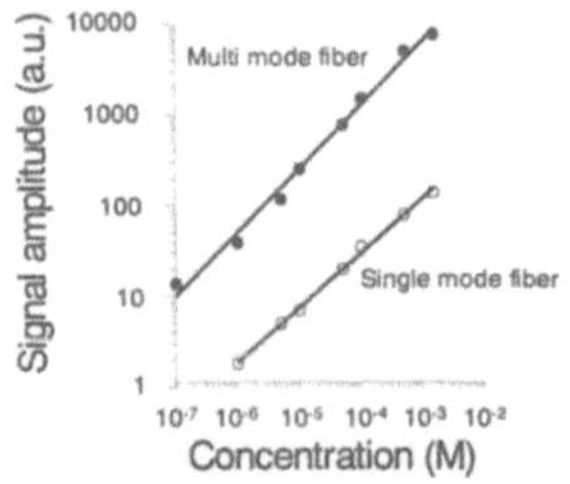

Fig. 5 Detected relative signal amplitude as a function of fluorescein concentration in a sodium borate buffer solution.

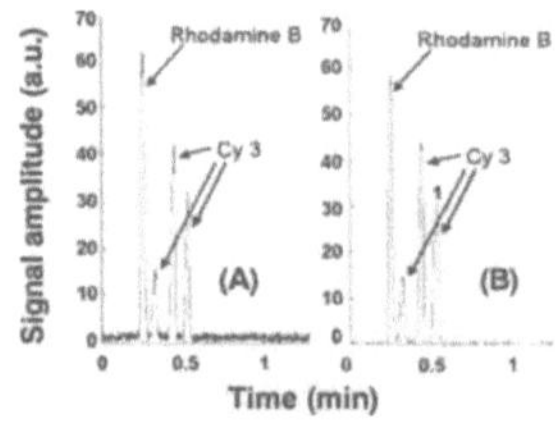

Fig. 6 Electropherogram of a mixture of Rhodamine B and Cy3 fluorescein using the developing microchip (A) and conventional method (B).

4. Conclusions

In this paper, we have described an innovative micro capillary electrophoresis device integrated with buried SU-8/SOG optical waveguides. No expensive photo detector will be disposed and no complicated optical alignment procedure is required using this simple approach. The detectable concentration fo`r Rhodamine B could be lower than 10^{-7} M for multi-mode optical fiber as the connection between the buried optical waveguide and the photo detector. Experimental results showed the developed microchip could separate and detect the mixture of Rhodamine B and Cy 3 fluorescent dye successfully. A miniaturized CE system could be realized with this simple approached.

Acknowledgements

The author would like to thank the National Science Council in Taiwan for financial support of the project. (NSC 90-2323-B006-011)

References

1.J. R. Webster, M. A. Burns, D. T. Burke, C. H. Mastrangelo, *Monolithic Capillary Electrophoresis Device with Integrated Fluorescence Detector*, Anal. Chem. 73, pp, 1622-1626, (2001)

2.M. L. Chabinyc, D. T Chiu, J. C. McDonald, A. D. Stroock, J. F. Christian, A. M. Karger, G. M. Whitesides, *An integrated fluorescence detection system in PDMS for microfluidic applications*, Anal. Chem. 73, pp. 4491-4498, (2001)

3.Z. H. Liang, N. Chiem, G. Ocvirk, T. Tand, K. Fluri, D, J. Harrison, *Microfabrication of a planar absorbance and fluorescence cell for integrated capillary electrophoresis devices*, Anal. Chem. 68, pp. 1040-1046 (1996)

4.C. H. Lin., G. B. Lee, Y. H. Lin, G. L. Chang, *A Fast Prototyping Process for Fabrication of Microfluidic Systems on Soda-Lime Glass*, J. Micromech. Microeng. 11, pp. 726-732, (2001)

DYNAMICS OF BUBBLES IN MICROCHANNELS

Mads Jakob Jensen, Goran Goranović, and Henrik Bruus
Mikroelektronik Centret, Technical University of Denmark, DK-2800 Lyngby
(group web-page: www.mic.dtu.dk/research/MIFTS)

Abstract
We present a thorough theoretical study of quasi-static motion and dynamics of bubbles in microchannels. We investigate the effects of geometry and surface physics on the behavior of bubbles, and we propose a set of design rules to minimize the clogging of the channels by bubbles.

Keywords: bubble dynamics, microchannels, CFD simulations

1. Introduction

Due to the large surface to volume ratio, surface effects become very pronounced and even dominant in microchannels. One important example is the formation of gas bubbles in microfluidic systems. Such bubbles present a significant problem: they can block the microchannels, disrupt the flow, and disturb measurements in an uncontrolled manner. For that reason, the sample liquids are usually pretreated by various methods before they enter a microfluidic chip. Examples of pretreatment methods are degassing, control of surface tension or high-frequency sound vibrations. However, in spite of such efforts gas bubbles may still be present, and once there it often requires large driving pressures to push them out of the system [1].

In this paper, based on the interplay between geometry, surface tension, and contact angle, we propose and analyse two designs: one that eases the motion of big bubbles through contracting microchannels, and one that traps the bubbles aside from the main fluid flow.

2. Theory of bubblemotion in a microchannel constriction

The first design concerns contraction of a channel. In order to push a bubble through the sudden and the tapered contractions shown in Fig. 1(a) and 1(b), respectively, a total pressure drop ΔP_{tot} is needed to overcome the sum of two pressure drops: (1) ΔP_{surf} arising from the solid-gas interface at the walls as well as from the curved liquid-gas surface, and (2) ΔP_{fric} due to friction. For the quasi-static motion studied here, the friction is vanishingly small, and the total pressure drop over the bubble in the sudden contraction and the tapered contraction is given by the surface contribution alone.

Y. Baba et al. (eds.), Micro Total Analysis Systems 2002, Volume 2, 733–735.

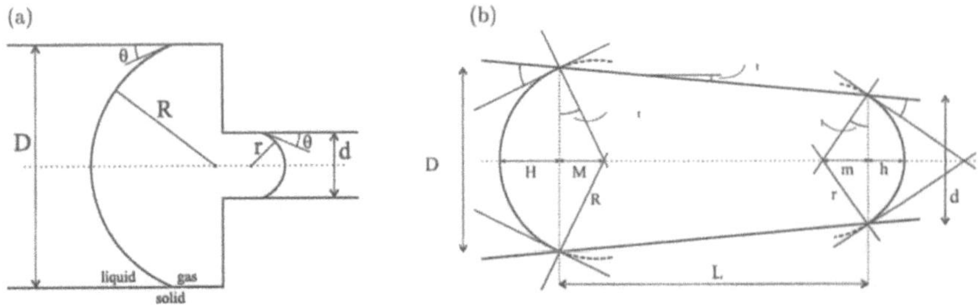

Figure 1: (a) A bubble passing through a sudden contraction. The curvatures at both ends are different and constant while the bubble is in the contraction. (b) A bubble in a tapered channel. Appropriate names have been given to different lengths and angles in use. The curvatures depend on the position of the bubble because of the tapering.

Using two different analytical methods, total energy considerations and the Young-Laplace equation, respectively, we have derived the following expressions for the pressure drop over bubbles in the two constrictions:

$$\Delta P_{\text{surf}}^{\text{sudden}} = 4\sigma \cos\theta \Big(\frac{1}{d} - \frac{1}{D}\Big), \tag{1}$$

$$\Delta P_{\text{surf}}^{\text{tapered}} = 4\sigma \Big(\frac{\cos(\theta + \theta_t)}{d} - \frac{\cos(\theta - \theta_t)}{D}\Big). \tag{2}$$

Here σ is the gas-liquid surface tension and θ is the contact angle describing the influence of the walls on the bubble dynamics. For any taper angle θ_t the driving pressure $\Delta P_{\text{surf}}^{\text{tapered}}$ for the tapered channel is smaller than the driving pressure $\Delta P_{\text{surf}}^{\text{sudden}}$ for the sudden contraction, hence lowering the threshold pressure to drive bubbles out of the contraction. It may thus be beneficial to use tappered constrictions in microfluidic networks where the presense of bubbles poses a problem.

3. A bubbletrap design

Sometimes it is beneficial to collect the bubbles in one place away from the main flow. For this purpose a special passive bubble trap is developed, using a combination of geometry and surface tension effects. Figs. 2a-2d show a 2D CFD simulation of a bubble-trapping process allowing unhindered flow. In a real system, several bubble traps could be placed at locations where there is a tendency of bubble creation. Using special semi-permeable materials (e.g. Gore-Tex), the bubbles could then be extracted from the region and out of the system.

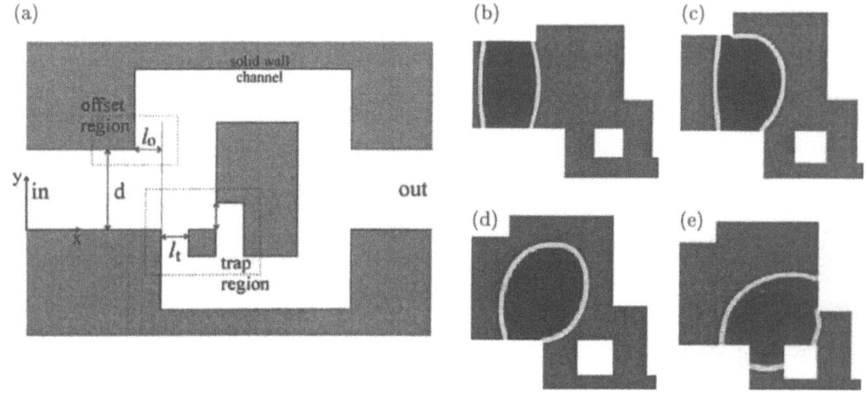

Figure 2: (a) A sketch of the trap geometry where the central length scales are marked: the width d of the main channel, the size l_0 of the offset between the upper and lower wall, and the width l_t of the narrow trapping channels. (b)-(e) Four snapshots of the simulated trapping procedure near the offset and trap regions. (b) A bubble moving towards the offset region. (c) Detaching of one side of the bubble by the offset effect. (d) The bubble is pushed towards the trapping region. (e) The bubble gets trapped.

4. Results and discussion

In our study of bubble dynamics in microchannels, theoretical concepts based on surface tension, geometry, and contact angle are considered. Numerical simulations are performed using commercial software CFD-ACE+ ver. 6.6 containing a specialized Volume-Of-Fluid (VOF) module. They are in good agreement with analytical results and are thus a reliable tool for analyzing bubble behavior in microchannels. The simulation approach is particularly useful for optimization and development of more complicated designs aiming at diminishing the problem of bubbles.

Acknowledgments

We thank our μTAS-colleagues at Mikroelektronik Centret, especially Jörg Kutter, for support and stimulating discussions. This work is partly supported by the Danish Technical Research Council, μTAS Frame Program Grant No. 9901288.

References

[1] P. Gravesen, J. Branebjerg, and O. Søndergård Jensen J. Micromech Microeng., **3**, 168-182 (1993).

EXPERIMENTAL VERIFICATION OF A BI-DIRECTIONAL DRIVING SYSTEM FOR MICROFLUIDS

C.P. Jen[1], W.D. Wu[1], Y.C. Lin[1], G.G. Wu[2], C.C. Chang[2] and Y.C. Chung[3]
[1]Department of Engineering Science, National Cheng Kung University, Tainan, Taiwan
[2]Department of Mechanical Engineering, Kun Shan University of Technology, Tainan, Taiwan
[3]Electronics Research & Service Organization, Industrial Technology Research Institute, Hsinchu, Taiwan

Abstract

A bi-directional microfluidic driving system was developed experimentally in this work. The bi-directional driving module combines two individual components for suction and exclusion. The individual components can be combined using a T-shape connection as well as in parallel. Modular devices with a parallel connection, presented herein, can provide flexible, predictable and linear control of bi-directional microfluid pumping. This work demonstrated the driving system and investigated the influences of the characteristics of the working liquids. The driving force, both suction and exclusion, for the liquid with the lower viscosity is stronger than those with the higher viscosity. Furthermore, the required flow rate at the suction component is quite large due to the strong surface tension of water.

Keywords: microfluid, pneumatic, bi-directional manupulation, driving system

1. Introduction

The microfluid management devices, such as micropumps, microvalves, microsensors and micromixers, have been rapidly developed over the past few years [1]. Micropumps are categorized into two groups: mechanical and non-mechanical. The performance of micropumps depends strongly on the special characteristics of that particular actuators. In medical and biomedical applications, the special transport mechanism must be designed for the μTAS to move samples and reagents through the microchannels that connect the unit procedure components in the systems [2]. However, the relationships that exist between the pumping mechanisms, the conditions under which the devices operate and the behavior of the multi-component fluids transported in these channels are quite complicated. It is necessary to develop a driving system, which can simplify the entire system.

2. Design of the bi-directional driving module

The main purpose of this investigation was to implement a microfluidic driving system for bi-directional manipulation. The bi-directional microfluid driving module is composed of the suction and exclusion components and the different designs of the module are shown in Fig. 1. The driving module is 600μm in depth [3]. The airflow passes through the air gallery of the suction component with constriction and provides suction due to the Bernoulli's equation. The individual exclusion component produces

Y. Baba et al. (eds.), Micro Total Analysis Systems 2002, Volume 2, 736–738.

exclusion through the pneumatic structure. A driving module with bi-directional pumping provides the net effect of suction or exclusion, which can be applied on the reaction chip for specific uses.

3. Results and discussion

The servo system includes an air compressor, a buffer tank and conduits for airflows as depicted in Fig. 2. The droplet of liquid (1.5 μl) is injected to the microchannel (Fig. 3) and its velocity can be determined through the CCD connected to the microscope. Figure 4 illustrates the experimental results of methanol for the T-shape and intermediate parallel connection when the flow rate at the exclusion component (Q_E) equals 100 and 400 mLPM (mL/min). The parallel connected module exhibits better linearility than the T-shape module and the slopes of the former is larger. Larger slopes imply the wider operating range. The results of methanol for the suction and exclusion tendency modules are depicted in Fig. 5. The slopes of the suction tendency module are larger than those of the exclusion tendency module at the region of negative velocity where the suction occurs. The output velocities of methanol (kinematic viscosity $\nu_{methanol} = 7.56 \times 10^{-7}$ m^2/s) are larger than those of ethanol ($\nu_{ethanol} = 1.52 \times 10^{-6}$ m^2/s) because of its lower viscosity (Fig. 6). The required flow rate at the suction component (Q_S) is quite large due to the strong surface tension of water. When Q_S larger than the threshold value of 3000 mLPM (in the suction domain, i.e. negative output velocities) or smaller than the threshold value of 3500 mLPM (in the exclusion domain, i.e. positive output velocities), the curves of the experimental results in this figure become linear, as depicted in Fig. 6. This work demonstrated the driving system and investigated the influences of the characteristics of the working liquids.

4. Conclusions

This work demonstrated the driving system and investigated the influences of the characteristics of the working liquids. The feature that the final airflow outlet is fixed in one direction is most advantageous for preventing cross-contamination between the servo-system and the chip. This microfluid driving system is therefore particularly suited to micro devices for biochemical analysis.

Acknowledgements

This work was supported by the National Science Council of the Republic of China under grant No. NSC90-2323-B-006-007.

References

1. S. Shoji and M. Esashi, *Micro flow devices and systems*, J. Micromech. Microeng. 4, pp. 157-171, (1994).
2. M. Elwenspoek, T. S. J. Lammerink, R. Miyake and J. H. J. Fluitman, *Towards integrated microliquid handling systems*, J. Micromech. Microeng., 4, pp. 227-245 (1994).
3. C. P. Jen and Y. C. Lin, *Design and simulation of bi-directional microfluid driving systems*, J. Micromech. Microeng., 17, pp. 115-121, (2002).

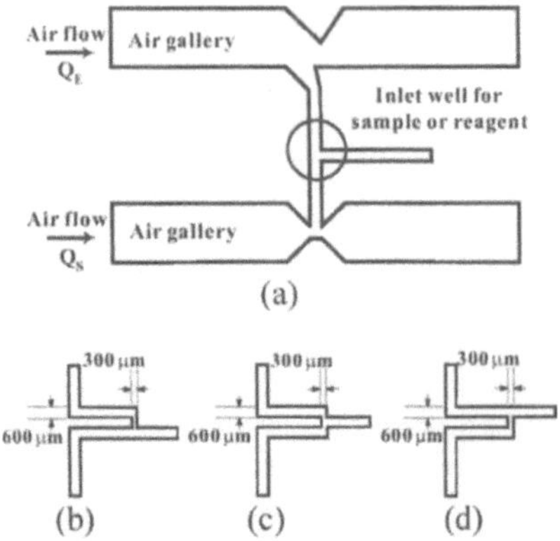

Figure 1: Schematic diagrams of the driving module: (a) T shape connection, (b-d) parallel connections in the dash line region of (a): (b) suction tendency, (c) intermediate and (d) exclusion tendency.

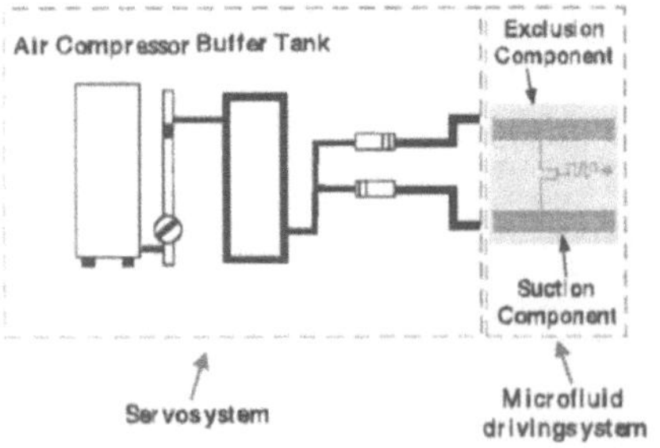

Figure 2: Schematic diagram of the entire bi-directional microfluid pumping system.

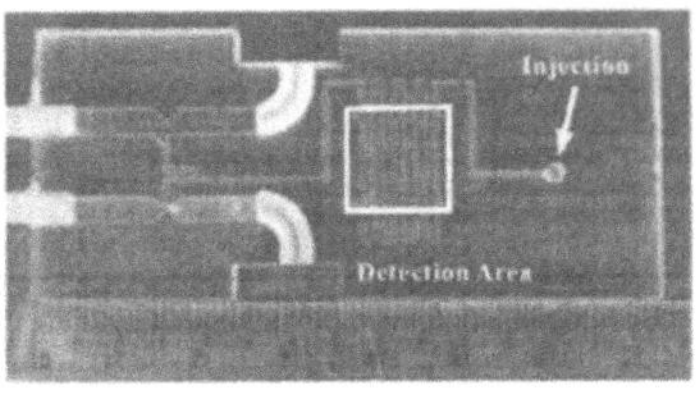

Figure 3: The driving module which used for the experiments.

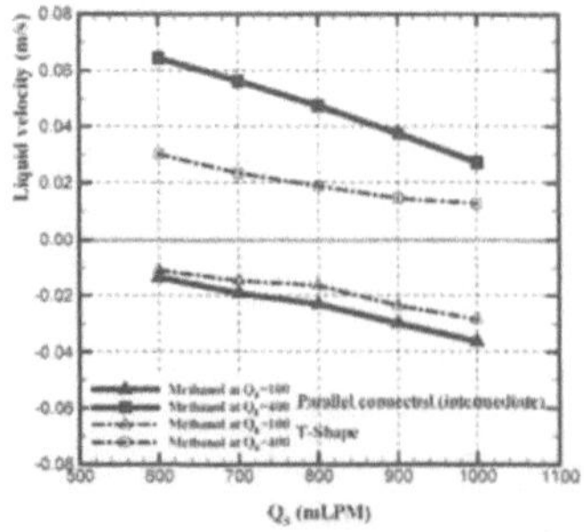

Figure 4: The experimental results for T shape and intermediate module.

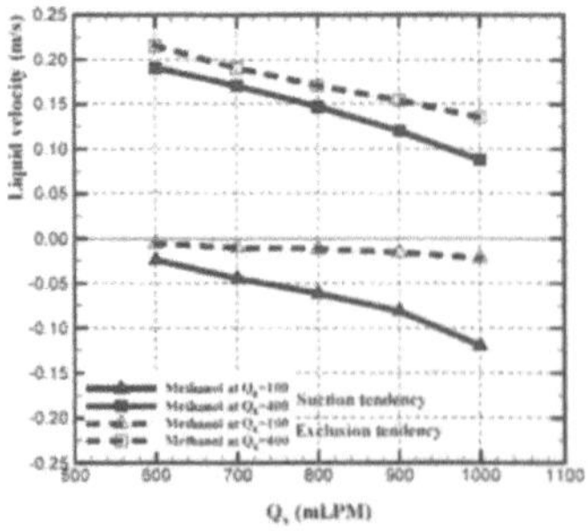

Figure 5: The experimental results for suction and exclusion tendency module.

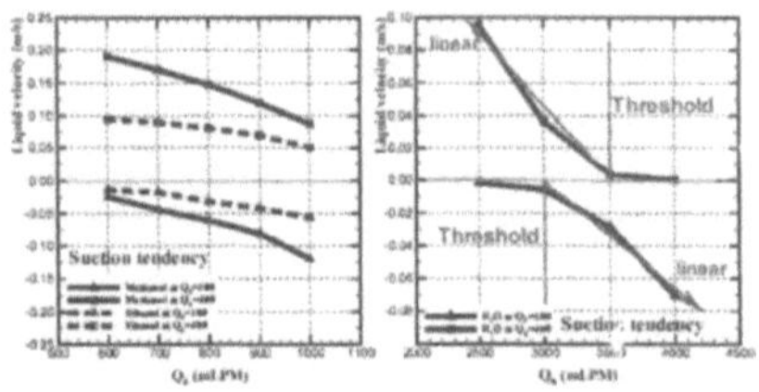

Figure 6: The experimental results for methanol, ethanol and water (in the suction tendency module). (surface tension: $\gamma_{water} = 6.79 \times 10^{-2}$ N/m; $\gamma_{methanol} = 2.25 \times 10^{-2}$ N/m; $\gamma_{ethanol} = 2.28 \times 10^{-2}$ N/m).

MICROFABRICATION OF A SI MESH STRUCTURE DEPTH FILTER

Yu Chen[1], Yubo Miao[1], Victor Samper[1], Fatimah Bte Mustafa[2], Qingxin Zhang[1], Chewkiat Heng[2], Huijen Lye[3], Titmeng Lim[3]
(chenyu@ime.a-star.edu.sg)

[1]:Institute of Microelectronics, 11 Science Park Road , Singapore Science Park II, Singapore 117685
[2]:Department of Paediatrics, National University of Singapore, Singapore
[3]:Department of Biological Sciences, National University of Singapore, Singapore

Abstract
The uTAS proposed the full incorporation of analytical procedures, which include sampling, sample preparation or pretreatment and actual analysis, into flowing systems. To incorporate a full uTAS, it is important to develop the individual microfluidic components. Microfilter is one of the microfluidic devices for sample preparation. This paper presents a fabrication method for a vertical silicon mesh structure depth filter as well as the application of white blood cell filtration. The results are also compared with another fabricated horizontal flow silicon filter.

Keywords: Micro-filter, microfabrication, silicon filter, microfluidic components, microfiltration

1. Introduction
Miniaturization of analytical techniques has become a dominant trend in research over the past decade. Research into miniaturization is primarily driven by the need to reduce costs, which is realized by reducing the consumption of expensive reagents and by increasing throughput and automation. Fundamentally, all analytical procedures follow a certain set of processes in a well-defined sequence. They usually includes sampling, sample preparation or pretreatment and actual analysis. The uTAS proposed the full incorporation of those analytical procedures into flowing systems. To incorporate a full uTAS, it is important to develop the individual microfluidic components. Microfilter is one of the microfluidic devices for sample preparation. The microfilter can be made by quartz [1] as well as by silicon material [2,3] and the silicon microfilter is proved the effective in isolation of white blood cells from whole blood as part of sample preparation for nucleic acid amplification. This paper presents a fabrication method for a vertical silicon mesh structure depth filter as well as the application of white blood cell filtration. The results are also compared with another fabricated horizontal flow silicon filter.

2. Fabrication
The vertical silicon depth filter [Fig.1] is fabricated by anodic bonding the microstructured silicon and pyrex glass wafers. The 3um wide mesh pattern [Fig.2] is formed on the silicon wafer firstly and followed by silicon DRIE [Fig.3] to make the 30um deep silicon vertical trenches. In order to form the silicon mesh structure

Y. Baba et al. (eds.), Micro Total Analysis Systems 2002, Volume 2, 739–741.

depth, the silicon that is under the mesh structure, needs to be fully released by SF_6. After silicon releasing, a 20um deep chamber is formed under the silicon mesh structure and the silicon mesh structure depth is fabricated [Fig.4]. The process of through holes for solution inlet and outlet is followed and will be done from backside of the silicon wafer. This process will be performed by KOH wet etching and SF_6 dry etching. Finally this processed silicon wafer needs to be anodic bonded with another pyrex glass wafer.

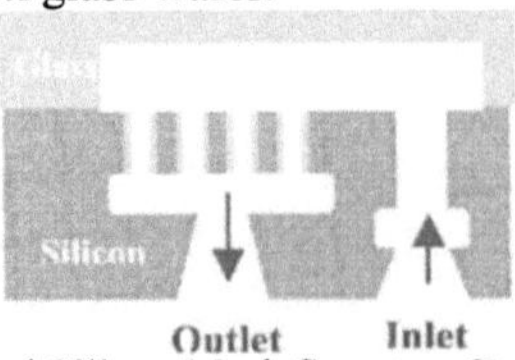

Fig.1 Vertical Silicon Mesh Structure Depth Filter

Fig.2 Mesh Structure

Fig.3 Mesh Structure Trench Etching

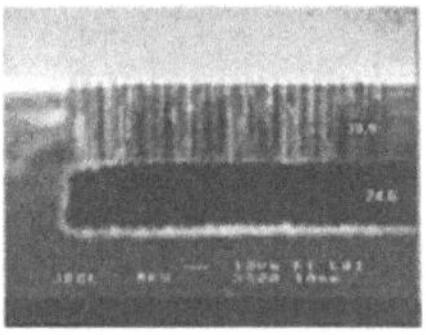
Fig.4 Mesh Structure Depth Filter

A 100um-depth cavity on the glass wafer is etched to form chamber pre-anodic bonding. The Cr/Au layer is the mask for glass etching and HF:HNO3 mixture is used as glass etchant. The dimension of the silicon mesh can be fine-tuned by depositing a layer of PECVD SiO_2 before bonding with the glass wafer. The sample solution will go through the inlet hole flowing to the glass chamber and pass through the silicon mesh depth for filtering, the elution solution will pass through the mesh structure silicon trench down to the bottom chamber and finally flow out through the outlet hole. The fabricated 6"wafer and diced filter devices are shown in Figure 5 and 6.

For comparison purpose, another horizontal flow designed filter is also fabricated. The Si is firstly deep etched to 100um to form the chamber and a group of 3um width channels by DRIE. The solution inlet and outlet holes are defined at wafer backside and are etched by KOH. Thermal oxide and LP nitride combined layer is used as mask for KOH wet etching. The processed silicon wafer finally bonded with another pyrex wafer. The sample solution will go through the inlet hole flowing into the silicon chamber and pass through the group of 3um width channels for filtering, then the elution solution will flow out through the outlet hole (Figure 7).

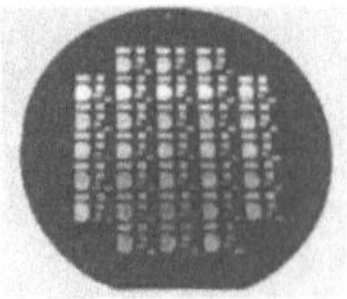
Fig.5 Fabricated 6"filter wafer

Fig.6 Filter devices

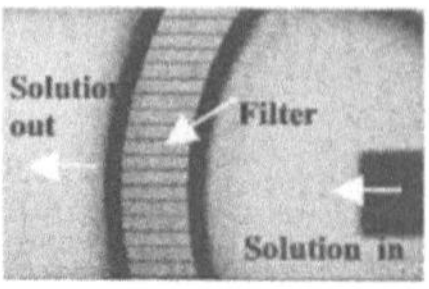

Fig.7 Horizontal flow filter

3. Results and discussion

Both vertical silicon depth filter and horizontal flow designed filter are initially tested for the white blood cells filtration from the whole blood sample. The 2ml of blood in 200ml of methylene blüe and 200ml of PBS are mixed as the testing sample, which is pumped into the filter through the inlet hole. The elution solution is collected from outlet. A haemacytometer is used to count the white blood cell for both testing sample and elution solution of the filter. The cells counted are based on 1ml of blood.

The filter efficiency is defined by the ratio of the number of white blood cells trapped by the filter to the total number of white blood cells from inlet sample and shows in Figure 8. The result shows that the fabricated silicon mesh structure depth filters are with a 72-85% filtration efficacy to trap white blood cells from whole blood sample. Three chips for each type of filters are tested and trapping capability for white blood cells are reproducible. The filtration area of the vertical mesh depth filter is from 4.5mm^2 to 64mm^2 and all show a high filtration efficacy. It is 10-20% higher than another fabricated horizontal flow silicon filter.

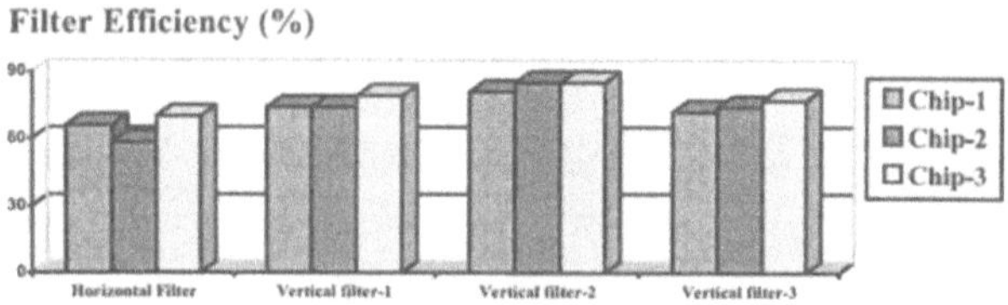

Fig.8. Efficiency of Filters

4. Conclusions

Both vertical mesh depth filter and horizontal flow silicon filter are fabricated successfully. The fabricated both vertical mesh depth filter and horizontal flow silicon filter can trap the white blood cells from whole blood sample more than 60%. The filter devices are with good reproducibility. The vertical mesh depth filter is with a 10-20% higher trapping efficiency and it is not significantly affected by the filtration area between 4.5mm^2 and 64mm^2. The further studies on filtration capability for the red blood cells from whole blood sample and microfluidic effect are needed.

Acknowledgements

The authors thank Biosensor Focused Interest Group in Singapore for providing financial support to DNA micro extraction system project. The authors also thank Dr. Narayanan Balasubramanian fully support for the DNA micro extraction system project.

References

[1] He B., Tan L., Regnier F., *Analytical Chemistry*, 71, 1464-1468, (1999)

[2] Wilding P., Kricka, L.J., Cheng, J., Hvichia G., *Analytical Biochemistry, 257,* 95-100, (1998)

[3] Cheng J., Kricka, L.J., Sheldon E.L., Wilding P., *Topics in Current Chemistry*, 194, 216-231, (1998)

In-situ Fabrication of Dialysis Membranes in Glass Microchannels Using Laser-induced Phase-Separation Polymerization

Brian J. Kirby and Anup K. Singh
Sandia National Laboratories, Livermore, CA (bjkirby@sandia.gov)

Abstract

Laser-induced phase-separation polymerization of a porous acrylate polymer is used for *in-situ* fabrication of dialysis membranes inside glass microchannels. A shaped 355 nm laser beam is used to define polymer membranes of 4-14 μm thickness, which bond to the glass microchannel and form a semipermeable membrane. Differential diffusion through the membrane is observed for fluorescein molecules versus 200 nm latex microspheres, showing the potential for this technique to integrate sample cleanup into chip-based analysis systems.

Keywords: phase-separated polymer, filtration, dialysis, sample preparation, polymer monolith

1. Introduction

Complex samples (*e.g.*, cell extract) often require extensive cleanup or pretreatment before introduction to analysis channels in a miniaturized device. These pretreatment steps are often performed off-chip using large volumes of sample and other reagents, and hence often add substantially to the total analysis time and cost. Dialysis, or size-based separation of species via selective diffusion through a semipermeable membrane, is a widely used technique for cleanup of biological samples [1,2]. We have developed a technique for fabricating thin (4-14 μm) porous polymer dialysis membranes within the channels of a glass microchip. UV laser-initiated polymerization (Figure 1) is used for controlled placement of the dialysis membrane in a chip for cleanup of complex or dirty samples; this technique is rapid and inexpensive and increases the potential functionality of integrated microfluidic devices. The semipermeable membrane allows diffusion of fluorescein molecules while preventing diffusion of 200 nm latex microspheres, showing the potential for extracting a small molecular weight analyte of interest from a complex matrix.

2. Substrate

Standard cross-shaped glass microchips were obtained from Micralyne; chemicals were obtained from Aldrich and used as received. To facilitate bonding between the membrane and the glass surfaces, glass surfaces were first exposed to a 2:2:1 (by volume) mixture of water, glacial acetic acid, and 3-(trimethoxysilylpropyl) acrylate for a period of 30 minutes, covalently linking the silane to the wall and exposing the acrylate group for polymerization.

Y. Baba et al. (eds.), Micro Total Analysis Systems 2002, Volume 2, 742–744.

3. *In-situ* Photopatterning of Polymer Membranes

Following surface treatment, channels are filled with a monomer/solvent/photoinitiator solution of 73:27 pentaerythritol triacrylate:1-propanol (by weight) with 10 mg/ml AIBN photoinitiator. Thin (4–14 μm) porous polymer membranes are fabricated *in-situ* by shaping and focusing the 355 nm output of a 12 kHz, 800 ps-pulse, 160 nJ-pulse, frequency-tripled Nd:YAG laser into a 1-2 μm sheet and using this sheet to generate photoinitiated phase-separation polymerization [3,4] in the irradiated region (Figure 2). The thickness of the membrane is determined by the laser sheet thickness, radical diffusion, solved-phase polymer diffusion, and bulk fluid motion [5]. Laser sheet thickness was minimized by spatially filtering the focused laser output with a 2 μm slit and imaging the resulting diffraction pattern at ~0.5 magnification onto the channel. Following polymerization, the system was flushed thoroughly with acetonitrile to remove residual polymer/monomer/solvent material and then filled with aqueous solutions for testing.

4. Demonstration of Differential Diffusion

Aqueous solutions of (a) fluorescein (Exciton) or (b) 200 nm, carboxylate-modified, dye-impregnated latex spheres (Molecular Probes) were alternately used to fill channels on one side of the polymerized membrane; the other side was filled with water. Solutions were allowed to come to rest and the diffusion of species (fluorescein or latex spheres) across the membrane was observed over several minutes using fluorescence excited at 488 nm. Figure 3 shows a schematic and images for both cases. For this porous polymer, with a nominal pore size of 30 nm (measured with Hg porosimetry, BET, and SEM), fluorescein diffuses across the membrane while the latex spheres do not. In future applications, monomers and solvents may be chosen to lead to a specific pore size distribution, which can be used to engineer the molecular weight cutoff of the porous membrane for a specific application.

Acknowledgements

The authors gratefully acknowledge the technical expertise of T.J. Shepodd, B.W. Wiedenman, R.N. Renzi, and D.A. Sunnarborg. This work was supported by the Laboratory Directed Research and Development program at Sandia National Laboratories. Sandia is operated for the United States Department of Energy under contract DE-AC04-94AL85000.

References

1. N. Xu, Y. Lin, S.A. Hofstadler, D. Matson, C.J. Call, R.D.A. Smith, *Anal Chem* **70** (1998).
2. J. Lichtenberg, N.F. de Rooij, E. Verpoorte, *Talanta* **56** (2002).
3. C. Yu, M.H. Davey, F. Svec, J.M.J. Frechet, *Anal Chem* **73** (2001).
4. D.J. Throckmorton, T.J. Shepodd, A.K. Singh, *Anal Chem* **74** (2002).
5. B.J. Kirby, T.J. Shepodd, E.F. Hasselbrink, *J. Chrom A*, in press (2002).

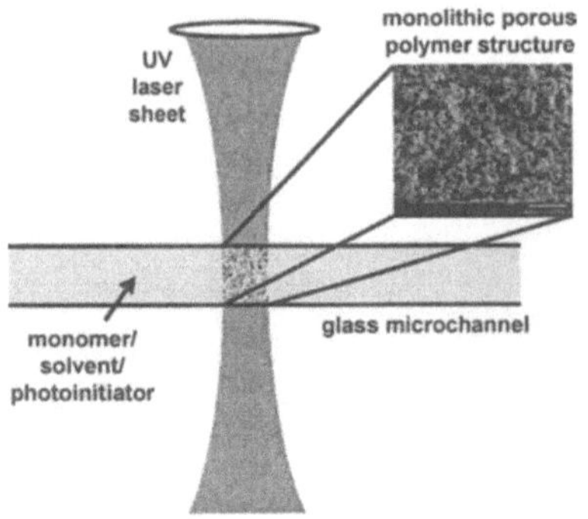

Figure 1. Phase-separation polymerization.

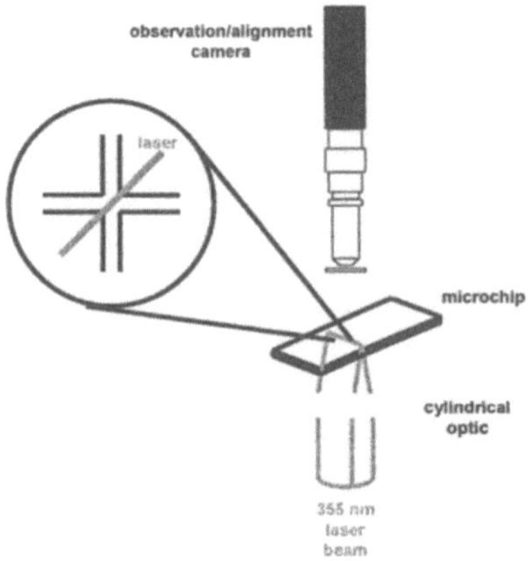

Figure 2. Beam-shaping optics for thin membrane polymerization.

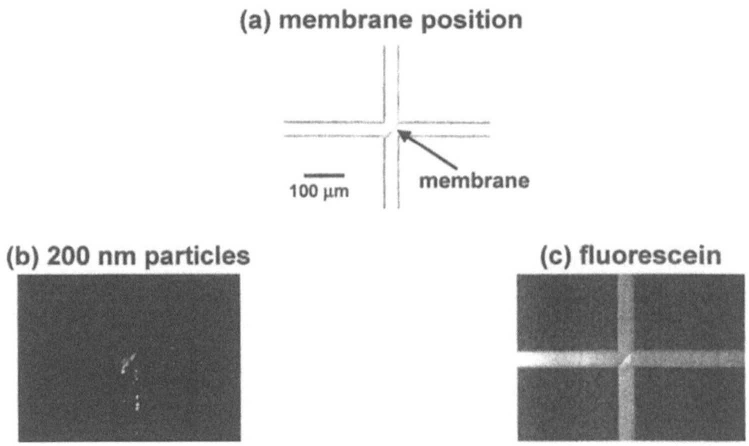

Figure 3. Images of selective diffusion across a 9 µm porous polymer membrane: (a) white light image of membrane location; (b) fluorescence image of 200 nm fluorescent spheres, which do not diffuse across the membrane; (c) fluorescence signal from fluorescein, which does diffuse across the membrane.

MICRO ULTRASONIC HOMOGENIZER CHIP MADE BY HYBRID MICROSTEREOLITHOGRAPHY

Koji Ikuta, Yasuhiro Sasaki, Hiroyuki Maegawa
Shoji Maruo and Tadahiro Hasegawa
Nagoya University, Furo-cho, Chikusa-ku, Nagoya 464-8603 Japan
(e-mail: ikuta@mech.nagoya-u.ac.jp)

Abstract

The world's first micro-ultrasonic homogenizer chip, a new type of biochemical IC chip [1], is proposed and developed successfully. The homogenizer chip utilizes ultrasonic cavitation to disrupt cells. Both a piezoelectric element and an exponential horn to generate ultrasonic cavitation are hybridized in the chip by using our original microstereolithography technique, the "hybrid-IH process [2]". We succeeded to disrupt cells (PC12) within only 3 seconds by using this chip. The efficiency of cell homogenization was almost 100%. The proteins extracted from PC12 cells were separated and detected by using SDS gel-electrophoresis and silver staining. The micro homogenizer chip is useful for pretreatment of various biochemical and biological analyses in microscale.

Keywords: micro homogenizer chip, biochemical IC chip, microstereolithography

1. Introduction

In general µ-TAS and lab-on-a-chips, several kinds of analytical apparatus such as electrophoresis and chromatography have been successfully miniaturized. However, few microdevices for pretreatment such as homogenization, fraction and filtration have been developed. We have been developing various types of biochemical IC chip to perform total chemical and biochemical processes including even pretreatment as well as analysis and fluidic control. We have already developed several types of biochemical IC chip such as a concentrator chip [2], reactor chip [3], optical detector chip, protein synthesis chip [4], micropump chip and switching valve chip[5].

In this paper, we developed a new micro homogenizer chip. This homogenizer chip utilizes ultrasonic cavitation to disrupt cells. Since ultrasonic cavitation is generated with a piezoelectric actuator, it is suitable for miniaturization unlike other methods such as blending and pressing.

2. Micro homogenizer chip

Figure 1 shows the schematic design of our homogenizer chip. A piezoelectric element and a micro exponential horn are integrated into a photocurable polymeric structure by using the hybrid IH process (microstereolithography) [2]. The chamber volume is 30 µl.

Y. Baba et al. (eds.), Micro Total Analysis Systems 2002, Volume 2, 745–747.

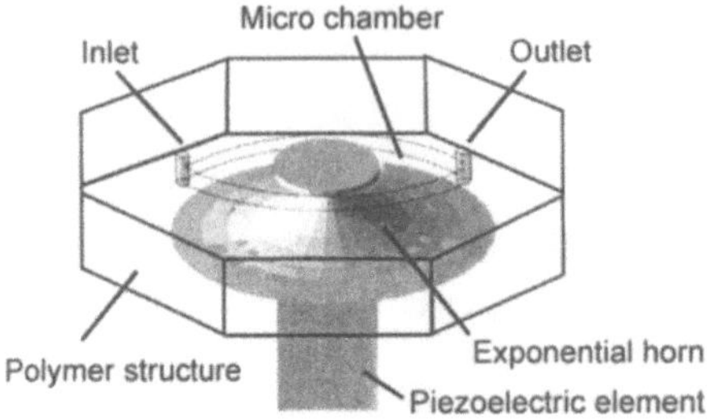

Fig. 1 Schematic design of micro-ultrasonic homogenizer chip.

Fig. 2 Prototype of micro-ultrasonic homogenizer chip.

By generating ultrasonic cavitation in the microchamber, cells are broken within a few seconds. A prototype of the homogenizer chip is shown in Figure 2. The cross-sectional shape of the chip is an octagon of 14 mm width, which is the current standard for biochemical IC chips. The exponential horn made from aluminum magnifies the amplitude of the oscillation of the piezoelectric element to generate ultrasonic cavitation.

3. Experimental Verification

Figure 3 shows an experimental setup to verify the basic performance of the homogenizer chip. In preliminary experiments, PC12 cells with HEPES buffer are introduced into the chamber with a micropipette. To generate ultrasonic cavitation, a rectangular wave of 300 VAC is applied within 3 seconds. We measured the dependence of homogenization efficiency on the driving frequency of the piezoelectric actuators. Figure 4 shows the experimental result of frequency dependence. This demonstrates that we can homogenize PC12 cells with 99.8% efficiency within 3 seconds at the optimal condition (frequency: 160 Hz, applied voltage: 300 VAC, cell concentration: 1x105 cells/ml). In this case, there were no harmful effects of heat generation on proteins by sonication. Typical photographs before and after homogenization are shown in Fig. 5. All of the cells were disrupted by sonication in the chip.

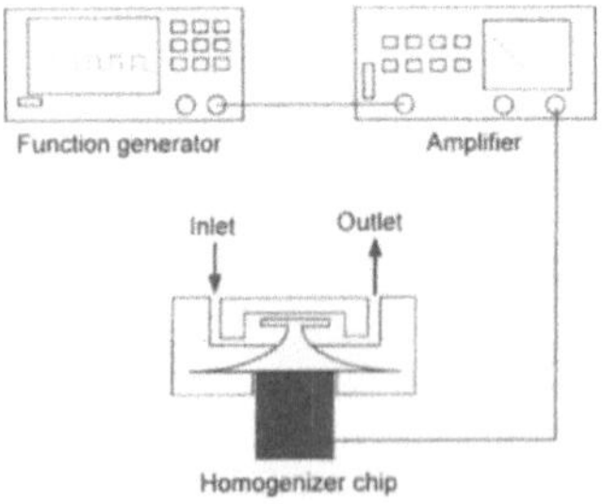

Fig. 3 Experimental setup for homogenization with the homogenizer chip.

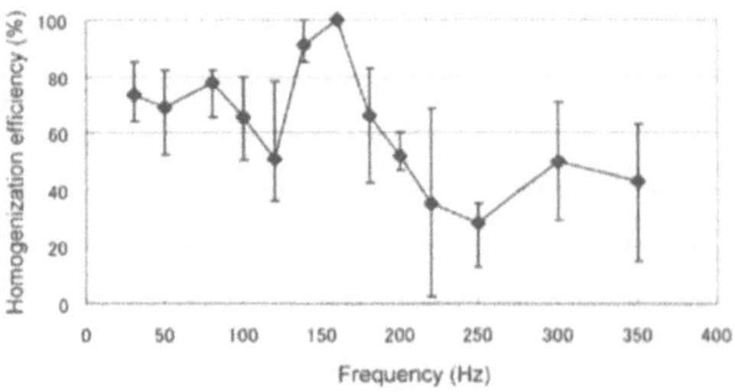

Fig. 4 Frequency dependence of homogenization efficiency.

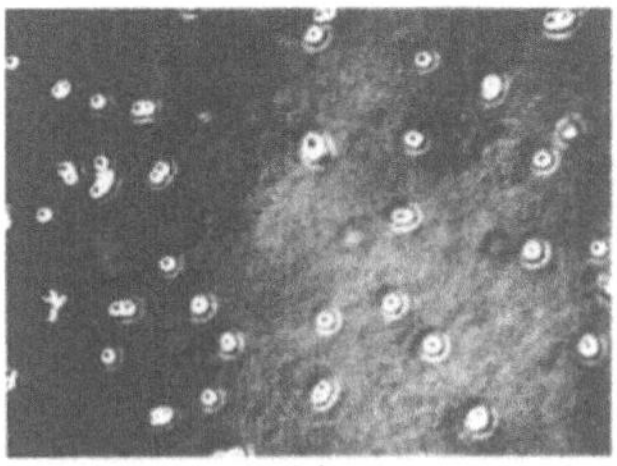

(a)

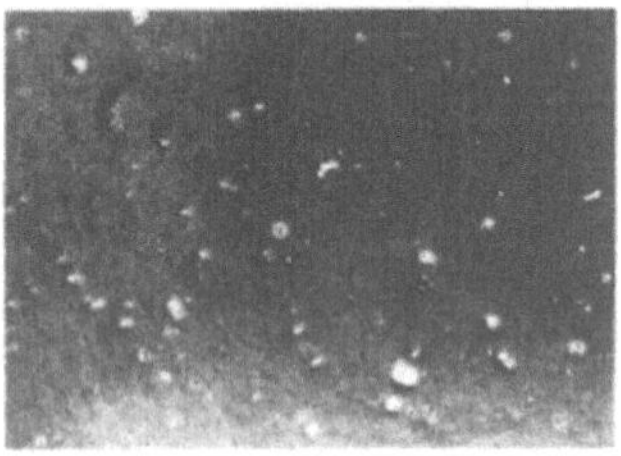

(b)

Fig. 5 Microscopic images of PC12 cells and homogenate. (a) Before homogenization, (b) After homogenization.

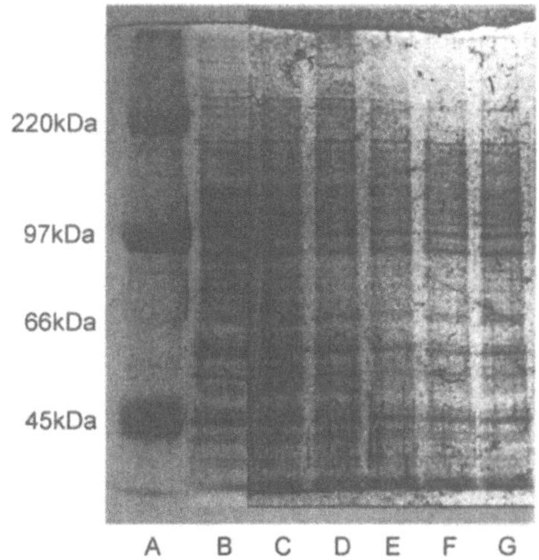

Fig. 6 Separation of proteins extracted from PC12 cells. A: Protein marker, B: Commercial homogenizer, C: Chip (30 Hz), D: Chip (50 Hz), E: Chip (80 Hz), F: Chip (138 Hz), G: Chip (160 Hz).

Finally we separated the homogenate of PC12 cells by SDS-gel electrophoresis. Figure 6 shows an experimental result of separation after silver staining. By using the homogenizer chip, we obtained a protein pattern similar to that of a commercially available large-size ultrasonic homogenizer (Microson XL2000). This result demonstrates that our homogenizer chip can disrupt cells without damaging proteins.

4. Conclusions

We have developed a micro homogenizer chip that can disrupt cells within a few seconds. The chip can be widely used for pretreatment of biochemical and biological applications. In the near future, our modular approach based on biochemical IC chips will enable a user-assembled, fully-integrated palm-top biochemical laboratory to be developed.

Acknowledgments

This work was partially supported by Grant-in-Aid for Scientific Research from the Ministry of Education, Culture, Science and Technology.

References

1. K. Ikuta, Artificial life V (MIT press), p. 17 (1997).
2. K. Ikuta, S. Maruo, T. Fujisawa and A. Yamada, Proc. of MEMS 99, p. 376 (1999).
3. K. Ikuta, S. Maruo, Y. Fukaya and T. Fujisawa, Proc. of MEMS 98, p. 131 (1998).
4. K. Ikuta, K. Ikeda, A. Takahashi and S. Maruo, Proc. of μ-TAS 2001, p. 66 (2001).
5. K. Ikuta, T. Hasegawa, T. Adachi and S. Maruo, Proc. of MEMS 2000, p. 739 (2000).

THE π-INJECTOR: A NEW ELECTROKINETIC DISPENSING ELEMENT FOR MICRODEVICE SEPARATIONS

Charles A. Emrich and Richard A. Mathies
Biophysics Graduate Group and Department of Chemistry,
University of California, Berkeley, CA 94720 USA

Abstract

We have developed a new electrokinetic injector for microfabricated separation devices that reduces the total number of reservoirs per channel from four to three while still allowing control of the plug size. Sample is introduced through two parallel channels, with the plug length determined by the distance between the channels. The π-injector is a straightforward solution to achieving increased channel densities in high-performance array devices.

Keywords: microfluidics, capillary electrophoresis, electrokinetic injection

1. Introduction

The high performance of microfabricated electrophoresis systems is due in large part to the extremely small and well-defined sample volumes that can be injected onto the separation column. These devices have traditionally used a cross injector to introduce the analyte in a direction orthogonal to the separation channel, and upon switching to the run configuration, injecting the sub-nL volume for analysis [1]. The cross injector has remained largely unimproved since its inception; however, with the development of microfabricated arrays to produce practical and useful microdevices [2,3], reducing the total number of reservoirs required for injection is crucial. The limiting features of a high-density array are the *macro*fabricated reservoirs, not the microfabricated channels. We recently showed that it is possible to create up to 384-lane systems by adopting a direct-inject format, though this gives resolution lower than the cross injector and introduces electrokinetic sample bias similar to conventional capillaries [4].

2. Theory

We present here a new design for a three-well cross injector, called the π-injector, with performance comparable to that of a standard four-well cross injector. A schematic of the π-injector is presented in Figure 1, showing two parallel channels leading from the sample reservoir to the main separation channel. A finite distance, *d*, separates these sample-loading channels and defines the final plug length. Operation of the π-injector, shown in Figure 2, occurs in three phases: i) injection, ii) excision, and iii) electrophoretic separation. During injection, analyte is moved electrokinetically from the sample reservoir (S) towards the cathode (C) and anode (A) reservoirs as presented

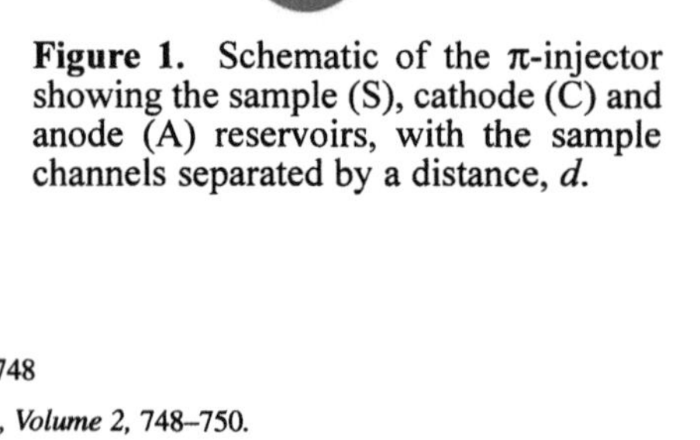

Figure 1. Schematic of the π-injector showing the sample (S), cathode (C) and anode (A) reservoirs, with the sample channels separated by a distance, *d*.

Y. Baba et al. (eds.), Micro Total Analysis Systems 2002, Volume 2, 748–750.

in Figure 2A. When the electric fields E_L and E_R are equal, the channel along d is equipotential, and electrokinetic transport will not occur into the volume, Δ, defined by the product of d and the area of the channel cross section. In this case, motion of the analyte into Δ occurs purely by diffusion. Diffusion will create a plug consisting of analytes of relative high electrophoretic mobility, μ, and particularly high diffusivity, D. A more homogenous plug may be created by performing the injection step at $E_L \neq E_R$, such that there exists some potential U_d across d that will drive analyte into Δ. Formation of the analyte plug in Δ is then accomplished by reversing the electric fields, such that the analyte in the channels is driven back to S (Fig. 2B). The final plug shape is determined by Δ, and excision is always performed with $E_L = E_R$. The electrophoretic separation is carried out in the standard fashion by applying an electric field between C and A (Fig. 2C).

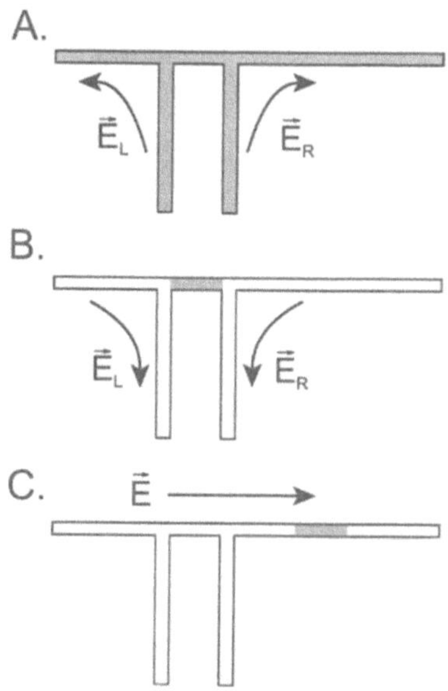

Figure 2. Schematic operation of the π-injector. A) injection, B) excision, C) electrophoretic separation. Electric field vectors are drawn assuming a positively charged analyte.

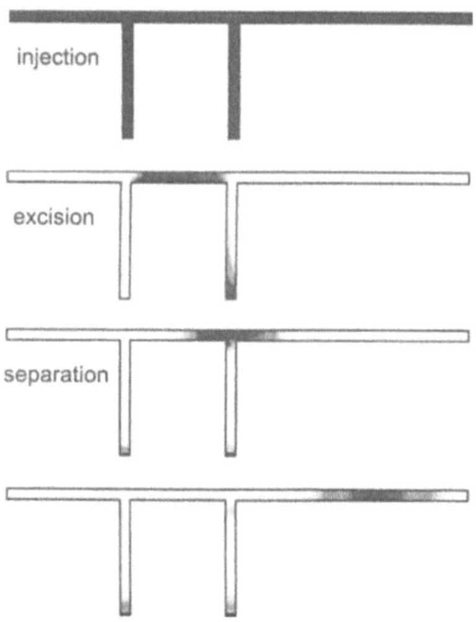

Figure 3. Finite element simulations with $E_L \neq E_R$ during injection, showing isolation of the excised plug followed by running the plug towards the anode (at right).

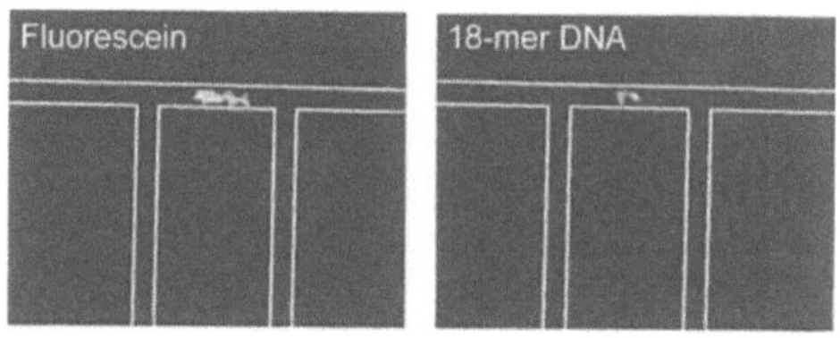

Figure 4. A diffusion-driven injection of fluorescein into a π-injector (d = 500 μm) showing the plug formed with a 15-s injection, 60-s diffusion, 45-s excision. C) A diffusion-driven injection of an 18-mer DNA primer labeled with the cyanine dye R110 using the same conditions. Runs were performed in 1% w/v HEC in 1X TBE excited at 488-nm. Stippling of fluorescence is caused by surface defects.

3. Results and Discussion

Finite element simulations of the π-injector with $E_L \neq E_R$ are presented in Figure 3; a plug is formed for a simulated 271-bp DNA fragment in a low-viscosity sieving matrix [5]. Diffusion-driven injections ($E_L = E_R$) of fluorescein dye give a higher plug intensity than an 18-mer DNA (Fig. 4), which was expected due to its higher diffusivity. Isolation of a plug of DNA is presented in Figure 5, accurately confirming results of the finite element simulation and demonstrating that the π-injector can be effectively used to define a narrow sample plug necessary for high-resolution separations.

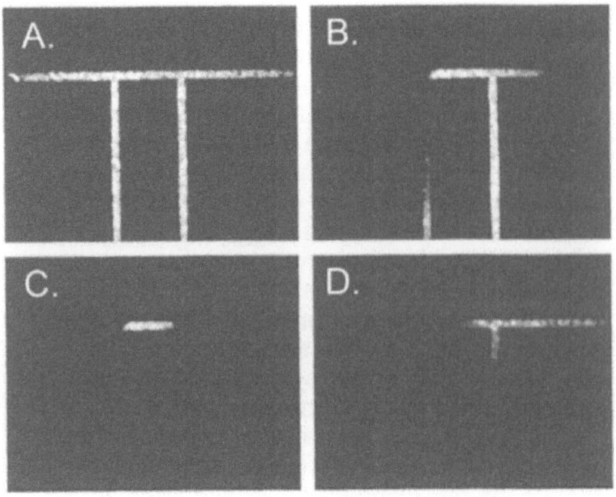

Figure 5. The π-injector A) filled with DNA with $E_L \neq E_R$ and B) excised leaving C) the isolated plug in Δ. The plug is then D) run down the column. DNA is an 18-mer labeled with R110. Run conditions correspond to simulation conditions as presented in Figure 3.

4. Conclusion

The high-quality separation permitted by the π-injector will enable the creation of higher density arrays by eliminating waste access wells and reducing the amount of electronic instrumentation required to drive the circuit. The injector's unique sensitivity to diffusivity could also be used as an orthogonal separation dimension to remove selectively the low-diffusivity, high-molecular weight contaminants often found in cell lysates and other biological samples.

References:

1. Harrison, D. J., Fluri, K., Seiler, K., Fan, Z., Effenhauser, C. S., Manz, A. *Science,* **261**, 895-897, (1993).
2. Medintz, I. L., Berti, L., Emrich, C. A., Tom, J., Scherer, J. R., Mathies, R. A. *Clinical Chemistry,* **47**, 1614-1621, (2001).
3. Paegel, B. M., Emrich, C. A., Wedemayer, G. J., Scherer, J. R., Mathies, R. A. *PNAS* **99**, 574-9, (2002).
4. Emrich, C. A., Tian, H., Medintz, I. L., Mathies, R. A. *Analytical Chemistry, in press*, (2002).
5. Coventorware v. 2001.3.

MANIPULATION OF SUSPENDED PARTICLES IN A LAMINAR FLOW

Andreas Nilsson[1], Filip Petersson[1], Hans W. Persson[1], Henrik Jönsson[2] and Thomas Laurell[1]
[1]Department of Electrical Measurements, Lund Institute of Technology, P.O. Box 118, S-222 10 Lund, Sweden
[2]Department of Cardiothoracic Surgery, Lund University Hospital, Lund, Sweden

Keywords: particle separation, suspended particles, radiation force, ultrasound

Introduction
For some time the art of separating suspended particles by applicating an ultrasonic standing wave field [1-3] has been investigated. The radiation force generated by the ultrasonic field causes the particles to move into the pressure nodal planes of the standing wave (Fig. 1). The radiation force F_r according to eqn.1 is proportional the acoustic pressure amplitude P_0, and the particle volume V_C and reciprocally proportional to the acoustic wavelenght in the fluid λ.

$$F_r = -(\frac{\pi P_0^2 V_c \beta_w}{2\lambda}) \cdot (\frac{5\rho_c - 2\rho_w}{2\rho_c + \rho_w} - \frac{\beta_c}{\beta_w}) \cdot \sin(\frac{4\pi z}{\lambda}) \qquad (1)$$

β_w and β_c are the compressibility for the fluid and the particle respectivly as are the densities ρ_w and ρ_c. The variable z is the distance between the particle and the pressure node. Previously reported work has mainly described and utilized this phenomena in macroscale fluidic systems.

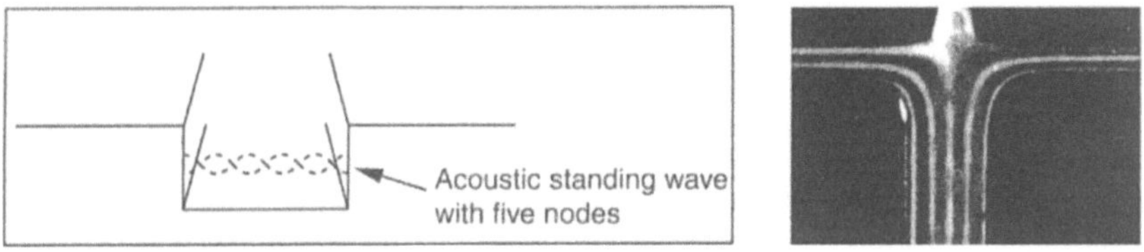

Figure 1. The radiation force causes the particles to gather in the nodes of the standing wave. Multiple bands of particle enrichment forms when a laminar flow is exposed to an ultrasonic standing wave.

This work reports acoustic particle separation in micro fluidic channels. As the trapping phenomena is utilized in a micro channel, the acoustic frequency has to be raised and thereby the trapping force increases proportionally. This in combination with the extremely laminar flow in the microdomain provides very stable streaming bands of particles, which according to the standing wave pattern positions the enriched particles at different lateral positions in the flow channel. By introducing a flow splitter at the end of the micro channel the particles can be separated from the rest of the fluid. Different designs of the microfluidic particle separation channel have been evaluated, with different channel widths and different orientations of the separations legs.

Y. Baba et al. (eds.), Micro Total Analysis Systems 2002, Volume 2, 751–753.

Materials and methods

The microfluidic structures were fabricated in <100>-silicon by means of double sided anisotropic wet etching. The etched structures were sealed on the channel side by a glass lid adheased by anodic bonding. To the inlets and outlets at the rear side of the chip, silicone rubber tubing was glued. The fluid used was water containing polystyrene particles of approximately 10μm diameter with a particle concentration of 2%.

Results and discussion

A cross-type design was initially made with the side channels 90 degrees perpendicular to the main input channel. The channel walls were defined by slowly etching <110>-planes, vertical to the wafer surface. Thereby an acoustic resonance cavity is obtained between the vertical channel walls (Fig. 2).

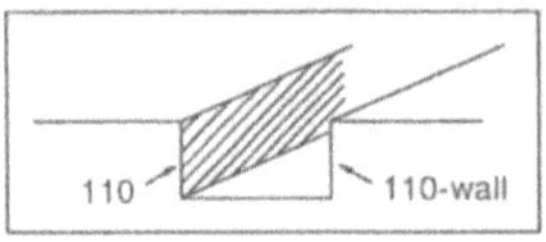

Figure 2. Microfluidic channel with vertical walls obtained by anisotropic wet etching. With a glass lid applied a resonance cavity is obtained between the walls.

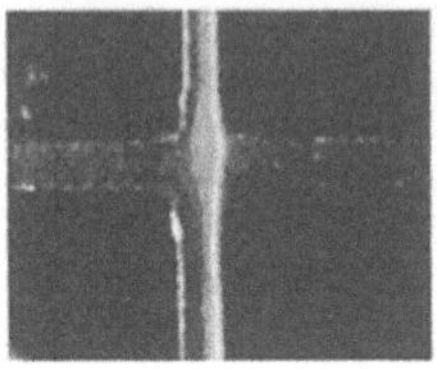

Figure 3a. A cross-type structure with a stading wave having one pressure node and particle exit in the central exit channel

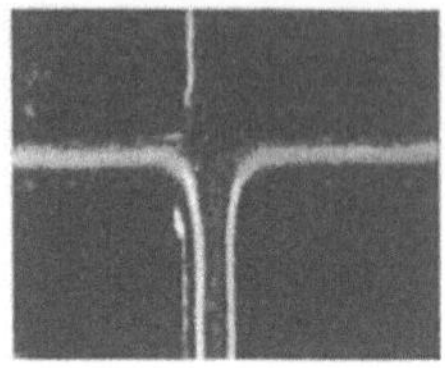

Figure 3b. A standing wave condition giving two pressure nodes and thus a particle exit at the side channels

When applying ultrasound to the suspended particles, a separation ratio of 90% was achived. Depending on where in the channel the enriched particles were located, the exit way of the particle flow became different. A single node standing wave caused the particles to gather in the middle of the channel and thus taking the central outlet (Fig. 3a). If a double node standing wave was applied the bands of concentrated particles was displaced towards the channel walls and leaving the chamber through the side outlets (Fig. 3b). The angle of the side channels was redesigned to 45 degrees (Fig. 4) and a separation ratio of 95% was achieved. Another advantage of changing the side angles to 45 degrees was that the stagnant zone as the fluid turned around the 90 degree corner was eliminated. This also reduced the bubble formation at these locations, as a more homogenuos flow was achieved across the channel.

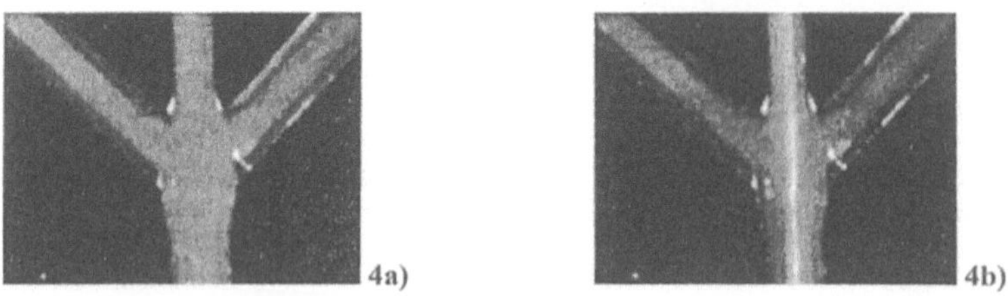

Figure 4a. Suspended particles in a 45 degree structure with no ultrasound applied. **b)** A 45 degree structure with a single band.

Since the 45 degree model achieved the highest separation ratio, a parallell stucture with eight separation channels were constructed (Fig. 5). With this new particle separator a throughput of 1ml per minute with a separation ratio of 95% was achieved. Since no negative effect was observed with the parallell structure, one can assume that it will be possible to link even more structures together to achieve an even higher throughput.

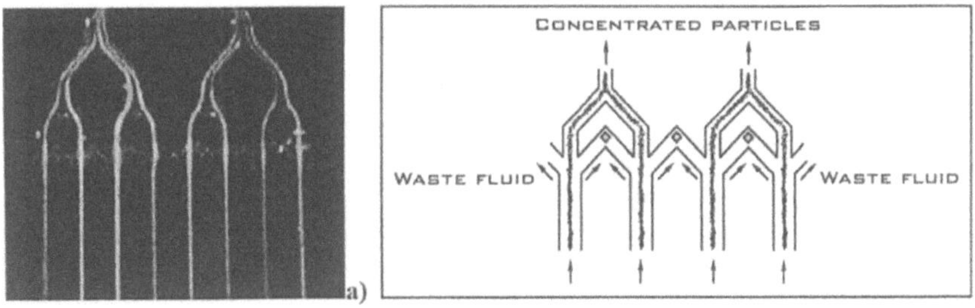

a) b)

Figure 5. **(a)** An eight channel multistructure demonstrating the posibility to scale up the system throughput without affecting the separation efficiency. **(b)** Schematics of the flow principle.

References

1. J.J. Hawkes, W.T. Coakley, A continuous flow ultrasonic cell-filtering method, Enzyme and Microbial Tech. 19 (1996) 57-62

2. M. Grötschl, Ultrasonic separation of suspended particles – Part II: Design and operation of separation devices, Acustica 84 (1998) 632-642

3. K. Yasuda, S-I. Umemura, K. Takeda, Concentration and fractionation of small particles in liquid by ultrasound, Jpn. J. Appl. Phys. Vol. 34 (1995) 2715-2720

POLYMER-BASED ACTUATORS INTEGRATED INTO MICROFLUIDIC SYSTEMS

Nikolas Chronis and Luke P. Lee
Berkeley Sensor and Actuator Center
Department of Bioengineering, University of California at Berkeley

Abstract
An in-plane polymer-based microactuator integrated into microfluidic systems is presented. Actuation is provided by thermal expansion of a polymer (SU-8) through resistive heating of a thin gold layer attached selectively at the bottom of the suspended polymer structure. The actuator can be easily integrated to any fluidic configuration, providing possibly a general tool for controlling and handling fluids and biological samples in microdevices.

Keywords: integrated in-plane polymer actuator, SU-8, chevron-type actuator

1. Introduction

The development of actuation mechanisms inside microfluidic environments is of major importance for advanced micro total analysis systems (μTAS), since such a technology can be used for a variety of applications such as fluid pumping, mixing, dispensing, handling or even mechanically lysing and separating cells as well as single cell manipulation for cellular bioassays and biophysical experiments.

Up to now, electrolysis-based actuators [1] and conductive polymers [2] have been suggested as possible solutions. The generation of a mechanical force in both cases is achieved through conversion of electrical to mechanical energy via a pathway that its biocompatibility is questionable. Alternatively, isolating the actuator [3] is a rather expensive solution that requires several fabrication steps.

Using a non-conductive photo-patternable polymer (SU-8) as a structural material, we fabricated a thermal microactuator that can operate – but not limited - in fluidic physiological environments with minimum interaction (Figure 1).

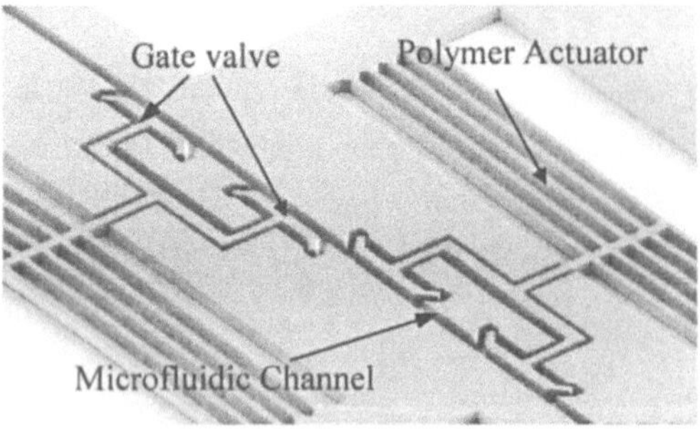

Figure 1. SEM of the microactuator in a two valve-type microfluidic system

Actuation is provided by thermal expansion of the SU-8 through resistive heating of a thin gold layer attached selectively at the bottom of the suspended polymer structure. The

Y. Baba et al. (eds.), Micro Total Analysis Systems 2002, Volume 2, 754–756.

bimorph SU8/gold structure moves in plane rather than out of plane when certain design rules are followed. Key point is the high thermal coefficient of expansion of SU-8 – 20 times higher than silicon- enabling the generation of the desirable displacement or mechanical force with small temperature increases at low voltages. Furthermore, the ability to activate the polymer actuator using alternating current (AC) – although not necessary -instead of direct current (DC) ensures that electrolysis and electrophoretic forces are absent.

2. Design and Fabrication Process

The microactuator consists of a V-shaped beam that is anchored through its ends to the substrate, allowing the apex to expand outwards upon thermal expansion of the polymer (Figure 2). Identical actuators are linked in series to increase the maximum output force. Similar actuators made of polysilicon already exist in the MEMS world and are known as chevron-type actuators.

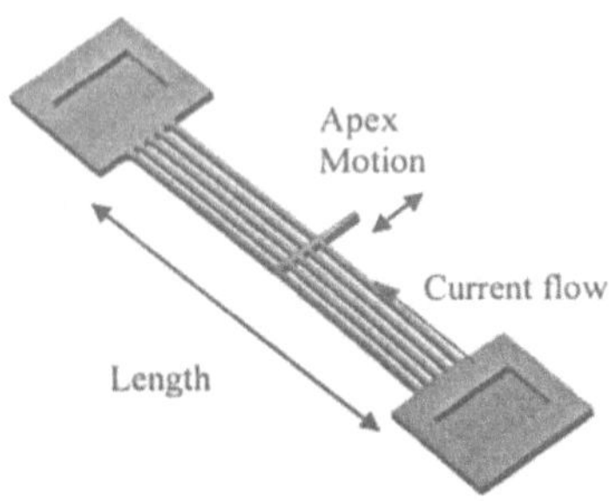

Figure 2. The chevron-type actuator schematically

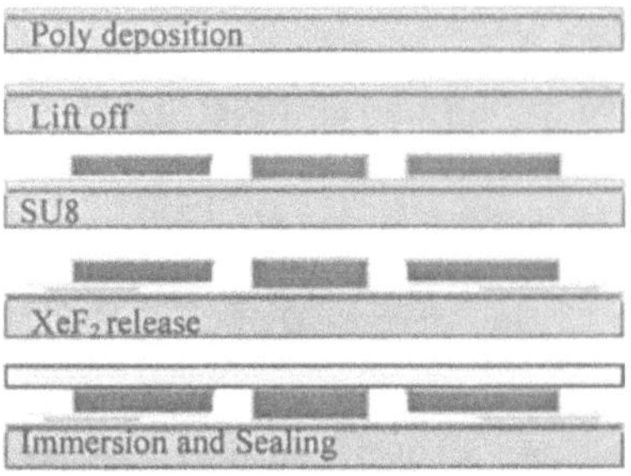

Figure 3. The process of integrating polymer microactuators in microfluidic systems

The fabrication is completed in a two-mask process (Figure 3). Starting with an oxidized silicon wafer, a layer of 4 μm undoped polysilicon, which serves as a sacrificial material, is chemically vapor deposited. Lift-off process is subsequently used to define the first thermally evaporated chrome/gold layer. The SU-8 is spun on the wafer and photolithographically patterned to form the structural layer. The structures are finally dry etch-released in xenon difluoride. The actuator is immersed in the solution and a cover glass slide is mechanically attached at the top of the structure.

3. Simulations and Experimental Results

We used ANSYS to estimate the displacement of the tip of the actuator for a given temperature elevation (Figure 4). Displacement of 15 μm is possible by increasing the temperature only 20 degrees for an actuator of total length of 1000μm. The temperature at the point where the actuator comes in contact with the fluid is expected to be much lower though.

The power consumption of similar microfabricated actuators having two V-shaped beams –operating in air- was experimentally measured versus displacement (Figure 5). The dependence is linear, while the shorter actuator needs less power than the longer one for the same amount of displacement.

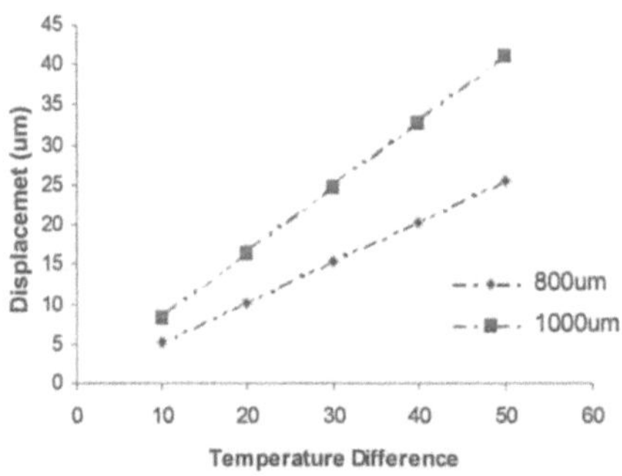

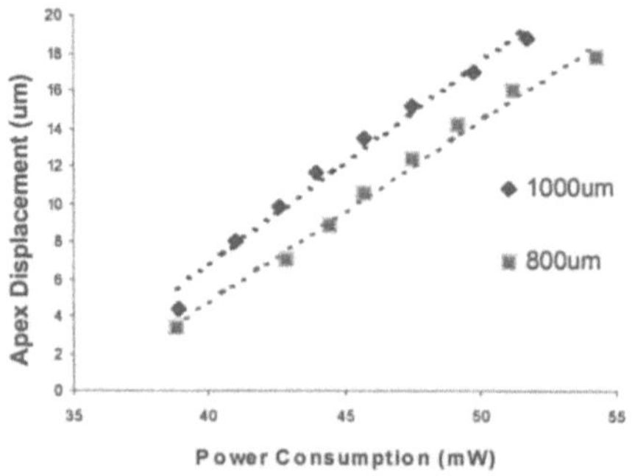

Figure 4. Displacement of the tip of the actuator at a given temperature elevation

Figure 5. Experimental data of two different lengths of actuators (800μm and 1000μm)

Figure 6 shows a valve-type configuration of the proposed microactuator. The gate valve is shown in open and closed position. The whole structure is immersed in a buffer solution containing cells.

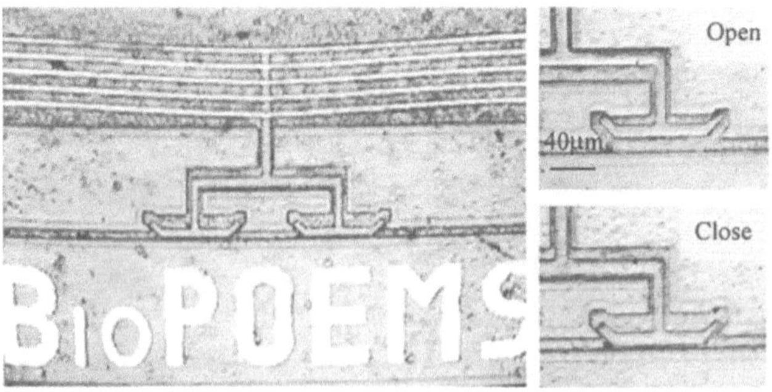

Figure 6. The actuator in action in a valve-type configuration driven at 1.2V at 400mA

4. Conclusions

A photo-patternable polymer-based thermal actuator that can operate in various physiological solutions with minimum interaction was developed for advanced microfluidic systems. Based on a two-mask process, it can be simply integrated to any fluidic network, providing an alternative actuation mechanism for fluid handling and single cell manipulation.

Acknowledgements

This work is fully supported by a DARPA grant under the BioFlips program.

References

1. A.Papavasilliou, D. Liepmann, and A. Pisano, Hilton Head Island, South Carolina, June 4-8, 2000.
2. E.Jager, O. Ingavas, I. Lundstrom, *Science,* Vol.288, June 2000.
3. M.Okandan, P.Galambos,S.Mani, J.Jakubczak, μTAS, Monterey, CA, 2001.

BARRIER EMBEDDED CHAOTIC MICROMIXER

Dong Sung Kim, Seok Woo Lee, Tai Hun Kwon[1] and Seung S. Lee
[1]Pohang University of Science and Technology, San 31 Hyoja-dong Nam-gu, Pohang, 790-784 Korea
(E-mail: thkwon@postech.ac.kr)

Abstract

In this paper, we present a *new* chaotic passive micromixer named *Barrier Embedded Micromixer (BEM). Chaotic mixing* in the microchannel has been *successfully* achieved by introducing periodic barriers on one microchannel wall which cause periodic perturbation in the velocity field of the helical flow.

Keywords: Barrier, Chaotic mixing, Hyperbolic point, and Periodic perturbation

1. Introduction

Mixing plays a very important role in micro-total-analysis-system [1]. Because of low Reynolds number (*Re*) of the flow in the microchannel, it is impossible to expect turbulent mixing of two fluids. So it is desirable to improve the mixing efficiency of micromixer by introducing a chaotic mixing mechanism. Chaotic mixing in Stokes flow regime can remarkably enhance the efficiency of mixing. In this regard, we developed a chaotic passive micromixer based on the helical flow perturbed by periodic barriers in the microchannel.

Ajdari and Stroock et al. [2,3] analytically and experimentally showed that oblique ridges can give rise to 3-dimensional helical flow in the microchannel. In addition to the helical flow, Stroock et al. [3] obtained the chaotic mixing by introducing an alternating velocity field by staggered herringbone structures on one microchannel wall [4].

2. Theory

In the case of just the helical flow, there exists one elliptic point in the cross-sectional flow, as schematically indicated in Fig. 1(a). With barriers on the surface, the helical flow in the barrier zone changes to have one hyperbolic point and two elliptic points in the cross-sectional flow as indicated in Fig. 1(b). Periodically alternating change of flow fields of Fig. 1(a) and Fig. 1(b) results in chaotic mixing with relatively low pressure loss to achieve the level of mixing efficiency of the BEM, which was proved by our group [5].

3. Experimental

In our chaotic micromixer design, we made slanted grooves on both upper and lower microchannel walls in order to enhance the helical flow as shown in Fig. 2. Width, thickness and groove thickness of the microchannel are 240μm, 60μm and 10μm,

Y. Baba et al. (eds.), Micro Total Analysis Systems 2002, Volume 2, 757–759.

respectively. Fig. 3 shows a BEM developed in this study. Barriers are 30μm wide, 40μm high and 600μm long. Micromixers are made by PDMS from SU-8 masters which are fabricated by conventional photolithography.

Fig. 4 shows schematic of mixing visualization experimental setup. Two fluid flows containing different dyes (blue (upper stream) and yellow (lower stream)) were driven by a syringe pump in the constant flow rate. In this experiment, *Re* was approximately maintained as 25.

4. Results and discussion

Figs. 5 and 6 show visualization experimental results without and with the periodic barriers, respectively. Fig. 5 clearly indicates the helical flow induced by slanted grooves in the microchannel. As far as the mixing efficiency is concerned, BEM achieves obvious mixing enhancement in comparison with just the helical flow case, as shown in Figs. 5(d) and 6(d) near the exit zones. It is because the hyperbolic point causes extreme stretching of material line and the periodicity of elliptic and hyperbolic points results in effective stretching and folding. This mechanism induces exponential interfacial area growth, consequently resulting in an efficient chaotic mixing in a passive micromixer.

5. Conclusions

In this paper, a *new chaotic* passive micromixer named *Barrier Embedded Micromixer* with successful experimental results is presented.

Acknowledgements

The authors would like to thank Korean Ministry of Science and Technology for the financial supports via the National Research Laboratory Program (2000-N-NL-01-C-148).

References

1. C.-C. Hong, et al., *A Novel In-Plane Passive Micromixer Using Coanda Effect*, Proc. Micro Total Analysis System 2001, pp. 31-33, (2001).
2. A. Ajdari, *Transverse electrokinetic and microfluidics effects in micropatterned channels: Lubrication analysis for slab geometries*, Phys. Rev. E, Vol. 65, 016301, (2002).
3. A.D. Stroock, et al., *Chaotic Mixer for Microchannels*, Science, Vol. 295, pp. 647-651, (2002).
4. J.M. Ottino, *The Kinematics of Mixing: Stretching, Chaos, and Transport*, Cambridge Univ. Press, Cambridge, (1989).
5. S.J. Kim and T. H. Kwon, *Enhancement of mixing performance of single screw extrusion processes via chaotic flows*, Adv. Polym. Tech., Vol. 15, pp. 41-69, (1996).

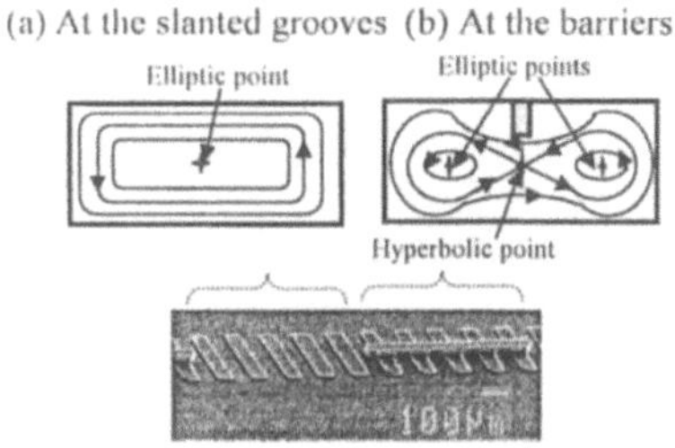

Figure 1. Cross-sectional flows: (a) with one elliptic point and (b) with two elliptic points and one hyperbolic point

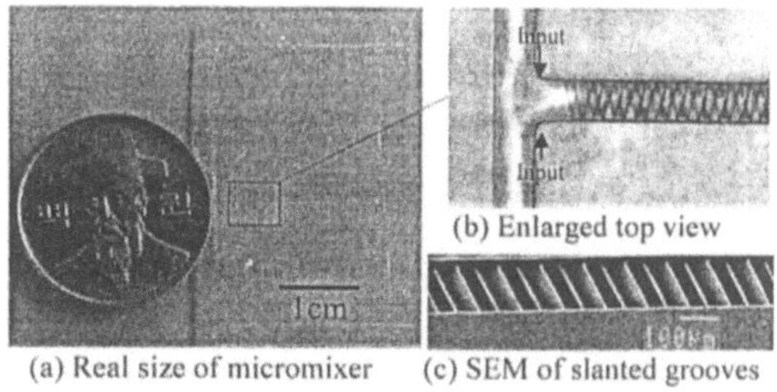

Figure 2. Microchannel containing slanted grooves: (a) and (b) top views of the microchannel and (c) SEM picture of slanted grooves at the bottom microchannel wall

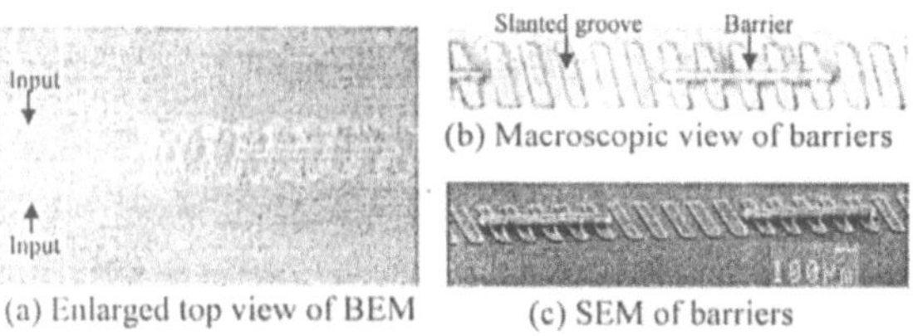

Figure 3. Barrier Embedded Micromixer (BEM): (a) and (b) top views of BEM and (c) SEM picture of barrier structures on slanted grooves at the upper microchannel wall.

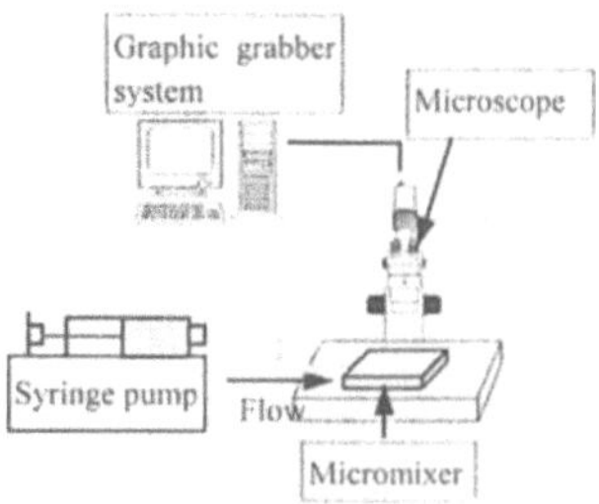

Figure 4. Schematic of experimental setup

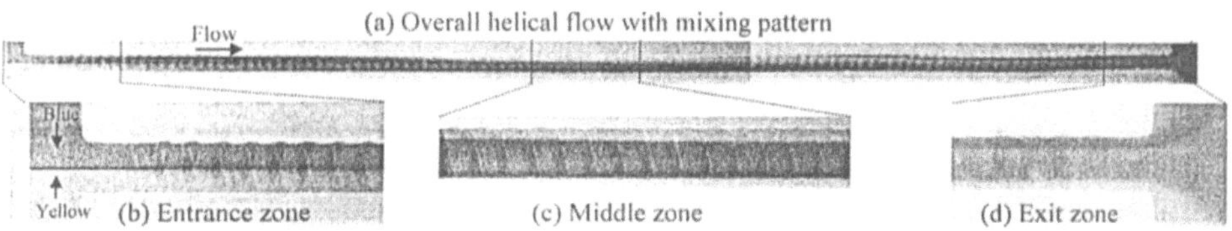

Figure 5. Mixing and helical flow behavior in the microchannel with just slanted grooves: (a) overall mixing pattern and (b)-(d) enlarged views of mixing patterns in three zones

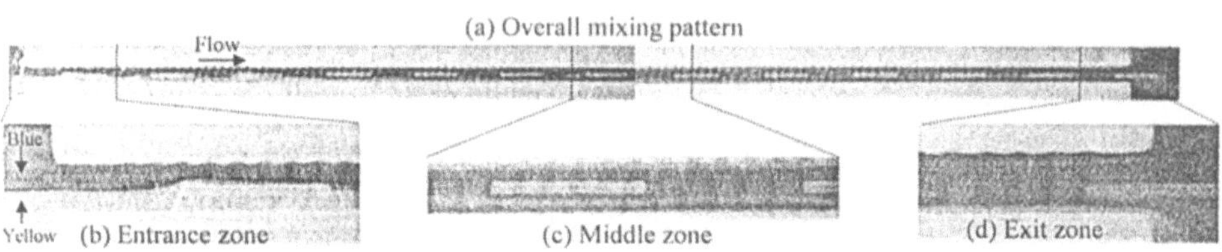

Figure 6. Chaotic mixing behavior in BEM: (a) overall mixing pattern and (b)-(d) enlarged views of mixing patterns in three zones

MAGNETICALLY ACTUATED COLLOIDAL CHAINS IN MICROCHANNELS

Sibani Lisa Biswal[1] and Alice P. Gast[2]
[1]Department of Chemical Engineering, Stanford University, Stanford, CA 94305
[2]Department of Chemical Engineering, MIT, Cambridge, MA 02139

Abstract

We report the use of chemically linked paramagnetic particles to manipulate microscopic flow in microchannels. In experiments with a rotating magnetic field linked chains show effective micromixing.

Keywords: paramagnetic particle, micromixing, rotating magnetic field

1. Introduction

Much of the current technology in uTAS relies on pressure-driven and electrokinetic flow control. In this work, we present the use of a magnetic field to control lab processes. The self-assembly of magnetic colloidal particles into permanent chains can provide a useful way to manipulate microscopic fluid flow. Once the chains have been linked chemically, they can be magnetically actuated to follow the magnetic field direction. We are designing a flow system that uses this unique mechanical property to perform micro-scale mixing and sieving.

2. Rotating Magnetic Field Theory

Several groups have studied the dynamics of paramagnetic particles in a rotating magnetic field. These particles come together to form rotating chains. However, at certain frequencies and lengths, these particles break apart to minimize their viscous drag. A linked chain is prevented from breaking up. Therefore, it will bend and flex to minimize its viscous drag.

We characterize the dynamics of this system by modeling the chains as rotating cylinders. There are two opposing torques dictating the conformation of the chain: a magnetic torque, T_m, and a viscous torque, T_v. The magnetic torque arises when the chain axis is not parallel to the magnetic field, H. The rotational friction of the chain through the surrounding fluid leads to viscous torque. This torque increases with increasing rotational speed.

The theory of kinetic momentum for a cylinder states:

$$I\frac{d^2\theta}{dt^2} = T_m + T_v \tag{1}$$

where I is the moment of inertia and θ is the angle between the direction of the external field and the orientation of the chain. Under stationary conditions, the two torques cancel and the chain rotates with constant angular velocity.

Y. Baba et al. (eds.), Micro Total Analysis Systems 2002, Volume 2, 760–762.

3. Experimental

We study two different means to link the paramagnetic colloids. The particles we use are .85µm diameter polystyrene lattices embedded with 10nm magnetic monodomains. The microspheres are coated with a tetrameric protein, streptavidin. The first linker we use is bis-biotin-polyethylene glycol (MW=735). The suspension is placed into a PDMS microfluidic channel 50µm in depth and 200µm wide. When the particles are placed in a magnetic field, the magnetic domains reorient, inducing a dipole in the particle to form chains. At this point the linker solution is added into the microchannel. The high affinity of biotin for streptavidin results in crosslinking of the particles. The polyethylene glycol chain length can be changed to provide an adjustable spacer length to the linker. The second linking agent used in these experiments is gluteraldhyde which binds to the carboxyl groups on the surface of the microspheres. This linking reaction results in a much stiffer chain that cannot bend or flex as easily as the PEG-linked chains.

In our experiment, we create magnetic fields in two perpendicular directions with two orthogonal pairs of coils to create magnetic fields. A multi-frequency synthesizer supplies AC current through the coils. The currents are in a sine wave shifted by 90° relative to each other. The coils consist of 0.4µm copper wire coiled around ferrite cores. With this configuration (Figure 1), we can achieve a magnetic field on the order of 10 kA/m. Observations were made using a 40x objective (Nikon). Images were recorded with a Hamamatsu CCD camera and digitized using SimplePCI.

Figure 1. Rotating magnetic field setup

Using a PDMS-glass fluidic chip, we create a mixing junction at the intersection of four imput channels (Figure 2). The chains are pumped into the junction from ports 1 and 4. Two different solutions are inserted in ports 2 and 3. Then, the rotating magnetic field is turned on to induce mixing of these two solutions.

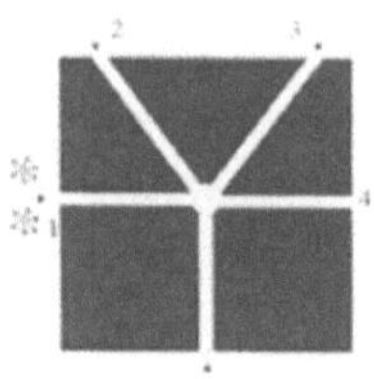

Figure 2: Micromixer Chip Layout

We functionalize the surface through silane chemisty. The glass side of the microchannel is placed in a silanizing solution: 2% [3-aminopropyl]triethozysilane

(ATES) in acetone for 5 minutes. Chains covalently attach to the functionalized regions of the glass channel (Figure 3).

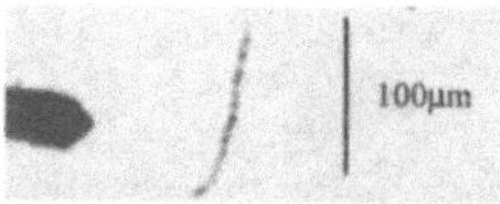

Figure 3: Chain attached to a surface

Chains attached to the surface bends in each direction of the field are unable to pivot. Patterns of silane patches can be created using a PDMS stamp. This substrate is then used to anchor the chains to create a device with patterns of chains.

4. Results and Discussion

The design of an effective mixer using linked chains requires an understanding of the properties of the chains. We have studied the flexability of these chains by micromechanics experiments and analysis of their thermal fluctuations [1]. For unlinked chains, alters the rotational dynamics. Sandre, *et al* [2], showed that for paramagnetic particles, a critical frequency exists at which the chains become unstable and break up. By linking the paramagnetic particles, the chains are prevented from breaking up. This can then result in interesting morphologies that depend on the chain length and frequency of the field. The following figure (Figure 4) shows typical morphologies observed with rotating chains of various lengths.

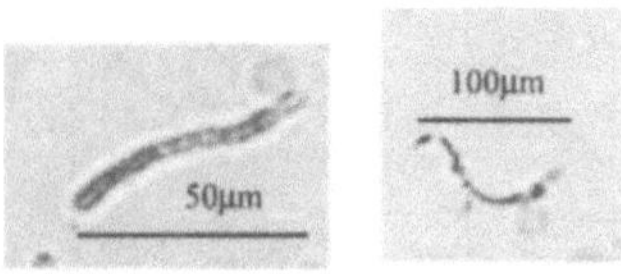

Figure 4: With Increasing Chain Length, there is an increase in flexibility

Shorter chains remain in a rod-like configuration as they rotate. However, longer chains begin to bend at the ends resulting in a S-shape configuration. Larger viscous drag creates a greater phase lag between the magnetic moment of the longer chains and the external magnetic field, leading to instabilities.

5. Conclusions

We have shown that permanently linked paramagnetic chains can be used as an effective tool in manipulating fluid flow. These structures display interesting dynamic properties.

References

1. Biswal, S and Gast, A.P., in preparation (2002).
2. Sandre, O., *et al* (1999). Assembly of microscopic highly magnetic droplets: Magnetic alignment versus viscous drag. *Physical Review E*. 59, 1736-1746.

SIZE EXCLUSION CHROMATOGRAPHY WITH PATTERNED NANO-PILLAR ARRAY

Masakazu Baba, Tohru Sano, Noriyuki Iguchi, Kazuhiro Iida, Toshitsugu Sakamoto, and Hisao Kawaura
Fundamental Research Laboratories, NEC Corporation
34 Miyukigaoka, Tsukuba, Ibaraki 305-8501, Japan
(e-mail: m-baba@aj.jp.nec.com)

Abstract

We report a new approach to performing biomolecule separation with the nano-fabricated channel. The channel has the alternating regions with space and stripe patterned with nano-pillar array. We can demonstrate the DNA molecule separation in the type of the size-exclusion chromatography for the first time. High resolution and multiple functions for biomolecule separation should be achievable with this approach because of the high controllability of pillar shape and the high designability of pillar pattern.

Keywords: Size exclusion chromatography, nanofabrication, pillar, electron-beam lithography, DNA separation

1. Introduction

There is wide interest in applying nanotechnology to biotechnology. Nanofabricated fluidics promise many advantages of high resolution and various functions for biomolecule separation because of its controllability of size and design [1-4]. Recently, some reports have shown DNA sieving with a pillar array [1-3] being used as a substitute for polymer gel in electrophoresis. In "sieve" type (S-type) separation, which relies on topological restrictions in a pillar array, small molecules move faster than large ones. Here we propose a different type of separation. We demonstrate the "size exclusion chromatography" type (SEC-type) separation of DNA molecules using a patterned nano-pillar array for the first time. Figure 1 describes the concept. The channel has alternating regions of stripes forming a pillar array and spaces. Large molecules (larger than the gap between pillars) cannot enter the stripe, but can move smoothly in the space. In contrast, small molecules (smaller than the gap) moved smoothly in the space and slowly in the stripe. Consequently, the larger molecules move much faster than the smaller ones, on average. This separation is the same as with conventional SEC. We fabricated the Si-based nano-structured

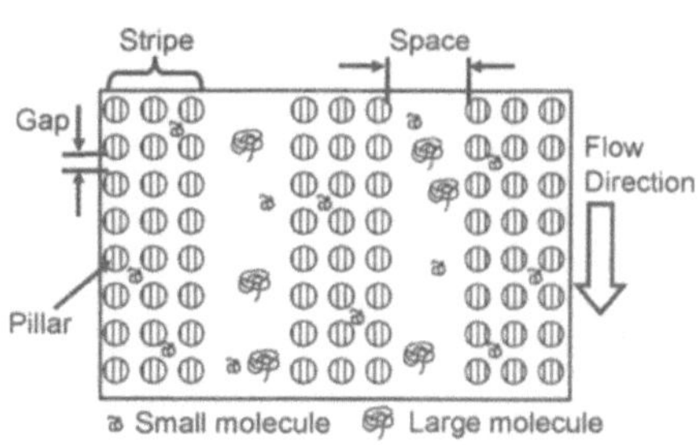

Figure 1. Principle of SEC-type separation with the stripe structure.

Y. Baba et al. (eds.), Micro Total Analysis Systems 2002, Volume 2, 763–765.

devices and demonstrated SEC-type DNA separation.

2. Experimental

Channels 80 μm wide and 360 nm deep were made on Si wafer by dry etching from a nano-pattern of the calixarene resist [5] fabricated by electron-beam lithography. For this demonstration, 5- and 10-kbp DNA molecules were observed simultaneously with YOYO-1 fluorescence dye and their motion was recorded as a video. An electrical potential of 0-80 V was applied across the 42-mm-long channel. The DNA velocities were measured by recording the time required for individual molecules to traverse an 80-μm section.

3. Results and discussion

We show the SEC-type separation with the stripe structures. In figure 2(a), we constructed the stripe with a Si nano-pillar array with a 350-nm pitch. Stripes were separated by a 1000-nm space. Figure 2(b) shows the velocity distribution of 100 DNA molecules of each size at 60 V. The 5-kbp DNA molecules were clearly slower than the 10-kbp ones. The video clearly revealed that 10-kbp DNA molecules moved only in the space, while the 5-kbp ones moved through both the stripe and space. This type of patterned nano-pillar array can thus act as an SEC.

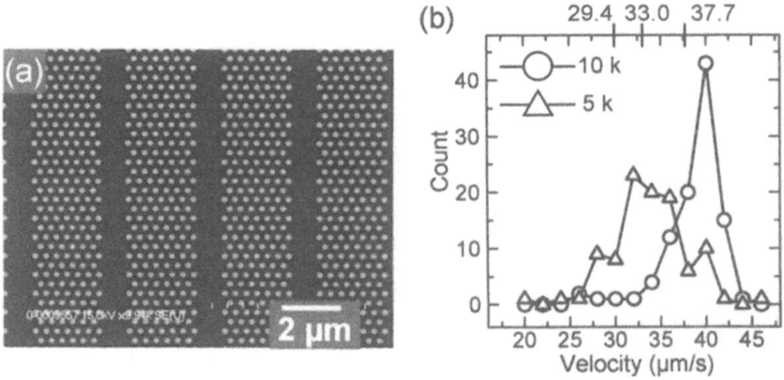

Figure 2. Stripe structure for SEC-type separation. (a) Scanning electron micrograph of the stripe structure with 350-nm pitch and 1000-nm space. (b) Velocity distribution of 100 DNA molecules with sizes of 5 and 10 kbp for applied voltage of 60 V. Average velocities were 37.7 and 33.0 μm/s for 10- and 5-kbp molecules, respectively. Except in the patterned region, the average velocities were the same (29.4 μm/s).

To determine the effect of the pillar pattern on separation, we also studied the S-type separation with simple array structures. As shown in figure 3(a), we constructed a square pillar array with a pitch of 250 nm. The pillars were 150 nm in diameter and separated by 100 nm in a square array. Figure 3(b) shows the velocity distribution of 100 DNA molecules of each size at 60 V. The 5-kbp molecules were faster than the 10-

kbp ones. Thus, S-type separation can be done with a simple array. This result is completely different from that of SEC-type separation.

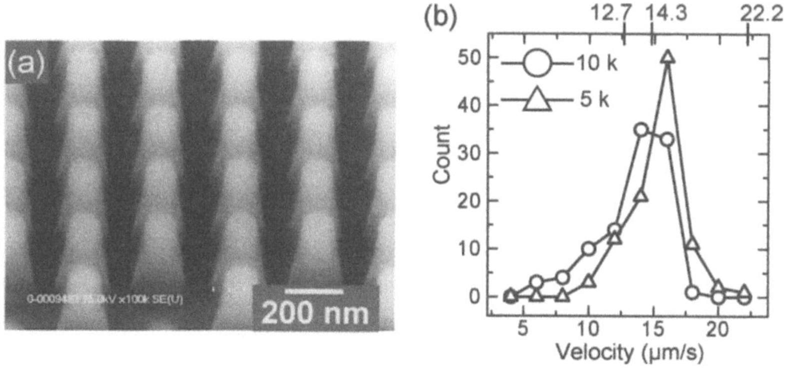

Figure 3. Pillar array structure for S-type separation. (a) Scanning electron micrograph of the pillar array with 250-nm pitch. (b) Velocity distribution of 100 DNA molecules with sizes of 5 and 10 kbp for applied voltage of 60 V. Average velocities were 12.7 and 14.3 μm/s for 10- and 5-kbp molecules, respectively. Except in the array region, average velocities were the same (22.2 μm/s).

4. Conclusion

We have demonstrated two types of the DNA separation with Si-based nano-structured devices. Size exclusion chromatography separation was done using a patterned nano-pillar array. The result was clearly different from that with S-type separation using a simple array structure. With both types of separation, 5- and 10-kbp DNA molecules were separated. The resolution was not so high because the DNA motions were observed in a section only 80 μm long and there was a little interaction between the DNA molecules and pillar array. The patterned nano-pillar array should be a good separation media for a micro fluidic system because the high controllability of the pillar size and shape produces high resolution and the high designability of the pillar pattern produces multiple functions for biomolecule separation. It is not limited to DNA analysis, but can also be applied to the separation of proteins, and cells by suitable adjustment of the structures of pillar pattern.

References

1. W. D. Volkmuth and R. H. Austin, Nature 358, 600 (1992).
2. S. W. Turner, A M. Perez, A. Lopez, and H. G. Craighead, J. Vac. Sci. Technol. B 16, 3835 (1998).
3. C.-Fu Chou, O. Bakajin, S. W. P. Turner, T. A. J. Duke, S. S. Chan, E. C. Cox, H. G. Craighead, and R. Austin, PNAS 96, 13762 (1999).
4. J. Han and H. G. Craighead, Science 288, 1026 (2000).
5. J. Fujita, Y. Ohnishi, Y. Ochiai, and S. Matsui, Appl. Phys. Lett. 68, 1297 (1996).

SIZE EXCLUSION CHROMATOGRAPHY USING SELF-ORGANIZED NANO-PORES IN ANODIC POROUS ALUMINA

Tohru Sano, Noriyuki Iguchi, Kazuhiro Iida, Hisao Kawaura, and Toshitsugu Sakamoto
NEC Corporation, Tsukuba, 305-8501 Japan
sano@ah.jp.nec.com

Abstract

We have proposed a new nano-biochip based on size exclusion chromatography (SEC), which uses anodic porous alumina. Since the SEC device does not require any filling material, such as polymer gel, it is highly integrable with other bioMEMS units for sequential bio/chemical reactions. With this device, we demonstrated separation of 165-kb from 10-kb DNA and confirmed that larger molecules move faster.

Keywords: anodic porous alumina, photopolymerizable resin, size exclusion chromatography, YOYO-1

1. Introduction

Methods for separating bio-molecules, such as chromatography, are key techniques in biotechnology. Artificial structures fabricated using nanotechnology are potentially much better separation carriers for chromatography than chemical compounds because we can design and control the shape and size of nanostructures. Here, we describe a nanostructured biochip that has nanoscale pores at the bottom of a microcapillary. This device works on the basis of SEC because small molecules easily run into the pores while large molecules are unlikely to find pores to enter, resulting the larger molecules moving more rapidly. We call these kind of chips patterned array chromatography (PAC). This separation mode differs from the capillary electrophoresis mode where smaller molecules move faster [1].

We have implemented this SEC-device concept in actual devices by using built-in anodic porous alumina with uniform nanoscale pores on its surface. Anodizing Al spontaneously forms nano-pores with diameters ranging from 5 nm (native protein scale) to 450 nm (large DNA scale) on the Al surface, and the diameter of the pores can be controlled (Figure 1). The anodic porous alumina ideally forms a honeycomb structure, and triangles and rectangles can be formed by using molds such as SiC [2].

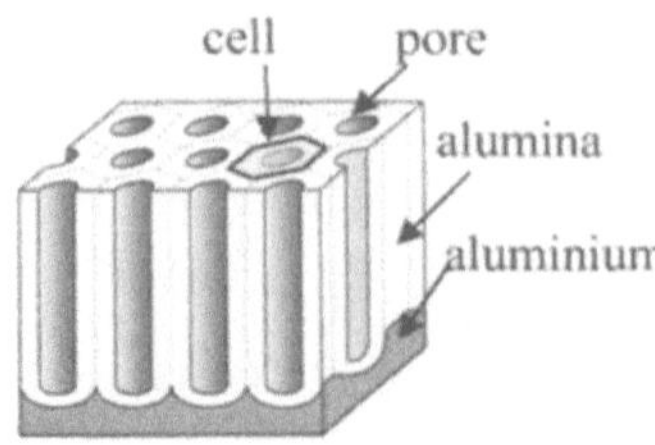

Figure 1. Structure of anodic porous alumina

Y. Baba et al. (eds.), Micro Total Analysis Systems 2002, Volume 2, 766–768.

2. Chip design and fabrication

We carried out a two-step-anodization. First, Al was anodized under a constant 120-V voltage in a 0.3-M H_3PO_4. Second, the anodic porous alumina was etched in 3-wt% H_3PO_4 at 30°C for 40 min. Anodizing the 500-nm-thick Al formed uniform pores with 120- to 150-nm diameters on the Al surface (Figure 2).

To prevent the sample solution from leaking, we spin-coated a photopolymerizable resin, Aronix UV-3750 (Toagosei Co.), onto the anodic porous alumina and applied direct patterning to the resin by applying UV irradiation through a mask. An 80-μm-wide and 40-mm-long channel was formed on the surface containing the nanoscale pores. The resin membrane was about 3 μm thick. After the polymerization, a glass cover with two holes was adhered to the resin at 100°C and a glass tube was attached at each hole (Figure 3).

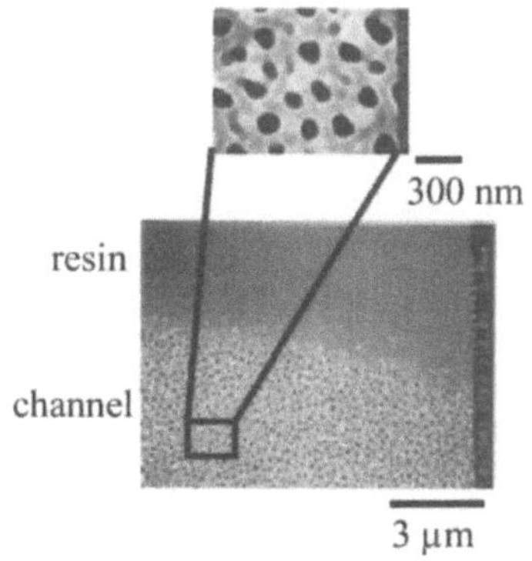

Figure 2. SEM micrographs of the channel showing pores ranging from 120- to 150-nm diameter.

3. Experimental

We prepared two kinds of DNA: the genome DNA of bacteriophage T4 (approximately 165 kb) and pVL1392 (approximately 10 kb) dyed with YOYO-1 (Molecular Probes Inc.). Electrophoresis was carried out in the channel after it was chemically coated with 0.5-wt% "Lipidure-HM" (NOF Co.). We observed the movements of the DNA molecules with a fluorescent microscope (ZEISS Co., Axioplan 2 Imaging) with an ICCD camera (Hamamatsu Photonics, C2400-21) and recorded them with digital VCR (SONY Co.).

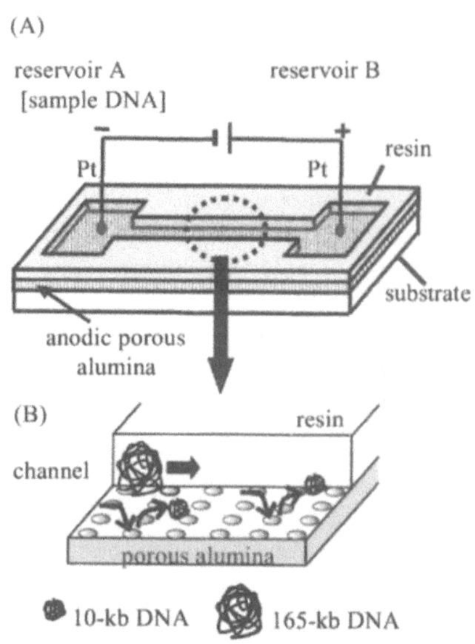

Figure 3. (A) SEC chip. The glass cover and reservoirs are not shown for clarity. Two reservoirs are connected to a power supply. The DNA runs from reservoir A to B in the nano-pore channel. (B) Schematic figure of the channel. The size of pore diameter is much smaller than the 165-kb DNA and approximately one third of the 10-kb DNA, so 10-kb DNA can contact nano-pores partially.

4. Results and discussion

We analyzed the running velocity for a 116-μm area of the viewfield. The obtained velocity distribution is shown in Figure 4. With these devices, we succeeded in separating 165-kb from 10-kb DNA and confirmed that the larger molecules moved faster. Our results thus show that this type of device can work on the basis of SEC. This is the first demonstration of nanostructure biochip technology. The longer-distance-run should improve the resolution of the separation due to increased interaction between molecules and holes.

Although Si and UV resin had an opposite directional electric osmotic flow (EOF), the Lipidure-HM depresses the EOF and the DNA was successfully moved into one direction. Since the radius of gyration (Rg) of 165-kb DNA is approximately 800 nm and that of 10-kb DNA approximately 460 nm [3], the smaller 10-kb DNA can make partial contact with the nanoholes and the larger 165-kb DNA has difficulty entering them. The structure still needs to be further optimized.

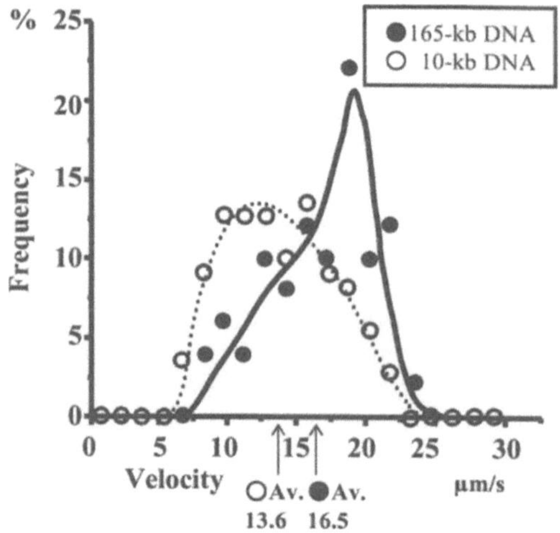

Figure 4. Velocity distribution of each DNA in the channel – solid line: 165-kb DNA, dashed line: 10-kb DNA. The average velocity was 16.5 μm/s for the 165-kb DNA and 13.6 μm/s for the 10-kb DNA. The number of measured samples was n=50 for the 165-kb DNA and n=110 for the 10-kb DNA. The window of velocity was 1.5 μm/s.

5. Conclusion

We have demonstrated a SEC-type separation using two kinds of DNA. The SEC device should be a better separation unit than the polymer-gel unit for future integrated nano-biochips because the device does not require filling material and is thus highly integrable with other bioMEMS units for sequential bio/chemical reactions. The chip should also reproduce data better and will be inexpensive to fabricate.

References

1. S. W. Turner, A. M. Perez, A. Lopez, and H.G. Craighead, *Monolithic nanofluid sieving structures for DNA manipulation,* J. Vac. Sci. Technol., B16, pp. 3835-3840 (1998).
2. H. Masuda, H. Asoh, M. Watanabe, K. Nishio, M. Nakao, and T. Tamamura, *Square and Triangular Nanohole Array Architectures in Anodic Alumina,* Adv. Mater., 13, pp. 189-192 (2001).
3. P. G. Righetti, *Electrophoresis in Analytical Biotechnology,* CRC Press, Boca Raton, FL (1996).

NANOCHANNEL ON FUSED-SILICA MICROCHIP AND LIQUID PROPERTIES INVESTIGATION BY TIME-RESOLVED FLUORESCENCE MEASUREMENTS

Akihide Hibara,[1] Takumi Saito,[1] Haeng-Boo Kim,[1] Manabu Tokeshi,[2] Takeshi Ooi,[3] Masayuki Nakao,[3] Takehiko Kitamori[1,2]

[1]Department of Applied Chemistry, University of Tokyo, 7-3-1 Hongo, Bunkyo, Tokyo 113-8656 Japan

[2]Kanagawa Academy of Science and Technology, 3-2-1 Sakado, Takatsu, Kawasaki, 213-0012 Japan

[3]Department of Engineering Synsthesis, University of Tokyo, 7-3-1 Hongo, Bunkyo, Tokyo 113-8656 Japan

(hibara@icl.t.u-tokyo.ac.jp)

Abstract

Nanochannels having widths of 330 nm and 850 nm were fabricated on fused-silica substrates and liquid samples were indroduced into the channels by utilizign capillary action. Liquid properties in the nanochannels were investigated by a time-resolved fluorescence measurements. The results suggested that water in the nanochannel has higher viscosity and lower dielectric constant than those in macro space.

Keywords: Nanochannel, capillary action, fused-silica, spectroscopy

1. Introduction

One of the most important investigation fields in microchip chemistry has been integration of flow chemical processing. The concepts of the continuous flow chemical processing (CFCP) and micro unit operations (MUO)[1] will advance integration density of microchip chemistry. It naturally follows smaller and thinner channels in nm (sub-μm) scale. Investigations into the mesoscopic liquid provide informative discussions on basic science of liquid and solution. In this presentation, we will demonstrate fabrication of nm-sized channels, liquid introduction, hydrodynamic analysis, and time-resolved measurements of aqueous solutions in the channels. The observed liquid properties in the nm-sized channels differed from bulk liquid properties and the difference was discussed in terms of statistical thermodynamics state of hydrogen network.

2. Fabrication and liquid introduction

330-nm- and 850-nm-sized channels were fabricated on fused-silica substrates by a fast atom beam etching method [2](Figure 1) and cover plates were sealed by a hydrofluoric bonding method. The channel layout around the nm-sized channels is shown in Figure 2. Liquid introduction process utilizing capillary action was observed under an optical microscope and the liquid velocity during the introduction was

Y. Baba et al. (eds.), Micro Total Analysis Systems 2002, Volume 2, 769–771.

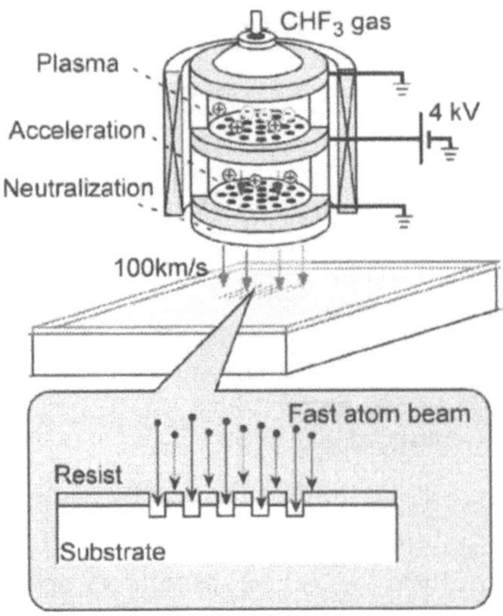

Figure 1 Fast atom beam method.

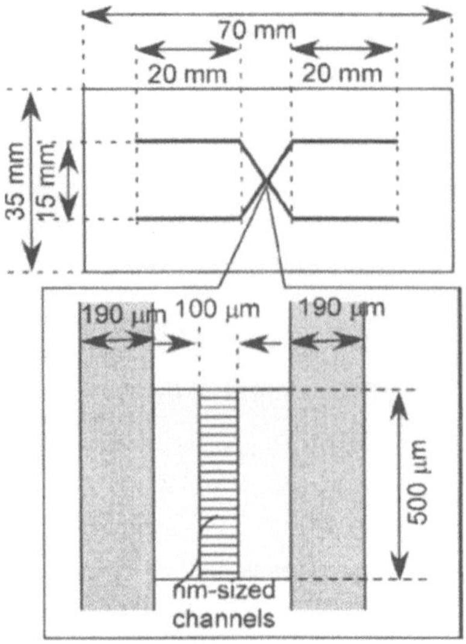

Figure 2 Layout and dimensions.

analyzed by surface energy and macro-scale hydrodynamics. In order to estimate liquid introduction velocity to the nm-sized channel, liquid was assumed to be driven by Laplace pressure ΔP as

$$\Delta P = \frac{\gamma}{S/l}, \tag{1}$$

where γ is surface tension and S and l are a cross section and periphery length of the section, respectively. When macro-scale fluid dynamics is assumed, pressure drop ΔP_d can be expressed as

$$\Delta P_d = \frac{32\mu v L}{D^2}, \tag{2}$$

where μ and v are viscosity and velocity of fluid and D is hydrodynamic diameter of the channel. L is distance between two points where ΔP_d is considered. Here, we assumed $\Delta P_d = \Delta P$ in order to estimate theoretical introduction time $\langle t \rangle$ and mean velocity $\langle v \rangle$. In case of channel length of L', $\langle t \rangle$ can be formulized as

$$\langle t \rangle = \frac{16\mu L'^2}{D^2 \Delta P}. \tag{3}$$

The averaged velocity $\langle v \rangle$ was calculated by $\langle v \rangle = L'/\langle t \rangle$. The observed velocity in the nm-sized channels seemed more than 2 times slower than the estimation, which corresponds to higher viscosity.

3. Fluorescent decay time analysis

Aqueous solutions of rhodamine 6G (R6G), sulforhodamine 101 (SR101) and rhodamine B (RB) in the channels were measured by the time-resolved fluorescence spectroscopy; spectra of the same solution in a 250-μm-sized channel were also measured as reference for macro space. While the fluorescence spectra in the 330-nm-, 850-nm- and 250-μm-sized channels agreed one another, the fluorescent decays in the nm-sized channels were faster in case of R6G and SR101 and slower in case of RB than that in the 250-μm-sized channels (Figure 3). The results suggested lower dielectric constant and higher viscosity in the nm-sized channels. The collective hydrogen-bond state of the water in the nm-sized channels is maybe enhanced compared with that of bulk water and affects the macroscopic parameters.

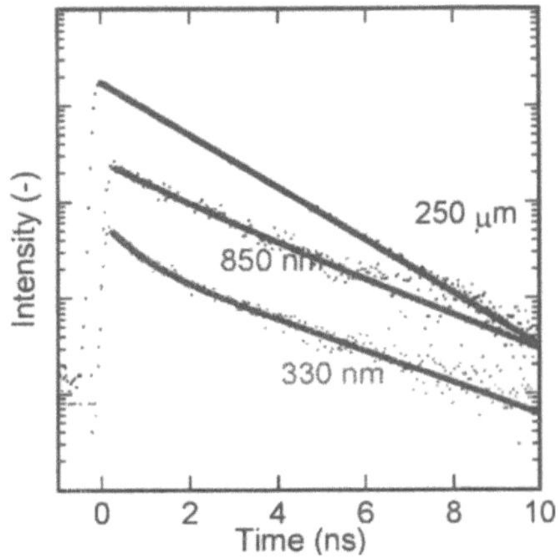

Figure 3 Size dependence of fluorescent decay of 1 μM RB aq.

4. Conclusions

We have fabricated nm-sized channels, and demonstrated a liquid introduction method into the channels utilizing capillary action. Liquid properties in the nm-sized channels were investigated by time-resolved fluorescence measurements of aqueous solutions of R6G, SR101 and RB. The results suggested that the aqueous solutions in the nm-sized channels have lower dielectric constants (polarity) and higher viscosities. In order to confirm the size effects, further experiments on liquid properties such as measurement of dielectric constant, low frequency Raman scattering spectroscopy, and flow visualization should be performed. From the viewpoint of analytical applications of the nm-sized channels, we will also investigate pressure-driven and electrophoretic flows in them.

References

1. M. Tokeshi, T. Minagawa, K. Uchiyama, A. Hibara, K. Sato, H. Hisamoto, and T. Kitamori, *Continuous-Flow Chemical Processing on a Microchip by Combining Microunit Operations and a Multiphase Flow Network*, Analytical Chemistry, 74, pp. 1565-1571, (2002).
2. M. Fujinami, M. Tokeshi, T. Odake, T. Kitamori, K. Sato, T. Sawada, K. Matsumoto, M. Nakao, T. Ooi, Y. Hatamura, *Microfabricated Channels and Fluid Control Systems for Integrated Flow Injection Analysis*, Proc. Micro Total Analysis Systems '98, Kluwer Academic Publisher, Dordrecht, The Netherlands, pp. 339-342, (1998).

CONTROLLED QUANTUM DOT SYNTHESIS WITHIN MICROFLUIDIC CIRCUITS

Joshua B. Edel, Robin Fortt, John C. de Mello and Andrew J. de Mello
Department of Chemistry, Imperial College of Science, Technology and Medicine, Exhibition Road, South Kensington, London SW7 2AY.
joshua.edel@ic.ac.uk, a.demello@ic.ac.uk

Abstract

Nanocrystalline semiconductors are of considerable scientific and commercial interest owing to their tuneable optical and electronic properties, and potential applications in a wide range of electronic devices. We show that the use of microfluidic synthesis procedures offers distinct advantages over bulk scale fabrication procedures for the production of nanoparticles. The small size of the microreactor ensures that thermal and chemical homogeneity pertain throughout the entire reaction volume, leading to well controlled reaction conditions.

Keywords: semiconductor, nanoparticle, microreactor, quantum dots

1. Introduction

In the last decade semiconductor quantum dots (QD) have become the subject of one of the most rapidly growing branches of chemistry and physics. QD's or nanoparticles (NPs) are in the transition regime between molecules and the bulk material. Due to quantum confinement effects NPs show unique physical and chemical properties such as size-dependent band gap shifts and size dependent photoluminescence. Currently there is considerable interest in synthesizing nanoparticles of well-defined sizes [1]. One of the primary challenges associated with the synthesis of NP's is the ability to limit the size of growing clusters and to prevent them from interacting with each other. To date, techniques for producing highly monodisperse nanoparticles have been relatively complex and involve multiple processing steps. Herein we show that by using a continuous flow microreactor [2], operating within Low Reynold's number regimes, we are able to synthesise and control the size and monodispersity of certain nanoparticles.

2. Experimental

To generate CdS nanoparticles stock solutions of $Cd(NO_3)_2{\cdot}4H_2O$ and Na_2S were prepared at concentrations of 4×10^{-4} M and hydrodynamically delivered into a microfluidic channel network at various flow rates (1-600 μL/min). The outlet flow from a microfluidic reactor chip was coupled to a quartz flow cell and absorption spectra were taken. The microfluidic chip used is a two-layer device made up of a glass/silicon/glass sandwich and has an internal volume of 600 nL. Two inlet flows are split into a series of separate multichannel streams (16 partial flows). This is achieved

Y. Baba et al. (eds.), Micro Total Analysis Systems 2002, Volume 2, 772–774.

by repeated splitting of the channels in such a way that an array of symmetrical elements results. Wafer-through nozzles connecting the two fluidic layers allow the two liquid streams to converge and mix. Channels are then sequentially combined in a reverse network until all partial flows are united in one broad outlet channel.

3. Results and discussion

Absorption spectra of CdS colloids obtained at flow rates ranging from 0.025 – 0.30 ml/min are shown in **Figure 1**. The maximum gradient of absorption profiles increases as a function of volumetric flow rate, indicating a reduction the dispersity of the produced NPs [3].

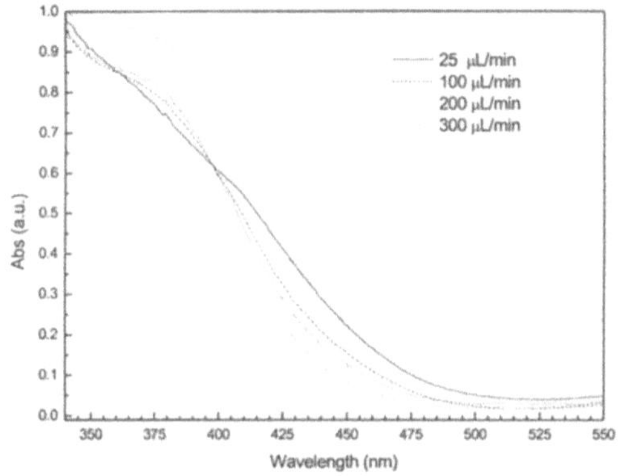

Figure 1. Absorption spectra of fluid stream containing quantum dots exiting the micromixer chip

Figure 2 shows absorption spectra of CdS using various initial cadmium concentrations at two different flow rates. Two key observations can be made. First, as the cadmium concentration is decreased absorption spectra shift to higher energies. Second, at higher flow rates the polydispersity of the produced NP populations is reduced.

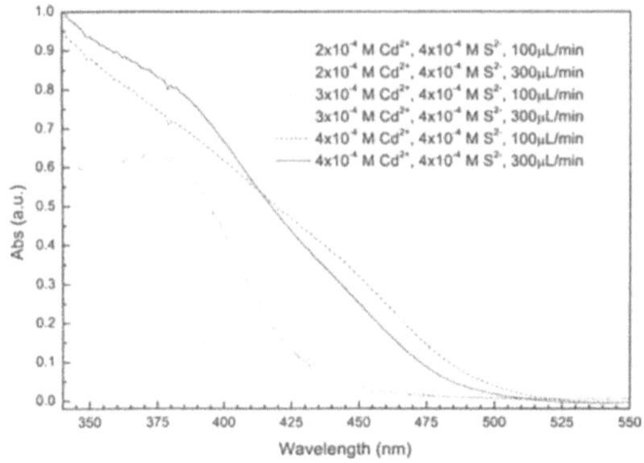

Figure 2 Absorption spectra of CdS colloids using various cadmium ion concentrations

An interesting feature of the data presented in Figure 1 is the existence of an approximate isosbestic point at 3.12 eV. This may be rationalized theoretically to be directly related to the polydispersity of the solution [3]. Let *N(E)* be defined as the relative occurrence of particles with bandgap *E*, and $A(E,E_G)$ being defined as an absorption spectrum of a given set of monodisperse nanopartiles with a given bandgap E_G. The average absorption spectrum of the population $\langle A(E)\rangle$ may then be approximated by $\langle A(E)\rangle = A(E,E_G)\otimes N(E)$ ($\otimes$ denotes a convolution). It should be noted that various absorption spectra will yield an isosbestic point only in special circumstances: *underlying population distributions must be of similar form – with the same modal energy gap & skewness – but different variances.*

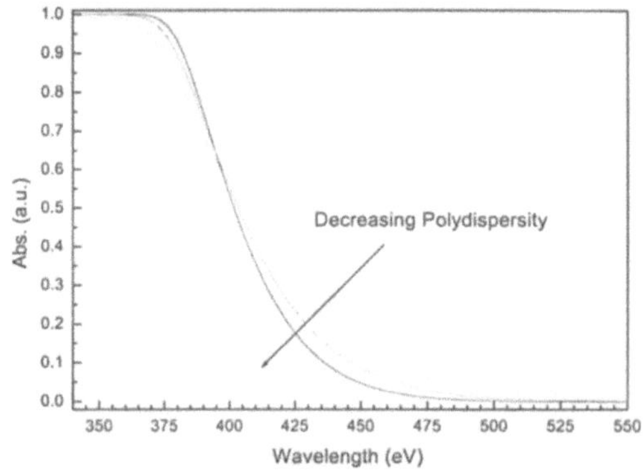

Figure3. Theoretical absorption spectra for samples of nanoparticles obtained by convolving an arbitrary single-particle absorption spectrum with three crystallite distributions of similar shape.

4. Conclusions

By adjusting experimental parameters such as concentration and flow rate both monodispersity and particle size of synthesised nanoparticles can be controlled.

Acknowledgements

J.B.E. acknowledges the receipt of an Overseas Research Studentship from the UK Government. R. F. is grateful to SmithKlineBeecham for an EPSRC CASE studentship.

References

1.M.A Hines, and P.J. Guyot-Sionnest,*J. Phys. Chem.*, 1996,**100**: 468 - 471.
2.F.G. Bessoth, A.J. de Mello, and A. Manz,*Anal. Commun.*,1999,**36**, 213 - 215.
3.J.B. Edel, R. Fortt, J.C. de Mello, A.J. de Mello,*Chem. Commun.*,2002,**10**, 1136-1137.

PDMS NANOSTRUCTURES FOR BIOMOLECULES ANALYSIS

Eric Leclerc[1, 2], Takatoki Yamamoto[2] and Teruo Fujii[2]

[1]LIMMS/CNRS-IIS, University of Tokyo
[2]Institute of Industrial Science, University of Tokyo,
4-6-1 Komaba Meguro-ku Tokyo, 153-8505, Japan
({rico, takatoki,tfujii}@iis.u-tokyo.ac.jp)

Abstract

To develop a platform for studying the dynamics of biomolecules, nanotowers with a pyramidal shape are developed. The nanotower structure is fabricated by micromolding of PDMS (PolyDiMethylSiloxane). As a demonstration, ribosomes obtained from *E.Coli.* are successfully immobilized at the top of nanotowers.

Key words: single molecule observation, PDMS, nanotower, ribosomes

1. Introduction

In order to study the time-resolved dynamics of biomoleculer reactions, visualizing the single molecule's behaviors is one of the strongest analytical methods. The most important point in this method is how to immobilize the target molecule on the solid substrate. To focus on a single molecule, the target molecule must be immobilized at pre-determined positions, preferably without non-specific adsorption to the other part of substrate to improve the ratio of successful observations. To realize this immobilization, the immobilizing area should be limited as small as possible. In this paper, a 3-D nanoscale tower structures are developed and the result of immobilization of ribosomes at the top of the nanotower is reported.

2. Fabrication of the Nanotower

It has been proposed and experimentally tested some basic idea to fabricate nanotowers for biological applications [1]. In this research, the nanotowers with pyramidal shape structure are fabricated by micromolding of PDMS, which mold master was fabricated by conventional anisotropic KOH etching of Si (Figure 1).

Y. Baba et al. (eds.), Micro Total Analysis Systems 2002, Volume 2, 775–777.

Figure 2 a) shows a SEM photographs of nanotowers covered with thin Al layer to visualize electrically insulative PDMS structure. The magnified view is shown in Fig. 2 b) revealed that the top surface area can be estimated around 100 nm by 100 nm.

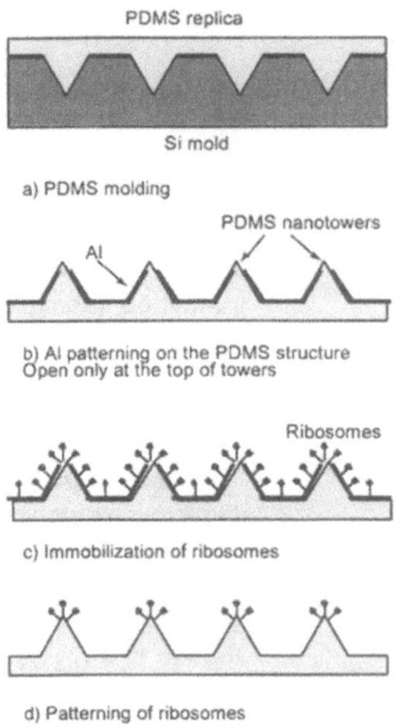

Figure 1. The process of nanotower fabrication and ribosomes immobilization

a) Array of nanotowers

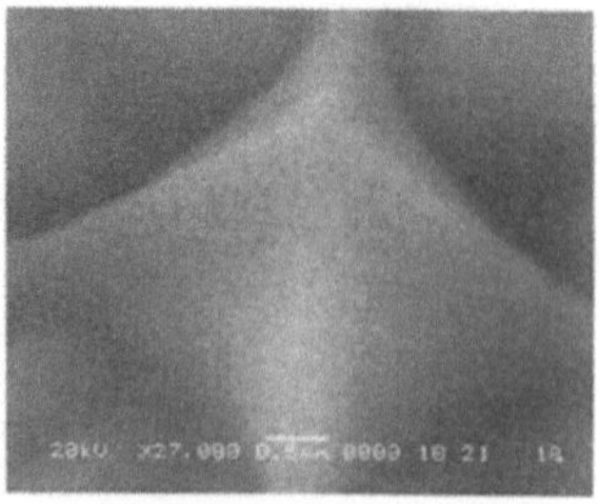

b) Magnified view of a nanotower

Figure 2. SEM view of pyramidal shape PDMS nanotowers

3. Bio-patterning

Ribosomes of *E. Coli* were immobilized at the top of the nanotower. At first, ribosomes were labelled with fluorescent dye of Alexa Fluor 488 (Molecular

Probes Inc.) which wavelength of excitation/emission is 494/517 nm, and then were immobilized using a cross-linker of Sulfo-SANPAH (Vector Inc.). The immobilized ribosomes are shown in Figure 3. As can be seen, there is a bright spot in the middle of the nanotower while the other part is dark. The background outside of the nanotower is caused by the non-specific adhesion of ribosomes. At the moment, it is difficult to analyze how many ribosomes are immobilized. However the photograph apparently shows some ribosomes are successfully immobilized at the top of the nanotower.

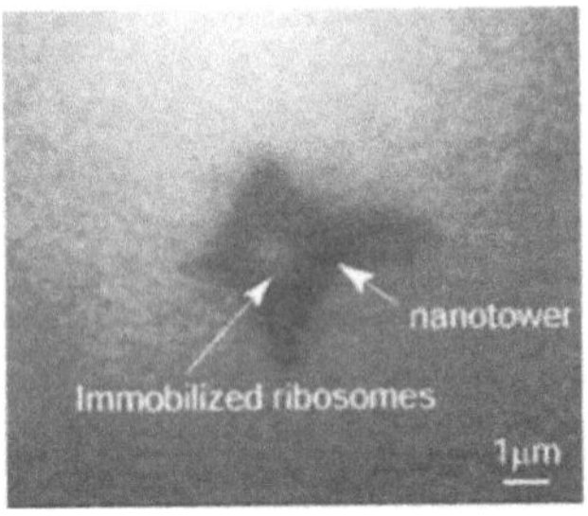

Figure 3. Immobilized ribosomes at the top of the nanotower

4. Perspectives

A nanoscale of pyramidal shape tower structures to study biomolecular dynamics was fabricated and it is demonstrated that ribosomes are successfully immobilized at the top. As a next step, we will observe the process of protein synthesis on the immobilized ribosome in the microreactor chip we developed before[2]. This result would supply a platform of "chip-based" single or a small number of molecules analysis and reaction methods.

References

[1] E. Leclerc and T. Fujii, *PDMS(PolyDiMethylSiloxane) Nanotowers for Biological Observations*, IEEE Trans. Nanotech. (2002) submitted

[2] T. Yamamoto, T. Nojima, and T. Fujii, *Cell-free Protein synthesis in PDMS-based Parallel Microrectors*, Proc. μTAS 2001, Monterey USA (2001) pp. 69-71

NANOSTRUCTURED ELECTRODES FOR NEURAL CHIP APPLICATIONS

Karen Cheung, Yang-Kyu Choi, Tim Kubow, and Luke P. Lee
University of California Berkeley, 497 Cory Hall #1774, Berkeley, CA, 94720, USA

Abstract

For the analysis of cultured neural networks and cells in on-chip biosensors, microelectrodes must be optimized for impedance and selectivity. We present a new method of increasing the effective electrode surface for decreased impedance. Here, high aspect ratio pillar-like polysilicon nanostructures are created in a reactive ion etch. The pillars are 0.3 μm in height and less than 100 nm in diameter. When compared to a control substrate, these nanostructures increase the surface area by 12-15 times and result in an order of magnitude decrease in impedance.

Keywords: nanostructures, impedance, microelectrode

1. Introduction

The interface between neurons and electrodes is one of the key issues in implantable microdevices or on-chip bioelectronics. The ideal electrode has maximum selectivity and minimum impedance. Selectivity refers to the ability to select a single neuron from a multitude of interconnected cells. High electrode impedance attenuates and filters the measured signal. In addition, a sufficient signal-to-noise ratio is required for data analysis; low electrode impedance gives high signal gain [1]. However, an increase in electrode selectivity with smaller geometric electrode size results in increased impedance and noise. To optimize the electrode, the impedance per unit of geometric surface area must be decreased by altering its nanostructure. Because the total electrode impedance is approximately proportional to the effective surface area, a 10 times increase in surface area should give an order of magnitude decrease in impedance [2]. Applications of these nanostructures extend beyond electrodes, because they may improve integration of tissue with an implanted microdevice as cells anchor onto the increased surface area that they provide.

2. Nanostructure Fabrication

The substrate is a 1 μm film of heavily n-type in-situ doped polysilicon, which was deposited by low pressure chemical vapor deposition (LPCVD) onto a quartz wafer. Electrode pads are defined by conventional I-line lithography and a two-step anisotropic plasma etch (Lam Research 9400 TCP): a breakthrough step removes the native oxide and an etch step patterned the polysilicon (Fig.1). After stripping the photoresist, nano-pillars are formed on the electrodes, also by using a two-step anisotropic plasma etch (Fig.2). The density and diameter of the nano-pillars is controlled by the ratio of HBr and O_2. A larger HBr:O_2 ratio is etch dominant, while a smaller HBr:O_2 ratio is passivation dominant and gives more densely packed nanostructures.

Y. Baba et al. (eds.), Micro Total Analysis Systems 2002, Volume 2, 778–780.

In this device, an array of nanostructured electrodes in the center of the substrate are connected to polysilicon lines which lead out to bond pads at the edge (Fig. 3a). The electrodes at the center are squares with widths of 20, 50, 100, and 200 μm. A 35 mm diameter tissue culture dish with the bottom drilled out was glued to the quartz substrate using 100% silicone adhesive (Fig. 3b). The bond pads are ultrasonically bonded to hand-made connectors for data collection.

3. Experimental

Impedance was measured from the electrodes at 1 kHz (Fig 4). The nanostructures increase the surface area by 12-15 times and result in an order of magnitude decrease in impedance.

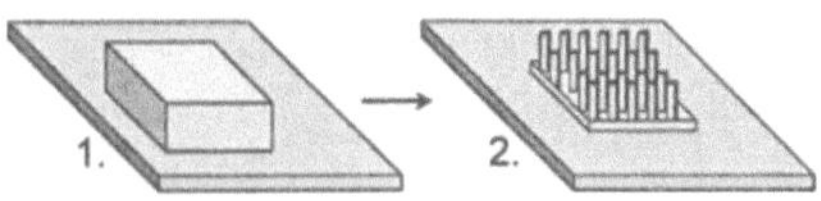

Fig. 1. Nanostructure fabrication.
1. A polysilicon film is deposited on a quartz wafer and patterned.
2. The film is etched back in a highly selective reactive ion etch, which produces the nanostructures.

Fig. 2. The nanostructures are less than 100 nm in diameter.

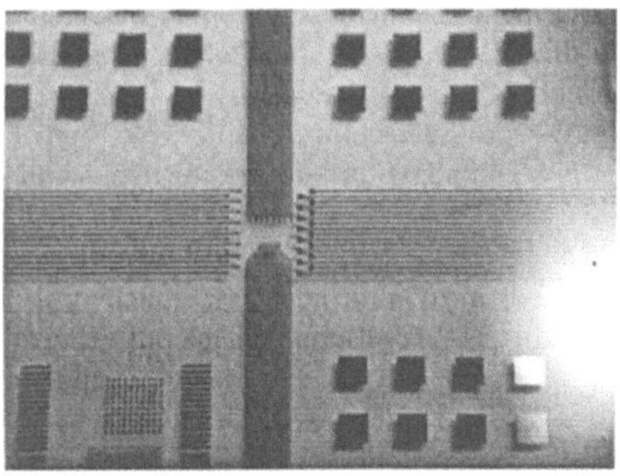

Fig. 3.(a) An array of electrodes in the center are connected to polysilicon lines.

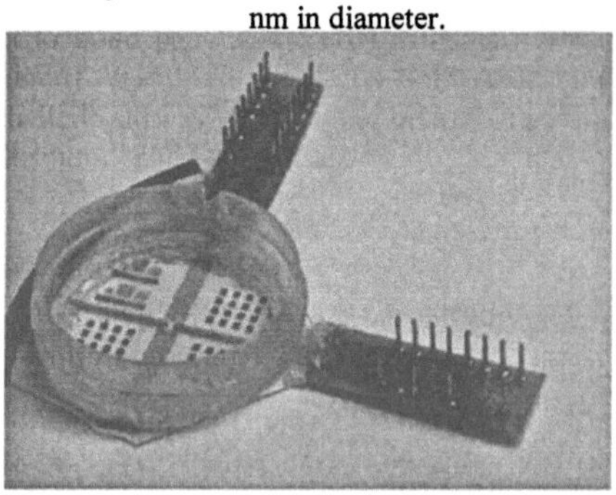

Fig. 3.(b) View of the entire assembly.

4. Results and discussion

High aspect ratio pillar-like polysilicon nanostructures were fabricated. Their diameters ranged from 20 to 100 nm; diameter and packing density are controlled by the ratio of HBr and O_2 in a highly selective silicon RIE. The nanostructures decrease the impedance by an order of magnitude.

Table 1. Nanostructure etch gas ratio affects the final surface area ratio

Wafer	*Nanostructure etch gas ratio*	*Surface area ratio*
Control	-	1
1	200 HBr : 6 O_2	12.5
2	200 HBr : 9 O_2	15

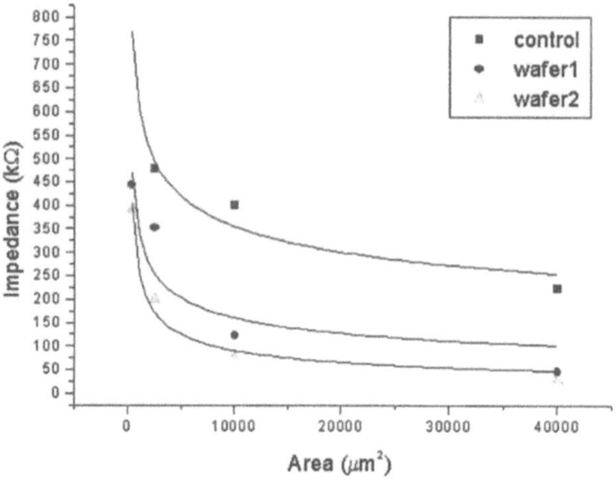

Fig. 4. Impedance recordings. Wafers 1 and 2 have nanostructures on the electrode pads. The nanostructures decrease the impedance by an order of magnitude.

5. Conclusions

High aspect ratio pillar-like polysilicon nanostructures were fabricated. The nanostructures increase the surface area by 12-15 times and result in an order of magnitude decrease in impedance. More studies are necessary in which these electrodes are used to record neural signals.

Acknowledgements

This work was supported by a grant from the Whitaker Foundation.

References

[1] A. Blau, Ch. Ziegler, M. Heyer, F. Endres, G. Schwutzgebel, T. Matties, T. Stieglitz, J. U. Meyer, and W. Goepel, "Characterization and optimization of microelectrode arrays for in vivo nerve signal recording and stimulation," *Biosensors & Bioelectronics*, 12(9-10):883:892, 1997.

[2] D. A. Stenger and T. M. McKenna, editors. *Enabling Technologies for Cultured Neural Networks*. Academic Press, New York, 1994.

Molecular filter-nanosieve

Terrell D. Neal and Axel Scherer
Department of Electrical Engineering,
California Institute of Technology,
Pasadena, CA 91125

Abstract

Demonstrated in this paper is a molecular sieve. The focus of this work is making a "smart" filter. This filter has electrochemical and electromechanical properties that would allow both sorting and filtering of solutions. Prior sifters and sieving structures have been at the mercy of a static material structure, whereas the convention proposed here allows dynamic control of filtering size during an assay.

The filter mechanism simply relies on a sandwiched layer of oxide, grown in a RF Sputtering system. This allows for almost Angstrom level control of the thickness. An oxide layer is sandwiched between two layers of silicon with e-beam written holes 100nm in diameter, one offset from the other. Dry etching is used to carve these holes in the silicon layers, and wet oxide etching defines a void in the sandwiched layer. The distance that the two silicon layers are separated dictates the smallest filtering size.

Analysis is available through several methods. Mechanically, a fluid can be passed through the holes carrying along with it the particles or molecules that are small enough to make it through the oxide layer. Electrochemically, a charged particle can be placed on one side of a container and separated by the filter, while having the tendency to get to the other side, with different concentration, by passing through the filter [1].

Keywords: microfluidic, filter, sieve, molecular, nanopore, nanofluidic

1. Introduction

The analysis of small volumes of solutions is extremely important for many biological applications. The goal of this work is to define small cavities that can be used to filter sub-100 nm particles and even molecules. Here we present the design and fabrication details of such filters, as well as preliminary experimental results.

2. Theory

Sputtering and etching are common processes that are used in microelectronics microfabrication, but can be applied to the nanofabrication of fluidic devices. Using these processes for new areas offers the opportunity for combining the excellent dimensional control from the microelectronics industry with the massive integration of miniaturized devices on a monolithic substrate. That is why we seek to fabricate our nanosieve lithographically. However, even with the excellent capabilities of high-resolution lithography, it is very difficult to define hole sizes below 10 nm with adequate uniformity for high quality filtration. Therefore, it is desirable to define filters in which the geometric size of the constriction is controlled by growth rather than by lithography. Here we demonstrate such a geometry, in which we use the vertical control over layer thickness to determine the filter size, and provide access to this constriction by

Y. Baba et al. (eds.), Micro Total Analysis Systems 2002, Volume 2, 781–783.

lithographically defined holes. To define the filter, we use two steps of electron beam lithography and dry etching with multiple alignments to obtain off-set holes within two thin silicon membranes, which are connected through a sacrificial silicon dioxide layer that ultimately controls the filter size. Since the layer thickness of a sacrificial layer can be controlled to within less than 1 nm during the construction of the filter, it is possible to accurately set the filter geometry and control the size of particles or molecules that can pass through the filter.

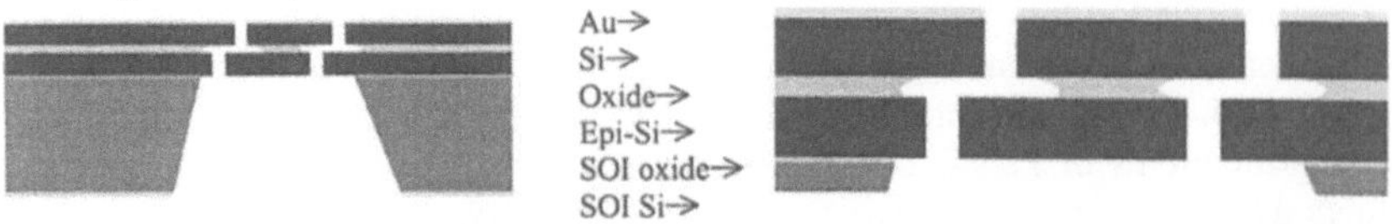

Figure 1. Illustration of nanosieve, left, and close-up on the right, with a thin gold film evaporated on both sides of the substrate. (Layers not drawn to scale.)

3. Experimental

The fabrication sequence is summarized in Figure 1, where we show the steps which are needed for defining our lithographic filter. First, we use electron beam lithography to define an array of holes on a silicon on insulator (SOI) sample. The thickness of the silicon layer is approximately 200 nm, whereas the SOI oxide thickness is approximately 400 nm. We use polymethylmethacrylate (PMMA) as an electron beam resist, as well as an etch mask. Using a chemically assisted ion beam etching system (CAIBE) with XeF2 reactive gas, the pattern written using e-beam lithography, illustrated in Figure 2, is etched through the first silicon layer. The beam voltage and other parameters are set appropriately in order to maintain an acceptable etch ratio between the etch mask layer and the silicon. Following this procedure, the etched holes are filled with gold, and the PMMA is removed in acetone. Then, a precisely controlled silicon dioxide sacrificial layer is sputtered onto the sample, followed by a thicker polycrystalline silicon layer. Another lithography procedure is then used to define a similar array of holes, with a small offset, on top of the original hole structure. Finally, hydrofluoric acid and gold etch are used to define the connection between the lithographically defined holes through the silicon dioxide sacrificial layer.

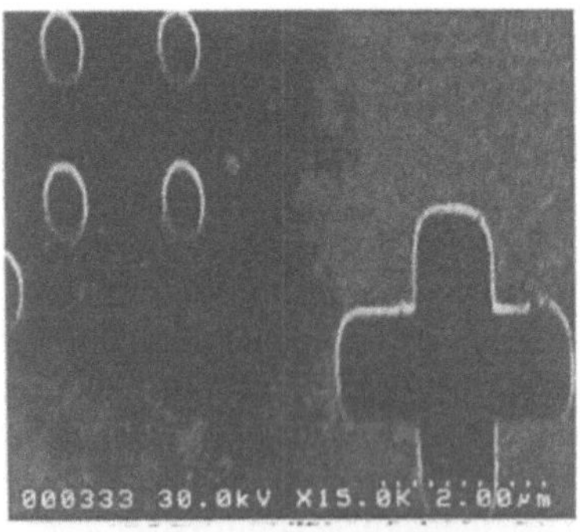

Figure 2. SEM view of e-beam written, etched holes and alignment mark used for offsetting the top layer.

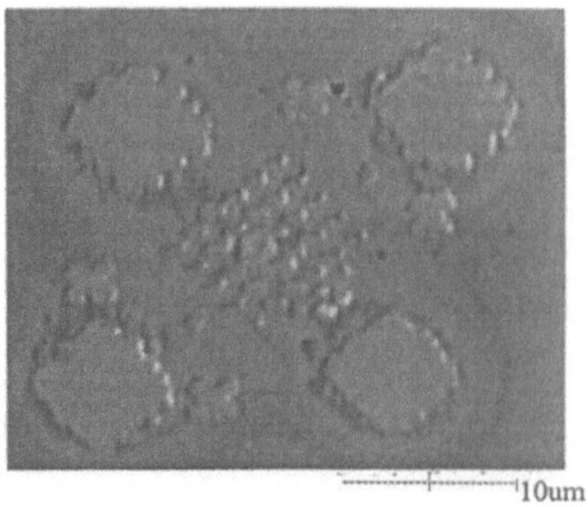

Figure 3. Suspended epi-Si membrane with e-beam written holes in the center.

We observe the hole size increase over the lithographic pattern by a few tens of nanometers in diameter after the etching process. This increase in pattern size can be adjusted for during the design of the filter by spacing the holes appropriately further apart. Thus, the sputtered oxide thickness and wet etching, using a buffered HF solution, creates the geometric structure of the filter. Objects exceeding the size of the void are not allowed to pass through, and the accuracy over the filter size is determined only by the control over the deposited thickness of the oxide.

In our device, gold is used to fill in the holes. Often, chromium or titanium layers are used to allow for adhesion to the substrate. However, adequate sidewall adhesion can be achieved in our structures by evaporating at a low rate <10 Angstroms/sec during the first several tens of nanometers. In addition, we have observed that timing is essential in ensuring that the gold lift-off process is successful.

4. Results and discussion

Fabrication of a lithographically defined filter is achieved through careful processing. More fabrication issues are encountered during the analysis. One convention looks to allow molecules or colloidal beads to pass through by way of a backside etch. Another convention, illustrated in Figure 3, would better suit the needs for mass scale production by allowing access to the bottom side of the filter from the top of the substrate. Undercutting the buried oxide layer in the SOI substrate allows access to the backside of the e-beam written holes.

Both conventions will require sub-micro-scaled fabrication on top of the surface. Whereas the primer would necessitate maneuvering of the sample in order to access the backside. These techniques are currently being explored.

5. Conclusions

Fabrication limits are approached when seeking to demonstrate filtration through a surface machined silicon on insulator fluidic channel. The high uniformity and large areas needed for filter action require the use of new geometries to ensure accurate definition of filter elements which can be used for biological assays. Use of fluorescent beads and fluorescently labeled molecules will be used to determine the quality of our filter. Meanwhile, the fabrication approach towards producing a filter in which the control over geometry is assured by deposition has been demonstrated. These devices will be able to be integrated into microfluidic analysis systems.

Acknowledgments

Terrell D. Neal was supported by the David Lucille Packard Foundation Fellowship. This work was supported by the National Science Foundation under grant BES-0119493, and by the Army Research Office under contract DAAD19-00-1-0392. We also wish to thank Jim Hone and Ali Husain for their thoughtful suggestions.

Reference

[1] Jiali Li, Derek Stein, Ciaran McMullan, Daniel Branton, Michael J. Aziz, Jene A. Golovchenko. "Ion-beam sculpting at nanometer length scales" **Nature**. 412. 166-129 (2001)

DEVELOPMENT OF MICROFLUIDIC SHEAR ASSAYS FOR QUANTITATIVE ANALYSIS OF CELL ADHESION

Hang Lu[1], Lily Y. Koo[1], Linda Griffith[1,2], Klavs F. Jensen[1]
Department of Chemical Engineering[1], Biological Engineering Division[2], Massachusetts Institute of Technology, 77 Massachusetts Ave. 66-560, Cambridge, MA 02139, USA (kfjensen@mit.edu)

Abstract

We report a series of microfluidic shear devices for studying cell adhesion on biological substrates. The method provides a tool for performing high throughput experiments with small amounts of sample and number of cells. Furthermore, it allows access to regimes of large adhesion strength that are difficult for conventional cell adhesion assays. We also demonstrate long-term cell culturing and experiments with medium perfusion in such microfluidic devices. These microfluidic devices could provide useful insight for tissue engineering and cell mechanotransduction studies.

Keywords: Cell adhesion, shear, high throughput

1. Introduction

Adhesive interactions between cells and their substratum are central issues in developmental biology, tissue maintenance and engineering, and biotechnological processes. Traditional methods for studying cell adhesion usually apply centrifugal or shearing forces to detach cultured cells from surface of interest [1]. These techniques are, however, inadequate for controlled and high throughput experiments, and sometimes unable to deliver large enough forces to detach cells from substrates. To overcome these limitations, we have developed a series of microfabricated devices for high-throughput microfluidic cell adhesion assays.

2. Designs of Devices

Our designs take advantage of laminar flow in microfluidic systems to induce steady shear stresses on adherent cells. Because of the small Reynolds number ($Re<100$), the entrance length is very small in microfluidic devices so that a steady shear field applies in most of the device. The new microfluidic devices can either accommodate multiple surface modifications (Fig 1A), or study multiple shear rates (Fig 1B), or both through a single device, hence achieving high throughput performance. The microfluidic channel height is kept small (25 μm ~ 100μm) compared to the width (250 μm ~ 1000 μm) to mimic Poiselle flow between two large parallel plates and minimize edge effect. The designed micro shear devices can achieve shear stresses (>10,000 dynes/cm^2), much higher than conventional equipment.

Figure 1C illustrates the device design for long-term incubation experiments. Separate bifurcated inlets perpendicular to the cell growth channel serve to infuse

Y. Baba et al. (eds.), Micro Total Analysis Systems 2002, Volume 2, 784–786.

medium or stimulation (e.g. growth factors) for prolonged periods. These side channels imply that the shear stress on the cells and the pressure drop are much smaller compared to the flow along the channel (e.g. by the shearing fluid). Moreover, variations of the nutrient and stimulus delivery to the cells are minimized.

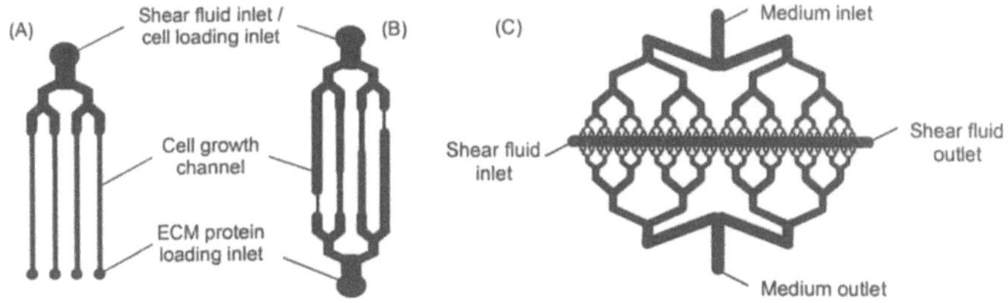

Figure 1. (A) Multiple substrates; (B) multiple shear stresses; (C) long term incubation device.

3. Experimental

The devices were fabricated using rapid-prototyping techniques of PDMS [2] with SU-8 on silicon as the master, and bonded to microscope slides using oxygen plasma surface treatments. Fibronectin of various concentrations was coated onto the glass, and BSA was used to block non-specific binding as a negative control. Murine WT fibroblasts (NR6) cells and Chinese Hamster Ovary cells were cultured and harvested before being seeded onto the microfluidic device. After incubation, the device was mounted onto a motorized stage and cell detachment was captured using time-lapse video-microscopy.

4. Results and discussion

Fig 2A illustrates NR6 cells on fibronectin-coated surfaces during a cell-detachment experiment. Both quantitative adhesion data and cell morphology observations are obtained – the former by comparing the number of cells on the surface before, during, and after the application of shear force. In fluorescence mode (Fig 2B), it is also possible to label and track structural or signaling proteins of interests. In the shear experiments, we have observed a strong fibronectin-concentration dependence on cell adhesion (Fig 2C). Moreover, by comparing two sets of experiments (Fig 2C and D), we have observed shear induced strengthening of cell adhesion on the fibronectin surfaces, which opens up opportunities for exploring cell signaling events

In a separate experiment, we have demonstrated long-term cell culturing in the device shown in Fig 1C. The cells were successfully maintained for 12 hours until the shearing experiments. Fig 3A and 3B demonstrate the necessity of medium perfusion (without which, cells disintegrate overtime). Similar to the short-term experiments, cell adhesion again demonstrated the substrate concentration dependence (Fig 3).

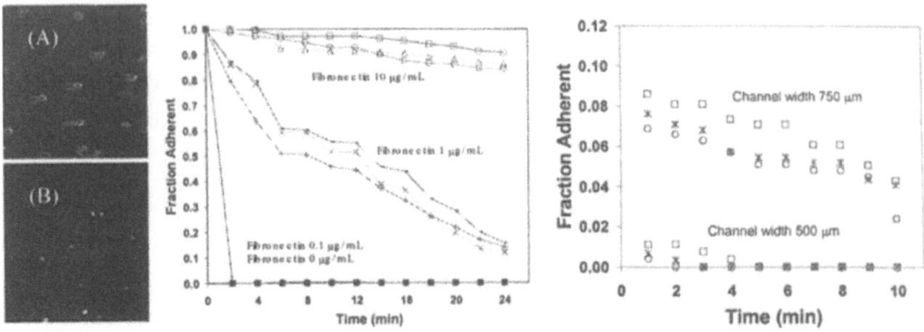

Figure 2. Sample images to study cell morphology upon the application of shear: (A) transmission mode; (B) fluorescent mode (vital stain). Quantitative shear assay: (C) Using device in Fig 1A coated with different ECM proteins; (D) Using device in Fig 1B coated with 10 μg/ml Fibronectin.

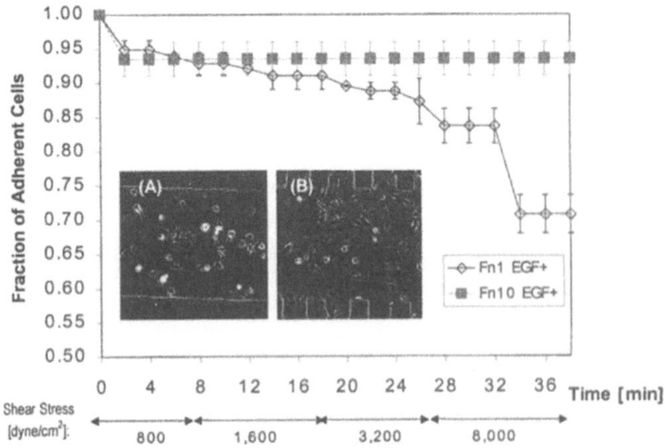

Figure 3. Long-term cell culture and stimulation experiment; (A) Without medium perfusion, cells disintegrate; (B) With medium perfusion, cell proliferation is apparent.

5. Conclusions

We have designed, fabricated, and tested a set of microfluidic devices for quantifying cell adhesion and related cell morphology. These devices could also be extended to studies of other biological phenomena, e.g., cell migration and chemotaxis.

References

1. S. Usami, H. Chen, Y. Zhao, S. Chien, and R. Skalak, "Design and Construction of a Linear Shear Stress Flow Chamber", *Annals of Biomed. Eng.* **21**, pp. 77-83 (1993).
2. D.C. Duffy, J.C. McDonald, O.J.A. Schueller, and G.M. Whitesides, "Rapid prototyping of microfluidic systems in poly(dimethylsiloxane)" *Anal. Chem.* **70**, pp. 4974-4984 (1998).

A MICROFLUIDIC BIOSYSTEM FOR METABOLIC MONITORING OF HUMAN ISLET CELLS WITH INTEGRATED BIOSENSORS

Chuan Gao, Horacio L. Rodriguez Rilo*, Jin-Woo Choi, and Chong H. Ahn
MicroSystems and BioMEMS Lab
Department of Electrical and Computer Engineering and Computer Science
*Center for Cellular Transplantation, Department of Surgery
University of Cincinnati, Cincinnati, OH 45221-0030, USA

Abstract

This paper presents a microfluidic biosystem for metabolic monitoring of human islet cells with an integrated biosensor. The polymeric microfluidic system has been designed and developed for real time handling and monitoring of human islet cell metabolism with simultaneous glucose measurements using the integrated glucose sensor. The microfluidic system, integrated with a glucose sensor for simultaneous glucose measurements, entraps and sustains pancreatic islet cells in a micro reaction chamber for monitoring cell metabolism. A single cluster (group of cells) of Islets of Langherans was exposed to different concentrations of glucose for 8 hours and the secreted insulin were transported and collected for further investigation on diabetes.

Keywords: microfluidic systems, glucose sensor, metabolic engineering, insulin

1. Introduction

"Genetic engineering of pancreatic islet cells" is an innovative approach that can provide a potential solution to the scarcity of organs available for clinical islet transplantation. The possibility of bioengineering islets can eliminate the need of two or more pancreata per recipient as is currently needed with the "Edmonton Protocol" [1]. Different glucose concentrations cause human islet cells to secret insulin or glucagons to maintain a normal euglycemic level through two feedback loops [2] as described in Figure 1. In conventional devices, hundreds of islet cells/clusters are gathered in one reaction tube for testing. Each cell may have variant extracellular glucose environments due to the coagulation and position in tube. The breakdown of any cell membrane may dramatically change the insulin measuring result. In addition, microfluidic biosystems for study on islet cells at single cell level are highly desired.

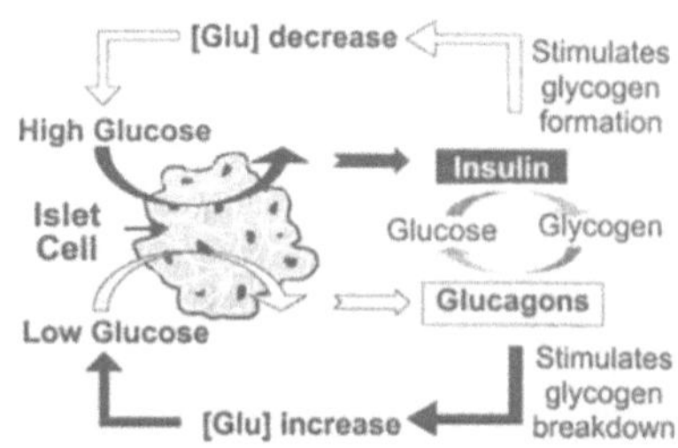

Figure 1. Islet cell function and performance in chemical environments.

For these reasons, we have developed a microfluidic biosystem for metabolic monitoring of human islet cells at single cell level with a capability of simultaneous

Y. Baba et al. (eds.), Micro Total Analysis Systems 2002, Volume 2, 787–789.

measurements of extracellular glucose concentrations using an integrated biosensor. Since islet cells smaller than 50 μm are treated as semi-functional cells or fragments and will not be counted in the cell activity measurement, pillars at a distance of 40 μm from each other have been positioned to avoid the migration of the cells from the chamber and to filter the unpreferred particles as illustrated in Figure 2. The secreted molecules are transported by the out flowing suspension and collected. Furthermore, the polymeric micro reaction chamber prevents the loaded and entrapped islet cells from the breakdown of cell membranes.

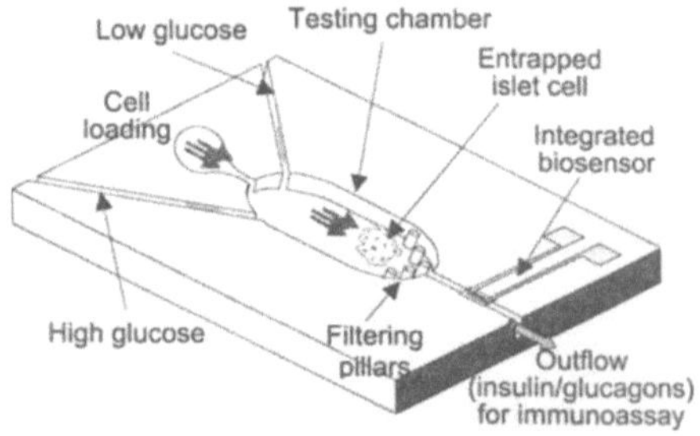

Figure 2. Schematic view of the biosystem for handling and monitoring of human islet cells at single cell level.

2. Fabrication and Experiments

Precisely designed polymeric (PDMS) channel/chamber structure casted from SU-8 mold is used for entrapment of human islet cell clusters. The novel multi-pillar structure holds the islet cell in the reaction chamber without blocking the fluid passing through. Solutions containing different glucose concentrations are injected by programmable syringe pumps. The hormones insulin, glucagons, somatostatin and pancreatic polypeptide secreted in response to different secretagogues such as glucose, caffeine, etc. are produced by diverse cell subpopulations trapped in the micro reaction chamber.

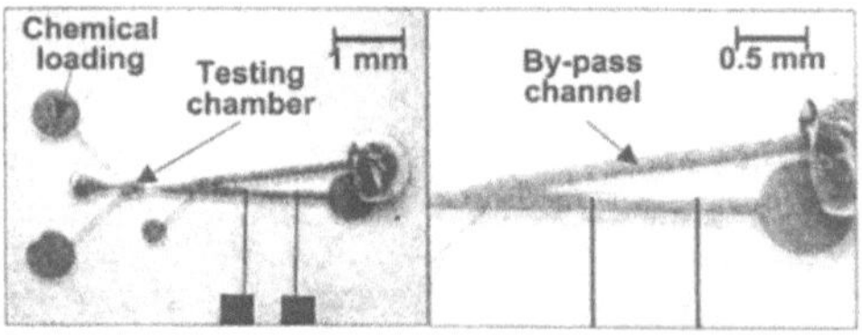

Figure 3. Microphotographs of the polymeric microfluidic biosystem integrated with biosensor. Two microvalves are used for switching fluidic pathways between by-pass channel and sensing channel.

The insulin/glucagons secreted by the trapped cell are transported by the outflow and collected in Eppendorf tubes. The outlets of the sensing channel and the by-pass channels are controlled by active microvalves to select the fluidic pathways as shown in Figure 3. The integrated glucose sensor simultaneously measures the extracellular glucose concentrations and decides suitable time of secreted sample collection at outlet in Eppendorf tubes for clinical evaluation.

3. Results and Discussion

Islet cells were preloaded into a customized chamber and a testing solution was perfused via a syringe pump. Different glucose concentrations were simultaneously monitored downstream and suitable time of sample collection was decided for accurate cell activity measurement. Figure 4 shows the monitoring curve during the switching between low glucose and high glucose concentrations. The injection and entrapment of human islet cell clusters during monitoring of cell metabolism under different environmental challenges are shown in Figure 5.

Generally, the real time monitoring of human islet cells requires immobilizing the cell for several hours without damage to the cells. During preliminary experiment, islet cells have been entrapped in the micro reaction chamber of the biosystem for up to 8 hours and no damage to the cells was observed. Infused glucose solutions alter the extracellular environment and transport the cell secretion to collection tubes. One of the advantages of the PDMS multi pillar design is that when the fluidic direction is reversed, cells previously studied in the chamber can be retrieved. The DNA quantifying analysis of the recycled islet cells can provide the efficiency of insulin generation by islet cells in different chemical environments.

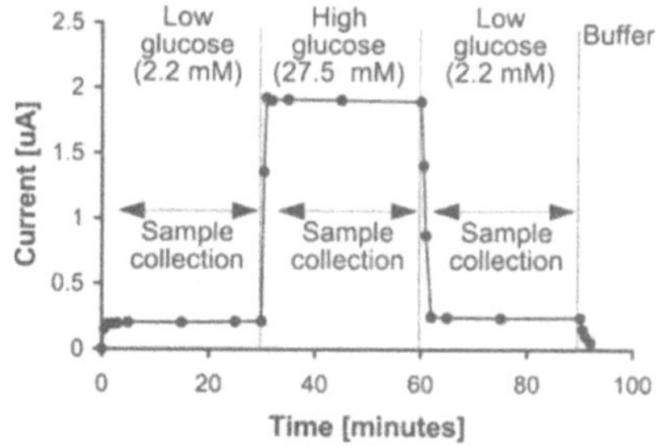

Figure 4. Simultaneous monitoring results of the extracellular glucose level over 90 minutes in the developed biosystem with the integrated glucose sensor.

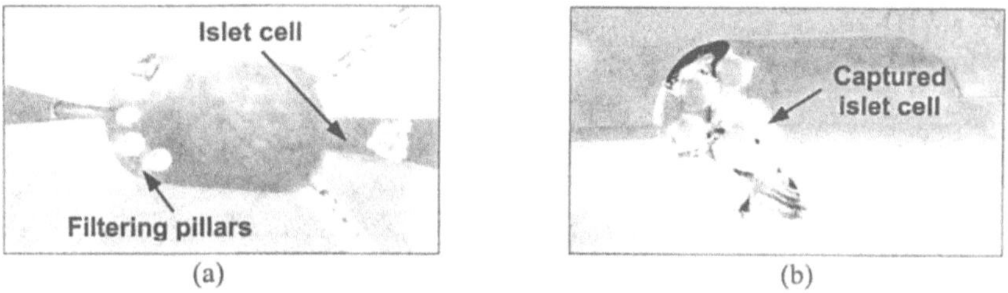

Figure 5. Injection and sustaining of human islet cell into testing chamber: (a) Loading of islet cell and (b) sustaining of islet cell. The islet cell is injected into the testing chamber and entrapped by the multi-pillar structure.

4. Conclusion

This newly developed biosystem showed the feasibility of real time monitoring and/or imaging in a continuous flow system and allows the study of multiple variables such as glucose concentration, secretory drugs, antioxidants, etc. The performance of cell metabolism can be observed by analyzing the collected outflowing fluids. The cell secretion under variant glucose concentrations is collected and investigated, which gives the clue for human diabetes research.

References

1. A. M. Shapiro, J. R. Lakey, E. A. Ryan, G. S. Korbutt, E. Toth, G. L. Warnock, N. M. Kneteman, and R. V. Rajotte, "Islet Transplantation in Seven Patients with Type 1 Diabetes Mellitus Using a Glucocorticoid-free Immunosuppressive Regimen," N Engl J Med, 343(4), pp. 230-238, 2000.
2. C. K. Mathews, K.E. van Holde, and K. G. Ahern, "Biochemistry," Addison Wesley Longman Inc, CA, USA, pp. 835-837, 2000.
3. C. Gao, J.-W. Choi, M. Dutta, S. Chilukuru, J. H. Nevin, J. Y. Lee, M. G. Bissell, and C. H. Ahn, "A Fully Integrated Biosensor Array for Measurement of Metabolic Parameters in Human Blood," Proc. of the 2nd Annual International IEEE-EMBS Special Topic Conference on Microtechnologies in Medicine & Biology, Madison, Wisconsin, USA, May 2-4, 2002, pp. 223-226.

A LabChip® Cell Sorting and Deposition System Based on a Rapid Functional Response to Stimulation of Purinoceptors

Jeffrey Wolk, Anhubav Tripathi, Roy Sandberg, Yue Suo, Andrea Chow, and Gary Wada
Caliper Technologies Corp., 605 Fairchild Drive, Mountain View, CA 94043.
(e-mail address: jeff.wolk@calipertech.com)

Keywords: cell sorting, microfluidic, receptors

1. Introduction

Cell separation technology has become an important tool in cell biology, allowing for the analysis and subsequent cultivation of specific cell subsets. This technique is especially applicable to transfected cells, where only a small proportion of the cells may be expressing the protein of interest. Several techniques have been investigated to perform efficient and rapid sorting, including fluorescence-activated cell sorting (FACS) [1]; magnetic-activated cell separation (MACS) [2]; and micro-fabricated fluorescence-activated cell sorting [3]. We have created a chip-based system that combines the functions of 1) cell sorting based on rapid functional response to an on-chip reaction, and 2) cell deposition in microplate wells.

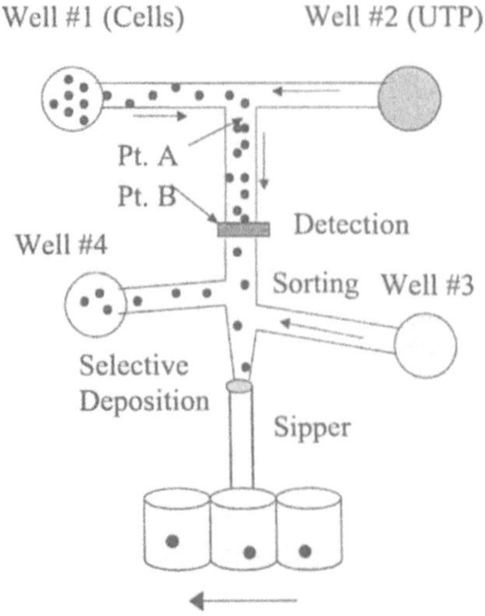

Figure 1-Cell Sorter Chip Schematic

2. Experimental

The sorting chip we have constructed is shown schematically in Figure 1. The chip configuration in Figure 1 shows that four wells and a sipper capillary are connected through micro-channels. THP-1 cells, present in well #1, are pre-loaded with the dye Fluo-4, which is sensitive to the presence of Ca^{2+} ions. THP-1 cells have a purinoreceptor that responds to uridine triphosphate (UTP) by triggering a signal transduction pathway that results in an

Y. Baba et al. (eds.), Micro Total Analysis Systems 2002, Volume 2, 790–792.

increase in intracellular Ca^{2+} ions. When a THP-1 cell comes into contact with UTP, there is a transient increase in fluorescence lasting roughly 30s. A photodiode detector is used to monitor the laser-activated fluorescence of cells flowing in the main channel. The cell fluorescence is shown in Figure 2 with the detector mounted initially near Point A (as labeled in Figure 1) and subsequently near Point B, which correspond to cell-UTP interaction times of 0 and 5s, respectively. The advantage of doing both the chemistry and analysis on chip is apparent, as the signal forming the basis for sorting would otherwise disappear before an analysis could be accomplished.

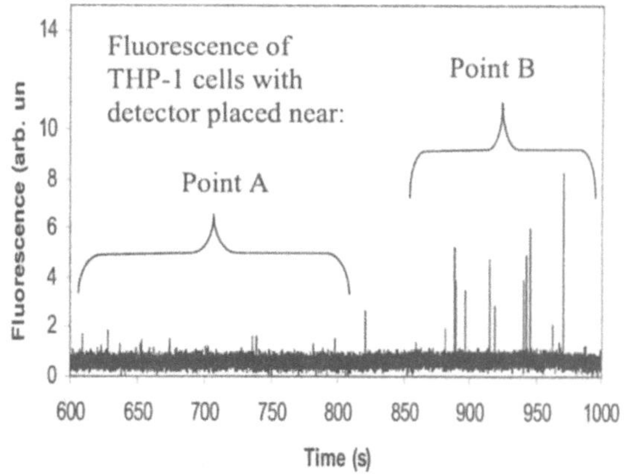

Figure 2 - THP-1 Cells treated with 33 uM UTP. Fluo-4 loaded cells show calcium flux response after 5 s incubation.

To test our ability to distinguish reactive and non-reactive cells, we prepared a mixture of Jurkat cells (which lack a purinoreceptor) with THP-1 cells in a ratio of 9:1. The Jurkat cells should show low fluorescence. As expected, we see in figure 3 that 11% of the cells we observed in our system showed fluorescence above the threshold level, which had been determined in a separate experiment using THP-1 cells.

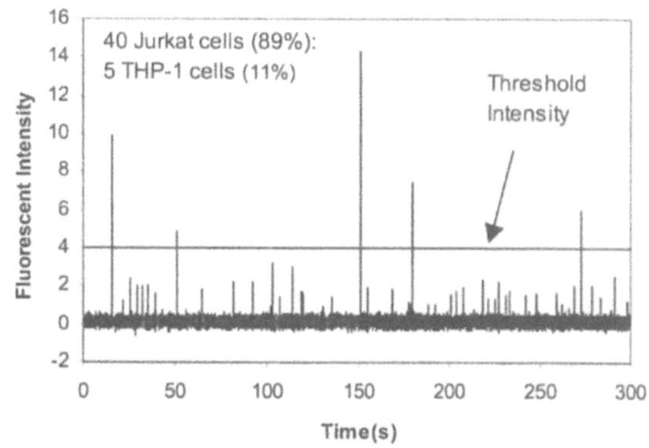

Figure 3 – THP-1 cells show fluorescence above the threshold value after stimulation with UTP, while the Jurkat cell fluorescence remains below threshold.

The cell sorting and deposition is initiated when a fluorescence signal threshold is exceeded. While the signal is below threshold, pressure is applied at the wells to drive the cells to well #4. When a highly fluorescent cell is detected, a microcomputer processing the optical signal passes a voltage signal to the pressure controller that diverts the cells to the capillary by momentarily changing the pressures at the wells. A voltage signal is also sent to an xyz robot positioning system that moves a microplate well under the capillary for single cell collection. In Figure 4, we show the change in pressure associated with a well on the chip in response to above threshold cell fluorescence. Figure 5 is a picture of a single THP-1 that has been sorted into a microtiter plate well.

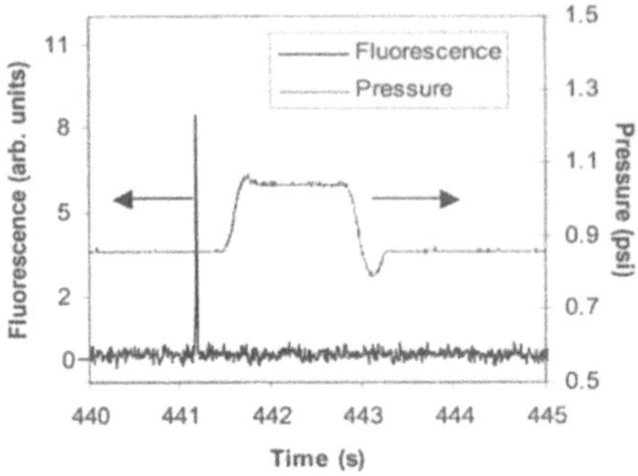

Figure 4 – Demonstration of pressure pulse applied to a well changing in response to above-threshold cell fluorescence.

Figure 5 – Fluorescent image of an isolated THP-1 cell that has been sorted into a microplate

3. Conclusion

We have thus successfully demonstrated a system that can both 1) sort populations of cells according to a rapid functional response to an on-chip reaction, and 2) deposit single cells in a microtiter plate wells for subsequent analysis and/or growth. This system is well suited for integration into a complete cell cloning system.

References

[1] W. A. Bonner, H. R. Hulett and R. G. Sweet, Rev. Sci. Instrument., 19, 813-816 (1972).
[2] S. Miltenyi, et al., Hematopoietic stem cells: the millhouse manual, Dayon: Alpha Med Press, 201-213 (1994).
[3] A. Y. Fu, et al., Nature Biotechnology, 17, 1109-1111 (1999).

POSITIONING AND IMMOBILIZATION OF DROSOPHILA EMBRYOS IN 2-D ARRAYS FOR DRUG INJECTION

Ralph W. Bernstein[1], Xiaojing Zhang[2], Stefan Zappe[2], Matt Fish[3], Matthew Scott[3], and Olav Solgaard[2]

[1]SINTEF Electronics and Cybernetics, Department of Microsystems, Norway
[2]Department of Electrical Engineering, Stanford University, USA
[3]Department of Developmental Biology, Stanford University, USA

Abstract
A technique for positioning and immobilization of fruit fly embryos in 2-D arrays is developed for use in high throughput microinjection of genetic material or drugs. The method is based on fluidic microassembly, and the results indicate that over 90% of the immobilized embryos would be injected with a matching array of microinjectors. The adhesion force of the immobilized embryos was estimated at 36 μN ± 22 μN.
Keywords: microinjection, array, embryo, positioning, Drosophila

1. Introduction

Technology for handling, sorting, and positioning of embryos is increasingly important in biomedicine. Applications include *in vitro* fertilization, cloning, and microinjection of genetic material or drugs for developmental biology studies. Demonstrations of micro technology for handling and immobilization of mammalian embryos [1], thermal branding for embryo labeling [2], and DNA injection [3,4] have been published. These reports are focused on handling of *individual embryos*, either in fluidic channels or by advanced robotic micro manipulation systems. In this paper we present technology for positioning and immobilization of embryos at specific sites by using surface tension. This microassembly technique enables realization of 2-D arrays of immobilized embryos, and opens for treatment of a large number of embryos in parallel. The technology can be utilized in many applications, but our focus is on high throughput injection of genetic material or drugs into fruit fly (Drosophila) embryos in order to investigate their influence on development. In this application the embryo arrays will be aligned with matching arrays of microinjectors [5], as shown in Fig. 1.

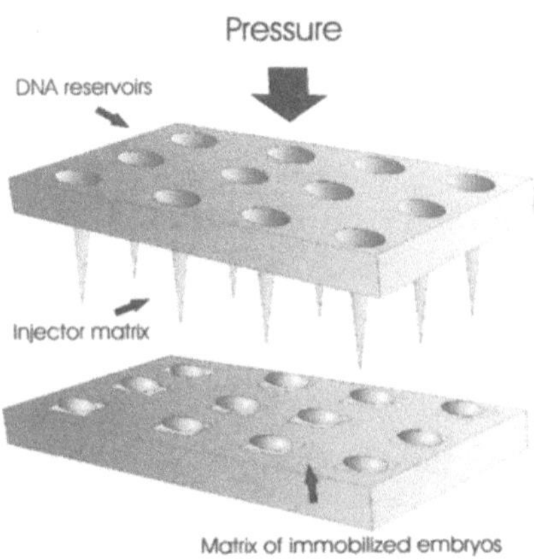

Fig. 1. Schematic drawing of high throughout DNA injection into an array of embryos immobilized and positioned by the technique described in this paper.

Y. Baba et al. (eds.), Micro Total Analysis Systems 2002, Volume 2, 793–795.

2. Experimental, results and discussion

The applied positioning and immobilization scheme is based on fluidic microassembly [6]. Arrays of 250 µm x 400 µm Cr/Au pads were patterned on silicon substrates using photolithography and wet etching. The dimensions of the pads correspond roughly to the size of Drosophila embryos. Samples containing 221 pads where etched for 15 min in 30% H_2O_2 in order to clean the gold areas and oxidize the silicon surface. The samples were then immersed in a 1mM octadecanethiol in ethanol solution and kept overnight. This treatment is adapted from [6] and results in formation of self-assembled monolayers on the gold surfaces and renders them hydrophobic, while the oxidized silicon is kept hydrophilic. After a subsequent rinse in ethanol and storage in methanol the samples were covered with a film of polychlorotrifluoro-ethylene based oil. This oil is known to be inert, have low toxicity, and it is regularly used to cover Drosophila embryos during injection experiments and hatching. Oil covered samples were immersed in DI water, leaving oil only on the hydrophobic gold areas. The oil film will act as a lubricant and adhesive for the immobilization of embryos in arrays. A gold pad covered with oil film is shown in Fig. 2a. Drosophila embryos were dispensed under water and distributed over the sample surface. An embryo positioned at a gold pad is shown in Fig. 2b. Embryos left on the substrate between the pads can be easily be rinsed off the sample when submerged in water.

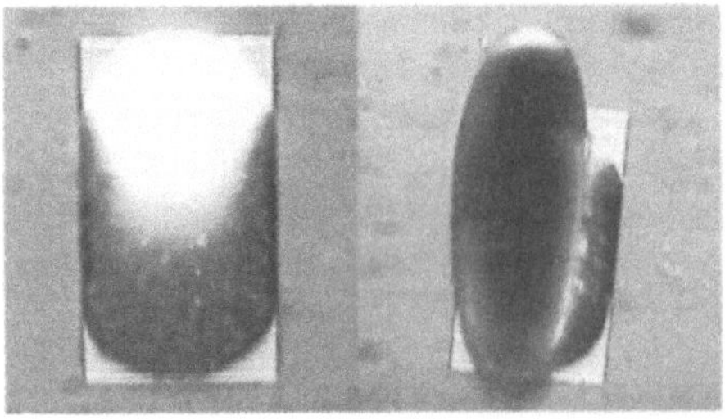

Fig. 2a and 2b.
A polychlorotrifluoroethylene oil drop on a hydrophobic gold pad immersed in water (left). A Drosophila embryo positioned and immobilized at the pad (right).

Although their orientation varies, the embryos tend to position and align themselves to the gold pads. In these first experiments more than 75% of the embryos were aligned within 30 deg. of the orientation of the gold pad. This level of alignment is sufficient for many injection applications.

Some clustering of embryos is observed. Without clusters close to 100% of the immobilized embryos would be hit by a corresponding array of injectors. This number is reduced to about 75% by the clustering observed in our experiments.

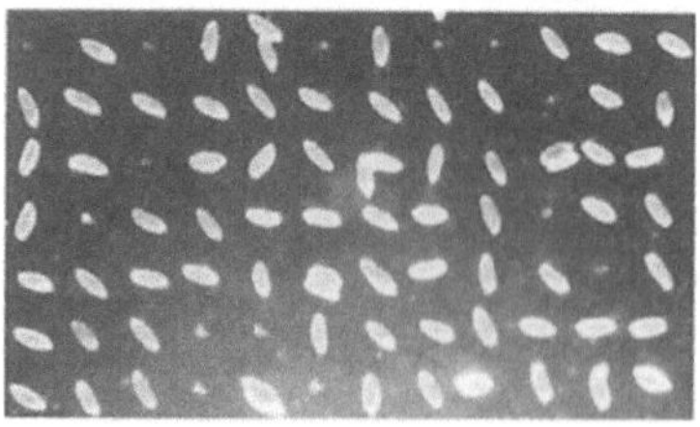

Fig. 3. A sample with immobilized Drosophila embryos. Note that although their orientation varies, the embryos would be injected by a needle array aligned with the center of the pads, except for a few cases of clustering.

To prevent clustering alternative designs with reduced size of the immobilization sites were tested. The results indicate that the clustering can be reduced dramatically. For the sample showed in Fig. 3 more than 90% of the immobilized embryos would be injected.

In order to assess the adhesion force of the embryos a micropipette with a controlled flow of water was used. It was moved into proximity of individual embryos, and the volumetric flow rate at which detachment of the embryos took place was measured. The corresponding drag force was calculated approximately using the approach given in [7]. The adhesion force was estimated at 36 μN ± 22 μN for the immobilized embryos, whereas embryos left at the substrate detach at a drag force of approximately 6 μN ± 3 μN. For comparison the mass of an embryo is approximately 13 μg.

3. Conclusion

A technique for positioning and immobilization of fruit fly embryos in 2-D arrays for drug injection is successfully demonstrated and characterized. In the future the method will be evaluated for high throughput injections of genetic material into embryos using a matching array of microinjectors.

Acknowledgements

The project was kindly sponsored by DARPA program Bio:Info:Micro, The research Council of Norway, SensoNor ASA, and Norchip ASA.

References

[1] I. K. Glasgow, H. C. Zeringue, D. J. Beebe, *Handling Individual Mammalian Embryos Using Microfluidics*, IEEE Trans. Biomed. Eng., vol. 48, pp. 570-577, May 2001.

[2] L. Wang, D. J. Beebe, A. R. Williams, and K. D. Easley, *Electrothermal Branding for Embryos Labeling*, IEEE Trans. Biomed. Eng., vol. 44, pp. 1128-1138, Nov. 1997.

[3] S. Yu, B. J. Nelson, *Microrobotic Cell Injection*, Proc. 2001 IEEE Int. Conf. On Robotics & Automation, Seoul, Korea, May 21-26, 2001, pp. 620-625.

[4] F. Arai, A. Kawaji, P. Luangjarmekorn, T. Fukuda, K. Itoigawa, *Three-Dimensional Bio-Micromanipulation under the Microscope*, Proc. 2001 IEEE Int. Conf. On Robotics & Automation, Seoul, Korea, May 21-26, 2001, pp. 604-609.

[5] K. Chun G. Hashiguchi, H. Toshiyoshi, and H. Fujita, *Fabrication of Array of Hollow Microcapillaries Used for Injection of Genetic Material into Animal/Plant Cells*, Jpn. J. Appl. Phys, vol. 38, pp. L279-L281, 1999.

[6] U. Srinivasan, D. Liepmann, R. T. Howe, *Microstructure to Substrate Self-Assembly Using Capillary Forces*, J. Micromech. Systems, vol. 10, pp. 17-24, March 2001.

[7] C. D. Cooper and F.C. Alley in *Air pollution control: a design approach*, Waveland Press; ISBN: 0881337587; 2nd edition (January 1994), pp. 91-93.

TOWARDS SINGLE-CELL-CONTROLLED ELECTROPORATION IN A MICROFLUIDIC DEVICE

Urban Seger[1], Shady Gawad[1], Alexander Scheer[2], and Philippe Renaud[1]
[1] STI-LMIS, EPFL, CH-1015 Lausanne, Switzerland
[2] Serono Pharmaceutical Research Institute, CH-1228 Geneva, Switzerland
urban.seger@epfl.ch

Abstract

We are developing an integrated microfluidic device that combines electroporation with impedance spectroscopy flow cytometry. We here report on the kinetics of fluorophore intake by individually lysed HEK cells and on the transfection of CHO cells confirmed by functional protein expression.

Keywords: Electropermeabilisation, Impedance Spectroscopy, Dielectrophoresis

1. Introduction

Electroporation (EP) is a method for transfecting biological cells. It is used in research as well as in industrial biotechnology. Typically some 10^5 to 10^7 cells in suspension are exposed to a single or a series of electric pulses of several kV/cm. Concepts of miniaturized EP-devices have been proposed [1, 2].

Electroporating a cell population at once results in very heterogeneous transfection and survival rate distributions [3]. Any method improving the quality of electroporation is of interest for handling small cell populations or rare cells. We are currently studying an approach of individually controlled single-cell electroporation.

We propose a microchip in which cells are flowing through a channel and characterized one-by-one by means of Impedance Spectroscopy Flow Cytometry (ISFloC) [4]. Each cell can be guided towards and optionally held at EP electrodes by negative dielectric forces. Electroporation is carried out with parameters (field strength, pulse duration, repetition rate) that are optimised for each cell according to its size and its cytoplasmic conductivity.

2. Chip Fabrication

The fluidic microchips consist of channels with integrated Ti/Pt electrodes. The channels (typ. 20 x 200 μm^2) are formed by polyimide processing on glass substrates and flip-chip bonding. The chips are mounted in a combined fluidic and electric interconnection block.

Y. Baba et al. (eds.), Micro Total Analysis Systems 2002, Volume 2, 796–798.

3. Experimental

Human Embryonic Kidney (HEK) and Chinese Hamster Ovary (CHO) cells are cultured in DMEM with 2.2% FCS at 37 °C. Negative dielectrophoresis is generated with sinusoidal 1 MHz fields of 1 kV/cm. Bipolar, rectangular electroporation pulses with variable field strengths from 2.5 to 10 kV/cm are delivered as 20-cycles pulse trains of 100 kHz, repeated at 50 to 250 Hz.

The functionality of our EP set-up is validated in two ways: (1) HEK cells are suspended in PBS buffer containing 0.1 mg/ml of the nucleic acid stain Propidium Iodide (PI). PI florescence increases when in contact with DNA. The cells are injected into the microchannel and held in place against a steady liquid flow by a dielectrophoretic trap, see Fig 1. (2) Exponentially growing CHO cells are suspended in DMEM containing Penicillin and Streptomycin (PenStrep). The solution is loaded with 10µg/ml pEGFP-N1, a plasmid coding for enhanced Green Fluorescent Protein (GFP). The cells are then flowed through the microchannel, passing between two parallel plate electrodes of 200 µm length. Flow speed is about 1 mm/s. The cells are finally incubated for 36 h

4. Results

(1) EP pulses were applied and the increase in fluorescence of the cells - indicating membrane permeabilisation - was optically monitored. Typical kinetics of PI intake for a given concentration is shown in Fig 2.

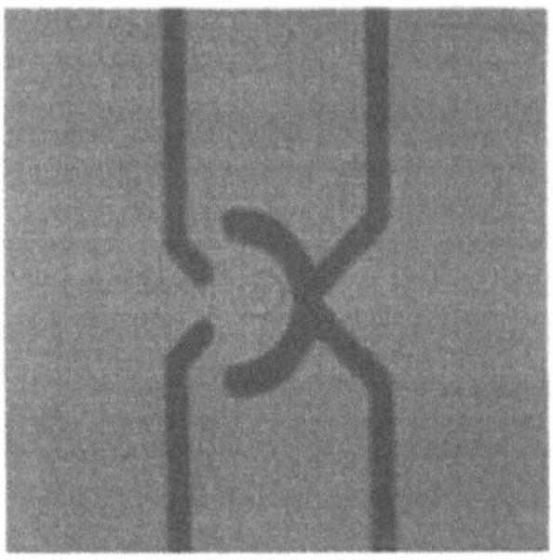

Figure 1. Optical top-view of a HEK cell held by a pair of arc-shaped DEP electrodes against a constant liquid flow from left to right. The cell is kept close to the EP electrodes.

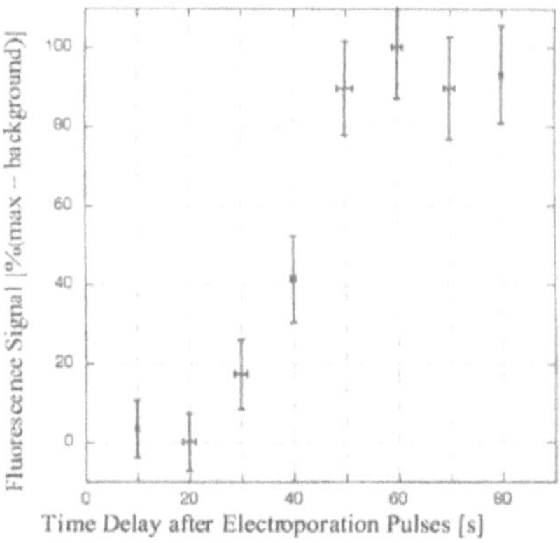

Figure 2. Kinetics of increase in fluorescence due to PI intake induced by permanent EP. (Duration: 300 ms, f: 100 kHz, E: 2.5 kV/cm)

(2) Green fluorescence was observed when plated CHO cells were inspected under a confocal microscope. This indicates successful electroporation and cell survival (Fig 3).

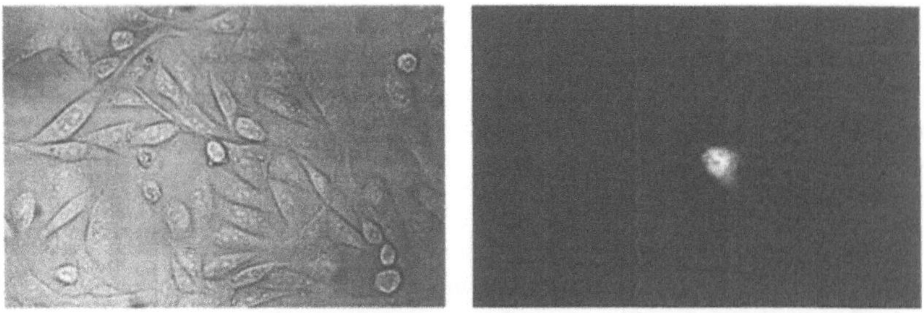

Figure 3. Bright and dark field photographs of a CHO cell expressing GFP. The cells were suspended in pEGFP-loaded medium, exposed to a bipolar pulsing field of 10 kV/cm for less than 0.2 s and incubated at 37 °C for 36 h.

5. Conclusion

After first phenomenological validations, on-chip electroporation is to be characterized by measuring the induced changes to the impedance spectrum $Z = Z(t)$ of each cell. Future steps will include cell viability studies as well as further protein expression assays of small cell populations.

By combining electropermeabilisation with electrical impedance monitoring we are developing an integrated microfluidic tool, which will permit a single-cell electroporation with both improved transfection and survival rates.

Acknowledgements

The authors are grateful to P. Pascoal, E. Ilegems, and J.-B. Perez (all of LCCPM, EPFL) for their help in cell culturing, DNA transfection and confocal microscopy.

References

1. Y. Huang and B. Rubinsky, *Micro-Electroporation: Improving the Efficiency and Understanding of Electrical Permeabilization of Cells*, Biomedical Microdevices 2:2, Kluwer Academic Publishers, pp. 145-150, (1999)
2. Y.-C. Lin and M.-Y. Huang, *Electroporation Microchips for Continuous Gene Transfection*, Sensors and Actuators B 79, Elsevier Science, pp. 137-143, (2001)
3. M.J. O'Hare, M.G. Ormerod, P.R. Imrie, J.H. Peacock, and W. Aschee, *Electro-permeabilization and Electrosensitivity of Different Types of Mammalian Cells*, Electroporation and Electrofusion in Cell Biology, Plenum Press, pp. 319-330, (1989)
4. S. Gawad, L. Schild, and Ph. Renaud, *Micromachined Impedance Spectroscopy Flow Cytometer for Cell Analysis and Particle Sizing*, Lab on a Chip 1, The Royal Society of Chemistry, pp. 76-82, (2001)

ENHANCED VELOCITY GRADIENTS WITHIN MICROFLUIDICS FOR CELLULAR MANIPULATION

Dino Di Carlo and Luke P. Lee
Berkeley Sensor and Actuator Center and Department of Bioengineering, University of California, Berkeley, CA 94720, USA
ddicarlo@socrates.berkeley.edu

Abstract

A model has been developed to quantify forces experienced by cells undergoing flow through integrated microstructured geometries within microfluidic devices. At sudden contraction points velocity gradients lead to enhanced shearing forces. Membrane tensions are modelled based on the inertial deformation of the cell with the fluid flow and membrane poration and lysis are predicted as a function of flow rate and geometry. The model is compared with the cell lysis results from an integrated sample preparation module for cell sorting, mechanical lysis, and mRNA purification, which was microfabricated using silicon bulk micromachining.

Keywords: Lysis, mechanical poration, microfluidics, membrane rupture model

1. Introduction

For micro total analysis systems (μTAS) where some analyte within cells must be probed or some foreign material injected, manipulation of cells and their membranes is required. This includes reversible opening of cell membranes (poration) or irreversible membrane rupture (lysis). Gene delivery microsystems require poration of cells while diagnostic microdevices often require complete lysis to access intracellular proteins or nucleic acids for pathogen detection or gene expression profiling.

We have developed a shear-based mechanical lysis device for functional genomics applications (Fig. 1a) [1] that consists of microchannels etched into a bulk silicon wafer and enclosed by a bonded Pyrex wafer. There is a sudden contraction filter within the microchannels (Fig. 1b) that creates large velocity gradients useful for mechanical manipulation of cells flowing through. A model has been developed to predict membrane tensions and thus lysis or poration in cells within the specific device and for similar geometries. Such analysis is useful in developing novel cell manipulation devices for mechanical poration and gene delivery or mechanical lysis for point-of-care diagnostics and functional genomics microsystems. It is also useful in determining maximum flow rates within microfluidic devices when minimal cell damage is a concern.

Y. Baba et al. (eds.), Micro Total Analysis Systems 2002, Volume 2, 799–801.

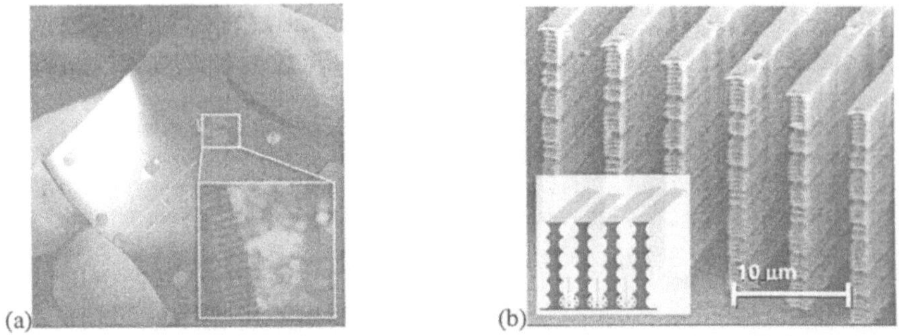

Figure 1. (a) Sample preparation module for gene expression analysis with inset of the mechanical lysis filter with HL-60 cellular debris fluorescing. (b) SEM of the lysis filter with 3-micrometer gap size and 15-micrometer depth. Micro-knife structures are seen, fabricated by increased scalloping with a modified deep reactive ion etch recipe.

2. Theory

To model forces on cell membranes the cells were considered to deform with the bulk solution which is a good assumption when the fluid-flow induced membrane tensions are greater than the elastic tension of the membrane and its associated cytoskeletal proteins [2]. The Navier-Stokes equation was used with the assumption of incompressibility and negligible viscous terms. Eq. 1 was developed considering flow and velocity gradients in the $\underline{e}_1$ direction in the microchannel frame, which are the main gradients that streamlines intersect (Fig. 2). L is defined as dv_1/dx_1 and is a function of x_1.

$$\frac{\partial p}{\partial x_1} = \rho(\frac{\partial v_1}{\partial t} + L(x_1)v_1) \tag{1}$$

The time derivative of velocity will drop out for a steady flow and if we integrate (1) over a streamline from *0* to *A* in the center of a filter gap in Fig. 2 we arrive at Eq. 2.

$$\int_{x_0}^{x_a} L(x_1)v_1 dx_1 = \frac{p_0 - p_a}{\rho} \tag{2}$$

Because of an induced elastic tension (T_{elas}) within the membrane a non-continuous pressure drop is observed from points *A* to *B*. Balancing $T_{elas} = (p_a - p_b)r_g$ where r_g is the gap size between obstacles and combining with (2) gives us Eq. 3, an expression for flow induced tension (T_{in}). Using r_g is accurate if the cell fills the obstacle gap during flow, which is a good assumption for most cells, which have diameters of 15-20 µm.

$$T_{in} = \rho r_{gap} \int_{x_0}^{x_a} L(x_1)v_1 dx_1 \tag{3}$$

Rupture tensions of 10 mN/m have been reported for monounsaturated phosphatidylcholine (PC) lipid bilayers for ramp rates of ~0.1 mN/m/s [3]. Too lengthy to report here, it is also apparent that exposure times to different tensions has an effect on membrane rupture as well.

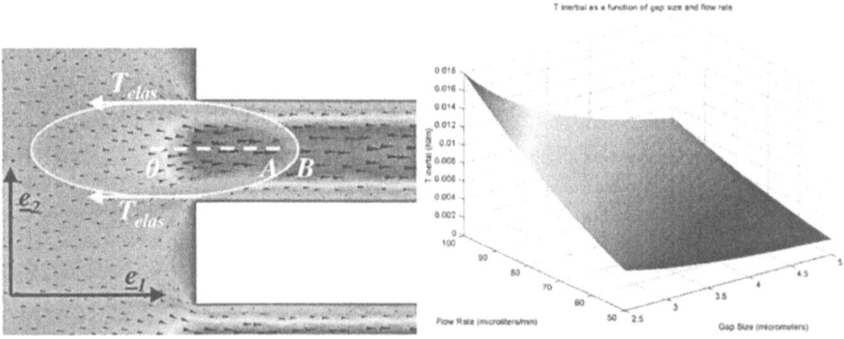

Figure 2. (Left) Diagram of cell flowing through obstacles, and critical regions for force calculations. Arrowheads are velocity vectors with size based on velocity magnitude. (Right) Max T_{in} as a function of flow rate (μL/min) and obstacle gap size.

3. Results and Conclusions

Using CFD-RC™ software to determine velocity fields maximum T_{in} was determined as a function of flow rate and obstacle gap (Fig. 2). T_{in} is a quadratic function of flow rate Q and power law in gap size. For our device geometry of 3-micron gaps a calculated flow rate of 84.8 (μL/min) is required to reach rupture tensions, which agrees with preliminary experimental data [1]. Below this critical flow rate, reversible poration without lysis is prevalent.

The analysis of the forces cells experience in microdevices is essential for developing new cell manipulating microsystems for controlled poration and lysis. Also, it is important to understand maximum flow rates when viable cells are desired to flow through specific geometries. The theory developed in this paper is a first step in understanding forces and manipulating cells on the microscale in a controlled manner.

References

1. D. Di Carlo and L. P. Lee. *Proceedings of the 2nd Annual International IEEE-EMBS Special Topic Conference on Microtechnologies in Medicine & Biology, Madison, Wisconsin, USA* (2002) 527
2. M. Lokhandwalla and B. Sturtevant. *Phys. Med. Biol.* **46** (2001) 413-437
3. K. Olbrich, W. Rawicz, D. Needham, and E. Evans. *Biophysical Journal.* **79** (2000) 321
4. M. Winterhalter. *Colloids and Surfaces A* **149** (1999) 161

CELL ANALYSIS IN VALVED MICROFLUIDIC DEVICES

Aaron R. Wheeler, Andrew M. Leach, and Richard N. Zare
Department of Chemistry, Stanford University, Stanford, CA

Abstract
A novel μTAS methodology for chemical cytometry is presented. Microfluidic chambers defined by microvalves are used to isolate and lyse cells, followed by fluorescent derivatization and separation of the cellular contents.

Keywords: cells, valves, PDMS, MEKC, electrophoresis

Introduction
The recent development of microfluidic valves[1] made from elastomeric polymers has created new and exciting possibilities for the field of micro Total Analysis Systems (μTAS). Valves are formed at the intersection of two channels separated by a thin poly(dimethylsiloxane) (PDMS) membrane. As pneumatic pressure is applied to the upper "control channel", the membrane is pinched, and the lower "fluidic channel" is effectively closed to flow (Figure 1).

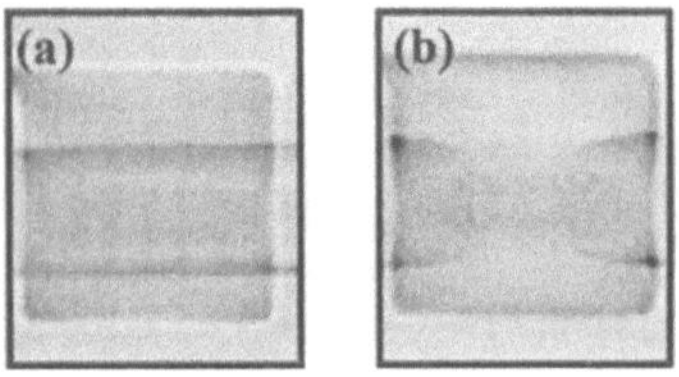

Figure 1. Microvalve (a) open, and (b) closed. The fluidic channel stretches from left to right; the valve (slightly out-of-focus) actuates when ~10 psi of nitrogen gas is applied.

We report application of microvalves for novel chemical cytometry analyses of mammalian cells. Previous μTAS experiments involving mammalian cells have proven difficult, often relying on high electric fields[2, 3] (which may damage the cells) for manipulation. Residual hydrostatic pressure imbalances and buffer incompatibilities have also proven to be problematic.[4] Valved devices allow us to isolate cells rapidly and reproducibly with pressure-driven flow, and subsequently lyse, label, and separate the cellular contents.

Experimental
Our methodology relies on a "two-chamber" design. Cells are loaded into a "cell chamber" with pressure driven flow; simultaneously, reactants (including lysing and

Y. Baba et al. (eds.), Micro Total Analysis Systems 2002, Volume 2, 802–804.

labeling reagents) are loaded into an adjacent "reactant chamber" (Figure 2). The chambers are then opened to each other, allowing the contents to mix and react.

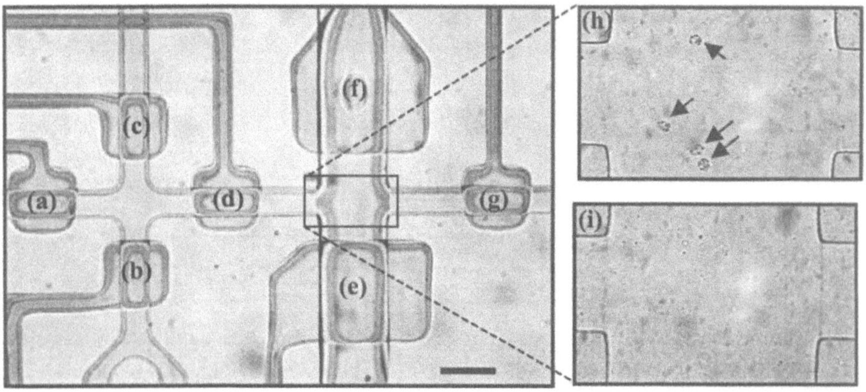

Figure 2. Images of valved cell analysis chip (bar 200 μm). Valves (a-d) define the reactant chamber. Reactants are loaded from (b) to (c). Valves (d-g) define the cell chamber. Cells are loaded from (e) to (f). Inset (h) of the cell chamber with four isolated *Cath.a* cells. Inset (i) same cell chamber ten minutes after opening valve (d) – as sodium dodecyl sulfate (SDS, 3%) in the reaction chamber diffuses into the cell chamber, the cells lyse. After reaction, sample is separated on a serpentine column to the left of (a).

Kinetics experiments demonstrate that the mixing contents of the combined chambers (~1.5 nL) react completely in less than 5 min (Figure 3).

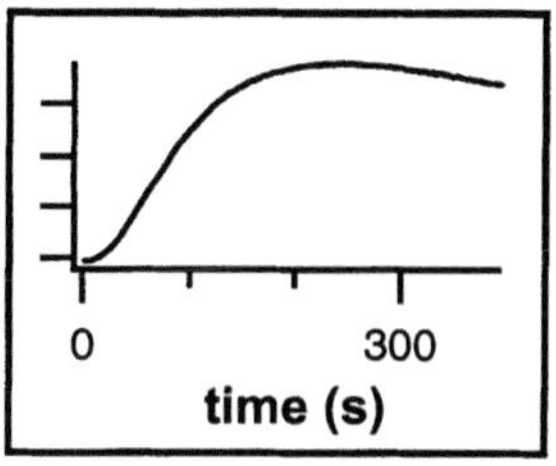

Figure 3. Reaction profile of two-chamber system. Labeling dye, naphthalene-2,3-dicarboxaldehyde (NDA), loaded in reactants chamber; amino acid standards (20 μM) loaded in cell chamber. Fluorescence monitored by focusing laser into cell chamber after valve (d) is opened (see figure 2). The reaction is complete in < 5 min, after which the signal photobleaches.

After reaction, the valves are opened into a serpentine separation column, high voltage is applied, and the mixture is separated by micellar electrokinetic chromatography (MEKC). Labeled cell contents are detected with epifluorescence, and the data are collected as MEKC chromatograms (Figure 4).

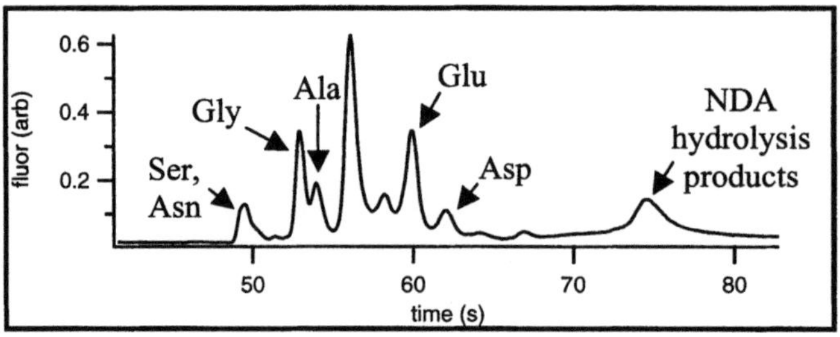

Figure 4. MEKC chromatogram of the contents of *Cath.a* cells. Cells were lysed and reacted in a solution of SDS (1 %), NDA (1 mM), and KCN (1 mM). The separation buffer was borate (10 mM, pH 9) SDS (1 %). The applied voltage was 5 kV over a three-turn serpentine column (15 cm).

Discussion and Conclusions

Complex μTAS cell analyses require strict control of a variety of different solutions. These include labeling reagents, lysing reagents, "cell-friendly" buffers, and separation buffers. Previous implementations of such experiments have been difficult, relying on off-column labeling, and constant, precise voltage control.[2-4] Our new method overcomes this problem by using pneumatic microvalves to effect complete, straightforward control over cell manipulation and the different solutions involved.

Acknowledgements

The microfluidic chips used in this project were manufactured by Fluidigm Corporation through their foundry services program. This work is supported by Beckman Instruments.

References

1. M. A. Unger, et al., *Monolithic Microfabricated Valves and Pumps by Multilayer Soft Lithography*, Science, pp. 113-116, (2000).
2. P. C. H. Li and D. J. Harrison, *Transport, Manipulation, and Reaction of Biological Cells On-Chip Using Electrokinetic Effects*, Analytical Chemistry, pp. 1564-1568, (1997).
3. C. T. Culbertson, et al., *Rapid Cellular Assays on Microfluidic Devices*, Proc. Micro Total Analysis Systems 2001, Kluwer Academic Publishers, Dordrecht, The Netherlands, pp. 285-286, (2001).
4. A. Wheeler, et al., *CATH.A Neuron Cell Analysis on a Chip with Micellar Electrokinetic Chromatography*, Proc. Micro Total Analysis Systems 2001, Dordrecht, The Netherlands, pp. 311-312, (2001).

A Multi-purpose Micro System for Electrophysiological Analyses of Single Cells

Arum Han[1], Eileen Moss[1], Richard D. Rabbitt[2], and A. Bruno Frazier[1]
[1]School of Electrical and Computer Engineering
Georgia Institute of Technology, USA
[2]Department of Bioengineering, University of Utah, USA

Abstract

A micro analysis system for multi-purpose electrophysiological analyses of single cells has been designed, fabricated, and characterized. The micro system contains multiple extracellular electrodes and opposing electrode pairs that interface with a single cell analysis cavity. These electrodes can be used for general singular electrode extracellular stimulation / recording studies, dual electrode applications such as impedance spectroscopy, and for applications requiring a greater number of electrodes such as impedance tomography. Additionally, the analysis cavity contains a sealing / clamping mechanism and a backside extracellular reference electrode for patch clamping studies. The analysis cavities were designed in an array format for parallel analyses. Proof of concept experiments were carried out using bovine chromaffin cells. The cellular analysis system was demonstrated by performing electrical impedance spectroscopy on chromaffin cells, air, and phosphate buffered solution (PBS) over a range from 40.0Hz to 1.0MHz. Furthermore, the cellular analysis system was demonstrated in patch clamping mode. Using the integrated glass sealing ring, giga-seals were obtained on a population of chromaffin cells. The micro system generated average giga-seal resistances of 0.23 GΩ with non-sealed average cavity-to-backside resistances of 10.5 kΩ.

Keywords: Electrical impedance spectroscopy, patch clamping, chromaffin cell

1. Introduction

Micro analysis systems capable of electrically / chemically interfacing with and analyzing single or small populations of cells have been under development for applications ranging from basic research on cellular activity to drug discovery [1,2]. The ability to interface single cells with micro analysis systems enables researchers to perform complex analyses requiring precise delivery of biochemicals to a microenvironment, integration of electrical functionality, and integration of optical functionality in a parallel fashion. Previous work demonstrated that interfacing single biological cells with microfluidic system in a flow channel can be used for analysis of red blood cells, white blood cells, and eukaryotic cells [1,3,4]. Using this technique eliminates the need for

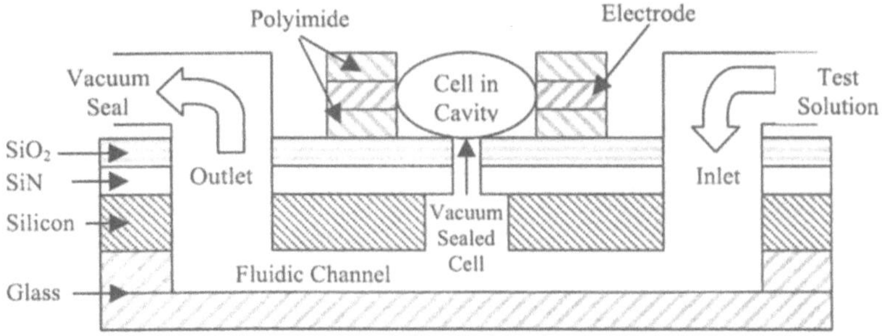

Figure 1. Schematic cross-section of the micro analysis system for electrophysiological study.

Y. Baba et al. (eds.), Micro Total Analysis Systems 2002, Volume 2, 805–807.

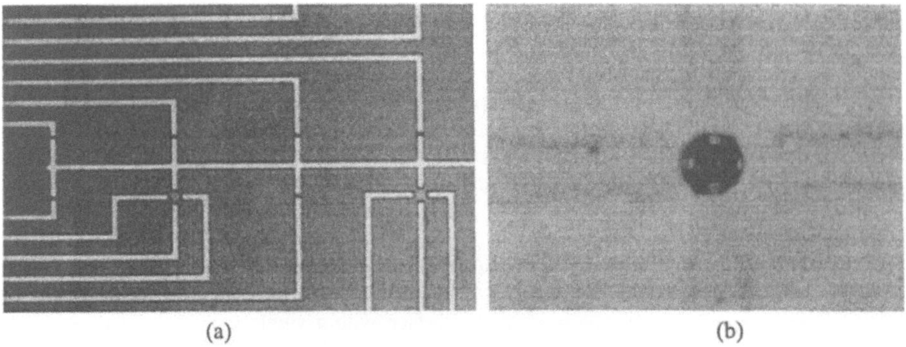

Figure 2. Fabricated system showing (a) array of cavities and (b) enlarged view showing the 16.0 μm polyimide cavity with 2.0μm vacuum via and four electrodes.

developing and integrating functionally equivalent conventional cell sorting techniques, such as fluorescent tagging, into the microdevice format.

In this work, a micro analysis system for electrophysiological study of cells has been designed, fabricated, and characterized using bovine chromaffin cell, a hormonal cell found in the adrenal gland. Chromaffin cells are interesting due to the electrically excitable characteristics. This system has the capability to perform whole cell patch clamping, impedance spectroscopy, and general extracellular stimulation / recording using integrated, multi-electrode configurations.

2. Methodology

The system is composed of analysis cavities that house a single cell, multiple electrodes surrounding the circumference of the cell to measure the electrical characteristics of the cell, and a vacuum via to create giga-seals for patch clamping and to hold the cells in position during the analysis operation. Arrays of such devices were fabricated on 3-inch silicon wafers with SiN and SiO_2 films deposited on the surface using plasma enhanced chemical vapor deposition. Backside vacuum ports were etched using 20.0 wt% KOH solution at 75°C. A 2.0 μm diameter vacuum via was etched through the SiN/SiO_2 film in the center of the analysis cavity. A 3.0 μm thick polyimide layer was spin-coated, followed by deposition and patterning of one or two pairs of opposing Ti/Au electrodes for each analysis cavity. A second layer of 3.0 μm thick polyimide was spin-coated over the patterned electrodes. Next, 10.0 μm and 16.0 μm diameter cavities were etched in polyimide using reactive ion etching (RIE) in a 90% oxygen, 10% CHF_3 environment. Using a 3-inch glass wafer, a network of fluidic channels was patterned and etched as a backside vacuum/fluid interface for the analysis cavities. The glass wafers were subsequently aligned to the overlying silicon wafer and anodically bonded. The glass microchannels were used to pull vacuum simultaneously through all the analysis cavities in the array. Reference electrodes were attached 300 μm from the backside of the vias to characterize the device in patch clamping mode. Figure 1 shows the schematic cross-section of the device.

3. Results and Discussions

Figure 2(a) shows arrays of 8 cavities with two and four opposing electrode configuration interfacing with each analysis cavity. Figure 2(b) shows a 16.0 μm diameter cavity with a 2.0 μm vacuum hole and four 4.0 μm wide opposing gold electrodes.

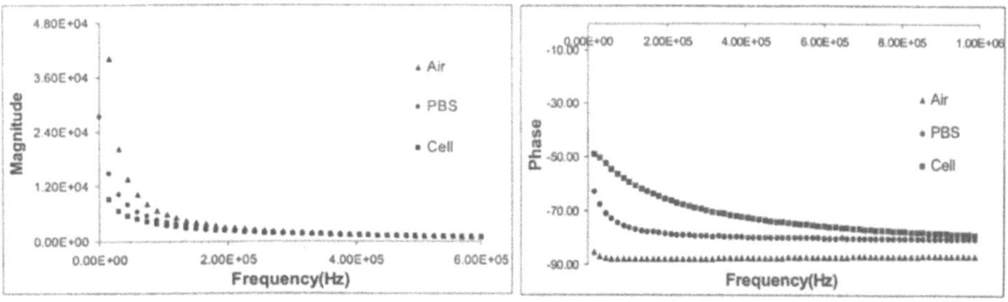

Figure 3. Electrical impedance spectrums for air, PBS, and chromaffin cells over a frequency range of 40.0 Hz to 1.0 MHz.

The sealing capability of the device was characterized with phosphate buffered solution (PBS) and bovin chromaffin cells. The device was filled with PBS and the impedance was measured between one of the integrated electrodes located inside the analysis cavity and the backside reference electrode. With PBS present and without a cell, the impedance showed an average of 10.5 kΩ at DC. When the vacuum was applied and a cell was trapped in the analysis cavity, the impedance increased to an average of 0.23 GΩ at DC.

In a second experiment, the impedance between two opposing electrodes in the analysis cavity was measured to obtain the impedance spectrums for air, PBS, and chromaffin cells. In the case of PBS, the resulting impedance was 9.67×10^5 (magnitude) and -58.2° (phase) at 40Hz. When the cell was trapped in the cavity due to vacuum, the impedance changed to 4.70×10^5 (magnitude) and -38.6° (phase) at 40.0 Hz. The device was then used to compare the impedance of air, PBS, and chromaffin cell over a frequency range of 40.0 Hz to 1.0 MHz (Figure 3).

4. Conclusions

A multi-purpose micro systems for biological cellular analysis was designed, fabricated and demonstrated. The feasibility of using the cellular analysis system for electrical impedance spectroscopy of single cells and for achieving giga-seals for patch clamping was successfully demonstrated on bovine chromaffin cells.

5. Acknowledgements

This work was supported by the National Institute of Health (NIH)/National Institute on Deafness and Other Communication Disorders (NIDCD) under Grant DC04928.

6. References

1. S.Gawad, S. Metz, L. Schild, and Ph. Renaud, "Impedance Spectroscopy Cell Analysis in Microchannels", *Micro Total Analysis Systems* 2001, pp. 253-255, 2001.

2. Axon Instruments, Inc., www.axon.com.

3. H.E. Ayliffe, A.B. Frazier, and R.D. Rabbitt, "Electrical Impedance Spectroscopy using Microchannels with Integrated Metal Electrodes", *IEEE Journal on Microelectromechanical Systems*, 8, pp. 50-57, 1999.

4. L.L. Sohn, O.A. Saleh, G.R. Facer. A.,J. Beavis, R.S. Allan, and D.A. Notterman, "Capacitance Cytometry: Measuring Biological Cells One by One", *Proceedings of the National Academy of Science*, vol. 97:20, pp. 10687-10690, 2000.

TOPOLOGICALLY INDUCED BACTERIAL AGGREGATION IN A MICROFABRICATED RANDOM MAZE

Sungsu Park[1], Emil Yuzbashyan[1], Peter Wolanin[2], Jeffry Stock[2], Pascal Silberzan[3] Robert Austin[1]

[1]Dept. of Physics, Princeton University, Princeton, NJ 08544, USA

[2]Dept. of Molecular Biology, Princeton University, Princeton, NJ 08544, USA

[3]Institut Curie, Paris, France

Abstract

We have used techniques of soft lithography to fabricate complex but well controlled topological microenvironments in order to study the collective behavior of *Escherichia coli.* We have found that starting from a homogeneously distributed suspension of bacteria, they self-aggregate into confined geometries. This was first observed in PDMS mazes designed to mimic complex environments, then more clearly in a simple geometry consisting of a large open area surrounding a 250x250 micron square with a narrow 40 micron channel. Our results show that bacteria chemotactically communicate amongst themselves and search out collectively highly confined environments when they are under stressed conditions caused by local bacteria densities.

Keywords: Chemotaxis, aggregation, maze, PDMS.

1. Introduction

Bacteria communicate each other by producing attractants and repellents. This, under certain circumstances, leads to a collective behavior such as elaborated geometrical patterns [1, 2]. We constructed PDMS mazes to probe under more realistic topological conditions than simple agar gels the communication of bacteria, under assumption that they could find the most suitable microenvironment for the colony survival within the maze. To test this prediction further, we constructed a simple geometry consisting of a large open area surrounding a square with a narrow opening. These two fabricated microenvironments should present a reasonable challenge to bacteria to be adequate to their complex behavioral strategies.

Y. Baba et al. (eds.), Micro Total Analysis Systems 2002, Volume 2, 808–810.

2. Experimental

Biocompatible and well controllable microenvironments were fabricated using soft lithography in which PDMS elastomer is casted to a rigid master made of SU-8 10. PDMS cast removed from the master was treated with oxygen plasma for 1 min to improve adhesion to a microscope glass slide. Chemotactic and nonchemotactic *Escherichia coli* strains were transformed with plasmids encoding green fluorescent protein (GFP) for ease of visualization. Bacteria were grown in LB or tryptone broth at 30°C with shaking at 150 rpm before being filled into each PDMS chip.

3. Results and Discussion

The fundamental phenomenon seen in Figure 1 is the aggregation of chemotactic *E. coli* into confined areas of the maze, although we cannot predict the actual region where the local clusters form. This clustering only occurs for chemotactic bacteria, a nonchemotactic "smooth swimmer" *E. coli* strain (all *che* genes deleted) does not produce such patterns (data not shown). The complex structure of the maze and the fact that it is a true random environment makes it difficult to predict where the clustering will occur.

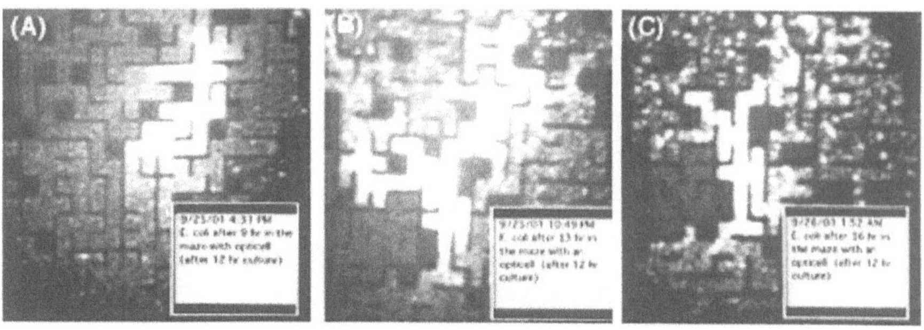

Figure 1. Epifluorescence images of *E. coli* aggregation in a random maze. (A) x100 image of dynamical clustering of the bacteria 9 hours after inoculation with early-log phase chemotactic *E. coli* in LB broth. (B) clustering of the bacteria 13 hours after the inoculation, showing an early stage of collapse on the aggregated bacterial structure. (C) heavily collapsed clustering 16 hours after the inoculation.

A simpler topology than the maze that still probes the essential elements of this bacterial communication and instability in the local density is an enclosed square that has a single inlet channel. This simple topology probes the ability of the bacterium to communicate with other bacteria, since the enclosed 2-D volume will provide a natural signalling center for the chemotactic signal. Figure 2 and 3 show that in a microfabricated environment it is possible to observe this prediction.

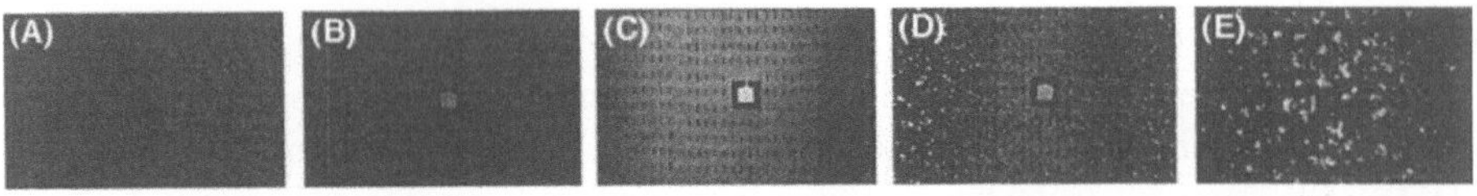

Figure 2. Collective behaviors of GFP-expressing chemotactic *E. coli* in a simple geometry consisting of a large open area surrounding a 250 x 250 micron square with a 40-micron opening. Bacteria are inoculated from both ends into a dry device, so the initial bacteria density is uniform. The small black squares are rectangles of PDMS to support roof of the device. (A) 10 min after filling with mid-log phase *E. coli* in LB broth. (B) 1 hour after filling. (C) bacteria aggregated around the square 2 hours after filling. (D) collapse in the boundary of the aggregated region into spherical colonies 5 hours after filling. (E) completed collapse of the aggregation 12 hours after filling.

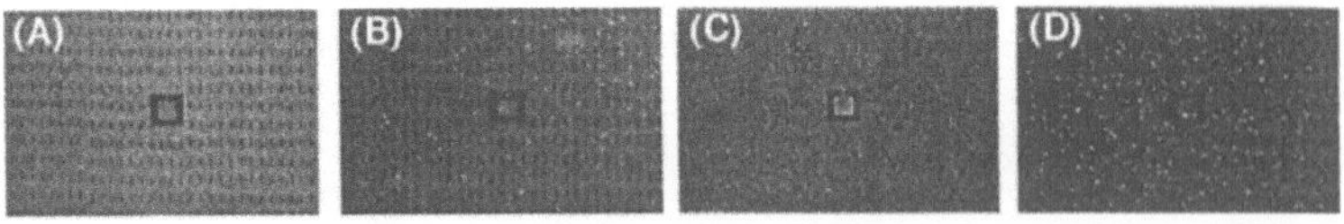

Figure 3. Nonchemotactic behavior of GFP-expressing *E. coli* in the simple geometry. (A) homogeneously dispersed bacteria 2 hours after filling with chemotactic *E. coli* in tryptone broth (B) 12 hours after filling with the chemotactic *E. coli* in tryptone broth. (C) homogeneously dispersed bacteria 2 hours after filling with early stationary-phase smoothly swimming *E. coli* in LB broth. (D) 12 hours after filling with the nonchemotactic *E. coli* in LB broth.

4. Conclusions

We have shown that bacteria are able to seek each other and effectively hide in confined places by their chemotaxis system. Techniques of microfabrication are feasible for constructing well-controlled microenvironments that test these predictions. Further analysis using mutant strains which knockout known aspects of the chemotaxis system can be used to understand the biological reason(s) lies on the aggregation phenomena.

References

1. E. O. Budrene and H. C. Berg, Nature '91, Vol. 349, pp. 630-633.
2. E. O. Budrene and H. C. Berg, Nature '95, Vol. 376, pp. 49-53.

MONITORING OF HUMAN CHEMILUMINESCENT REPORTER CELL LINES EXPRESSING G-PROTEIN-COUPLED RECEPTORS IN A MICROFLUIDIC SYSTEM

Richard Davidsson[1], Jesper Bristulf [2], Knut Kotarsky[2], Björn Olde[2], Christer Owman[2], Martin Bengtsson[3], Thomas Laurell[3], Jenny Emnéus[1]

[1]Department of Analytical Chemistry, Lund University, Sweden
[2]Division of Molecular Neurobiology, Lund University, Sweden
[3]Department of Electrical Measurements, Lund Institute of Technology, Lund University Sweden
(richard.davidsson@analykem.lu.se)

Abstract

A microfluidic system for monitoring genetically modified HeLa cells immobilized on silicon microchips is presented. Stimulation of cell surface receptors results in an increase of luciferase expression in the reporter cells. The enzyme activity is assayed in the microfluidic system by injection of luciferin and ATP. The following chemiluminescent reaction is monitored using a photomultiplier tube, mounted right above the microchip with immobilized HeLa cells.

Keywords: microfluidic, microchip, immobilized cells, reporter gene assay, chemiluminescence

1. Introduction

Cellular assay systems are of growing importance in drug discovery and high-throughput screening (HTS) for studying cell-cell communication, externally triggered protein expression and release of exogenous compounds, toxicity and bioavailability. Microfluidics offer one great advantage over traditional plate assays in the possibility to monitor dynamic processes, *e.g.*, real time monitoring of cellular events. Additionally, the small dimensions (*e.g.*, shorter diffusion paths), low reagent consumption are other positive aspects.

The microfluidic system described here uses the cell line HF1, based on HeLa cells and constitutes a highly efficient cell-based reporter system specially designed for HTS of various heterologously expressed membrane receptors [1].

The reporter system is based on an optimized reporter vector, which includes a synthetic enhancer composed of multiple TPA-responsive elements (TRE) fused to a minimal cytomegalovirus (CMV) promoter. This promoter controls the expression of a novel chimeric reporter gene based on enhanced green fluorescent protein (EGFP) and *Photinus* luciferase. Upon ligand activation of a given membrane receptor, Ca^{2+} mobilization and signaling through the mitogen-activated protein kinase cascade takes place. This leads to a strong activation of the reporter promoter and concomitant expression of the chimeric reporter protein (Figure 1).

Y. Baba et al. (eds.), Micro Total Analysis Systems 2002, Volume 2, 811–813.

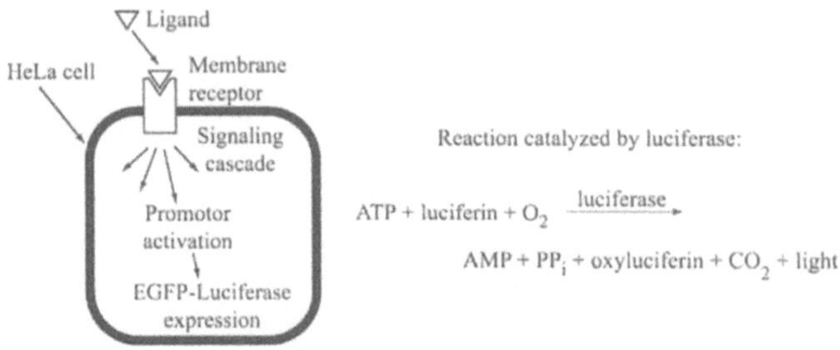

Figure 1. Schematic figure of the receptor-mediated signal transduction, which will result in the formation of luciferase and EGFP.

2. Experimental

The silicon microchips were fabricated by chemical wet etching described by Laurell et al. [2] and Drott et al. [3]. The microchips had an overall dimension of 13×3.1 mm, consisting of 28 parallel V-grooves, 10 mm long, 100 μm wide at the top and 71 μm deep, with each end falling into inlet and outlet basins The total volume of a microreactor was approximately 1.9 μL.

The microchip surfaces were modified by silanization followed by glutaraldehyde activation and attachment of poly-L-lysine [4]. Finally, the HF1 cells were immobilized to the poly-L-lysine coated microchips by adsorption.

As a receptor-stimulating agent we used 1 mM adenosine 5'-triphosphate (ATP) which stimulates the endogenous expressed ATP receptors present in HeLa cells.

Figure 2a shows the microfluidic system for assaying immobilized HF1 cells, and Figure 2b the specially designed plexi glass flow-cell for incorporating the microchip into the flow system.

The luciferase activity was assayed by either injection of substrate mixed with a detergent that lysed the cells, or in a non-invasive way by injecting the substrate at different pH and using a luciferin derivative.

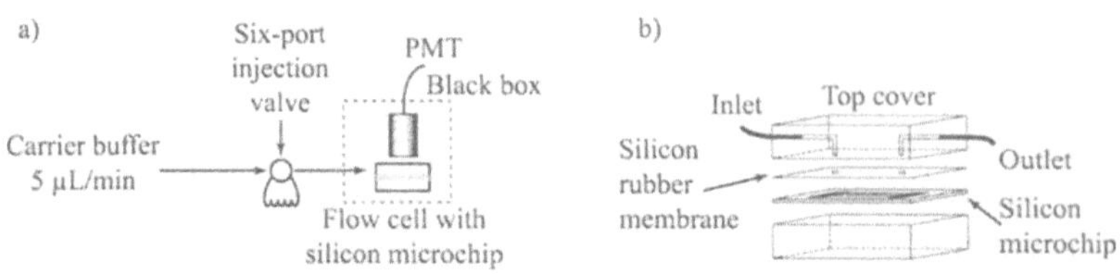

Figure 2. a) The microfluidic system used for measuring the luciferase activity of HF1 cells, immobilized on silicon microchips; b) Detailed figure of the flow-cell for incorporating the microchip in the flow system.

3. Results and discussion

Figure 3 shows a scanning electron microscopy (SEM) picture of HF1 cells immobilized on a silicon microchip.

Figure 3. SEM picture of immobilized cells.

The luciferase activity in immobilized HF1 cells (after ATP stimulation) is shown in Figure 4, at consecutive injections of luciferin mixed with ATP at pH 6.0 or 7.0. Luciferin, which is uncharged at pH 6.0, reaches most likely higher intracellular concentrations at this pH, favoring the light production.

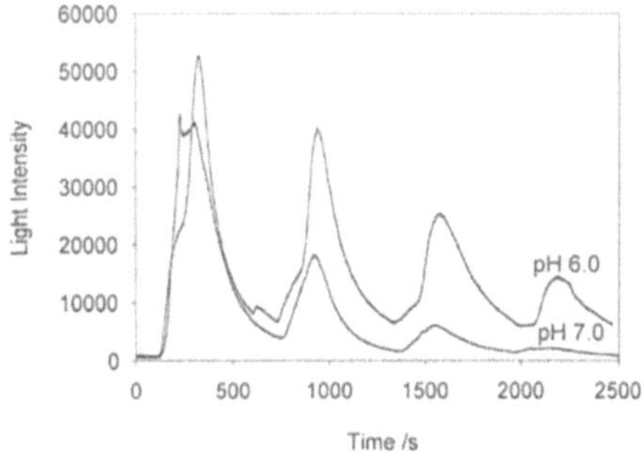

Figure 4. Luciferase activity of HF1 cells for two different pH's of the substrate solution.

4. Conclusions

These initial results clearly show that human cells could successfully be immobilized to poly-L-lysine coated silicon microchips. The production of luciferase was triggered via stimulation of the endogenous expressed ATP receptors, and the enzyme activity was determined by injection of luciferin with the resulting signal measured by a photomultiplier tube (PMT) in the microfluidic system.

References

1. Kotarsky, K.; Owman, C.; Olde, B. *Anal. Biochem.* **2001,** *288*, 209-215.
2. Laurell, T.; Drott, J.; Rosengren, L.; Lindström, K. *Sens. Actuators B* **1996,** *31*, 161-166.
3. Drott, J.; Lindström, K.; Rosengren, L.; Laurell, T. *J. Micromech. Microeng.* **1997,** *7*, 14-23.
4 Weetall , H. H. *Methods Enzymol.* **1976,** *44*, 134-148.

DEVELOPMENT OF NOVEL BIOASSAY SYSTEM USING MICROPATTERNED MAMMALIAN CELLS

Hirokazu Kaji, Matsuhiko Nishizawa, Tomokazu Matsue
Department of Biomolecular Engineering, Graduate School of Engineering, Tohoku University, Sendai, Japan
(matsue@bioinfo.che.tohoku.ac.jp)
(nishizawa@bioinfo.che.tohoku.ac.jp)

Abstract

The patterning of cardiac myocytes on a micron scale was achieved by microcontact printing (μCP) of fibronectin (FN) onto a hydrophobically pretreated glass sbstrate. The patterned myocytes conjugated with each other by forming a gap junction, as judged from synchronized Ca^{2+} transient over the pattern, and thus simultaneously contracted. Furthermore, it was shown the chemical stimulus with 1-octanol rapidly and reversibly decreased the activity of gap junctional communication of the patterned myocytes. The patterning of cardiac myocytes on an indium tin oxide (ITO) microelectrode array was also attempted for developing novel bioassay devices.

Keywords: Patterning, Microcontact printing (μCP), Cardiac myocytes, Confocal Ca^{2+} imaging, Microelectrode array

1. Introduction

In vitro assays using cultured cells have been commonly carried out for evaluating the cytotoxic effects as an alternative to animal experiments. On the other hand, the patterning of living mammalian cells has been extensively investigated in recent years, since the spatial control of cellular adhesion and growth in vitro is critically important in many areas of biotechnology as well as for the fundamental studies of cell biology. [1] In the field of biosensors, it will be undoubtedly required to place cells in predetermined locations on the device chip with defined shapes and sizes. Also, in order to make further progress in sensors which afford much sophisticated information, it will be indispensable to evaluate "network activity" of micro-tissue resulted from patterning of nerve and muscle cells.

2. Experimental

As schematically described in Fig. 1, the pattern of fibronectin (FN) was prepared by μCP method using the poly(dimethylsiloxane) (PDMS) microstamp. [2] The stamp was fabricated using a template of a photoresist pattern on a glass slide. First, the PDMS stamp was treated with O_2 plasma (60 W, 1 min) to make its surface hydrophilic, followed by the application of a FN aqueous solution (2.8 mg ml^{-1}) onto the stamp surface. After drying for 20 min under ambient conditions, the FN-inked stamp was put on a hydrophobic glass substrate preteated with n-octadecyltriethoxisilane and pressed at ca. 6 g cm^{-2} for 1 h. After thoroughly rinsing with vigorous

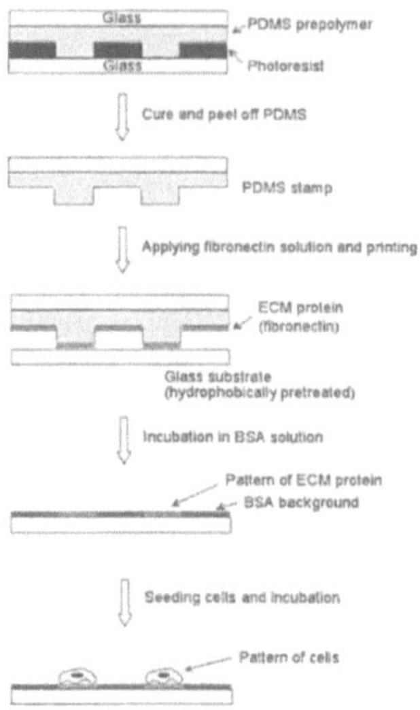

Figure 1.
Schematic diagram describing the procedure used to prepare the FN patterns on a glass substrate.

Y. Baba et al. (eds.), Micro Total Analysis Systems 2002, Volume 2, 814–816.

stream of PBS, the substrate was further incubated for 30 min in a PBS solution containing 3.0 mg ml^{-1} of bovine serum albumin (BSA) to cover the remainder with BSA molecules. Subsequently, the cardiomyocyte-rich suspension (1 x 10^6 cell ml^{-1}, from chick embryos) was seeded onto the protein-patterned substrate and incubated to get cellular micropatterns. [3]

The intracellular Ca^{2+} imaging was performed using a Nipkow-type high-speed confocal microscopy system (CSU21, Yokogawa Electric Company). The cultured myocytes were loaded with 10 μM fluo-3 AM for 30 min at 37℃, and studied in a Hepes buffered isotonic solution consisting of (in mM) NaCl 137, $CaCl_2$ 1.3, KCl 5.4, $MgCl_2$ 1.0, glucose 5.5, and Hepes 5.0 (pH 7.4). Cells were excited with a 488-nm argon laser, and fluorescence at 530 nm was detected with a digital CCD camera (HiSCA, C6790-81, Hamamatsu Photonics) through a barrier filter. Images were recorded on a computer and analyzed with Aquacosmos software (Hamamatsu Photonics).

3. Results and discussion

Figure 1 shows (a) the optical micrograph of the PDMS stamp used for the μCP of FN in order to make the band arrays of 5, 10, 20, 50, and 100 μm widths and (b) the phase contrast micrograph of the cardiac myocytes cultured on the FN-patterned substrate for 24 hrs. As can be seen in Fig. 2b, the band pattern of the PDMS is satisfactorily represented as a cell pattern during the 24-hr incubation. It is of interest that the myocytes in the narrow patterns (5 μm and 10 μm widths) are spreading along the line with a spindle shape, which is in contrast to the random orientation of cells in the wider line patterns (20 μm, 50 μm, and 100 μm widths). Furthermore, most of the cells in the narrow patterns spontaneously contracted in the direction along the lines at a beating frequency of ca. 1~2 Hz, indicating that these cells are myocytes and the number of fibroblasts contained in the patterns is negligible. It is also important that these cells are electrically connecting to each other through a so-called "gap junction", as proved below with fluorescent imaging. These results indicate that the restriction in the shape and orientation by patterning would not have a significant influence on the fundamental features of cardiac myocytes such as the formation of the gap junction and the synchronized contraction. It should be noted that, in Fig. 1b, some of myocytes grew out from the FN-patterned area. Such partial collapse of the cellular pattern would be due to the gradual exchange of albumin with serum proteins contained in the culture medium. [4]

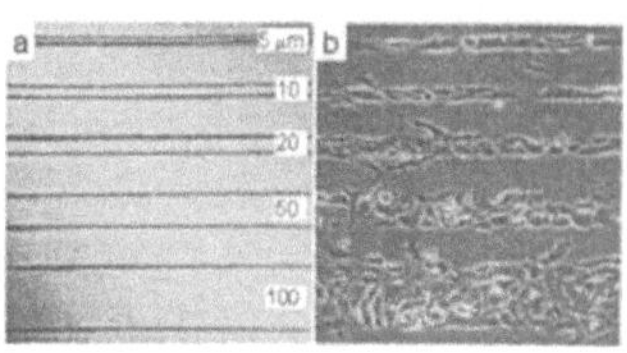

Figure 2.
(a) The optical micrograph of the surface of the PDMS stamp. (b) The phase contrast micrograph of the cardiac myocytes cultured on a FN-patterned substrate for 24 hrs.

With the view to analyze the cellular events in the patterned cardiac myocytes, fluorescent Ca^{2+} imaging was performed. The patterned cells were loaded with 10 μM fluo-3 AM for 30 min at 37℃, and studied in a Hepes buffered isotonic solution. Figure 3 shows (a) the phase contrast micrograph, (b) the confocal fluorescent Ca^{2+} image of the cardiac myocytes cultured on the 5-μm width FN pattern, and (c) the fluorescence time-courses of the cytosolic Ca^{2+} observed at the points denoted in the image (b) as C1 and C2. The line pattern in Fig. 3 consisted of at least seven cardiac myocytes. As can be clearly seen in Fig. 3c, a spontaneous Ca^{2+} transient for each measured area

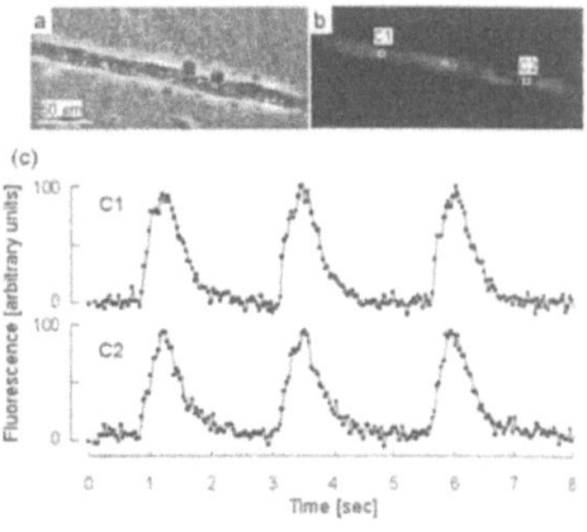

Figure 3.
(a) The phase contrast micrograph and (b) the confocal fluorescent Ca^{2+} image of the cardiac myocytes cultured on the 5-μm width FN pattern. (c) Time course of cytosolic Ca^{2+} fluorescence measured in the points denoted in the image (b) as C1 and C2.

simultaneously occurred, according to the Ca^{2+}-induced Ca^{2+} release (CICR) mechanism. Moreover, the profiles of the Ca^{2+} transients at C1 and C2 are identical with peaks at 1.20, 3.49 and 6.00 s. These results indicate that the seven cardiac myocytes are electrically conjugating each other to form a cardiac tissue. It would be also worth noting that the fluorescence intensities at C1 and C2 seem to be equal, indicating that the sarcoplasmic reticulums are uniformly dispersed over the linear tissue. Such a patterned cardiac tissue will provide a suitable model for the pharmacological studies of the gap junction itself and of the mechanism of heart disease such as arrhythmia. For example, we have shown that the micropatterned myocytes rapidly and reversibly lost their activity of gap junctional communication by 1-octanol.

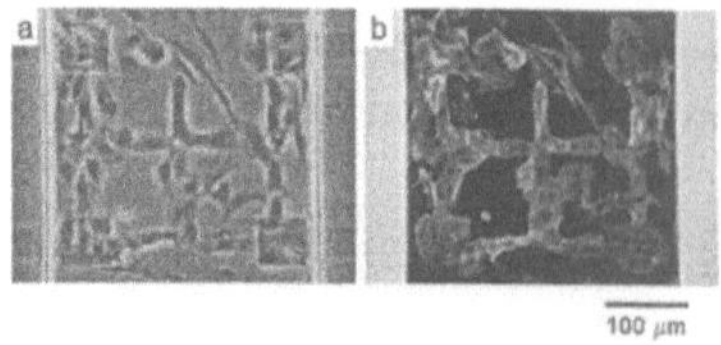

Figure 4.
(a) The phase contrast micrograph of the cardiac myocytes cultured on a FN-patterned electrode array. (b) The confocal fluorescence image of actin filaments.

The integration of cellular patterns with a microelectrode array has also been attempted, which enables us the stimulation and recording at the desired parts of the pattern. Figure 4 shows (a) the phase contrast micrograph of the cardiac myocytes cultured on the FN-patterned electrode array substrate for 24 hrs and (b) the confocal fluorescence image of F-actin. Since the microelectrode array was fabricated from indium tin oxide (ITO), the pretreatment was carried out in the same way as glass plates and a variety of fluorescent measurements were also applicable in the usual manner. It can be seen, although the cellular pattern is partially broken, the stamp pattern is roughly reproduced as a cell pattern (Fig. 4a). The staining of F-actin reveals that the muscle fibers grow along the pattern and the myocytes connect each other (Fig. 4b). These results satisfactorily demonstrate the integration of the patterned cardiac myocytes onto the microelectrode array substrate.

In addition to the micropatterning of cardiac myocytes for bioassay applications, we have also investigated the patterning of neuronal cells, HeLa cells, and embryonic stem cells. Such a cellular pattern would provide a suitable system for the pharmacological investigation of a gap junction and a synaptic junction, and for the studies of the mechanism of cellular event in a micro-tissue level, e.g., the arrhythmia generation and the synapse formation.

4. Conclusions

We have applied the μCP-based patterning technique to cardiac myocytes, and achieved the formation of the line-patterned myocytes as narrow as 5 μm in width. By means of confocal Ca^{2+} imaging, it was proved that the patterned myocytes conjugated with each other by forming a gap junction and thus contracted simultaneously. Although we focused on the patterning of cardiac myocytes herein, a cellular pattern of other kinds of cells such as neurons should be attractive for the whole-cell bioelectronics.

Acknoledgements

This work was supported by Grants-in-Aid for Scientific Research on Priority Area (No. 11227201) and for Scientific Research B (No. 13450348) from the Ministry of Education, Science and Culture, Japan.

References

1. Y. Xia and G. M. Whitesides, *Angew. Chem. Int. Ed.*, 37, 550 (1998), and references therein.
2. M. Nishizawa, K. Takoh, T. Matsue, *Langmuir*, 18, 3645 (2002).
3. H. Kaji, K. Takoh, M. Nishizawa, T. Matsue, *Biotechnol. Bioeng.*, in press.
4. M. Nishizawa, A. Takahashi, H. Kaji, T. Matsue, *Chem. Lett.*, in press.

DNA EXTRACTION FROM BACTERIAL CELLS BY REVERSE ELECTROPORATION AND SPLITT METHODS ON A MICROFABRICATED DEVICE

Melanie Fennah* & Andreas Manz**
*Department of Biological Sciences, Imperial College, London, UK, SW7 2AY.
**Department of Chemistry, Imperial College, London, UK, SW7 2AY.

Abstract
A miniaturised DNA extraction and separation device has been microfabricated using PDMS and glass. The device uses reverse electroporation to extract plasmid DNA from bacterial cells and SPLITT (split flow-thin cell flow fractionation) to separate the extracted DNA from the cell sample. The chip has been termed a Poration-SPLITT (PS) device.

Keywords: DNA, extraction, electroporation, SPLITT, microfluidics.

1. Introduction

Many of the conventional methods used to carry out genetic analysis have been transferred to microfabricated devices, decreasing analysis time dramatically. These include methods for amplification [1, 2, 3], separation and sequencing of DNA fragments [3, 4, 5, 6, 7]. The rate-determining step for genetic analysis is the often lengthy extraction of DNA samples from cells. Until now, miniaturised methods and devices for these procedures have not been developed. We report the microfabrication of a miniaturised DNA extraction and separation device (Figure 1).

Figure 1. Photograph of PS chip.

This device carries out two functions, (a) extraction of DNA from cells in the sample using reverse electroporation and (b) separation of DNA from the other sample components using electrical SPLITT. A schematic showing the operation of the PS chip is shown in Figure 2. A buffer stream is fed into the chip at the top right and the cell sample enters at the top left, these pass through the electroporation region where

Y. Baba et al. (eds.), Micro Total Analysis Systems 2002, Volume 2, 817–819.

DNA is released, then pass through the SPLITT separation region where DNA is transferred to the buffer stream on the right. Extracted DNA exits at the bottom right, cell waste exits at the bottom left. Chip-based electrotransfection has already been reported [8] but we believe that ours is the first report of chip-based electroextraction of DNA.

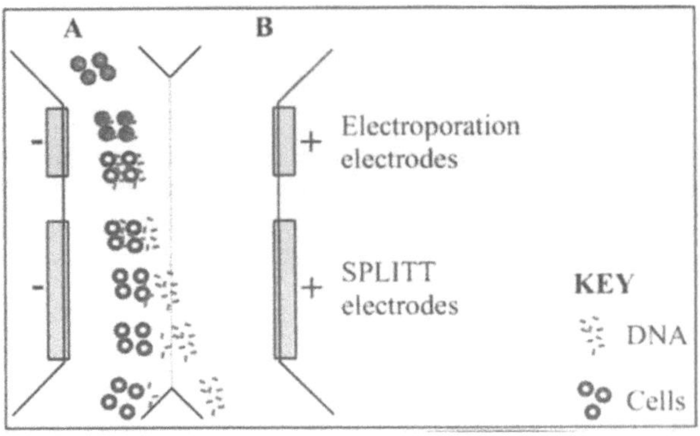

Figure 2. Schematic showing PS chip operation.

2. Experimental

E. coli XL-1 Blue cells containing the target plasmid (ME2) were prepared for electroporation by washing and concentrating in 1mM Tris-HCl, 50mM sucrose solution to remove traces of the growth media. These electrocompetent cells and 1mM Tris-HCl, 50mM sucrose solution were introduced to the chip at flow rates of 23.43nl/s using computer-controlled pumps. Pulsed fields of 200μs duration were applied across the pulsing electrodes at 20ms intervals. A separation voltage of 350V/cm was also applied. Voltages and pulse shape were monitored using a digital multimeter and an oscilloscope. Samples were collected from both outlets. Aliquots were plated on ampicillin-agar plates and incubated overnight to determine whether cells had crossed streams. An aliquot of the electrocompetent cells was also plated to compare cell numbers at outlets A and B relative to the original cell sample. The samples were then filtered through sterile filters and PCR was performed to determine the amount of DNA transferred from stream A to stream B. PCR products were analysed by capillary gel electrophoresis using the chip-based Agilent Bioanalyser.

3. Results & Discussion

The effect of the pulse field strength on DNA release was examined by varying the applied pulse voltage. Pulse field strengths of 0, 2, 4, 6, 8 and 12.5kV/cm were compared. The overnight cultured ampicillin-agar plates showed that no cells crossed from stream A to stream B, the cells collected from outlet A were still viable but there were slightly fewer colonies on this plate than on the original cell sample plate. The percentage concentration of DNA in stream B compared to the total DNA concentration at both outlets was determined after PCR using the Agilent Bioanalyser. The results of this experiment, normalised with respect to 0V, are shown in Figure 3.

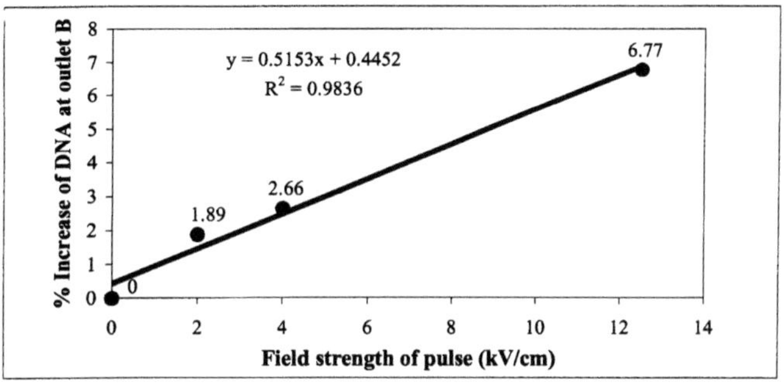

Figure 3. Dependence of DNA release on pulse voltage at a SPLITT separation voltage of 350 V/cm and a flow rate of 23.43 nl/s.

4. Conclusions

The PS device has been shown capable of extracting plasmid DNA from bacterial cells. There appears to be a linear relationship between pulse strength and DNA yield.

Acknowledgements

The authors would like to acknowledge the financial support of the Biotechnology and Biological Sciences Research Council and Amersham Pharmacia Biotech.

References

1. M.U. Kopp, A.J. de Mello & A. Manz, *Science*, **280**, pp.1046-1048, (1998)
2. P. Wilding, L.J. Kricka, J. Cheng, G. Hvichia, M.A. Shoffner & P. Fortina, *Anal. Biochem.*, **257**, pp. 95-100, (1998)
3. L.C. Waters, S.C. Jacobson, N. Kroutchinina, J. Khandurina, R.S. Foote, & J.M. Ramsey, *Anal. Chem.*, **70**, pp. 158-162, (1998)
4. A.T. Woolley, G.F. Sensabaugh & R.A. Mathies, *Anal. Chem.*, **69**, pp. 2181-2186, (1997)
5. C.S. Effenhauser, A. Manz & H.M. Widmer, *Anal. Chem.*, **67**, pp. 2284-2287, (1995)
6. S. Mouradian, J.W. Skogen, F.D. Dorman, F. Zarrin, S.L. Kaufman & L.M. Smith, *Anal. Chem.*, **69**, pp. 919-925, (1997)
7. P.L. Ross, P.A. Davis, & P. Belgrader, *Anal. Chem.*, **70**, pp. 2067-2073, (1998)
8. S. Attiya; E. Majid; D.J. Harrison, *HPCE 2001 symposium, 14th International Symposium on Microscale Separations and Analysis*, 2001

CELL SEPARATION WITH TRANSIENT ADHESIONS

Wesley C. Chang, Luke P. Lee, Dorian Liepmann
Berkeley Sensor and Actuator Center, University of California, Berkeley, 94720
(e-mail: wcchang@me.berkeley.edu)

Abstract

A biomimetic method for cell separation based on adhesive rolling and transient tethering has been demonstrated in microstructured fluidic channels coated with an E-selectin IgG chimera. Using E-selectin-ligand adhesions, capture and several hundred-fold enrichment of HL-60 cells on channel surfaces under continuous sample flow was achieved. Additionally, the difference in flow-driven cell transit speeds through the fluidic channel (HL-60 vs. U-937 cells) provides the basis for fractionating the different types of cells.

Keywords: cell separation, sample preparation, microfluidics, cell adhesion

1. Introduction

A basic step in many biological assays is separating and isolating different types of cells from raw samples. Normally, this function is performed by various macroscale techniques based on differences in physical, biochemical, or immunological properties among different cells [1]. Development of miniaturized cell separators has focused on shrinking conventional methods to accommodate small samples. Such methods, however, still require incubations for cell labeling and other preparatory steps that can hurt yield. MicroTAS systems by contrast must handle small volumes while limiting preparation steps and minimizing loss of sample.

To better meet these requirements, an alternative method for separating cells has been devised by mimicking the physiological process of leukocyte recruitment to blood vessel walls: adhesive cell rolling and transient tethering [2]. Reproducing these interactions for cells on surfaces of microstructured fluidic channels can serve to *capture* and *concentrate* cells and even to *fractionate different cell types* from a continuously flowing sample in a manner suited to the requirements of micro-assay systems. Targeted cells adhere to channel surfaces and then transit through the channel at slow speeds via adhesive rolling and transient tethering, while fluid and unaffected cells elute through at much higher speeds. Among captured cells, differences in transit speeds can permit different cell populations to segregate chromatographically.

2. Methods

Two different microstructures were fabricated in silicon via DRIE (Fig. 1). Each consisted of a 10mm long by 500μm wide by 40μm deep trench containing micro-pillars and reservoirs with through-substrate access holes at either end of the channel. Flow structures were then sealed with anodic bonding of pyrex coverslide.

A 1μg/mL solution of E-selectin IgG chimera (Glycotech, Inc.) in PBS was perfused into the sealed microfabricated channels for 2 hours to allow the proteins to physisorb

Y. Baba et al. (eds.), Micro Total Analysis Systems 2002, Volume 2, 820–822.

on to the channel surfaces, covered with native silicon dioxide. A 3% solution of fetal bovine serum (FBS) in PBS was used to "block" remaining sites on the surfaces. Cell binding to the Fc segment of the chimera protein was blocked by 2μg/mL of anti-Fc $F(ab')_2$ antibody, which was incubated with the fluidic channel for 30 minutes prior to testing and included in cell suspensions used during flow experiments.

Suspensions of HL-60 and U-937 myeloid cell lines (raised in RPMI-1640 with 10% FBS), at about 100,000 cells/mL, were pumped at 1μL/min through the test structures by a syringe pump. The flow was viewed under a microscope, and average transit speeds of the cells were determined by dividing the transit distance of 700μm by the transit time. Cell capture rate and tethering durations on the rear surfaces of pillars were also recorded as cells traversed this length.

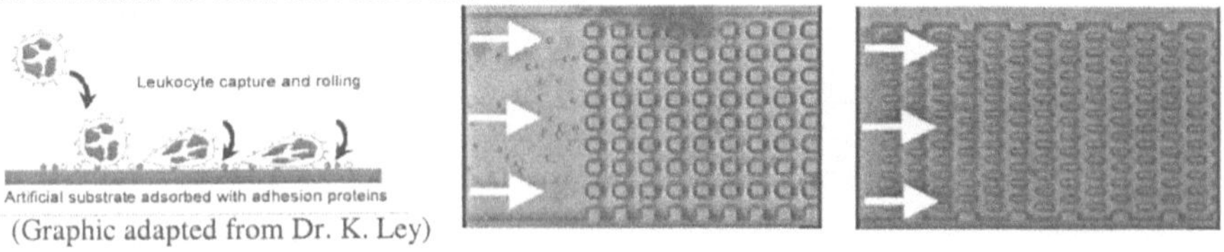

(Graphic adapted from Dr. K. Ley)

Fig.1: Cell capture (left); Top views: "Square" (center) and "Offset" (right) designs.

—100μm

3. Results and Discussion

Cell capture and enrichment (for HL-60 cells only). The "offset" geometry more effectively captured cells flowing into the array, with 95% of cells captured over the observed 700μm interval, while only 73% were captured in the "square" channels. This suggests that cell capture is favored by more circuitous flow fields and the curved, non-linear streamlines of the "offset" geometry. Cell concentrations were greatly enriched from the original concentration in both devices, but the "square" geometry was able to collect a higher concentration of cells. In the "square" geometry, cell concentrations reached a steady state of 377±91 times, compared to 229±42 in the "offset" geometry.

Cell transit. Under continuous flow, cells captured by the micropillar surfaces transited slowly down the length of the fluidic channel, driven by the fluid flow and occurring under three distinct regimes: 1) shear driven rolling along the channel surfaces; 2) transient immobilization of the cell on the trailing surface of micropillars, where adhesion bonds undergo tension, instead of tangential shearing; and 3) transport in the free stream following dissociation of the cell from the trailing surface and before recapture downstream. Overall transit speeds of cells are determined by the nature of repeated progressions through these regimes. Cell rolling speeds scale directly with shear stress at low shears but begin to plateau beyond 4dyne/cm^2. Rolling speeds for U-937 are about 50% more than those of HL-60 at all shears (Fig. 2a). Tethering on the trailing surface of pillars was also related to magnitude of shear stress due to the dependence of bond dissociation kinetics with mechanical loading [3]: tension of around 5.6pN for cells tethered on a "square" pillar resulted in just over twice the median tethering duration (12.5 vs. 6.5 sec.) as that for cells under 59pN of tension on an "offset" pillar, which exposed cells to greater fluid shears (Fig. 2b). Following dissociation, cells in the "offset" array generally spent more time in the free stream before recapture downstream. Therefore, in the "offset" geometry shorter tethering durations (on the trailing surface) and longer and more arbitrary distances carried in the free stream resulted in wider distributions of transit speeds. In the "square" geometry, however, cell transit was slowed more and the transit speeds were more tightly distributed, allowing the difference between the transit of the two cell types to be discerned (Fig. 2c) and providing the mechanism for fractionating the two populations.

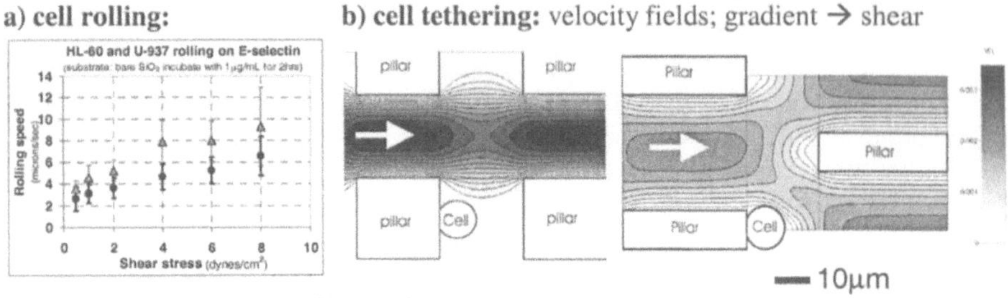

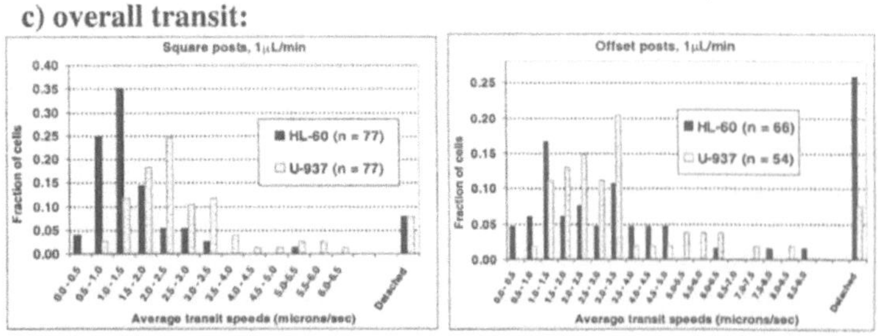

Fig. 2: Cell rolling (*above, left*) and transient tethering (*above, right*) govern overall transit (*lower*).

4. Conclusion

This work demonstrated the use of selectin-coated surfaces for capture of cells from rapid flow in microfluidic devices. The use of protein-mediated, transient adhesions for cell separation—capture, enrichment, and chromatographic fractionation—can provide a new and convenient alternative to more cumbersome techniques for microTAS by reducing or eliminating the need for sample preparation and post-processing and the storage of processing reagents, saving time and maintaining high process yields. In this demonstration, it was shown that the design and arrangement of micropillar arrays greatly affected the performance of adhesion-based cell separation. Future designs for micropillar arrays must explicitly address the three regimes of cell motion in order to achieve desired cell transit, capture, and enrichment characteristics.

References

1. D. Recktenwald and A. Radbruch, ed., *Cell separation methods and applications*, New York, NY: Marcel Dekker (1998)
2. Lawrence, M.B. and T.A. Sringer, *J Immunol.* **151**(11): pp. 6338-46. (1993)
3. Orsello, C.E., *et. al.*, *TRENDS in Biotech.* **19**(8): pp. 310-316. (2001)

GENE MANIPULATION SYSTEM ON INTEGRATED POLYMER MICROCHIPS

Morio Fukui, Masumi Yamada, Takaaki Saeki, Yasuhiro Kakigi, Nae Yoon Lee and Minoru Seki
Department of Chemistry and Biotechnology, School of Engineering, The University of Tokyo, 7-3-1 Hongo, Bunkyo-ku, Tokyo 113-8656, Japan
(e-mail: mseki@chembio.t.u-tokyo.ac.jp)

Abstract

We have developed the microchips for gene manipulation: DNA extraction and purification microchips based on the alkaline-SDS lysis method, and a microfluidic DNA- transfection device via electroporation. These chips allow contamination free, high throughput, rapid and automatic gene manipulation.

Keywords: Gene manipulation, integrated microchip, microfluidic system, alkaline-SDS method, electroporation

1. Introduction

In recent years, the opportunities for gene manipulation such as DNA extraction, enzyme restriction, ligation, amplification, transfection etc. have been increasing more and more with the development of biotechnology. However, these procedures for gene manipulation are time- and labor-consuming, and difficult for continuous operation. Therefore we developed the two types of microfluidic device that make possible to extract plasmid DNA from bacterial cells and a continuous DNA transfection via electropolation. These devices would be applicable for various kinds of gene manipulation by connecting in tandem together or with other types of microfluidic devices.

2. Experimental

Firstly, we developed microdevices for each unit operation of alkaline-SDS method; cell lysis, protein elimination and DNA recovery. The cell lysis microchip has large cylindrical chambers with 2 mm both in diameter and in depth for gaining enough retention time to complete cell lysis. The microchip for protein elimination has a microfilter with a sedimentation tub so that the channel should not be clogged up. The DNA recovery microchip contains glass powder for the adsorption and recovery of

Y. Baba et al. (eds.), Micro Total Analysis Systems 2002, Volume 2, 823–825.

plasmid DNA deposited by isopropyl alcohol. Next, we fabricated microchips integrated each unit operation. Figure 1 shows the schematic diagram of an example of the integrated microchip for DNA extraction and purification.

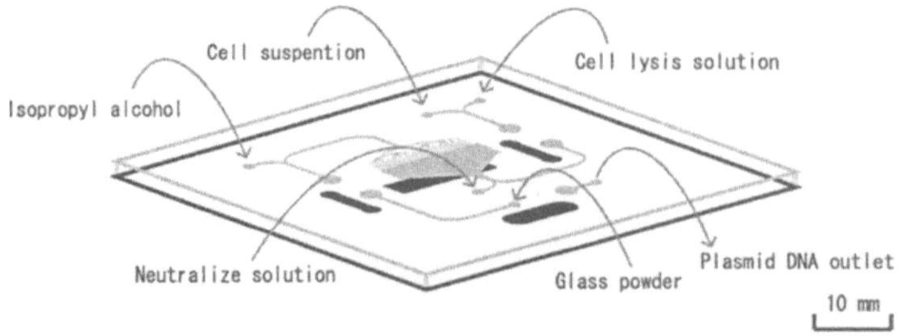

Figure 1.Schematic diagram of the integrated microchip for DNA extraction and purification.

We have also developed a microfluidic device for DNA-transfection via electroporation. Figure 2 is the photograph of this microfluidic device. The diameter of platinum wire electrodes is 200 μm and the width of the channel is 100 μm. *E.coli* DH5α cells and pBluescript plasmids were used for DNA transfection.

Although electroporation is the fastest and most efficient technique for gene transfection [1], it usually requires pulse electric field with high voltage. The microfluidic device of electro -poration can lower the necessary strength of applied voltage because of the narrower space between two electrodes. The advantages of this electroporation microdevice are the reduced use of cells and genes. Moreover, we can apply pulse electric field to each cell simply using a DC power supply because it takes a very short and controlled retention time for each cell to pass through between two electrodes in a flow system.

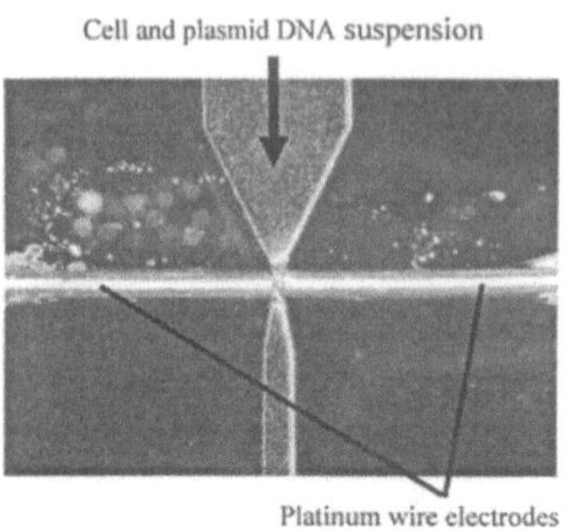

Figure 2. The photograph of the microfluidic device for electroporation.

3. Results and discussion

Figure 3 shows the quantity of the protein that remained after the treatment of alkaline-SDS method both in conventional scale and in the microchip for protein elimination. This result shows that the microchip with a microfilter have the potential for removing protein equivalent to alkaline-SDS method in a conventional scale.

Plasmid DNA recovery and purification using the DNA recovery microchip was successfully achieved under the similar recovery efficiency to that in conventional scale as shown in Figure 4.

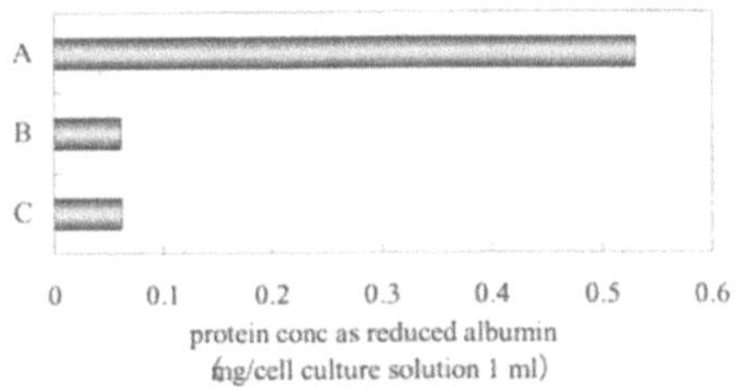

Figure 3. Quantity of residual protein. A is that before removing protein. B is in conventional scale and C is on the microdevice.

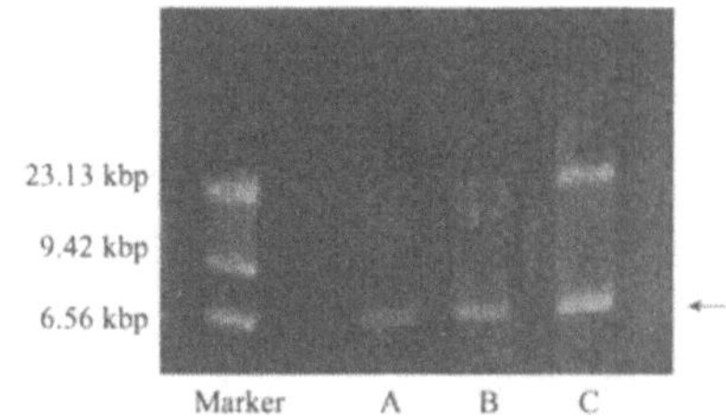

Figure 4. Electrophoretic analysis of plasmid DNA: lane A; recovered by glass powder in conventional size. B; on microchip. C; recovered by centrifugation.

Figure 5 shows the relationship between transfection efficiency, applied voltage and flow rate in the electroporation microdevice. It can be seen that transfection was successfully achieved in a rather lower voltage by a simple DC power supply compared to conventional systems.

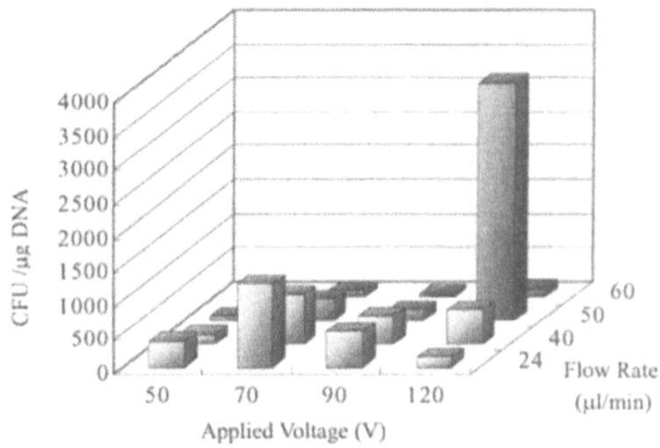

Figure 5. Relationship between transfection efficiency, applied voltage and flow rate in a electroporation microdevice.

4. Conclusion

We have successfully achieved some unit operations for gene manipulation in microfluidic devices. These devices will make it easy and reproducible to carry out various procedures of gene manipulations that are labour- and time-consuming in a conventional method.

Acknowledgement

This research was supported in part by Grants for Scientific Research on Priority Areas [A] (No. 13025216) and Millennium Project (No. 12319) from the Ministry of Education, Culture, Sports, Science and Technology in Japan.

Reference

1. R.A. Meldrum, M. Bowl, S.B. Ong and S. Richardson, *Biochem. Biophys. Res. Comm.*, **256**, 235-239, (1999).

THREE-DIMENSIONAL PDMS MICROSTRUCTURE FOR 3T3-L1 FIBROBLAST AND HEP G2 CELLS PERFUSION CULTURE

Eric Leclerc[1, 2], Yasuyuki Sakai[2], and Teruo Fujii[2]

[1]LIMMS/IIS-CNRS, University of Tokyo,
[2]Institute of Industrial Science, University of Tokyo, 4-6-1 Komaba Meguro ku Tokyo,153-8505, Japan
({rico, tfujii}@iis.u-tokyo.ac.jp)

Abstract

In this paper, we present the results of 3T3-L1 and Hep G2 cells culture in a network of multidimensional microchannels made in PDMS microdevices. The cells were kept living and some tissue like structure could be observed during the perfusion culture. The protocol of cells culture was optimised for microdevices, which could be applied in future to tissue engineering.

Key words: cells culture, tissue engineering, PDMS

1. Introduction

For the tissue engineering applications, some studies are performed to achieve cells culture in biocompatible materials with the MEMS technology [1,2]. Indeed, using the miniaturization processes, it is expected in future to be able to build artificial organs inside microdevices for medical or chemical utilizations, like implantation or toxicology studies. The material investigated to build the microdevice is a silicone elastomer, PDMS (PolyDiMethylSiloxane).

2. Fabrication of the microdevices

A protocol to establish the cells culture in a PDMS microdevice with a fully closed perfusion system was then developed and the detailed procedure of the fabrication process of the microdevice is presented in a previous paper [1]. Briefly, a multilayer structure of SU-8 is fabricated as a negative master of desired microstructures. Then, the PDMS polymer is poured and molded. The

Y. Baba et al. (eds.), Micro Total Analysis Systems 2002, Volume 2, 826–828.

channels are designed to achieve a mono-directional flow to get uniform repartition of the culture medium inside of the microdevice as shown in Figure 1. Then, the microdevice is connected to a perfusion circuit, and the whole circuit is autoclaved for sterilization.

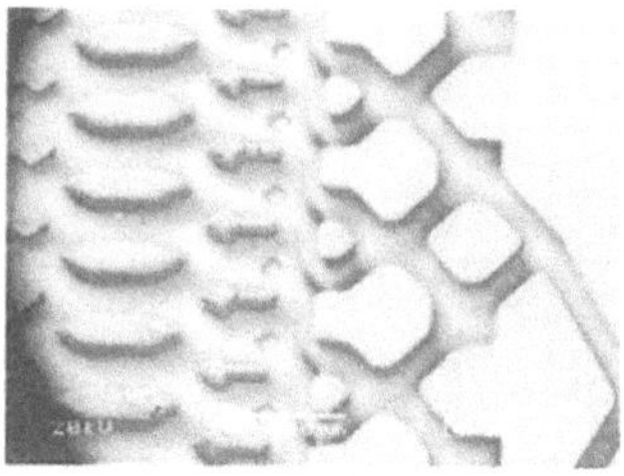

Figure 1:PDMS microchannel network for cells culture

3. Cells culture

We used Hep G2, human hepatocarcinoma cells, as a model liver cells and 3T3-L1, a mouse fibroblast as a model of connective-tissue-derived cells. Due to a collagen coating, the cells are properly attached inside the device. The cells were successfully kept alive for two weeks and the culture was stopped with living cells. Hep G2 cells aggregations and fibroblast elongations were observed and show some tissue like structures as shown in Figures 2 and 3. As an indicator of the cells' activity, the concentration of albumin in the perfusion circuit is shown in Figure 4, which is produced by the cells during the perfusion culture.

4. Conclusions

Cells culture was successfully performed in PDMS based microdevices placed in a closed perfusion system. Some tissue like structure could be observed inside the network of three-dimensional microchannels. The present method could be extended in future to realize the microdevice-based tissue engineering.

References

[1] E. Leclerc, Y. Sakai, and T. Fujii, *Cells culture in a three-dimensional network of PDMS*, Biomedical Microdevices (2002) submitted.

[2] K.R. King et al., *Microfluidics for tissue engineering microvasculature: endothelial cell culture*, Proc. µTAS 2001, Monterey USA (2001) pp.247-249

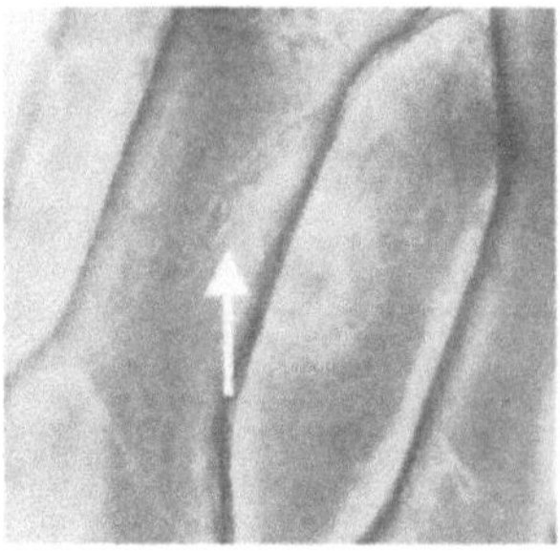

Figure 2. typical view of 3T3-L1 cells elongation in the microdevice after one day of culture

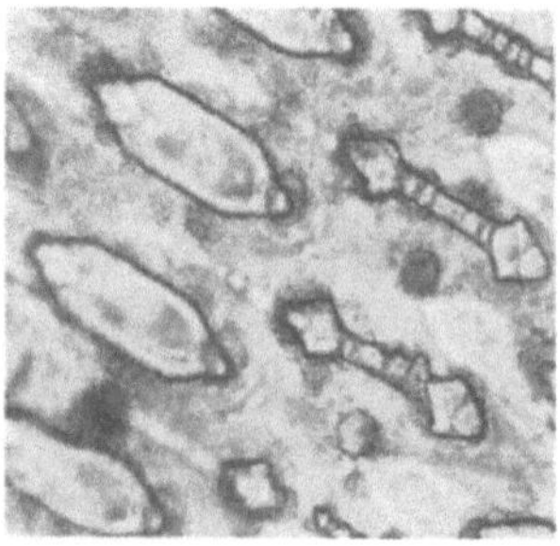

Figure 3. Hep G2 cells aggregations in the network after six days of culture.

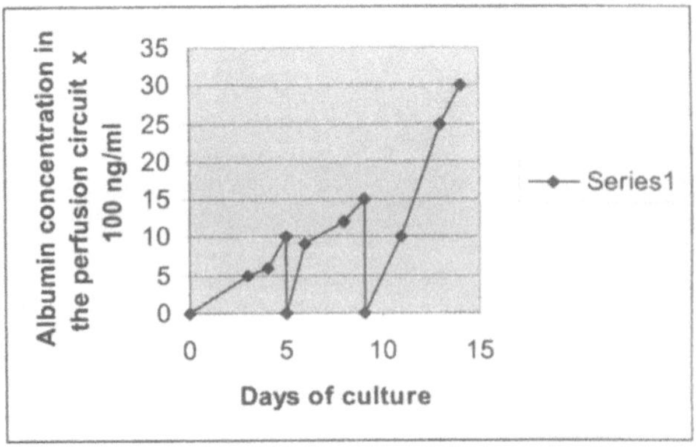

Figure 4. Albumin production during HepG2 experiments. Culture medium was changed on 5th and 9th days of culture. The number of cells increased, leading to the higher concentration of albumin in the perfusion circuit.

DIELECTROPHORETIC CHROMATOGRAPHY OF CELLS

David Holmes and Hywel Morgan
Bioelectronics Research Centre, Department of Electronics and Electrical Engineering, University of Glasgow, Glasgow, G12 8LT, U.K.
d.holmes@elec.gla.ac.uk, h.morgan@elec.gla.ac.uk

Abstract

We present a cell separation device that uses a combination of dielectrophoresis (DEP) and pressure driven fluid flow to separate differing cells types along the length of an electrode array within the device. Cells are initially focussed into a thin sheet mid-way between the upper and lower walls of the flow channel using negative-DEP. As they flow further through the device they pass over a second electrode array and experience a positive-DEP force pulling them onto that electrode array. The distance along the second electrode array at which a specific cell type is held is defined by the cells dielectric properties and size, with cells of similar properties collecting in a defined band.

Keywords: dielectrophoresis, particle separation, electrokinetic focusing.

1. Introduction

Dielectrophoresis (DEP) is the motion induced in a polarisable particle when it is placed within a non-uniform ac electric field [1,2]. In this paper we describe a novel dielectrophoretic cell separation system. The system consists of two separate electrode arrays integrated into the one flow channel. Each of the electrode arrays has a different function; the first uses negative-DEP to focus particles into a narrow sheet in the mid-plane of the flow channel and the second array uses positive-DEP to pull particles out of the solution and onto the electrode array where they are held under positive-DEP (see figure 1).

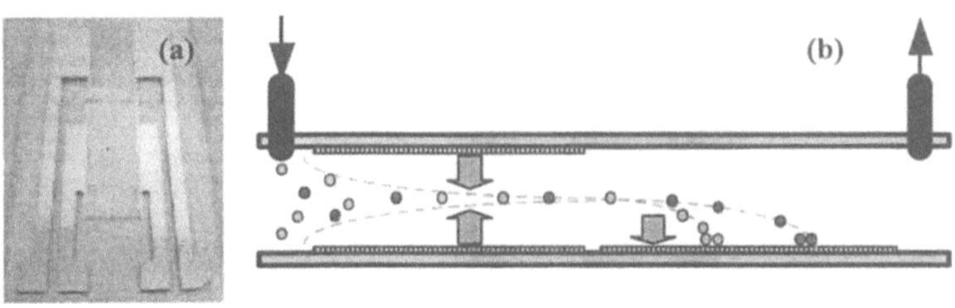

Figure 1. (a) Electrodes, (b) Schematic of DEP chromatography chip.

Y. Baba et al. (eds.), Micro Total Analysis Systems 2002, Volume 2, 829–831.

The position at which a particle ends up at along the separation electrode array depends upon the forces acting upon that particle (DEP, gravity, fluid flow). For a given set of experimental conditions (e.g. the frequency and magnitude of the applied electric field, fluid flow rate) particles with the same size and dielectric properties will follow a similar trajectory and band in the same position along the electrode array. Particles with differing size and/or dielectric properties will follow distinctly different trajectories and band at different positions along the array.

2. Experimental

The devices were fabricated on glass using SU8 to define the flow channel [3]. Cells are suspended in a low ionic strength media and pumped via a syringe pump into the device. Upon entering the device they are distributed randomly throughout the height of the channel. The first electrode array pushes the cells away from the upper and lower walls of the chamber into the centre of the fluid stream before they enter the separator. The separator comprises the second electrode array (down stream from the first) patterned onto the lower channel wall. This attracts particles under positive-DEP to the electrode surface and retains them their against the fluid flow. Since all the particles are positioned at the central plane of the channel as they enter the separator similar particles will follow similar trajectories and collect forming a band.

3. Results and discussion

The device is capable of fractionating cells or other particles depending upon their dielectric properties and size. Cells with different dielectric properties are retained at different distances along the electrode array. Figure 2 shows simulated particle trajectories for the four major peripheral blood mononuclear cell sub-populations. The differences in their trajectories is due to their different sizes and dielectric properties.

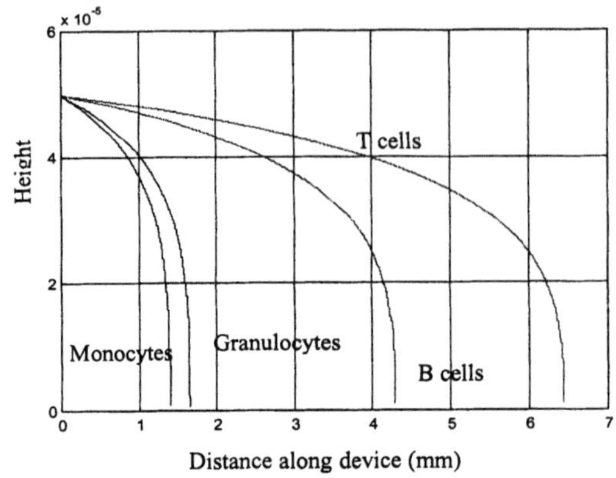

Figure 2. Simulated separation trajectories for the different peripheral blood mononuclear cell sub-populations.

Experiments carried out using the THP-1 monocyte cell line demonstrates the formation of a band as these cells are pulled out of the flow stream and collect on the separation electrode. Variation of the applied forces results in altered particle trajectories and subsequently different banding position along the electrode array. Figure 3(b) shows an example where variation in the fluid flow rate result in movement of the banding position. The theoretical position of the band is in close agreement with the experimental results.

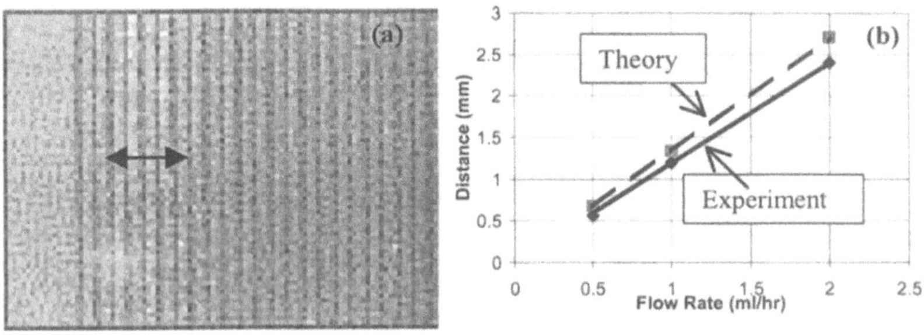

Figure 3. (a) Banding of THP-1 monocytes on the separation electrode. (b) Comparison of theoretical position and experimental.

4. Conclusions

This technique is potentially applicable to the fractionation of a variety of particle types. We are currently investigating the possibility of using the system to perform differential blood counts.

Acknowledgements

The authors would like to thank Dr M.Thomas at Glasgow Royal Infirmary for use of the THP-1 cell line. DH was funded by an EPSRC studentship.

References

1. Pethig, R., *Dielectrophoresis: Using inhomogeneous AC electrical fields to separate and manipulate cells.* Critical Reviews in Biotechnology, 1996. **16**(4): p. 331-348.
2. Pohl, H.A., *Dielectrophoresis.* 1978, Cambridge: Cambridge University Press.
3. Holmes, D., M. Thomas, and H. Morgan, *Dielectrophoretic separation/isolation of rare particles/cell types form a heterogeneous suspension within a microfluidic system*, in *Micro Total Analysis Systems 2000, Proceedings.* 2000, Kluwer Academic Publ: Dordrecht. p. 115-118.

CELLULAR ANALYSIS SYSTEM OF HISTAMINE RELEASE ON A MICROCHIP

Takahito Tokuyama, Shin-ichiro Fujii, Mitsuru Abo and Akira Okubo
Department of Applied Biological Chemistry, Graduate School of Agricultural and Life Sciences, The University of Tokyo
1-1-1 Yayoi, Bunkyo-ku, Tokyo, 113-8657 Japan

Abstract

The article describes the analysis system of histamine release reaction using mast cells on a microchip. In this study, the elastmetric material, poly(dimethylsiloxane) (PDMS), was employed to fabricate micro channels and chambers. Rat peritoneal mast cells were retained in the microchip, and were stimulated with chemical releaser. Histamine released from the cells was derivatized with *o*-phthalaldehyde (OPA) in the microchip, and its fluorescence was detected. This system consists of two sections, histamine release section and histamine derivatizing section, which were integrated into one chip. The calibration curve for histamine exhibited linearity at the concentration range from 60 to 1000 μM. By this integrated system, 2.8 nmol of histamine were detected using 15000 cells.

Keywords: histamine release, animal cell, fluorescence derivatization, PDMS, microchip

1. Introduction

Microchip applications have drawn much attention in the biochemical analysis system using cells, proteins and microbes[1]. The advantages of analysis on microchip are small consumption of samples and reagents, short analysis time, and experimental accuracy and so on. It is expected that the cellular analysis microchip is the ideal system for experiments on stimuli and response of cells. Histamine, which is primary found in mammalian mast cells, is the representative mediator in connection with allergies[2], and is released by some stimuli. Therefore tracing its behavior is important in physiology. Moreover, in the analysis of biological mediators it is important to minimize the diffusion because they are secreted in a small amount and tend to be degradated in a short time. In this study we try to retain mast cells on microchip for the rapid and sensitive analysis of histamine.

2. Experimental

Fabrication of microchip: Micro channels (700 μm wideth and 100 μm depth) were fabricated using poly (dimethylsiloxane) (PDMS) with glass template. As a fluidic device, the upper layer with connecting tube was just laid on the lower layer with the channels. Syringes were set on the syringe pump and were connected with the channel *via* fused-silica capillary tubes (250 μm i.d.; 350 μm o.d.). This cellular analysis system consists of two sections. One is designed for the histamine release reaction layer and the other is the histamine fluorescence derivatizing reaction layer. Schematic image of the

Y. Baba et al. (eds.), Micro Total Analysis Systems 2002, Volume 2, 832–834.

histamine release section (A), and the histamine fluorescence derivatizing section (B) were detailed in Fig. 1A and 1B respectively.

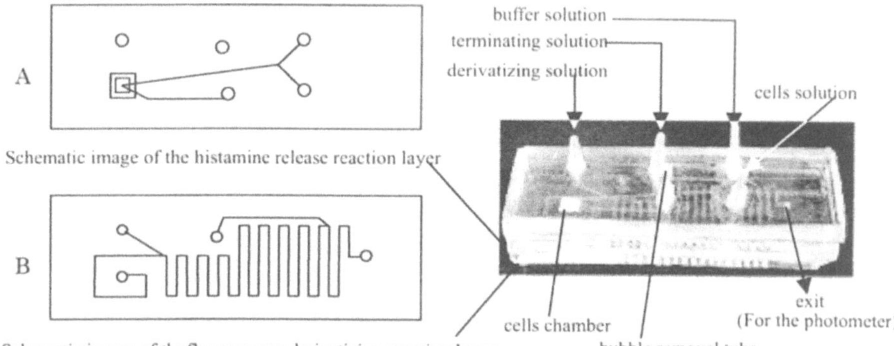

Fig. 1. Integrated microchip for the determination of histamine release using mast cells

Histamine release section: As shown in Fig. 1A, histamine release layer has a chamber (3 μl) retaining the mast cells. Cells were retained by the membrane filter which was put between this layer and the lower layer. Consequently less than 1 x 10^5 cells were retained in the chamber. Polyethylene tube (1 mm i.d.), which was thicker than other tube(250 μm i.d.), was used for the injection of the cells. The air bubbles were removed by the additional channel that located upstream of the cell chamber.

Rat peritoneal mast cells were collected from the abdominal cavity of male Wister rats, weighting 250-300 g. Chemical releaser (C48/80; 1 μg/ml) was used to stimulate the cell. All solutions were degassed sufficiently before introducing into the chip. Firstly, all channels were filled with the buffer and the chip was dipped in a water bath. The temperature of the whole system was kept at 37 °C. Next, cells were introduced through the channel toward the chamber. Then the chemical releaser solution was introduced through the other channel. After 10 min incubation, histamine eluted from the chip was derivatized by *o*-pthalaldehyde (OPA), and the fluorescence intensity was measured using the fluorescence photometer. Fig. 2 shows that the relationship between the number of the inserted cells and the fluorescence intensity. The roughly linear relationship was observed when the 3 – 30 x 10^3 cells were inserted in the chip. The influence of free amino acids was negligible (data not shown). The obtained intensity was mostly derived from the fluorescence of the histamine, and it was about 60 - 70%

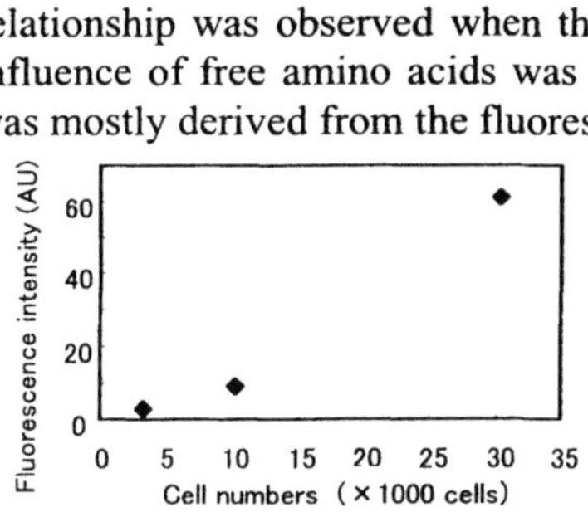

Fig. 2. Relationships between the introduced number of the cells and fluorescence intensity.

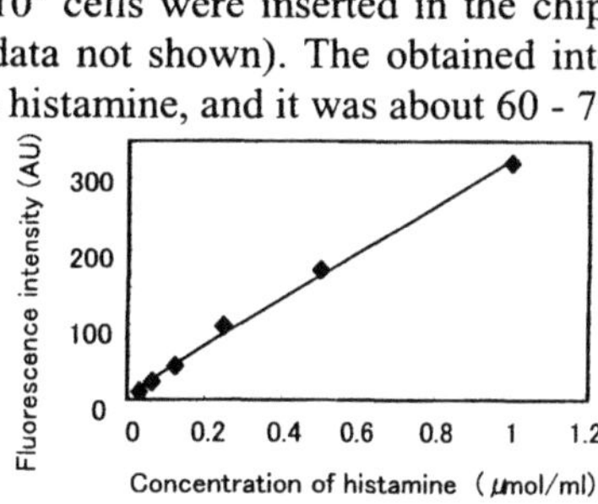

Fig. 3. Calibration curve of histamine fluorescence derivative

of the value that was expected by the conventional test-tube size assay[3] performed on the same condition.

Fluorescence derivatizing section: Fig. 1B shows the schematic image of the fluorescence derivatizing reaction chip. Three channels were fabricated for the sample, OPA and terminating solution (citric acid) respectively. Firstly, the sample was mixed with OPA at the first junction, and the reaction time was adjusted by the length of the winding channel. Next, the reaction was terminated by addition of citric acid at the second junction. After the addition of the citric acid, the histamine fluorescence derivative flowed toward the exit of the chip. The fluorescence of the elutant was detected by the fluorescence photometer.

The fluorescence intensity is related with the flow rate in the channel. When the flow rate was more than 10 μl/min, the fluorescence intensity decreased sharply. Thus, the flow rate was settled at 10 μl/min. The reaction time was shortened as compared with conventional assay. The calibration curve of the histamine was shown in Fig. 3. The calibration plot displays a good linearity from 60 nmol/ml to 1 μmol/ml. It is assumed that this concentration range is equal to the quantity of the histamine emitted from about 2000 to 20000 mast cells.

Integrated analysis system: Above two functional sections were integrated into one chip system. These sections were performed by the same procedure as the histamine release section and fluorescence derivatizing section. When 15000 cells were injected in the system, 2.8 nmol of the histamine was detected by the above calibration curve.

3.Conclusion

We have developed the integrated cellular analysis system of histamine release on the PDMS microchip. This system consists of the histamine release layer and the histamine fluorescence derivatizing layer. Less than 1×10^5 cells were retained in the cell chamber. Flow rate was employed 10 μl/min, and the reaction time was shortened as compared with conventional assay. The calibration curve of the histamine using this derivatizing layer exhibited good linearity at the concentration range from 60 nmol/ml to 1 μmol/ml. When 15000 cells were introduced, 2.8 nmol of the histamine was detected in this system. To our knowledge, this is the first application of the histamine release reaction using primary mast cell on the microchip, and the released histamine was detected certainly. This integrated system will be suitable for the rapidly analysis of the biological mediators, which are secreted in a small amount and degradation in a short time.

References

[1] A. Manz, F. Bessoth, Y. Xu, P. Monaghan, Miniaturizing conventional analytical methods: Micro reactors for bioassays and cell growth, BIOSENSORS2002, London UK, pp.110 (2002).

[2] M.J. Rothe, M. Nowak, F.A. Kerdel, The mast-cell in health and disease, J Am Acad Delmatol, Florida USA, 615-624, (1990).

[3] M. Akagi, N. Matsui, S. Mochizuki, K. Tasaka, Inhibitory effect of egualen sodium: a new stable derivative of azulene on histamine release from mast cell-like cells in the stomach, Pharmacology, Tokushima Japan, 203-209, (2001).

SYNCHRONIZATION OF BACTERIAL CULTURES IN A MICROSYSTEM

T. Fettah Kosar[1], Deirdre R. Meldrum[2], and Mark R. Holl[2]
[1]Department of Bioengineering, University of Washington, U.S.A.
[2]Department of Electrical Engineering, University of Washington, U.S.A.

Abstract

A method is presented to size-fractionate and hence synchronize a mixed population of the rod-shaped bacterium *Methylobacterium extorquens* AM1, which exhibit a direct relationship between length (size) and age. Preliminary studies have shown that *M. extorquens* AM1 cells readily respond to negative and positive dielectrophoresis (DEP) and align parallel to the field vectors in a uniform AC field. When these bacteria were placed in a liquid cell medium flowing through a microfluidic channel, they were observed to interact with physical obstacles present inside the channel. A novel MEMS-based device was designed, fabricated, and tested to combine the effects of electrokinetic alignment and steric hindrance for the size-fractionation (synchronization by age) of these bacteria. The method demonstrated in this device is suitable for integration into lab-on-a-chip instrumentation for the analysis of selected metabolites and intermediates in non-spherical non-aggregating microorganisms.

Keywords: BioMEMS, cell synchronization, *Methylobacterium extorquens* AM1, dielectrophoresis, electrokinetic alignment

1. Introduction

In microbiology, microorganisms are usually analyzed in large mixed populations where there are cells present at all possible stages of life cycle. The immediate disadvantage of such an approach is that the results are averaged over all the cells in the sample and, hence, are not representative of an individual cell. Detection of protein expression and metabolic activity at low relative levels for single bacterial cells is often not possible or infeasible using current techniques. Analysis of small, synchronized populations of the microorganism of interest to increase concentrations to detectable levels is proposed as a feasible alternative to single cell analysis under selected conditions. A method is presented to size-fractionate and hence synchronize a mixed population of the rod-shaped methylotrophic bacterium *Methylobacterium extorquens* AM1 [1], which exhibit a direct relationship between length (size) and age.

2. Motivation

Methylotrophic bacteria appear to play an essential role in the cycling of single-carbon compounds. They are one of the major sinks of methane, which has been identified as one of the significant greenhouse gases, in the global methane cycle [2]. Beside their ecologically important role, they are also excellent candidates to be utilized as biochemical factories for producing chemicals from cheap and simple substrates or for minimizing industrial hazardous waste. Hence, they carry both scientific and economic importance. Nevertheless, before they can be fully exploited, a thorough analysis and understanding of their physiology is imperative.

Y. Baba et al. (eds.), Micro Total Analysis Systems 2002, Volume 2, 835–837.

3. Experimental

Our preliminary studies have shown that AM1 cells readily respond to negative and positive dielectrophoresis (DEP) [3] and align parallel to the field vectors in a uniform AC field (Figure 1). When these bacteria are in a liquid cell medium flowing through a microfluidic channel, they were observed to interact with physical obstacles present inside the channel. A novel MEMS-based device was designed, fabricated using SU-8 and PDMS on a glass wafer, and tested to combine the effects of electrokinetic alignment and steric hindrance through an obstacle course for the synchronization of these bacteria [4] (Figure 2). It was shown that smaller bacteria travel faster through the device than larger ones, enabling their fractionation by size, hence, by age (Figure 3).

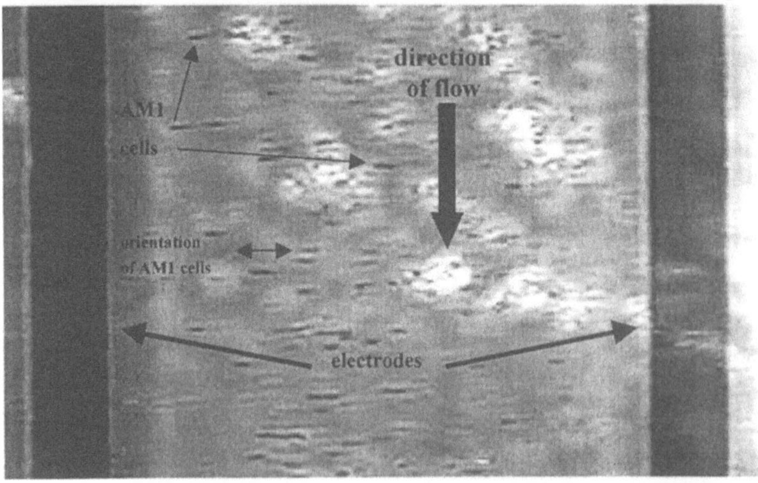

Figure 1. Electrokinetic alignment of AM1 cells in an AC electric field while flowing through a microfluidic channel between two parallel microelectrodes.

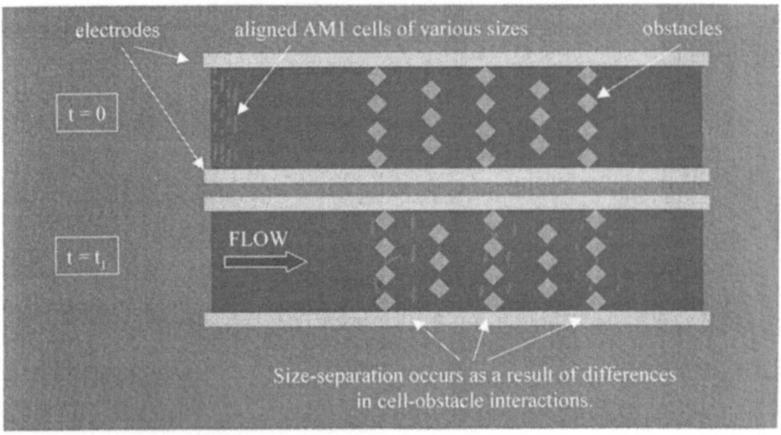

Figure 2. Synchronization principle of AM1 cells.

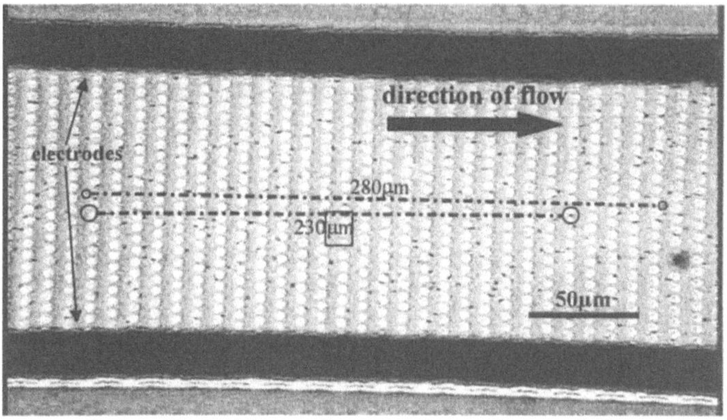

Figure 3. A 3μm cell (small circle) travels faster than a 6μm cell (large circle) through a section of the device during a 5-second period. The regularly repeating white shapes are the obstacles (3μm diameter). The black dots scattered inside the channel are the AM1 cells. The center-to-center distance between obstacles in the direction of flow is 10μm, 6μm in the transverse direction. The depth of the channel is 10μm.

4. Discussion and Conclusions

The preliminary development work presented here suggests the potential for a novel culture synchronization device that is simpler to use, less expensive to produce, faster, and smaller than other conventional synchrony devices and methods. Because of the flexible microfabricated design, this device can be fabricated with connections to other modules on the same chip and incorporated into a lab-on-a-chip instrument for the metabolic analysis of small cells populations. This analysis method has the potential to enable estimation of metabolic conditions in a single cell through synchronous amplification of biological moieties of interest. Although it was developed using *Methylobacterium extorquens* AM1 cells, this method can be adapted to any non-spherical non-aggregating unicellular organism.

Acknowledgements

We thank Prof. Mary E. Lidstrom and her laboratory staff for sharing their expertise with us and for providing the AM1 cultures. This research was supported though NIH grant 5 R01 HG01497-05 and NHGRI grant 5 P50 HG002360-02. We gratefully acknowledge this support.

References

1. M.E. Lidstrom, *The Aerobic Methylotrophic Bacteria,* in The Prokaryotes, Springer-Verlag, New York, (1992).
2. R.S. Hanson and E.V. Wattenberg, *Ecology of methylotrophic bacteria,* Biotechnology **18**: 325-48, (1991).
3. H.A. Pohl, *Dielectrophoresis : the behavior of neutral matter in nonuniform electric fields*, Cambridge University Press, New York (1978).
4. T.F. Kosar, *Synchronization of Bacterial Cultures in a Microsystem*, MS Thesis, Department of Bioengineering, University of Washington, Seattle, U.S.A. (2002).

Single Cell Analysis of Bovine Chromaffin Cells Using Micro Electrical Impedance Spectroscopy

Swomitra K. Mohanty[1], Surendra K. Ravula[1], Katherin Engisch[2] and A. Bruno Frazier[1]
[1]School of Electrical and Computer Engineering, Georgia Institute of Technology, Atlanta, GA, USA
[2]Department of Physiology, School of Medicine, Emory University, Atlanta, GA, USA

Abstract

Single cell characterization using electrical impedance spectroscopy (EIS) has been conducted on bovine chromaffin cells and red blood cells. Measurements were taken using a micro scale electrophysiological analysis system fabricated using micromachining (e.g. MEMS) technologies. Impedance measurements were taken over a frequency range of 40 Hz to 3.0 MHz. The data obtained for the cells was compared to data obtained for air, and phosphate buffered solution.

Keywords: electrical impedance spectroscopy, chromaffin cells, red blood cells, MEMS, microfluidic, cell sorting

1. Introduction

The study and development of microfluidic systems for biological and chemical analysis has been an area of increasing interest over the past decade. Of particular interest are micro biochemical analysis systems capable of analyzing and processing a variety of biological materials. Micro analysis systems and biosensing systems based on single cells are areas that have benefited greatly by the application of micromachining technologies to biological analysis. The ability to interface cells with microsystems makes it possible to perform analyses which otherwise would not be possible due to system complexity (e.g. multiple precisely defined electrodes, micro scale fluid control). Previous work has shown that interfacing cells with microsystems has made techniques such as electrical impedance spectroscopy (EIS) of single red and white blood cells possible [1,2]. Traditional cellular analysis techniques generally involve fluorescent or radioactive tagging. This is an effective method and yields important information, but requires additional sample preparation and modification of cell biochemistry. The use of micromachining technologies to fabricate micro systems for electrical impedance spectroscopy of single cells enables rapid, parallel analyses and collection of data when compared to conventional techniques for interfacing with cells. Initial data from Ayliffe et al. demonstrated that a characteristic impedance spectrum exists for different types of cells. Potential applications for EIS have already been demonstrated in cell sorting and cell-based diagnostic systems [3, 4]. The latter group used

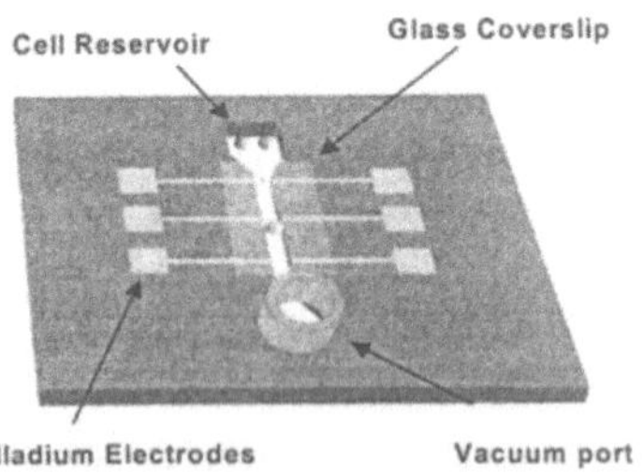

Fig 1. Schematic of a micro electrical impedance device. Cells are pulled into the fluid-filled channel via vacuum and positioned in between palladium electrodes.

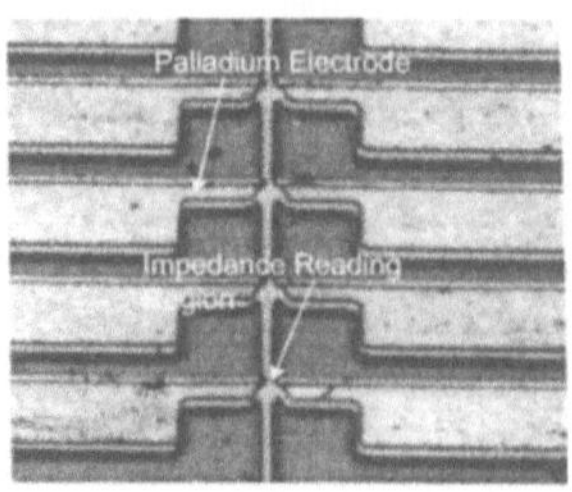

Fig 2. μ–EIS analysis system with 7 μm channels and electroplated palladium electrodes.

Y. Baba et al. (eds.), Micro Total Analysis Systems 2002, Volume 2, 838–840.

planar electrodes to analyze erythrocytes in wide channels followed by a sorting operation. Thus far, the majority of cell studies in microsystems have focused on red blood cells. Little work has been done on other cell types. This paper expands the use of μ–EIS to the characterization of bovine chromaffin cells, a hormonal cell found in the adrenal gland. Chromaffin cells are interesting because they are electrically excitable cells. Using EIS, membrane activity can be studied with regard to ion channel activity and response to chemical stimuli.

In this work a μ-EIS device was fabricated for impedance characterization of bovine chromaffin cells and red blood cells. Cells were separated based on the diameter and the viability of the cell. Impedance and phase measurements were compared to air, and phosphate buffer solution (PBS), to investigate the viability of using electrical impedance spectroscopy for cell identification, cell sorting and electrophysiological analysis applications.

2. Methodology

Figure 1 shows a schematic diagram of the μ-EIS system. The micro electrical impedance device was fabricated using standard MEMS fabrication technology. The system was fabricated on a silicon substrate. Initially a dielectric layer of Si_3N_4 was deposited using plasma enhanced chemical vapor, followed by a thin film of Ti (500Å) and Ag (1000Å) for an electroplating seed layer. The film was then patterned to form isolated pairs of opposing electrodes. Next, micro channels with a cross-sectional area of 7.0x7.0 μm^2 were constructed on the surface of the silicon wafer using photo-definable SU-8 material (SU-8-10C, MicroChem Corp.). The microchannels were used as flow channels for routing the individual cells and as micromolds to define the electroplated electrodes. Palladium was electroplated into the micromolds to a thickness of 7.0 μm, as shown in Fig 2. The fluid microchannels were sealed using a glass cover slip with input/output ports.

Red blood cells were taken from a human sample and diluted in a buffered solution for reading in the μ-EIS. Bovine chromaffin cells were separated using a percoll gradient, which separates

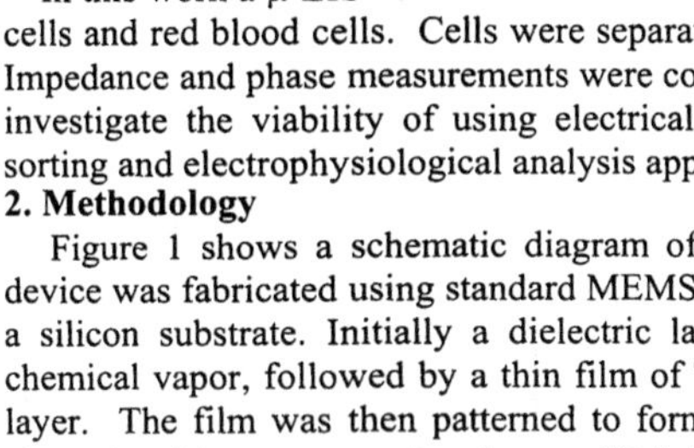

Fig 3. Impedance measurements of Air (top), PBS (middle) and most commonly measured chromaffin cells (bottom).

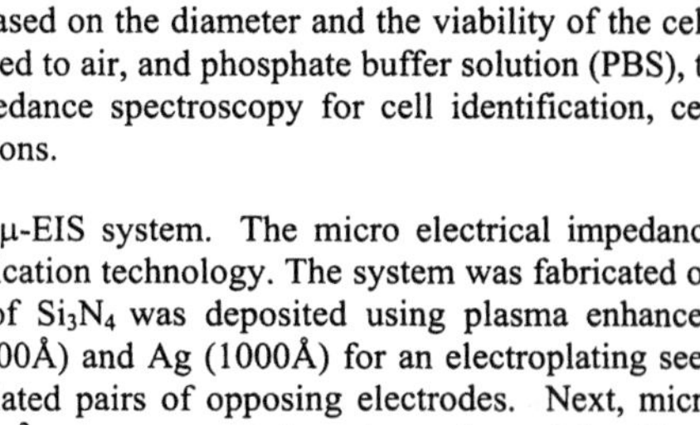

Fig. 4. Impedance measurements of all types of chromaffin cells measured and blood cells.

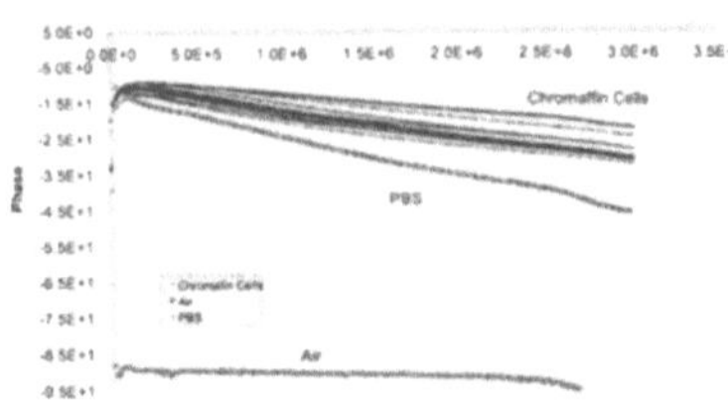

Fig 5. Phase readings for Air, PBS, and most common chromaffin cells.

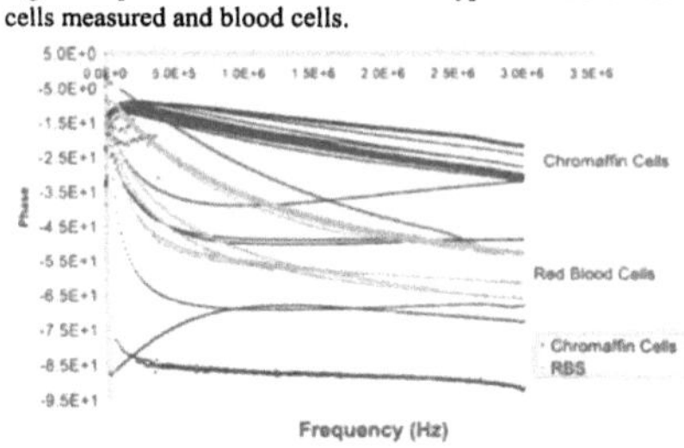

Fig 6. Phase readings of all types of chromaffin cells and red blood cells.

cells according to size and buoyancy. The adrenal gland was processed and spun at approximately 14,000 rpms in a test tube with percoll. Layers of cells develop with the smaller diameter cells on the lower layers. Cells were then cultured from the appropriate layer. Dead cells were determined by using a Trypan Blue. The cells were suspended in a recording medium (Dulbecco's PBS 1x, 1.00 mg/L , D-glucose with 36 mg/L sodium pyruvate, calcium).

The μ-EIS was initially characterized with air, and then PBS to provide a baseline for the impedance data recorded. The cells were brought into the channel via vacuum and placed in between opposing electrodes. Impedance readings were taken using an HP4294 impedance analyzer over a frequency range 40 Hz to 3.0 MHz. Magnitude and phase data were taken for each cell. Measurements were done on statistically relevant population of chromaffin cells and red blood cells.

3. Results and Discussion

The impedance and phase measurements for air, PBS, red blood cells and chromaffin cells are plotted in Fig 3, 4, 5, and 6. Large differences in cell impedance were observed at lower frequencies of the spectrum. At higher frequencies, the curves levelled out to a characteristic impedance value due to the elimination of the membrane capacitive component. Figure 3 shows the characteristics of air, PBS, and chromaffin cells. Air has the highest magnitude at approximately 5 MΩ over the frequency spectrum. It also had a 90 degree phase shift as expected since the device in air behaves like an ideal capacitor. PBS had a lower magnitude with an average reading just above 10 kΩ across the spectrum. This is expected since PBS is a conductive medium. Figure 4 shows magnitude measurements for red blood cells fell in the range between 10 kΩ to 100 kΩ. Chromaffin cells showed up commonly at three different levels. The first level measured between 1 kΩ and 10 kΩ and accounted for most of the cells observed. Several cells measured between the 10 kΩ to 100kΩ range and other cells fell in between the 100 kΩ to 1000 kΩ range. Distinct phase readings were observed for air, PBS, chromaffin cells, and red blood cells. This is shown in Fig 6 where the chromaffin cells with the lowest magnitude had the lowest phase shift, while red blood cells with a higher magnitude exhibited a higher phase shift.

The results have interesting implications in regards to cell characterization and identification. Red blood cells and chromaffin cells clearly have different impedance spectrums. Differences in magnitude between some cell types were not very large but the phase between them was very different (red blood cells and lowest magnitude chromaffin cell population). Therefore the phase data provides added sensitivity for cell identification. Three different groups of chromaffin cells were observed. The viable chromaffin cell population had two average sizes of 10μm and 20μm in diameter. The viability of chromaffin cells also declines soon after sample preparation. These size differences between cells, and whether the cells were viable or not, resulted in the recorded impedance spectrums.

4. Conclusions

Electrical impedance spectroscopy was successfully performed on bovine chromaffin cells and red blood cells with a micro device fabricated using MEMS technology. The μ-EIS device is sensitive enough to distinguish between chromaffin cells and red blood cells based on the gain and phase signatures of the impedance spectrum.

5. Acknowledgments

This work was supported in part by the National Institute of Health (NIH)/National Institute for Environmental Health Sciences (NIEHS) under Grant ES10846 and in part by the NIH/National Institute on Deafness and Other Communication Disorders (NIDCD) under Grant DC04928.

6. References

[1] H. Ayliffe, B. Frazier, and R. Rabbitt, *Am. Soc. Mech. Eng. Biomed. Eng. Conf.*, v. 35, pp. 485-486, June 1997
[2] H. Ayliffe, et al, *Int. Conf. Solid-State Sensors & Actuators*, pp. 1307-1310, Chicago, IL, June 1997
[3] H. Ayliffe, B. Frazier, and R. Rabbitt, IEEE *Journal of Microelectromechanical Systems*, Vol. 8, NO. 1, March 1999
[4] S.Gawad, S. Metz, L. Schild, and Ph. Renaud, *Micro Total Analysis Systems* 2001, pp. 253-255

ON-CHIP NEURAL CELL-CULTIVATION SYSTEM FOR LONG-TERM OBSERVATION WITH MULTI-ELECTRODE AND MICROCHAMBER ARRAYS

Kazunori Takahashi[1], Yoshihiro Sugio[1], Hiroyuki Moriguchi[1], Yasuhiko Jimbo[2] and Kenji Yasuda[1]

[1] Department of Life Sciences, Graduate School of Arts and Sciences, University of Tokyo, 3-8-1 Komaba, Meguro, Tokyo 153-8902 Japan

[2]Department of Materials Science, NTT Basic Research Laboratories, NTT, Ltd., Japan

(E-mail: cyasuda@mail.ecc.u-tokyo.ac.jp)

Abstract

A new type of multi-electrode array system for long-term neural-cell cultivation has been developed. The core of the system is a 9-room microchamber array, formed of thick SU-8 photoresist, with a 9-channel multi-electrode array connected to each chamber. The biotin-coated microchamber array plate is covered with a streptavidin-decorated cellulose semi-permeable membrane, fixed on the surface of the plate by streptavidin–biotin attachment, thus separating the microchambers from the nutrient medium circulating above. The advantages of this system are as follows: (1) Contamination-free cultivation of neural cells is achieved by sealing the semi-permeable membrane lid on the microchamber array. (2) The simple, flexible design of the microchamber array enables the positions of the cells to be fixed between thick-photoresist walls 1-5 micrometers wide and 25 micrometers high. (3) Optical microscope observation and electrical recording enable long-term simultaneous analysis of the cells. Our experiments showed that nerve cells cultured in the microchamber array were grown safely, without any contamination. These results demonstrate the potential usefulness of our multi-electrode array system.

Keywords: microchamber array, semi-permeable membrane, long-term recording, multi-electrode array, neural cell

1. Introduction

Studies of the communication among neural cells through connections like axons are a minimum requirement for developing models to understand brain function. Understanding the relationship between the numbers of neural cells and networks and the efficiency of the learning process requires a single-cell–based, long-term cultivation and analysis system for neural cells, in which we can cultivate the cells, with full control over their spatial arrangement and connections, stimulate them, and record their responses. To achieve this purpose, we have designed a novel on-chip microchamber array formed on a glass plate, with electrodes installed. Here we describe this on-chip neural-cell cultivation system for long-term cultivation and analysis.

Y. Baba et al. (eds.), Micro Total Analysis Systems 2002, Volume 2, 841–843.

2. On-chip Neural-cell Cultivation System

The microchamber array structure was designed to topologically control the number of cells, the spatial distribution of cells, and the network shape on the chip. The electrodes installed in each microchamber enable us to electrically stimulate each cell and record its response during long-term cultivation. Figure 1 shows a schematic of our on-chip neural-cell cultivation system. The multi-electrode array (MEA) chip is placed under a microscope. Micrographs of the neural cells can then be recorded periodically with a CCD camera and a computer (lap-time measurement) to measure the changes in network shape. The microchambers on the chip were formed of thick SU-8 photoresist and a semi-permeable membrane. The SU-8 forms enclosures for the neural cells (with walls about 25-μm high), enabling us to control the formation of the neural-cell network. The semi-permeable membrane

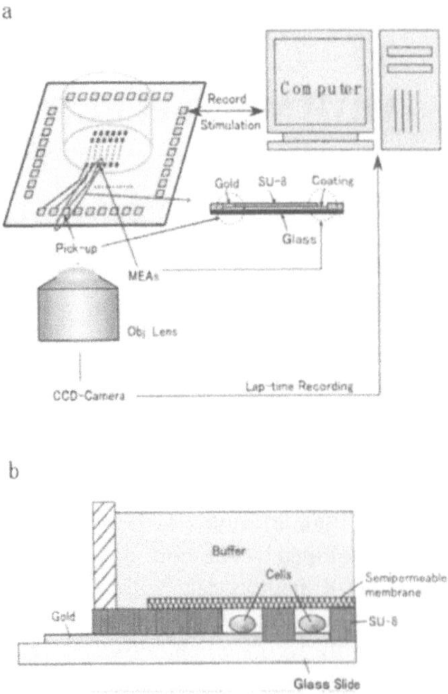

Figure 1. On-chip multi-electrode neural-cell cultivation system

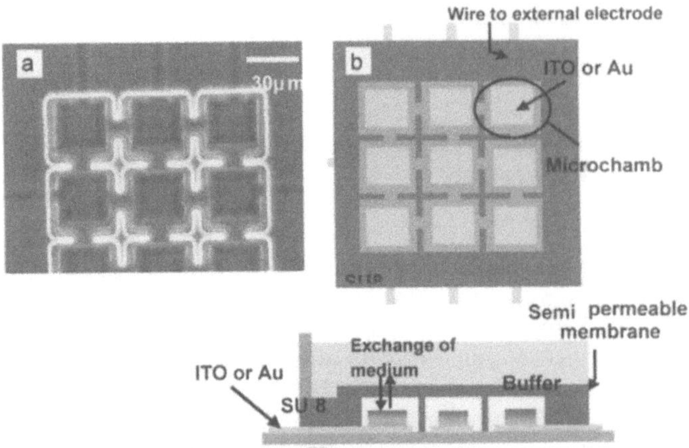

Figure 2. Structure of multi-electrode array chamber.

was strongly attached to the microchambers and functions as a lid to prevent uncontrolled changes in the formation of the network.

The lid also makes it possible to maintain contamination-free conditions inside the microchambers while still supplying fresh nourishment from outside. We communicate with the neural cells through electrical stimulation via the ITO or gold electrodes installed in each chamber. These electrodes can simultaneously be used to stimulate the cells and record their response signals. Figure 2 shows a micrograph (a) and schematic (b) of a 9-room microchamber structure with ITO electrodes. The wires connecting these ITO electrodes and the external electrodes are covered with SU-8 for disconnection of electric signals from other cells.

3. Experimental Results and Discussion

Figure 3 shows a micrograph of neural cells on MEAs. The size of each electrode was 30 µm x 30 µm. We confirmed that each neural cell escaped from its chamber by movement of those cells, forming the desired pattern for the neural-cell network. No bacteria contaminated the chambers during cultivation even when bacteria existed in the cultivation medium. Thus, we concluded that this system works appropriately well for long-term neural-cell cultivation. We also found that the neural cells were successfully cultivated and the network extended throughout the microchambers. However, the performance of the cells was not sufficient to be measured as we had expected. This might be caused by the incompleteness of our neural-cell cultivation technique, especially for single-cell cultivation.

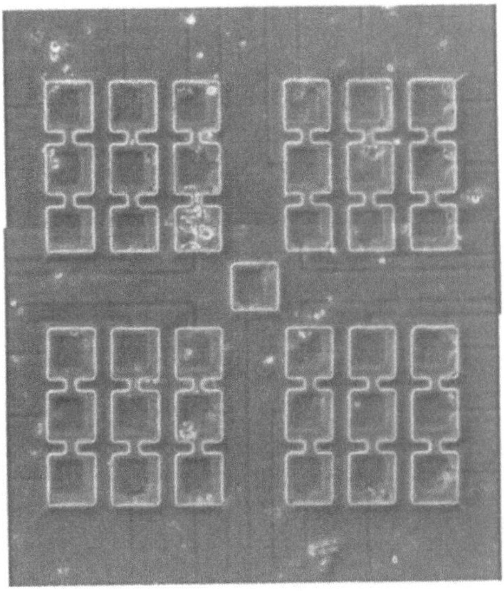

Figure 3. Phase-contrast micrograph of neural-cell cultivation on MEAs.

4. Conclusion

We have designed and tested a novel on-chip neural-cell cultivation system for long-term analysis, including topological control of the cells' network patterns and recording of their electrophysiological responses. Although we confirmed that our cultivation technique using a semi-permeable membrane as a lid prevented contamination, we still need to improve our single-cell cultivation technique with MEAs and microchambers. The next step will be to practically cultivate and control the neural network pattern.

References

1. S. M. Potter, T. B. DeMarse; *J. Neurosci. Methods,* **2001**, 110, 17-24
2. Y. Jimbo, T. Tateno, H. P. C. Robinson; *Biophys. J.* **1999**, 76, 670-678
3. T. Hirono, K. Torimitsu, A. Kawana, J. Fukuda; *Brain Res.*, **1988**, 446, 189-194

SELF-ASSEMBLED MAGNETIC BEADS FOR CELL SORTING

Cécile Goubault [1], Jean-Louis Viovy [1], Jérôme Bibette [2]
[1] Laboratoire PCC, Institut Curie (UMR CNRS / IC 168), Paris, France
[2] Laboratoire Colloïdes et Nanostructures, ESPCI, Paris , France
cecile.goubault@curie.fr

Abstract

We present an original matrix for cell sorting by affinity : magnetic beads self-organised into an array of columns. Data concerning the development of such matrices will be presented and the potential for cell sorting will be discussed.

Keywords : self-assembling, cell sorting, magnetic beads

1. Introduction

Magnetic separation is currently used to capture a specific cell population from a heterogeneous suspension. When added to a suspension, functionalised magnetic beads bind the desired target, forming a "bead-target complex" which can be easily isolated with a magnet. This technique is not so suitable to get a pure sample of rare cells. The drainage of the "bead-target complex" due to the magnet and the stress applied on cells in the phase collected by the magnet create non specific binding. Furthermore, the capture of all the targets requires an important quantity of beads which covers up cells and renders their study by histology difficult.

In the present study, we propose to use the self-organisation properties of magnetic beads : when placed between two planes, they form an array of columns when a homogeneous magnetic field is applied perpendicularly to the planes (Fig. 1). Matrices of this type have been previously applied with success to DNA separation [1]. Functionalisation of such matrices should enable cell sorting via specific interactions. This direct on-chip approach offers strong potential advantages with regards to the previous technique : it is very gentle to cells (no pressure applied) and the capture surface is larger than with direct functionalisation of channels.

H

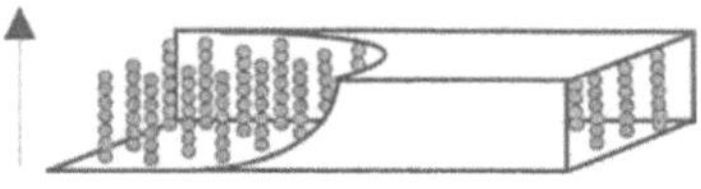

Fig. 1. Schematic of the array of columns formed by a suspension of magnetic beads

Y. Baba et al. (eds.), Micro Total Analysis Systems 2002, Volume 2, 844–846.

2. Study and optimization of matrices

Polydimethysiloxane (PDMS) microchannel devices were prepared by soft-photolithography. The magnetic matrix was self-assembled in situ by introducing the suspension of magnetic beads in the channel, and submitting it to a magnetic field perpendicular to the channel.

At first, we investigated reversible arrays as described in [2]. The characteristics of the array depend on two factors: (i) the distance (L) between the two planes : the higher the distance, the larger the column spacing (d) (ii) the volume fraction of beads : the higher the volume fraction (ϕ), the thicker the columns.

The parameters controlling column spacing were investigated with ours beads [3] in order to find the right spacing to avoid cell jamming and to obtain columns thick enough to get a robust array. Figure 2 shows a ln-ln plot of column spacing vs channel thickness at constant volume fraction. The dependence of d on L follows approximately a power-law relation $d=3L^{0.28}$, consistent with previous results on different particles [3].

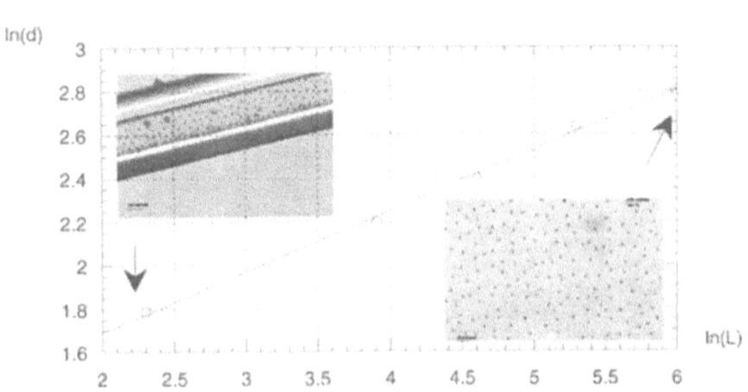

Fig. 2 : Column spacing as a function of channel thickness on a ln-ln scale. Pictures of the array in a 50 μm high channel (top left) and in a 400 μm high channel (right bottom)

. The influence of bead concentration was also investigated at constant channel thickness (L=100 μm). As ϕ increases from 0 to 15%, the average distance between column centers does not change significantly, but column thickness increases. Above $\phi \cong 15\%$, the network evolves into a labyrinth-like structure (Fig. 3). This structure is not suitable for cell sorting because the cells are trapped in dead-ends.

Finally, we prepared irreversible arrays (Fig. 4), following [4], optimizing concentrations of beads and reagents to avoid aggregation of the beads before the magnetic field is applied.

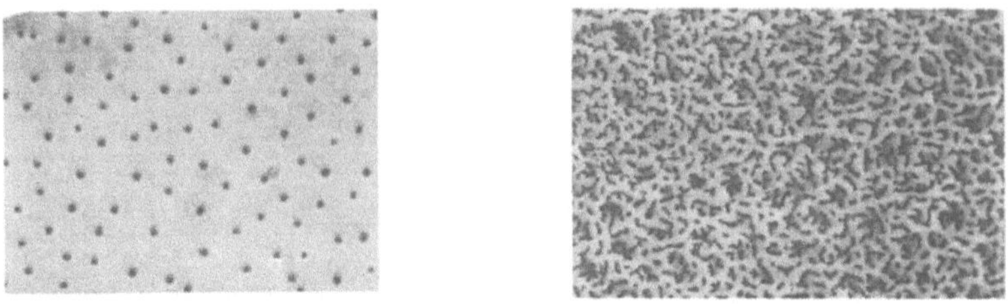

Fig. 3 : Array of columns ($\phi \cong 5\%$) [left]; Labyrinth-like structure ($\phi \cong 20\%$) [right]

3. Migration of vesicles and erythrocytes

Vesicles (as a model system to mimic cells) and erythrocytes were driven by an electric field in the pre-assembled arrays of magnetic beads. In a reversible array, we observe that erythrocytes or vesicles carry away the columns like a bowling ball in a skittle game (Fig. 5). Moreover, if the columns are charged (due to their functionalisation), they can migrate with the electric field. Anchoring of the columns to the wall facilitates the resistance of the array to the force exerted by cells when they flow through the channel.

Fig. 4 : Irreversible columns attached to the channel wall

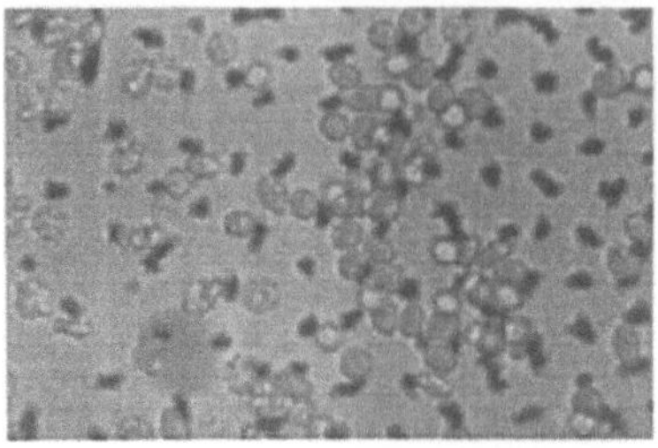

Fig. 5 : Cell jamming and deformation of a reversible array by erythrocytes

4. Conclusion

We developed an on-chip cell sorting method based on specific interactions of cells with self-assembled columns of magnetic beads. We demonstrated that column spacing compatible with cell migration can be achieved. Preliminary results suggest that permanent array will be more efficient than reversible ones. Moreover, we observe essentially no non-specific adsorption. The results obtained so far confirmed that self-assembled beads are a promising new medium for cell sorting in microfluidic system.

Acknowledgements

This work is supported in part by the French Minister of Research under the “Post-Genome” and “Nanostructures” programmes.

References

1. P. S. Doyle *et al.*, *Science*, **295**, 2237 (2002)
2. J. Liu *et al.*, *Phys. Rev. Lett.*, **74**, 2828 (1995)
3. provided by ADEMTECH S.A., Pessac, France
4. J. Philip *et al.*, *J. Phys. D : Appl. Phys.*, **30**, 2798 (1997)

OBSERVATION OF EXTREMELY LOW TRANSMEMBRANE POTENTIAL OF CELLS IN ELECTROPORATION USING MICROCHIPS

Yu-Cheng Lin, Ming-Yuan Huang, Min Li
Department of Engineering Science, National Cheng Kung University,
1 University Road, Tainan, Taiwan

Abstract

We develop a micromachined electroporation chip successfully performing cell electroporation under a very low applied electric field, which yielded an extremely low transmembrane potential to create membrane pores. The transmembrane potentials of four different cells obtained from the experimental results were one to two orders of magnitude smaller than 0.2 V, which is the lowest transmembrane potential reported in the current literatures since early 1980s. The four different cells were successfully transfected under multi-pulse or single pulse electric field stimulation. The lowest applied voltage used was 0.05 V or 0.1 V for four different cells which resulted in the estimated maximum transmembrane potentials lower than 20 mV in all cases. The lowest transmembrane potential was found at 6 mV for Huh-7 cells (human hepatocellular carcinoma cell line).

Keywords: MEMS, electroporation, microchip

1. Introduction

Electroporation is a technique that introduces foreign materials into cells by applying impulses with an extremely high electric field to create multiple transient pores in the cell membrane through a dielectric breakdown of the cell membrane. The dielectric breakdown is caused by electric field-induced transmembrane potential. Literature [1] shows that the transmembrane potential has to reach at least 0.2 V, usually 0.5-1 V, to cause the dielectric breakdown of lipid bilayer membranes in order to create membrane pores for transporting ions. An electroporation microchip for continuous gene transfection has been developed to handle a large number of cells in a continuous flow [2]. An *in vitro* electroporation microchip for gene transfection with surface electrodes has successfully performed electroporation to deliver genes into cells with an applied voltage of 10 V [3]. In most of the cases of this study, the applied voltages themselves are actually lower than 0.2 V, which is the lowest transmembrane potential reported in current literatures.

2. Design and Microfabrication of Electroporation Microchips

The electroporation microchip (Fig. 1) consists of two components, the well-defined cavity for cell accommodation and culturing as well as the interdigitated electrodes for providing electroporation electric power. The cavity region was designed to match with the electrode position and provide a precise cell culturing area, a 6 mm diameter circle. The electrodes are 100 μm wide with a 100 μm spacing.

Y. Baba et al. (eds.), Micro Total Analysis Systems 2002, Volume 2, 847–849.

Figure 1: Schematic drawings of an *in-vitro* electroporation microchip; 3D view (**A**) and topview (**B**), and the dimensions of the interdigitated electrodes on the microchip (**C**).

3. Plasmids Preparation and Cell Culture

The *pEGFP-N1* plasmids (Clontech, USA), coded with green fluorescent protein (GFP) gene, were extracted and purified from *E. coli.* Four different cells comprised of two different types, primary cell and cell line, are used in this study. Chang liver cells (human liver cell line), Hep-G2 (human hepatocellular carcinoma cell line) and Huh-7 cells were maintained at 37°C, under 5% CO_2, in Dulbecco's modified Eagles's medium (DMEM). The microchip was coated for 24 hrs with 0.01% polylysine in phosphate-buffered saline (PBS) for cell lines, and with gelatin for 6 hrs for primary cells. The primary cells, HUVEC, were maintained at 37°C, under 5% CO_2 in the endothelial growth medium 2 (EGM2). After washing with PBS, the cells were loaded into the cavity of the microchip for cell accommodation. The cells, at a concentration of 10^5 cell/ml in a total volume of 50 μl, were seeded onto the pretreated microchip. After 24 hours prior to electroporation, the cells on the microchip were then washed with PBS twice, added with 4 μg GFP plasmids and subjected to electroporation.

4. Electroporation setup and detection system

A lab-built power supply, able to deliver a series of square wave pulses with adjustable voltage (0.01-10 ± 20% V), duration (0.1-100 ms) and frequency was used in this study. The cells on the electroporation microchip were treated with one or multiple electrical pulse shocks at 25 Hz with pulse duration of 20 ms, and then incubated at 37°C, 5% CO_2 for 24 hours.

The imaging and detection system consists of a microscope (BX60, Olympus, Japan) and a cooled charge-coupled device (CCD) camera (Penguin 600CL, Pixera, USA). The cells were inspected under a fluorescent microscope 24 to 48 hours after electroporation. The successfully transfected cells emit a green fluorescence between 515 to 540 nm when they are excited using a 420 – 480 nm light source.

5.Results and discussion

In order to verify that the authenticity of extremely low transmembrane potential can create membrane pores for other kinds of cells, four cells were tested under applied voltage by either 0.1 V or 0.05 V using three-pulse stimulation. The cells, which were successfully transfected with GFP plasmids under the applied voltage as low as 0.05 V, exhibited green fluorescence (Fig. 2). Control experiments without using electroporation process show no transfection. The four different cells used in this experiment were electroporated using applied voltages as low as 0.05 V or 0.1 V, which resulted in estimated transmembrane potentials varying between 6 mV to 18 mV. The extreme case may approximately reduce the maximum transmembrane potential by a factor of two. However, using different models to calculate the transmembrane potential will not affect the major result of this study.

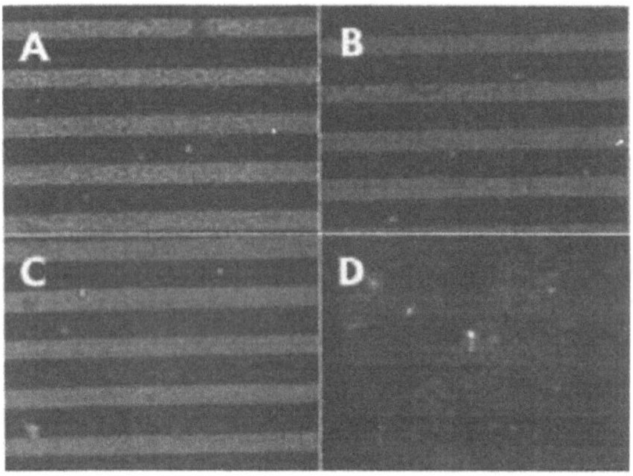

Figure 2: Images of successfully transfected different cells (displaying green fluorescence) electroporated by three pulses under 0.1 or 0.05 V applied voltages; (**A**) Chang liver cell line was electroporated using 0.1 V, (**B**) HUVEC cell was electroporated using 0.1 V, (**C**) Hep-G2 cell line was electroporated using 0.05 V, and (**D**) Huh-7 cell was electroporated using 0.05 V.

6. Conclusions

In this study, using very small amounts of cells and genes is enough for evaluating the gene delivery and expression. The transmembrane potentials of four different cells, categorized into cell line and primary cell, have been experimentally found to be as low as 7.5 mV, which is much lower than that reported in current literatures. It is important to note that the induced transmembrane potential cannot be greater than the applied voltage which, in this case, is 0.05 V. Herein the extremely low transmembrane potentials of cells in electroporation are observed using the electroporation microchips.

References

1. J.C. Weaver, Y.A. Chizmadzhev, *Theory of electroporation: a review*, Bioelectrochem. Bioenerg., 41, pp.135-160 (1996).
2. Y.C. Lin, C.M. Jen, M.Y. Huang, C.Y. Wu, X.Z. Lin, *Flow-type electroporation chips for gene transfection*, 79, Sens. Actuators B, pp. 137-144 (2000).
3. Y.C. Lin, M.Y. Huang, *Electroporation microchip in vitro gene transfection*, J. Micromech. Microeng., 11, pp. 542-548 (2001).

ON CHIP GENERATION AND REACTION OF UNSTABLE INTERMEDIATES: MONOLITHIC MICROREACTORS FOR DIAZONIUM CHEMISTRIES

Robin Fortt, Robert C.R. Wootton and Andrew J. de Mello
Imperial College of Science, Technology and Medicine, Exhibition Road, South Kensington, London SW7 2AY, UK.

Abstract

Monolithic microreactors for the safe and expedient continuous synthesis of products involving unstable intermediates were fabricated and tested. The optimisation of diazonium salt formation in anhydrous conditions and its subsequent *in situ* chlorination using the Sandmeyer reaction (Scheme 1) in a microfluidic network under hydrodynamic pumping regimes is presented. Significant enhancement of yields was observed due to enhanced heat and mass transfer in microfluidic systems. Direct detection by Raman spectroscopy is demonstrated as a tool for on-line dynamic optimisation.

Keywords: organic synthesis, microreactor, continuous flow, hazard reduction

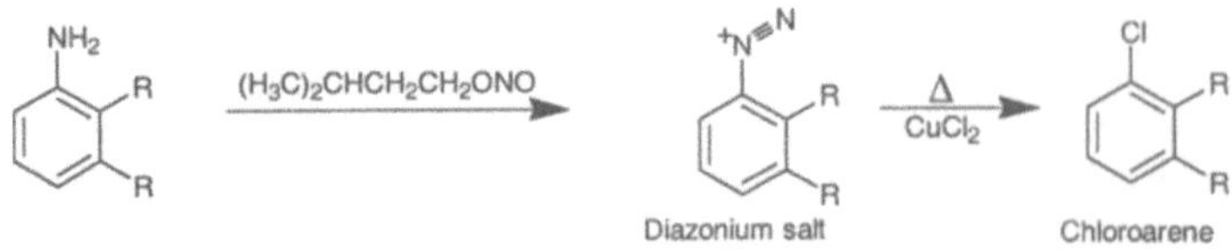

Scheme 1. Chloro-de-diazonation – the Sandmeyer reaction.

1. Introduction

The diazotisation of aromatic amines is an industrial process of great importance [1]; the range of functionalities that can be introduced via de-diazonation reactions being vast. The dangers of diazotisation are well known [2], diazonium salts being sensitive to physical agents such as heat, light, shock, static electricity, etc that can lead to rapid uncontrollable decompositions and explosions. Traditionally diazonium salts are prepared in aqueous conditions due to the stabilising effect of water on the diazonium salt and used without isolation limiting the applicability of the reaction to some substrates. Dry diazonium chemistry can be performed on a monolithic micro-fluidic device (Figure 1) in complete safety as at any given instant the total volume of dry diazonium salt is measured in nanolitres. Using micro-

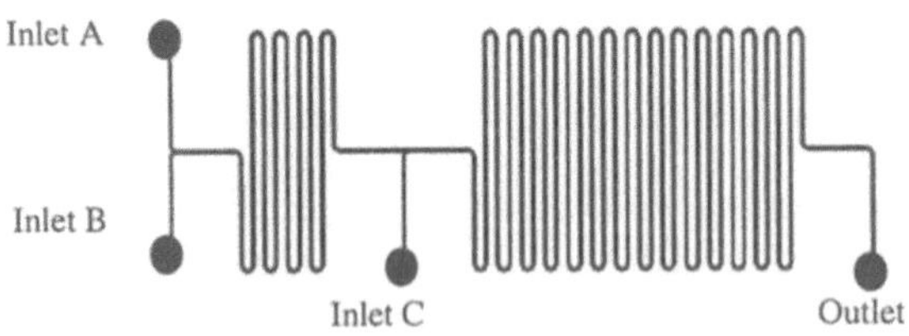

Figure 1. Schematic of microreactor channel. Channel depth 50μm, width 150μm, length 36cm.

Y. Baba et al. (eds.), Micro Total Analysis Systems 2002, Volume 2, 850–852.

fluidic devices in continuous flow regimes, real-time on-line optimisation is possible using direct detection methods. This is demonstrated by the use of Raman spectroscopy for on-chip detection and optimisation of flow rates.

2. Results and Discussion

Monolithic microreactor structures combine two or more steps into a single device. Incorporating reagent mixing, quench addition and heated decomposition sections into one device decreases experimental complexity, minimises dead volumes and the problem of precipitation is overcome. In devices incorporating heated sections a steep thermal gradient is established due to the large surface to volume area of the device. This results in two entirely different temperature environments with a separation of only 20mm without the need for the inclusion of heat exchangers. For a typical structure flow rates of $3.2\mu Lmin^{-1}$ through the first two inlets were used to allow for complete diffusive mixing and reaction over the first serpentine delay section.

By operating in continuous flow microreactors on-line and real-time optimisation of reaction conditions can be achieved by utilization of direct monitoring techniques. Raman spectroscopy utilises monochromatic light usually in the visible region of the spectrum and is an excellent real-time, on-line fingerprinting tool for chemical identification. Each channel can be addressed individually for facile reaction monitoring with little or no glass Raman signal background from the chip itself. To demonstrate the use of Raman spectroscopy for direct detection and dynamic optimisation of reaction conditions, solutions of aniline and isoamyl nitrite in N,N-dimethylformamide were infused into the device at a variety of flow rates (Figure 2). Both aniline and N,N-dimethylformamide have strong characteristic Raman bands between 200 and 1700 cm^{-1} providing ample information for sample identification.[3] Raman spectra can be obtained at millimolar concentrations or below and at data rates of up to five spectra per second allowing on-line real-time reaction monitoring (Figure 3). Utilizing optimized concentration conditions chloro-de-diazonation was performed

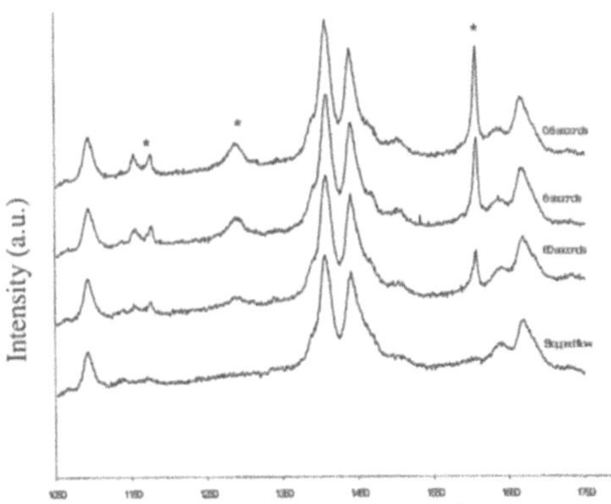

Figure 2. On-line reaction monitoring of aniline diazotisation by Raman spectroscopy

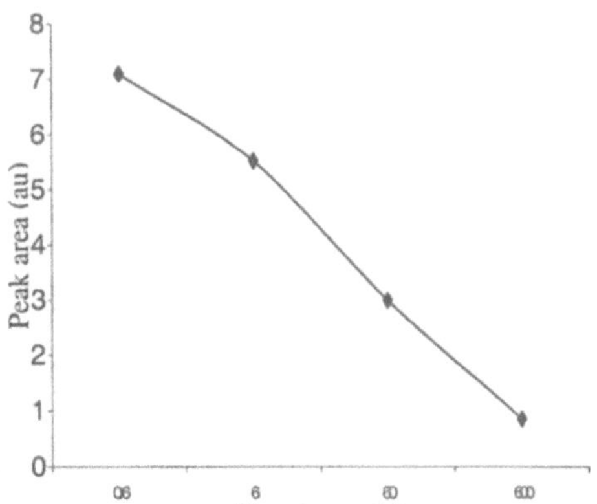

Figure 3. Dynamic optimization of aniline diazotisation by Raman spectroscopy

for three separate aryl amines on a microreactor each being diazotised by amyl nitrite following identical procedures. No attempt to optimize temperature conditions was made during this study nor was an acid catalyst employed. To affect diazotisation, isoamyl nitrite was chosen which is often used on a laboratory scale to perform diazotisations on water sensitive substrates.[4] Alkyl nitrites are thermally unstable and may readily decompose or explode on heating.[5] By using micro-reactor components, however, reagents such as alkyl nitrites can be utilised for industrial processes in relative safety. In addition utilizing alkyl nitrites often results in yields inferior to those utilizing sodium nitrite. Substrates containing alkyl groups such as the toluidines used in these experiments can also lead to disappointing yields. By operating in the micro-environment, microfluidic reactors may be used to increase yields of such reactions to more satisfactory levels. The advantages of microflow reactors can be clearly seen by comparison to a traditional bulk scale synthesis employing the same solutions and temperatures. In contrast to poorly mixed bulk solutions with little temperature control, yields of reactions performed on the microreactor showed significant improvement with increases of 15 - 20% being observed (Figure 4).

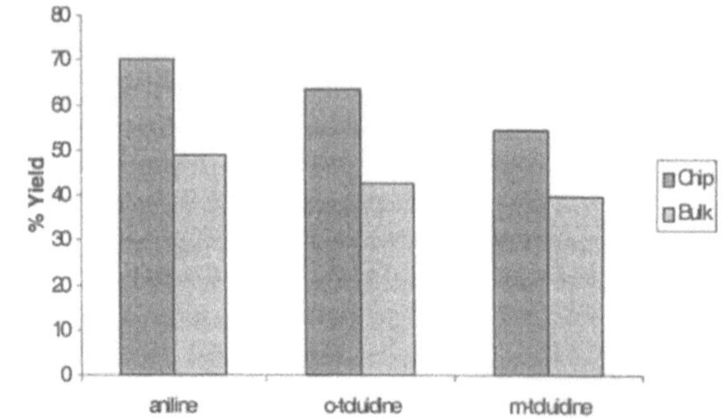

Figure 4. Comparison of nanoreactor vs bulk scale chloro-de-diazotisation.

3. Conclusion

The utility of a microfabricated microreactor to effect the safe, continuous flow multistage syntheses involving unstable, reactive intermediates has been demonstrated. The benefits of the technique include elevated temperature handling of thermally unstable intermediates and reagents in anhydrous conditions with increased safety over existing procedures. The simple structures described and the safe nature of the procedure show clear advantages over bulk scale syntheses with significant yield increases and facile optimisation offering a viable alternative to industrial routes.

4. References

1. The Chemistry of Diazonium and Diazo Groups; S. Patai Ed. Wiley, N.Y., (1978),
2. H. Zollinger, Diazo Chemistry I; VCH:Weinheim, pp. 24 (1994),
3. G. N. R Tripathi, J. Chem. Phys, 73, 11, pp. 5521-5530 (1980),
4. M. P. Doyle, B. Siegfried, J. F. Dellaria Jr., *Journal of Organic Chemistry, 42*, pp.2426-2431 (1977),
5. *SORB: safety relevant characteristic data of chemical materials,* pp.146. (1976).

MICROSYSTEMS TECHNOLOGY FOR CELL SCREENING IN NEW MEDICINES DISCOVERY

Norbert Klauke[1], Godfrey Smith[2], and Jon Cooper[1]
[1]Department of Electronics, University of Glasgow, Glasgow G12 8LT
[2]Biomedical and Life Sciences, University of Glasgow, Glasgow G12 8QQ

Abstract

A high-density array of assay sites for cell based drug profiling was micromachined on microscope glass coverslips. The shortening and intracellular calcium of individual ventricular muscle cells electrically stimulated in picolitre volume was recorded using video enhanced light microscopy. The electromechanical coupling of cells enclosed in microchambers decayed in contrast to cells permanently superfused in microchannels, indicating ischemic conditions in the restricted volume. This cell-based model of ischemia will enable the profiling of new protective compounds in heart failure.

Keywords: active cell assay, electrical stimulation, adult ventricular cardiomyocyte

1. Introduction

The pharmaceutical industry remains under increasing pressure to reduce costs, to improve the number of new chemical entities as lead compounds and consequently to increase the rate of drug discovery. Coupled with this, increased numbers of biological/drug targets, delivered by both the genome and proteome projects, has moved the current discovery bottleneck "down- stream" into cell based assay systems. Increased demands for high-through-put cell screening of new lead compounds has led to the development of a microsystems technology, with the potential to improve the rate and the quality of information.

A variety of tools have enabled cell profiling to be performed on-chip. These include separation technologies, to allow the purification of cell samples (including the removal of cell debris), the development of valved microincubators, to ensure continuous superfusion of metabolically active cells, the dispensation of picolitre volume drugs across high throughput chip based cell screens and the optical and/or electrochemical measurement of cell metabolism.

In this paper we describe the fabrication and use of electrode arrays integrated into arrays of picolitre chambers or microchannels on glass coverslips to enable electrical stimulation of isolated ventricular muscle cells.

2. Theory

The theoretical threshold for the hydrolysis of water is set by its oxidation/reduction energy given as

Anode:	$2\ H_2O$	$\rightarrow$	$4\ H^+ + 4\ e^- + O_2$	$E^0 = 1.230$ V
Cathode:	$2\ H_2O + 2e^-$	$\rightarrow$	$2\ OH^- + H_2$	$E^0 = -0.828$ V

Y. Baba et al. (eds.), Micro Total Analysis Systems 2002, Volume 2, 853–855.

The threshold for electrical stimulation of cardiomyocytes oriented parallel to the electrical field is 20 V/cm. This field strength can be achieved without hydrolysis by downscaling the distance between the electrodes (Fig. 1).

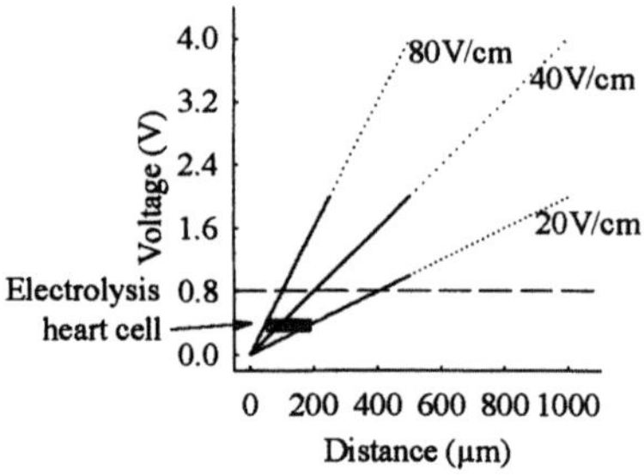

Fig. 1 Voltage/distance plot

3. Experimental

Hearts were removed from terminally anaesthetised rabbits and myocytes were isolated from the left ventricle by perfusion with collagenase solution. Individual cells were introduced into the chip-based microsystem, fabricated from glass-metal-PDMS hybrids using soft lithographic methods. Sample preparation and fluidic control were integral to the overall system. The optimum conditions for ion-channel stimulation (and control of membrane potential) were investigated, and in general, it was found that the threshold for contraction was ca. 20V cm-1. Use of a bi-polar stimulus pulses, presented to the cell using photolithographically defined microelectrodes prevented the accumulation of electrolysis products, enabling the cells to survive for up to 24 hours, on the chip, within perfused pL volume titre chambers.

4. Results and Discussion

Figure 1 shows an example of a ventricular heart muscle cell aligned to the stimulating electrodes. Field strength for excitation was 600 mV/200μm, well below the

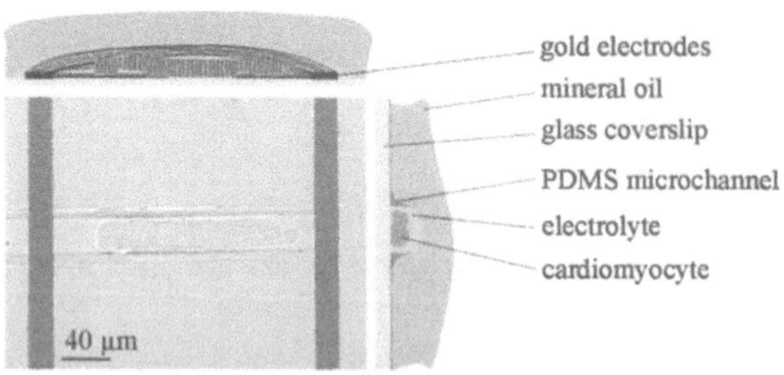

Fig. 1 Micrograph of single cardiomyocyte

threshold for hydrolysis experimentally determined as shown in Figure 2.

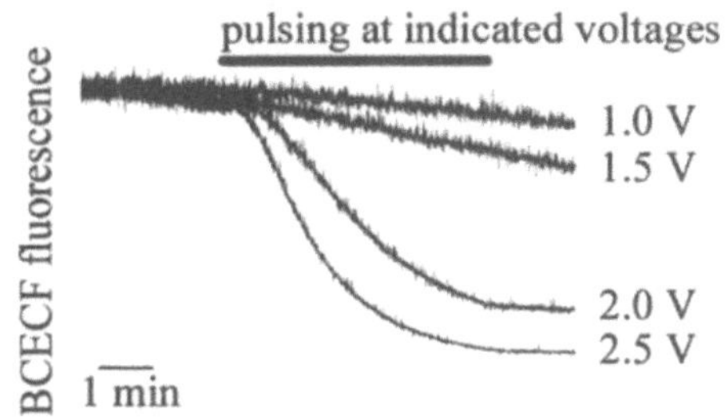

Fig. 2 pH change in microchamber

Figure 3 shows the amplitude of the contraction of a single heart muscle cell electrically stimulated either in the microchamber (left, without perfusion) or in the microchannel (right, continous superfusion)

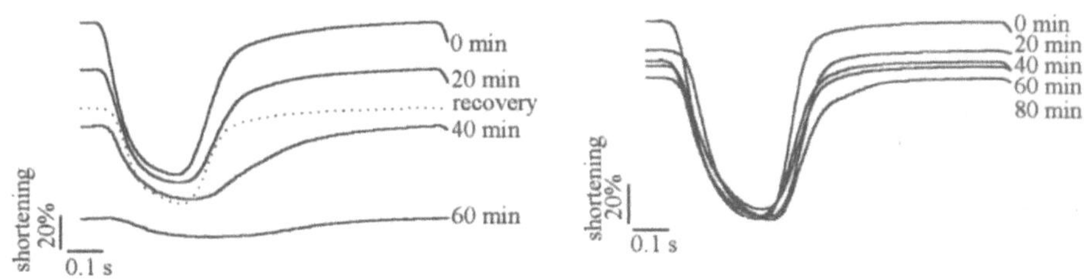

Fig. 3 Decay of amplitude of contraction in enclosed volume (left), no change in perfusion channel (right)

The amplitude of the intracellular calcium transient in heart cells stimulated over a period of 60 min in the microchamber did not change indicating impaired excitation contraction coupling due to ischemic condition within the restricted extracellular space.

5. Conclusion

The presented data indicate that microsystems/lab-on-a chip technology can be developed with associated stimulus electrodes to allow the study of arrays of individual cardiac cells. The chip can be readily incorporated into a confocal or multi-photon microscope to enable high spatial as well as temporal resolution of pharmaceutically induced sub-cellular processes.

Electrophysiological High Throughput Drug Screening System

Nobuhiko Ozaki[1], Ryuta Ogawa[1], Hirokazu Sugihara[1], Fumiaki Emoto[1], Makoto Tanabe[2], Satoshi Konishi[2], and Hiroaki Oka[1*]
[1]Matsushita Electric Industrial Co., Ltd., 3-4 Hikaridai Seika Soraku Kyoto 619-0237 Japan
[2]Ritsumeikan University, 1-1-1 Norohigashi kusatsu Shiga 525-8577 Japan
(*e-mail: oka@crl.mei.co.jp)

Abstract

A novel type electrophysiological biochip to realize efficient electrophysiological recording was designed and fabricated. This biochip enabled to record field potentials derived from living cells without cell damage.

Keywords: High Throughput Screening, Electrophysiological recording, and Biochip

1. Introduction

Progress in genomic science and combinatorial chemistry has dramatically increased the number of screening test required in drug development. High-throughput screening (HTS) method is used to identify new drugs from extremely large numbers of candidate. In conventional HTS, sample cells were stained with special dye, and effects of chemical compounds are evaluated by observing optical characteristic change of the dye. However, heterogeneous staining, dye breaching and cell toxicity of the dye prevent reliable and stable recordings. On the other hand, electrophysiological methods can measure effects of compounds more precisely, but low throughput capacity makes it difficult to apply it to HTS. In order to overcome these disadvantages, we developed a novel biochip and system for electrophysiological high throughput screening (E-HTS).

2. E-HTS system

We developed the E-HTS system consisting of three units: 1) biochip unit, 2) measuring unit and 3) data analysis unit (Fig.1).

Biochip unit:

E-HTS biochip has 16 recording chambers arranged in 4x4 array. Each chamber is shown by a white square, and recording electrodes and terminals corresponded to the chambers are patterned on the bottom side as shown from the arrow (Fig.2 A). Each chamber has 100 through-holes for capturing cell (Fig.2 B). These through-holes, of which upper and lower diameters are 20 and 5 μm respectively, are fabricated by RIE etchings for making a bowl shape (Fig.2 C). This structure makes it possible to capture a single cell in the through-hole by suction. The recording electrode acquires field potentials of 100 cells as an electrical signal in each chamber.

Y. Baba et al. (eds.), Micro Total Analysis Systems 2002, Volume 2, 856–858.

Measuring unit:

The measuring unit includes a head amplifier (HA), a main amplifier (MA), and filters. Total gain of HA and MA is 100 to 10,000 times, and unnecessary frequency component is removed through filters.

Data Analysis unit:

Effects of compounds are analysed statistically by using the method described in Katz et al.[1]. It is known that various chemical compounds affect open probability of ion channels, and thus, cause change of voltage fluctuation amplitude of a cellular electrical signal. Distribution of standard deviations (SD) of the voltage fluctuation amplitude is used as an index of the effects of compounds.

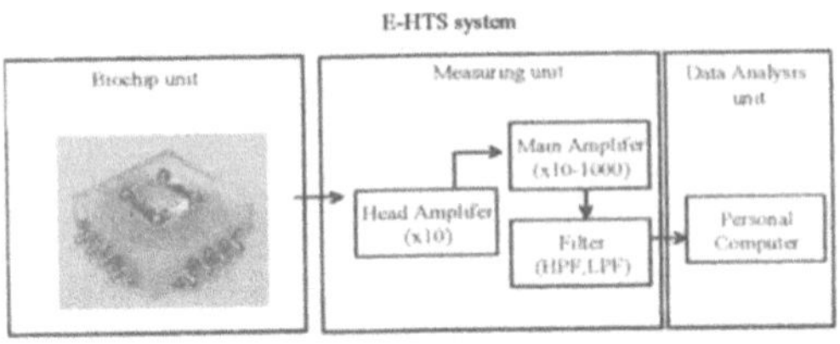

Fig. 1. A block diagram of E-HTS system

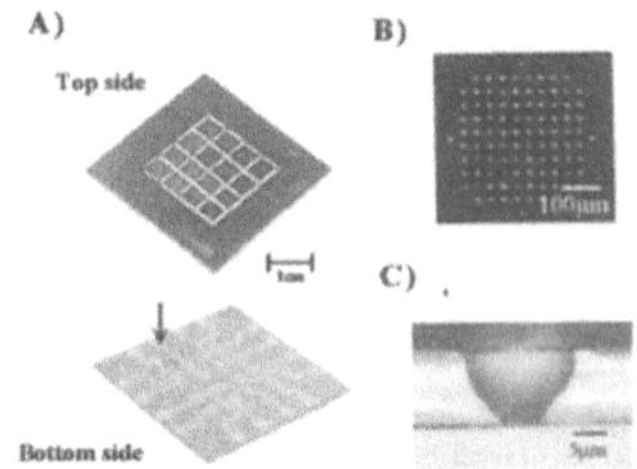

Fig. 2. E-HTS biochip
A) An entire image of the biochip. B) An enlarged image of one chamber. C) A cross sectional view of one through-hole.

3. Experimental

Fabrication of Biochip

We started with a 525 μm (15/2/508 μm) thick in (100) SOI wafer with a resistivity of 0.3 V cm. These bowl-shaped holes were etched by reactive ion etching (RIE). To make chambers below the through-hole, SOI wafer were immersed into TMAH/IPA etchant solution, 75℃ 24 hours, etched by anisotoropic wet etching. To make through-holes, the middle SiO2 layer was etched by dry etching process from the bottom of the wafer. Subsequently, we structured the electrical contacts by sputtering with Cr 2000 Å/Au8000 Å deposition.

Cell preparation

Lymnaea stagnails used for experiments had a shell length of 15-25 mm. The central nervous system (CNS) of *Lymnaea stagnails* was dissected quickly and ganglion cells in CNS were dissociated according to the method described in O. B. Nesic et al.[2].

Measurement

10 μM Glutamate (final conc.) was applied to each chamber by pipette. The voltage signals were amplified 100 to 10,000 times, and filtered at 5 kHz, and stored for 30 sec. The sampling frequency was 10 kHz. SDs of voltage fluctuation were calculated at every 20 msec.period of time.

4. Results and discussion

Fig.3 shows the effect of 10 μM glutamate (Glu) using the distribution of SDs of field potentials at the period of time. Before applying the Glu, the distribution of SDs is shown by open column. The median and 3σ were 7.11 and 1.02, respectively. After applying the Glu, the distribution of SDs is shown by closed column. The median and 3σ were 9.09 and 1.77, respectively. The distribution of SD of voltage fluctuation was clearly changed with applying Glu. These results suggest that E-HTS biochip enables to detect an excitatory effect of Glu on *Lymnaea* ganglion cells.

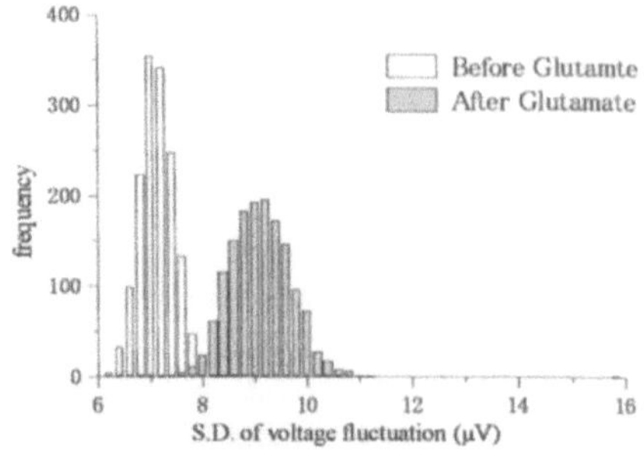

Fig. 3. A distribution of voltage fluctuation

Each distribution considering the deviation of histogram was evaluated numerically utilized Z factor. Z factor is used to confirm reliability of a HTS system. It is known that a HTS system has enough reliability when Z factor is between 0.5 and 1.0.[3].

$$Z = 1 - \frac{3\sigma_{control} + 3\sigma_{compound}}{|mean_{control} - mean_{compound}|} \qquad (1)$$

In consequence, Z factor shown in figure 3 was calculated 0.58. In this time, Z factor satisfied the difference of activated cell condition, however the sensitivity will have to be improved for high throughput drug screening.

5. Conclusions

E-HTS biochip, which had bowl-shaped through-holes on Si substrates, enabled to record the effect of Glu without any pre-treatment. The Z factor in *Lymnaea* ganglion cells when stimulated with Glu was 0.58. This value suggests that E-HTS biochip will realize an electrophysiological HTS.

Voltage fluctuation of field potentials around cell is reflected an essential change as a certain index induced by effects of chemical compounds. E-HTS system will play a crucial role not only at HTS of hit compounds but also at the optimisation of lead compounds, and will be expected to accelerate the drug development.

References

1. B. Katz and R. Miledi, The statistical nature of the acetylcholine potential and its molecular components, J. Physiol., 224, pp.665-699, (1972).
2. O. B. Nesic et al., Glutamate as a putative neurotransmitter in the mollusc, lymnaea stagnails, Neuroscience, 79, pp.1255-1269, (1996).
3. JH. Zhang et al., A simple statistical parameter for use in evaluation and validation of high throughput screening assay, J.Biomol. Screen, 4: pp.67-73, (1999).

A COMBINATORIAL LIBRARY OF PYRAZOLES PERFORMED IN AN AUTOMATED MICRO REACTOR SYSTEM

Eduardo Garcia-Egido,[1] Valerie Spikmans,[2] Stephanie Y. F. Wong[2] and Brian H. Warrington[2]
[1]GSK Cambridge Technology Centre, Cambridge Chemistry Laboratory, Lensfield Road, Cambridge, CB2 1EW, UK
[2]Glaxo SmithKline, New Frontiers Science Park (North), Third Avenue, Harlow, Essex, CM19 5AW, UK
(Eduardo_2_Garcia-Egido@gsk.com)

Abstract

This paper reports a combinatorial library synthesis, detection and analysis carried out in an automated micro reactor system. The reactions were performed in a non-well based glass micro channel reactor. The results demonstrate the ability of a single micro reactor to carry out combinatorial chemistry.

Keywords: Pressure-driven flow, automation, micro reactor, synthesis, combinatorial chemistry

1. Introduction

Micro channel flow reactors potentially provide a basis for carrying out ultra-high throughput chemical synthesis on a greatly reduced scale that is compatible with highly miniaturised modern screening techniques. Recently reported descriptions of micro-channel based synthesis of thiazoles [1] and Diels-Alder adducts [2] demonstrating time and scale economies are favourable. Of late, interesting approaches to the synthesis of combinatorial libraries using well-based micro reactors have been reported [3]. However, the number of compounds made '*in one shot*' is limited by the chip design and, in addition, the product analysis is carried out off-line. A non well-based micro reactor seems to be more suitable to perform synthesis and detection in an integrated system [4].

2. Objectives

The aim of this work is to show the capabilities of an automated micro reactor system to perform combinatorial chemistry, detection and analysis on line. Another objective is to demonstrate the ability of a single-channel micro reactor to carry out the synthesis of combinatorial libraries.

Y. Baba et al. (eds.), Micro Total Analysis Systems 2002, Volume 2, 859–861.

3. Experimental

Reactions were performed in a non-well based glass micro reactor with a serpentine-design channel (100 μm width) and integrated with the autosampler (Figure 1).

Figure 1. Integration of the micro reactor with the autosampler

The starting materials were introduced into the system using an autosampler. A μ-HPLC pump was used to mobilise reagents towards the two micro reactor's inlets. The reaction plugs were sequentially diluted by using a dilution pump (the dilution solvent is injected at the micro reactor's outlet) and detected by using an external UV detector. Only a small amount of diluted plug is introduced into an on-line LC-MS system. Products were identified and yields quantified by MS and UV analysis.

4. Results and discussion

The reaction selected to test the system was the double condensation of aromatic 1-3, dicarbonilic compounds with hydrazines to afford pyrazoles. A library of 21 pyrazoles (7 x 3) was synthesized (Figure 2).

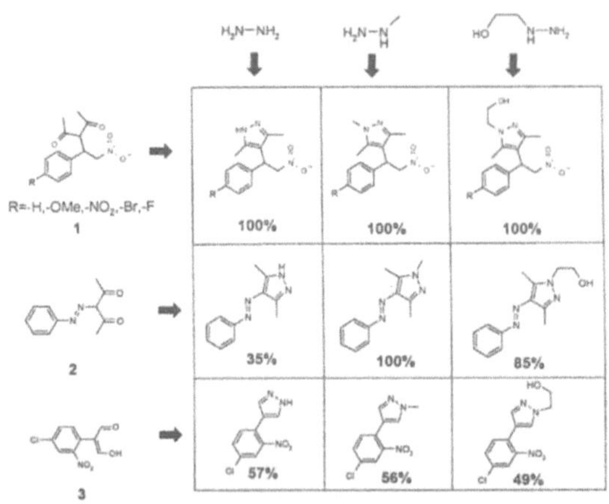

Figure 2. Library of pyrazoles performed in the automated system

The flow rate used to perform the library was 2 μl/min (210 s of residence time on chip). The total time needed to carry out the synthesis and analysis of the whole library (21 compounds) was 5h 15 min (15 minutes for each reaction). Most of the reactions worked until completion as no starting material was detected.

5. Conclusions

The results of this work indicate that the automated micro reactor system could be used to support the generic synthesis of combinatorial libraries detection and analysis in a short time. This work demonstrates also the suitability of a non-well single-channel micro reactor to carry out multiple syntheses in '*one shot*' under pressure-driven flow.

Acknowledgements

E. Garcia-Egido thanks the EU for a Postdoctoral Marie Curie fellowship (HMPI-CT-1999-00066).

References

1. E. Garcia-Egido, S. Y. F. Wong and B.H. Warrington, *A Hantzsch synthesis of 2-aminothiazoles performed in a micro reactor system,* Lab Chip, 2, pp. 31-33, (2002).
2. M. Fernandez-Suarez, S. Y. F. Wong and B. H. Warrington, *Spatial and temporal control in the synthesis of a series of cycloaddition products in a glass microchip with pressure driven flow*, Proc. LabAutomation 2002, Palm Springs, CA, Jan. 27-30, pp. 285 (2002).
3. Y. Kitukani, H. Hisamoto, M. Tokeshi and T. Kitamori, *Fabrication of a glass microchip with three-dimensional channel network and its application to a single-chip combinatorial synthetic reactor*, Proc. Micro Total Analysis Systems 2001, Kluwer Academic Publisher, Dordrecht, The Netherlands, pp. 161-162, (2001).
4. V. Spikmans, S. J. Lane, B. Leavens, A. Manz and N. W. Smith, *On-chip post-column derivatisation reactions in capillary liquid chromatography-mass spectrometry*, Proc. Micro Total Analysis Systems 2001, Kluwer Academic Publisher, Dordrecht, The Netherlands, pp. 222-223, (2001).

HIGH-THROUGHPUT SCREENING OF ANTICANCER DRUGS USING MICROARRAY BASED CELL CHIP

Sathuluri Ramachandra Rao, Yoshinori Akagi, Yasutaka Morita and Eiichi Tamiya
School of Materials Science, Japan Advanced Institute of Science and Technology, 1-1, Asahidai, Tatsunokuchi, Ishikawa 923-1292, Japan
(srrao@jaist.ac.jp)

Abstract

This paper describes fabrication of a microarray based cell chip and its suitability for cultivating human cancer cells (HeLa) in aqueous medium. Apart from this, the fabricated cell chip was also applied in high-throughput screening of anticancer activity against HeLa 229 cells of known standard carotenoids and carotenoid chemicals isolated from bacterial cultures. The anticancer or cytotoxic activity of these targeted chemicals was monitored by a fluorogenic substrate-calcein. The imaging was carried out using a chip scanner and the anticancer activity was expressed in terms of cell death. These fabricated microarray based cell chips play a profound role in high-throughput screening of potential anticancer drugs.

Keywords: High-throughput screening, microarray, cell chip, anticancer drugs

1. Introduction

High-throughput screening methods for drug discovery play a key role in the search for new potential drugs and therapeutic targets. Over the past decade, a variety of scientific advances and economic pressures have driven the need for improved drug discovery screening technology [1, 2]. These advances include the growing number of potential therapeutic targets emerging from the field of functional genomics, the rapid development of large compound libraries derived from parallel and combinatorial chemical synthesis techniques, and the ever increasing pressure to reduce developmental costs while enhancing commercial competitiveness in the pharmaceutical industry.

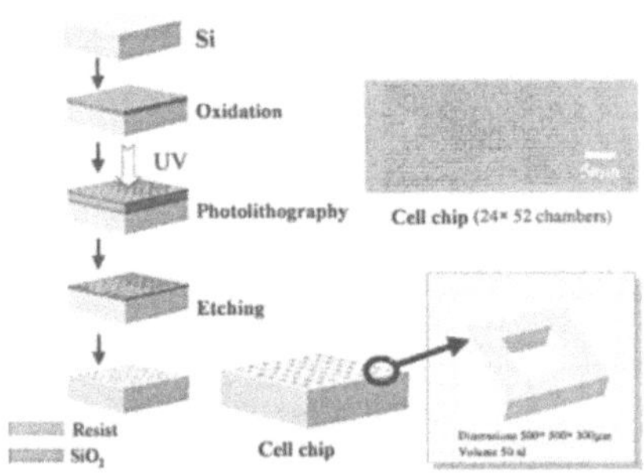

Fig. 1. Schematic representation of cell chip fabrication

Y. Baba et al. (eds.), Micro Total Analysis Systems 2002, Volume 2, 862–864.

Testing very small samples is attractive because it requires only minute amounts of both the component being assayed and the components of the assay itself, both of which may be quite expensive or available in limited amounts. The most convenient reactions to test in high-throughput are those involving chromogenic or fluorogenic substrates [3]. This paper focuses on high-throughput screening of anticancer chemicals derived from bacterial cultures. They were tested against HeLa cells, the proliferation of which was monitored by using a cell-chip.

2. Fabrication of a microarray based cell chip

The cell chip used in this study is made from a Si wafer. At first, the wafer was washed with acetone, then immersed in a 60 % (v/v) hydrogen fluoride solution, and finally allowed to dry at room temperature. Subsequent oxidation of Si wafer surface was achieved by heating at 1000°C for 8 hrs. A chip pattern (24 x 52) was then printed on the Si wafer surface by a photolithographic technique. Further, the Si wafer with chip pattern was subjected to etching with tetra methyl ammonium hydroxide (TMAH) solution for about 4-5 hrs, followed by heating treatment for about 8 hrs in order to obtain an hydrophilic surface, which allows cultivation of animal cells in an aqueous medium. This completed the fabrication of a microarray based cell chip with a pattern of chambers (24 x 52). Each chamber was parallelepiped of about 0.5 x 0.5 x 0.3 mm, and accommodated approximately 50 nl of sample volume (Fig. 1).

3. Experimental

HeLa 229, a cell line causing cervical cancer, was purchased from the Dainippon pharmaceutical company, Japan. The cells were maintained in a minimal essential medium (Dainippon, Japan) consisting of 10% (v/v) fetal bovine serum, 1% (w/v) glutamine and 1% (w/v) non-essential amino acid mixture. They were subcultured every 2 days. Lutein, a xanthophyll, was purchased from Sigma chemicals, USA, while β-carotene was obtained from Wako chemicals, Japan. A carotenoid extract mainly consisted of lutein, astaxanthin, and canthaxanthin etc. from *Stenotrophomonas* sp. in an ethanolic form was also used in this study.

The chemicals dissolved in 50 nl ethanol at different concentrations were distributed into cell chip chambers (0.5 x 0.5 x 0.3 mm) by a specially designed robot dispenser (Cartesian Technologies, USA) and allowed to dry inside the cell dispenser chamber. Then 50 nl volume of solution containing an average 165 HeLa cells were distributed in each chamber by the robotic cell dispenser while maintaining the chamber humidity between 75-85%, which allows cells to keep their native activity. The chips containing the cells were then incubated at 37°C in a 5% CO_2 atmosphere for one day. At the end of incubation, chips were processed with fluorogenic substrate-calcein (Promega, USA) according to the protocol. The chip was then analyzed using a chip scanner (Hitachi, Japan). The cell viability was measured from the intensity of the fluorescence signal emitted by live cells.

4. Results and discussion

It was found that cell proliferation was inhibited significantly in the chambers that were treated with lutein, β-carotene and carotenoid extract as compared to control cultures with the same volume of ethanol and number of cells. The anticancer activity of these chemicals was measured by subtracting the fluorescence intensity emitted by of all viable cells in each treatment on a chip from the intensity of control chambers. The difference in the intensity was a measure of anticancer activity of these chemicals against HeLa 229 cells (Figs. 2a, b). Cytotoxicity of β-carotene and lutein was higher than that of the carotenoid extract of bacterial cultures. Experiments are presently in progress to optimize both the cell number and the concentration of the chemicals.

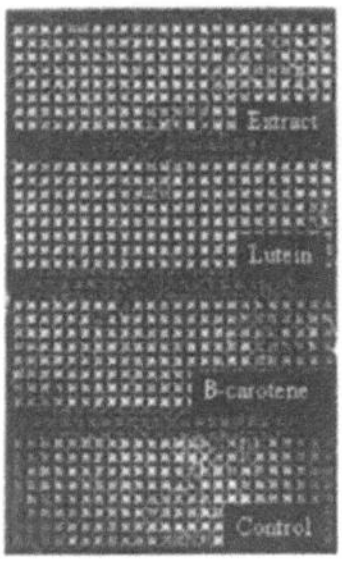

Fig. 2a. Cell chip showing emission of fluorescence signals from live cells

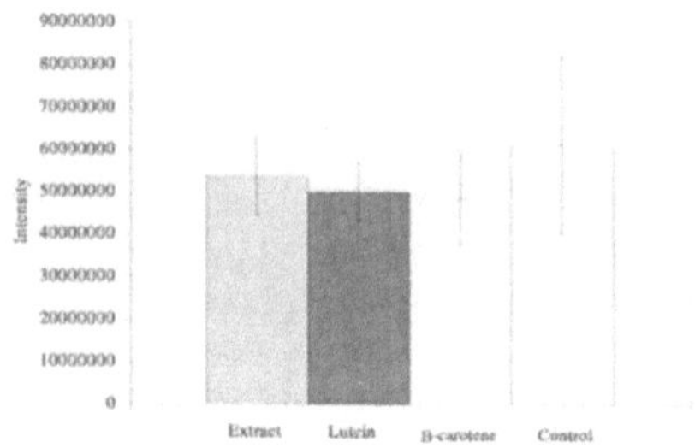

Fig. 2b. Anticancer activity of carotenoids and carotenoid extract

5. Conclusions

This study showed an example of successful cultivation of animal cells in tiny chambers of a microarray-based cell chip. Further, the fabricated microarray-based cell chip can be used in screening of novel potential anticancer drugs in a high-throughput assay.

Acknowledgements

SRR gratefully acknowledges the support of Japan Society for the Promotion of Science (JSPS) towards Postdoctoral fellowship.

References

1. J. Wolcke and D. Ullmann, *Miniaturised HTS technologies-μHTS*, Drug Discovery Today, 2001, 6, 637-646.
2. R.P. Hertzberg and A.J. Pope, *High-throughput screening: new technology for the 21st century*, Current Opinion in Biotechnology, 2000, 4, 445-451.
3. S.A. Sandberg, *High-throughput screening and Ultra-high-throughput screening: solution- and cell-based approaches*, Current Opinion in Biotechnology, 2000, 11, 47-53.

ELECTRIC FIELD ASSISTED EXTRACTION AND FOCUSING OF FINGERPRINT RESIDUES BY MEANS OF A MICROFLUIDIC DEVICE

Silvia Valussi* and Andreas Manz

Department of Chemistry, Imperial College London, South Kensington
London SW7 2AY, UK
(E-mail: s.valussi@ic.ac.uk)

Abstract

This paper presents a novel approach for the analysis of fingerprint components based on a microfluidic device. The method enables direct sampling of one single fingerprint, on-chip pre-concentration, separation, and analysis of targeted fingerprint components with on-chip fluorescence detection.

Keywords: fingerprints, Lab-on-a-Chip, sample pre-concentration, forensics

1. Introduction

Fingerprints are still one of the most important contact traces left during the commission of a crime, as their uniqueness places an individual at the scene.

Although the chemistry of fingerprints is far from being well-understood [1], significant inter-individual differences in the composition of sweat and sebum as well as in that of fingerprint residues have been reported [2,3]. These findings reveal the presence of biochemical markers in fingerprint residues and suggest the possibility of identifying personal traits (e.g. gender, age, ethnic origin, living habits, etc.) via the chemical analysis of components in fingerprints and/or skin. In this study, we targeted the amino acids present in fingerprint residues, which we visualized by labeling with fluorescamine.

2. Experimental

A glass microfluidic device was used to focus and separate fingerprint material laid on the top of an agarose gel substrate (Figure 1). Fuorescamine-labeled fingerprint components were extracted through the gel by means of an applied potential difference between a ring electrode placed on the top surface of the gel (ground) and another electrode (positive) located in a buffer reservoir (100mM TRIS/20mM borate, pH 9.2) at the end of the channel (Figure 2). During this step the extracted components focused at the gel-buffer interface inside the microchip channel, producing a concentration band.

Typically after 300s, the voltage polarity was reversed and, as a result, the concentration band was mobilized and its components separated and detected.

* on leave from The Forensic Science Service, Lambeth Site, London SE1 7LP, UK

Y. Baba et al. (eds.), Micro Total Analysis Systems 2002, Volume 2, 865–867.

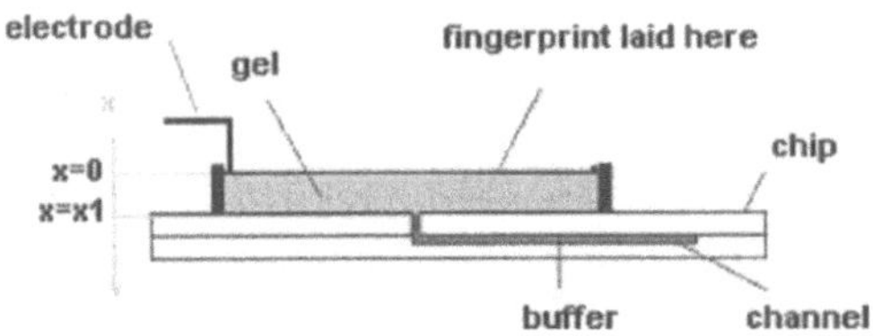

Figure 1. Experimental set up

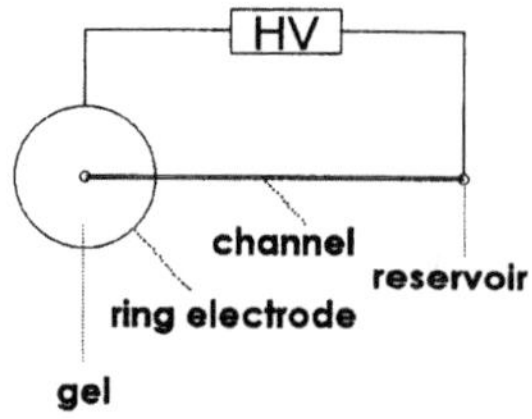

Figure 2. Schematics (top view)

3. Electro focusing: Theory

Within our experimental set up (Figure 1), the on-chip pre-concentration phenomenon arises at the point along the x-axis of the electric field where the flux of ions N_x is zero, *i.e.*

$$N_x = -D\frac{dc}{dx} + \left(u_x(x) + z\omega \frac{\sigma_1 E(x)}{\sigma(x)} \right) c = 0 \tag{1}$$

where D is the diffusion coefficient ($m^2\ s^{-1}$), u_x is the convective velocity ($m\ s^{-1}$), c is the concentration of the ion ($mol\ l^{-1}$), z is the charge, $E(x)$ is the potential gradient ($V\ m^{-1}$), $\sigma(x)$ is the conductivity ($\Omega^{-1}\ m^{-1}$), and ω ($m^2\ s^{-1}\ V^{-1}$) is the absolute mobility.

In eq (1) $u(x)$ and $E(x)$ vary with x due to the gel-buffer configuration and to the circular geometry of the electrode utilized on the top of the gel. $\sigma(x)$ also varies with x according to

$$\sigma(x) = \sigma_1 \text{ when } 0 < x < x_1 \text{ and } \sigma(x) = \sigma_2 \text{ when } x > x_2,\quad \sigma_1 << \sigma_2 \tag{2}$$

where σ_1 and σ_2 are the conductivities of the gel and the buffer respectively.

In order for an ionic analyte to focus it is necessary that the convective velocity $u(x)$ and the electrophoretic velocity $v(x) = z\omega E$ balance each other, *i.e.*

$$u_x(x) = -z\omega\sigma_1 \frac{E(x)}{\sigma(x)} \tag{3}$$

Due to the steepness of $u(x)$ and $E(x)$ there exists a well-defined focusing plane in proximity of the gel-buffer interface, in which the analyte ion is trapped and concentrated.

4. Results and discussion

Figure 3 shows the results obtained after mobilising the concentration band in the case of a model compound, 75 fmol of aspartic acid (A) and in that of a single fingerprint (B). Time t_0 marks the start of the mobilization phase. In Figure 3A, the plate number N was 136,900, with associated plate height H of 0.15μm, which is actually below the theoretical lower limit [4] as posed by longitudinal diffusion. At present, the reason for this surprising result is unknown. From these data, the estimated limit of detection for fluorescamine-labeled aspartic acid is about 1 fmol in terms of amount of substance laid on the gel (S/N=3). In Figure 3B, at least three peaks can be distinguished in the graph, possibly related to different amino acids. The plate number N for peak 2 was 6,944 with H equal to 2.9μm, implying that additional sources of band broadening are present. Considering that the migration time is shorter than for aspartic acid, which is twice negatively charged, the peaks could represent amino acids or other primary amines in fingerprint residues with a single or a partial negative charge.

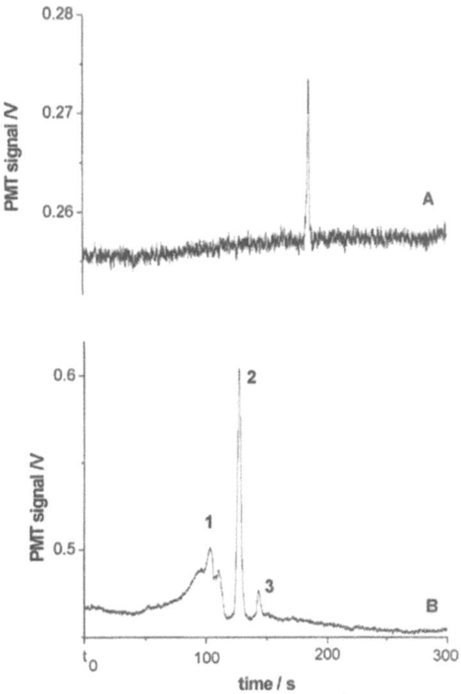

Figure 3. A) Aspartic acid, 75 fmol. S/N=14.
B) Single fingerprint from female donor.

Acknowledgements

The Forensic Science Service for the financial support and especially Dr. Julia Elliott. We would also like to thank Dr. Jan Eijkel for the useful discussions.

References

1. Bramble SK & Brennan SJ "Chemistry of print residue" in Encyclopedia of Forensic Sciences, 3rd vol., Jay A Siegel & Geoffrey C Knupfler Eds (2000),
2. Buchanan MV, Asano K & Bohanan AM, "Chemical characterization of fingerprints from adults and children, *SPIE*, **2941** (1996) 88-95,
3. Bohanan AM "Latents from pre-pubescent children versus latents from adults", *J. For. Ident.* **48**, 5 (1998) 570-573,
4. Effenhauser CS, Manz A, Widmer HM "Glass chips for high-speed capillary electrophoresis separations with submicrometer plate heights" *Anal. Chem.* **65** (1993) 2637.

DESIGN OF A MICROFLUIDIC FLOW INJECTION ANALYSIS SYSTEM

Andrew M. Leach, Aaron R. Wheeler and Richard N. Zare
Department of Chemistry, Stanford University, Stanford, CA 94305
amleach@stanford.edu

Abstract

A flow injection analysis system consisting of multiple peristaltic pumps, an injection valve and a mixing column has been constructed in a polydimethylsiloxane monolith. The performance characteristics of the device were evaluated with fluorescent dyes. Additionally, a microfluidic enzyme kinetics assay has been explored.

Keywords: flow injection analysis, valves, PDMS

1. Introduction

Flow injection analysis (FIA) is a technique based on the repetitive injection of a known sample volume into a liquid stream that contains one or more reactants. The length of a column following the injection valve controls mixing and reaction times. Flow rates of milliliters per minute and injection volumes of 10 to 100 microliters are common in FIA experiments. The recent development of microvalves formed in elastomeric polymer devices makes possible the design of a miniaturized FIA system.[1]

Although microfluidic FIA experiments have been widely described in the literature, most systems have consisted of full-size pumps and injection valves coupled with miniature mixing column. Recently, Veenstra and coworkers[2] reported a microfabricated FIA device that contained peristaltic pumps and fluidic channels within one monolithic unit. Our experiments further the development of microfluidic flow injection systems by including pumps, an injection valve, and a mixing column inside a single polydimethlysiloxane (PDMS) device.

2. Experimental

In this study multiple peristaltic pumps, an injection valve system, and a mixing column have been constructed within a single two-layer PDMS structure (Figure 1). Fluidic channels were 100 μm wide and 10 μm deep. Pumps are used to manipulate sample, carrier, and reagents solutions at flow rates of up to 40 nL/min. The injection valve system has a volume of 1.5 nL and operates in a similar manner to a standard 6-way 2-position valve used in macro scale FIA experiments. Sample and carrier solutions that continuously flow through the injection system are segregated by a series of microvalves. By altering the state of these valves, the carrier stream is redirected through the injection loop resulting in the movement of a sample plug onto the mixing column. A serpentine column, 88 mm in length, following the injection valve provides adequate mixing and time for reactions. Mixing is predominantly controlled by

Y. Baba et al. (eds.), Micro Total Analysis Systems 2002, Volume 2, 868–870.

molecular diffusion. Two samples can be simultaneously loaded onto the chip resulting in enhanced sample throughput and the potential for fast concentration calibration routines.

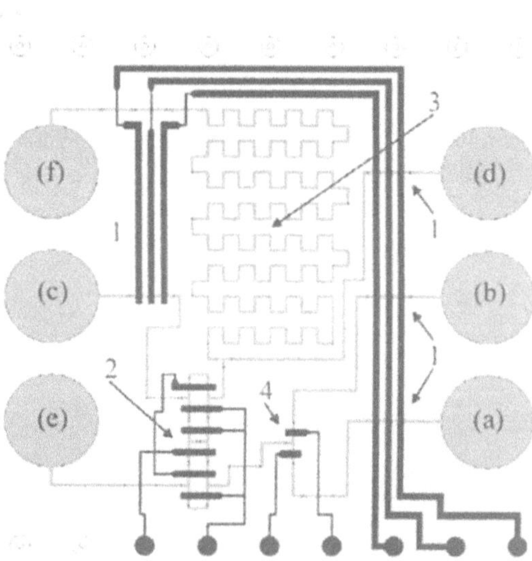

Figure 1. Diagram of the microfluidic FIA device. *Gray* lines represent fluid channels whereas *black* lines represent control channels. The device consists of (1) four peristaltic pumps, (2) an injection valve, (3) a mixing column, and (4) a sample selector. Fluid reservoirs contain (a,b) samples, (c) carrier, (d) reactant, (e) sample waste, and (f) column waste.

3. Results and Discussion

The microfluidic FIA system has been characterized with the fluorescent dye Rhodamine 110. Figure 2 shows the replicate injection of three concentrations of dye as measured at the end of the mixing column. As predicted by theory, peak dispersion was found to increase approximately as the square root of column length. Peak height

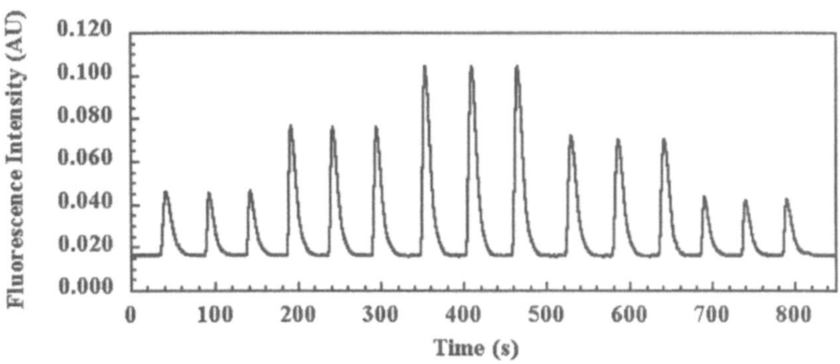

Figure 2. Replicate injections of 3.3, 6.6, and 10 μM Rhodamine 110 fluorescent dye solutions.

reproducibility was found to be better than 3% relative standard deviation. Memory effects in the injection valve were found to be minimal. It is estimated that 60 samples per hour can be analyzed with the current device.

The hydrolysis of fluorescein diphosphate by alkaline phosphatase (AP) seen in Figure 3a demonstrates the ability of the microfluidic FIA system to perform rapid enzymatic assays. The analysis was completed in 10 minutes with a total enzyme consumption of 200 nL, corresponding to approximately 420 attomoles of AP. A Lineweaver-Burk double reciprocal plot (Figure 3b) was used to determine enzyme kinetics values.[3] The maximum enzyme velocity was calculated to be 3.1 ± 0.7 nM/s, with a Michaelis-Menten constant of 4.0 ± 1.0 μM, and a turnover rate of 29 ± 7 molecules per second.

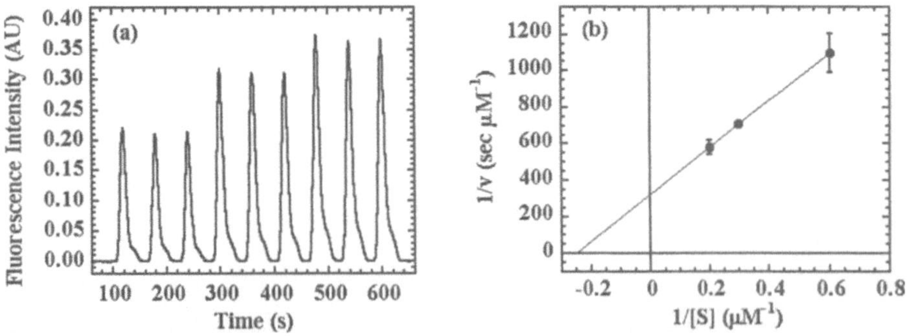

Figure 3. (a) Reaction of 3.3, 6.7 and 10.0 μM fluorescein diphosphate with 0.3 mg/L alkaline phosphatase. All solutions were in 10 mM borate buffer at pH 9.1 with a reaction time of 110 seconds. (b) Lineweaver-Burk double reciprocal plot of data seen in Figure 3a. Error bars represent the standard deviation of three discrete measurements.

Acknowledgements

The microfluidic chips used in this project were manufactured by Fluidigm Corporation through their foundry services program. This study is supported in part by Beckman Instruments. Additionally, AML is grateful to the NIH for a postdoctoral fellowship.

References

1. M.A. Unger, et al., *Monolithic microfabricated valves and pumps by multilayer soft lithography*, Science (Washington, D. C.), 288, pp. 113-116, (2000).
2. T.T. Veenstra, et al., *Monolithic versus modular integration of a micro-FIA system for ammonium determination*, Proc. Micro Total Analysis Systems 2001, Kluwer Academic Publishers, Dordrecht, The Netherlands, pp. 664-666, (2001).
3. R.A. Copeland, *Enzymes*, 2nd ed.; Wiley-VCH, Inc., New York, (2000).

BIOCHEMICAL REACTION ON A CHIP USING A QUANTITATIVE VOLUME CONTROL

Sang-Ho Lee[1], Chang-Soo Lee[2], Youn-Suk Choi[3], Kwang-Won Suh[3], Byung-Gee Kim[2], and Yong-Kweon Kim[1]

[1]School of Electrical Engineering and Computer Science,
[2]School of Chemical Engineering, Seoul National University, ENG 420-007, Seoul 151-742, Korea
[3]KyungWon Tech Co. Ltd.
sholee1@snu.ac.kr

Abstract

This study introduces a biochemical reaction on a nanofluidic chip by quantitative nanoliter volume control. The proposed nanoliter metering techniques showed accuracy with 2 % error at the criterion of 20-nL metering. A biochemical reaction on a chip was performed with β-galactosiadase rendering fluorescent product from fluorescein di-β-galactopyranoside (FDG) substrate.

Keywords: Enzymatic reaction, mixing, metering, PDMS, fluorocarbon

1. Introduction

Major operations in lab-on-a-chips (LOCs) for biochemical analysis involve preparation of chemical or biochemical reagents, separation and detection. The sample preparation step includes various fluidic operations such as injection, metering, mixing, and reaction [1-3].

A precise quantitative liquid handling is one of essential techniques among various fluidic operations on the LOCs for biochemical analyses [3-5]. This study introduces an in-situ monitoring of biochemical reaction on a chip conducted by a quantitative volume control.

2. Experimental

The quantitative control was performed by the proposed variable metering technique. Sample reagents were metered as a desired volume using the variable metering technique (figure 1(a)). Metered liquid droplets were manipulated into a mixing and region by surface guided control for reaction (figure 1(b)). The proposed fluidic handling methods were designed by a hydrophobic valving and a selective wettability patterning.

For the fabrication of the nanofluidic chip, polydimsilioxane (PDMS) microchannel networks were aligned and bonded on the glass wafer with the hydrophobic FC patterns.

Y. Baba et al. (eds.), Micro Total Analysis Systems 2002, Volume 2, 871–873.

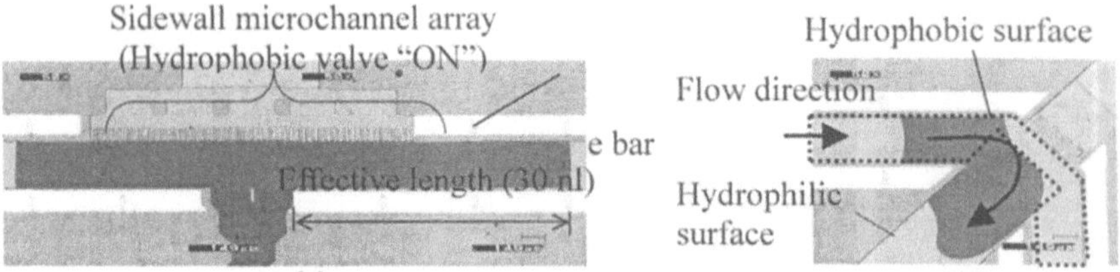

(a) (b)

Figure 1. Proposed fluidic manipulation methods; (a) Variable metering using direct measuring in microchannels and (b) Surface guided flow control.

3. Result and discussion

We made several numerical models to predict and find out the detail flow pattern of the liquid in surface guided flow control. A commercial CFD code (CFD-ACE+) solved the full Navier-Stokes equations using a finite volume approach. The unsteady motion of the free surface was solved on Eulerian grid using the Volume of Fluid (VOF) technique. Higher order surface reconstruction algorithms were employed to predict the free surface shape. To predict the surface tension effects, the Laplace-Young high-order nonlinear boundary conditions at the liquid surface were used to accurately model the liquid-gas interface and at the wall contact. Simulation results were well agreed with the experiments. Figure 2 shows the computational simulation results for the surface guide control.

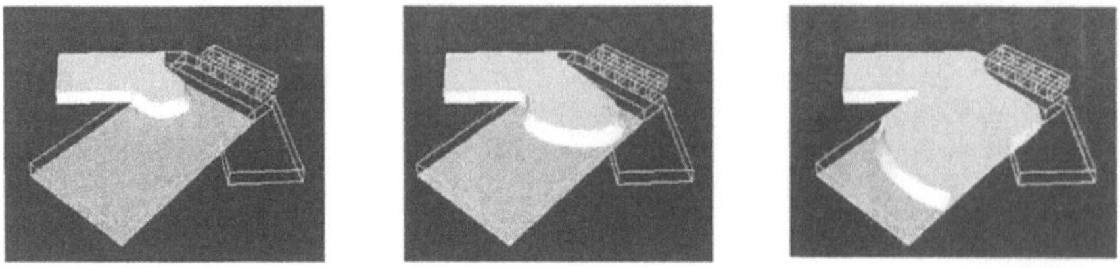

Figure 2. Computational simulation results for the surface guide control.

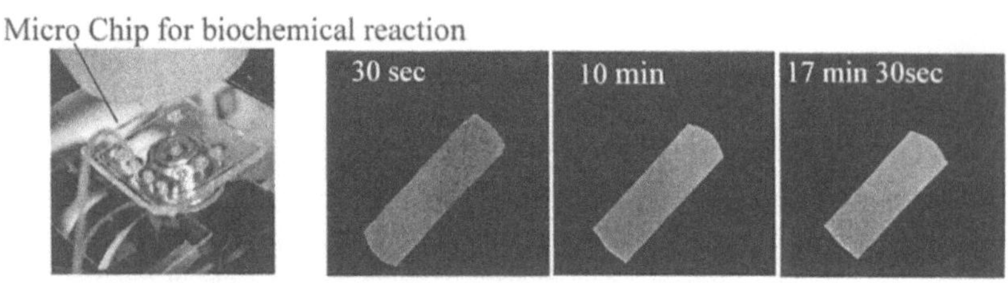

(a) (b)

Figure 3. Biochemical reaction on the nanofluidic chip; (a) the nanofluidic chip mounted on a confocal microscope, (b) the fluorescence images during the reaction enzyme/FDG (30nl/30nl) as a function of reaction time (at room temperature).

In this study, a biochemical reaction on a chip was performed with β-galactosiadase rendering fluorescent product from FDG (fluorescein di-β-galactopyranoside) substrate [6,7]. The substrate (FDG) and the enzyme (β-galactosiadase) solution respectively were manipulated on the nanofluidic chip at the desired volume and mixing ratio. In order to investigate the reactivity in different ratio the enzyme to the substrate, the fluorescence intensity was observed during the reaction as shown in figure 3.

4. Conclusions

The novel quantitative volume control method has been developed to perform an in-situ monitoring of biochemical reaction on a chip. A nanofluidic chip was fabricated with a glass wafer with hydrophobic patterns and a PDMS microchannel networks using the microfabrication techniques. The enzymatic reaction with the FDG and enzyme demonstrated a feasibility of the fabricated nanofluidic chip to the biochemical reaction on a chip. The FDG and enzyme solution were metered and mixed respectively as a desired volume (20~40-nL) by the proposed nanoliter metering techniques and surface guided flow control. Fluorescence intensity of fluorescein product was directly proportional to a ratio of the enzyme to the FDG.

Acknowledgements
This work was supported by the Nano Bioelectronics and Systems Research Center of Seoul National University.

References

[1] M. Krishnan, V. Namasivayam, R. Lin, R. Pal and M. A. Burns, *J. Current Opinion in Biotechnology* **12** (1), pp. 92-98, (2001).
[2] J. Khandurina and A. Guttman, *J. Chromatography A* **943**, pp. 159-183, (2002).
[3] K. Handique, D. T. Burke, C. H. Mastrangelo and M.A. Burns, *Anal. Chem.* **72**, pp. 4100-4109, (2000).
[4] K. Hosokawa, T. Fujii and I. Endo, *Anal. Chem.* **71**, pp. 4781- 4785, (1999).
[5] S. H. Lee, and Y. K. Kim, Micro TAS '02, pp 205-206, (2000).
[6] E. F. Tsung and R. D. Tilton, *J. Colloid and Interface Science* **213** (1), pp. 208-217, (1999).
[7] B. J. Bruke and F. E. Regnier, *J. Electrophoresis* **22** pp. 3744-3751, (2001).

INTEGRATED DEVICE FOR POINT-OF-CARE GENOTYPING

Eugen Ermantraut, Torsten Schulz and Thomas Ellinger
CLONDIAG® chip technologies GmbH, Loebstedter Str. 103-105,
D-07749 Jena, Germany

Abstract

A new integrated microdevice is presented which allows the performance of complete arraybased genotyping assays within a single reactor.

Keywords: integrated PCR & hybridization device, micro wet printing, genotyping

Diagnostics is expected to become very soon the major area of DNA array applications. Especially for medical point-of-care applications easy and robust DNA analysis instruments are required, which will allow to conduct a complete assay with sample preparation, target amplification and labeling, analysis of the target for specific markers and interpretation of raw data within a single system. Here we introduce test applications within our new integrated platform, the Assay Processor AP, that performs all of these steps within a single reactor, thus combining DNA array analysis with microfluidics performance. The AP cartridge is an integrated microfluidics PCR device with a synthesized probe array within the microreactor. Target amplification by PCR and analysis by array hybridization is accomplished in a single step reaction within the cartridge. Due to its robust performance and unprecedented simple structure, the AP is a real test strip format in pharma-associated diagnostics at the point of care.

Small array chips are the basis for the integration into the microreactor. The applied arrays are made by micro wet printing (μWP) on the four inch wafer scale. The chips are 3 mm x 3 mm in size with a density up to 10^6 probes/cm^2. A specific challenge to be mastered was hybridization of amplified targets in the amplification buffer. Probe length is typically between 12 and 30 bases. Appropriate feature dimensions are in the range of 4μm to 64μm. We believe that μWP is the ultimate method for large scale high yield synthesis of microarrays. Current data on synthesis yield, opportunities and limitations will be presented.

For highly variable and affordable AP applications, we developed a new type of universal array leading to a new class of assays. Based on this kind of array, a target is detected by a bispecific probe, which consists of two regions complementary to the target and the probe on the array, respectively. Bispecific probes are labeled during its interaction with the target and can be detected after hybridization to the array. We designed an universal DNA array of 256 probes showing low cross hybridization. The array has been used within the AP cartridge in an PCR-LDR-based assay. Point

Y. Baba et al. (eds.), Micro Total Analysis Systems 2002, Volume 2, 874–875.

mutations in the human P450 *cyp2D6* gene could be detected with high discriminatory power.

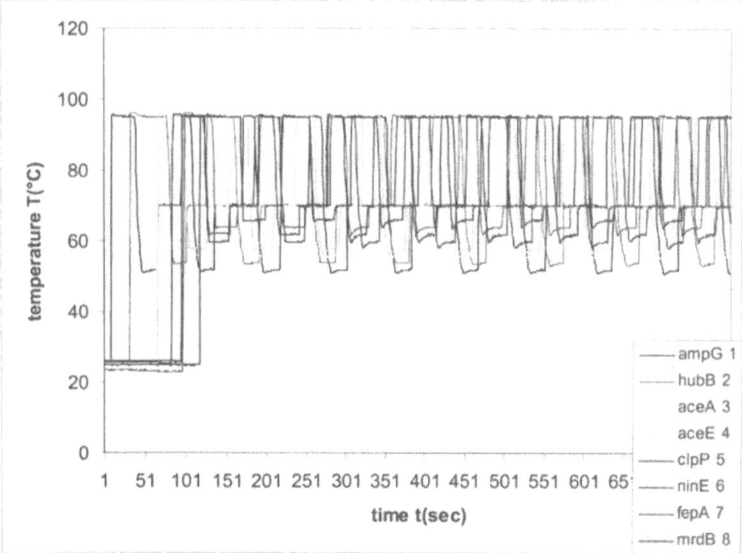

Temperature regime maintained in eight independent PCR amplification reactions in the Assay Processor Cartridge, actual temperature values are plotted against time.

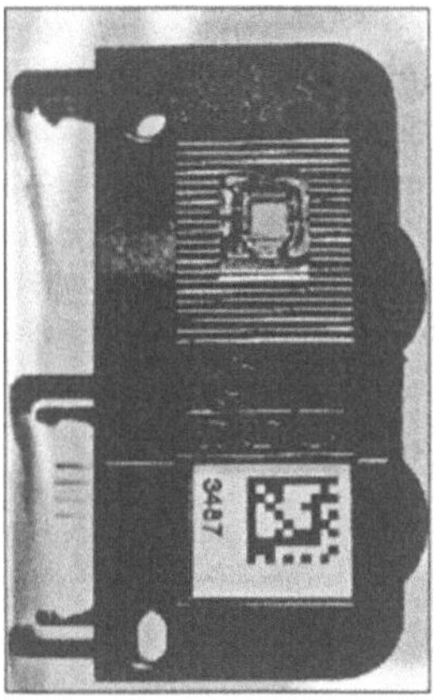

The Assay Processor Cartridge bearing the amplification machinery and the microprobe array (left). Multiplex PCR generated amplification product of cyp2D6 gene fragments detected by distinct hybridization pattern on pre-synthesized micro probe array (array size: 2 mm x 2 mm; feature size: 64 µm x 64 µm) in the Assay Processor Cartridge (right).

PHOTOCYANATION OF AROMATIC HYDROCARBONS ACROSS OIL/WATER INTERFACE IN POLYMER MICROFLUIDIC CHIPS

Fumihiko Kitagawa, Kosei Ueno and Noboru Kitamura
Hokkaido University, Kita-ku, Sapporo, 060-0810 Japan
(e-mail:kitamura@sci.hokudai.ac.jp)

Abstract

We explored an oil/water interfacial photocyanation reaction of pyrene (PyH) by using polymer microchannel chips: width of 100 μm, depth of 20 μm. As a typical experiment, an aqueous NaCN solution and a propylene carbonate solution of PyH and 1,4-dicyanobenzene were brought together into a Y-structured microchannel chip by pressure-driven flow. Light radiation onto the whole of the channel chip resulted in formation of 1-cyanopyrene (PyCN). The absolute PyCN yield as high as 78 % was attained under the optimum conditions.

Keywords : microreactor, oil/water interface, photoreaction, polymer microchip

1. Introduction

An experimental approach to conducting efficient and fast synthetic organic reactions in microchannel chips has received broad interests, since a chemical reaction in the chip is advantageous in respect to efficient heat management and mass transfer [1-3]. Nonetheless, it is known that laminar flow in a microchannel sometimes disturbs mixing of reactants (i.e., solutions), which is a main drawback for a synthetic reaction in a microchannel. One possible approach to overcome such a drawback is utilization of a liquid/liquid stream in a microchannel to conducting an interfacial reaction. In practice, we reported previously that a double Y-type polystyrol channel chip fabricated by an imprinting method could be applied to liquid/liquid extraction of an aluminum chelate [4]. We also confirmed that a stable oil/water interface was formed in the microchannel. Therefore, an application of an oil/water interface produced in a channel chip to an interfacial reaction would be very promising.

As an example of an interfacial reaction, on the other hand, we recently demonstrated that photocyanation of pyrene proceeded efficiently in an oil-in-water emulsion system via a photoinduced electron transfer (PET) mechanism [5]. In the present study, we fabricated two types of a polystyrol microchannel chip, double Y-structured type (two-phase chip) and three-phase channel chips (three-phase chip), and applied the chips to an interfacial photocyanation reaction of PyH.

2. Experimental

Microchannel chips were fabricated on polystyrol substrates by an imprinting method. The templates for imprinting were fabricated on silicon substrates by photolithography. The overall structural layouts of the microchannel chips used are shown in Fig. 1. One

Y. Baba et al. (eds.), Micro Total Analysis Systems 2002, Volume 2, 876–878.

of the chips allows two-phase (oil/water) flow in a microchannel, whose dimension was set 100 μm width, 20 μm depth, and 350 mm total length (from entrance to exit junction, Fig. 1a). The other is a three-phase type channel chip, which allows an oil/water/oil stream in the channel (Fig. 1b). A polystyrol substrate and the silicon template were fasten tightly between two flat glass plates and heated at 108 °C for 25 min. As a fluidic device, the channel substrate was covered and bonded with a flat polystyrol substrate by clamping both substrates between two flat glass-plates and heating at 108 °C for 18 min.

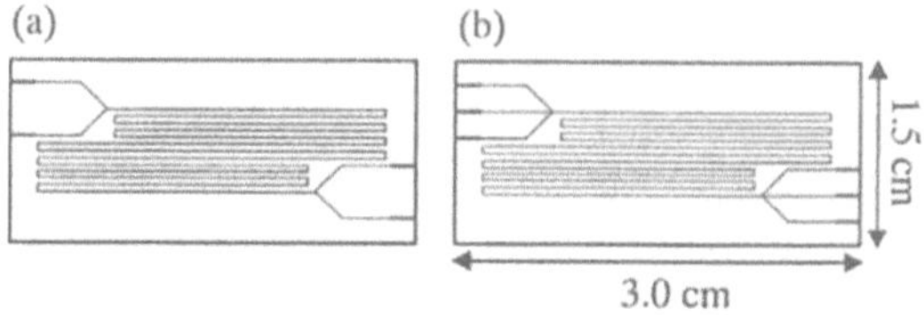

Fig. 1 Schematic drawing of the channel chip design; (a) two-phase chip and (b) three-phase chip.

Two 100 μl-volume syringes, each being filled with a propylene carbonate (PC) solution of PyH (20 mM) and 1.4-dicyanobenzene (DCB, 40 mM) or an aqueous NaCN (1 M) solution, were equipped with a dual-syringe pump and connected with the microchannel chip via capillary tubes. The two fluids were introduced to the channel chip with the same flow velocity.

For photocyanation of PyH, light from a high-pressure Hg lamp was irradiated onto the whole of the channel chip. For product analysis, the oil phase was collected from the exit of the microchannel and analyzed by GC-MS. The absolute yield of 1-cyanopyrene (PyCN) was determined by a calibration curve obtained by GC analysis using an authentic PyCN.

3. Results and discussions

According to our previous studies, we expected that the following photoreaction would proceed in a microchannel (Scheme 1). Under laminar flow of an aqueous NaCN solution and a PC solution containing PyH and DCB in a microchannel, PET between PyH and DCB takes place in the oil phase. The cation radical of PyH would be then subjected to nucleophilic attack by a cyanide ion at the oil/water interface or in the water phase. In the presence of O_2, the cyanated pyrenyl radical is oxidized to give PyCN. Since the solubility of PyCN in water is very poor, so that extraction of PyCN to the oil proceeds along solution flow in the microchannel.

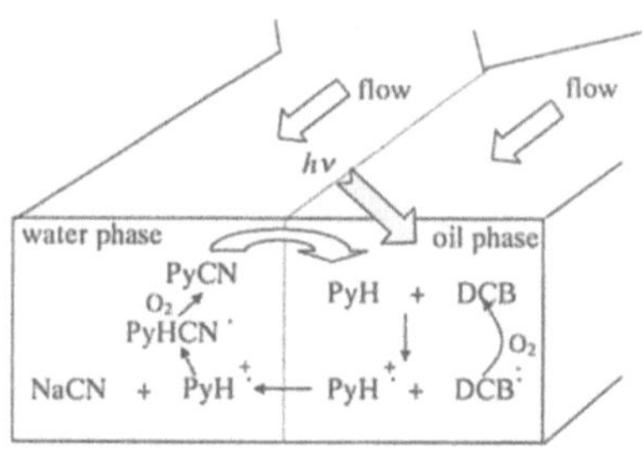

Scheme 1 Photocyanation reaction of PyH in the two-phase channel.

The absolute PyCN yields obtained by using two- and three-phase chips at several flow velocities are summarized in Table 1. In the present system, the reaction time is determined by the contact time (t) between the two phases; t = channel length (l = 350 mm) / average linear flow velocity (v mm/sec). In the case of two-phase chip, the reaction yield was 28 % with the reaction time of 210 s, while that was 7.5 %(t = 70 s) at u = 0.6 μl/min. These results demonstrate that the photocyanation reaction of PyH can be controlled by the solution flow rate in the microchannel.

	u = 0.6 μl/min t = 70 s	u = 0.4 μl/min t = 105 s	u = 0.2 μl/min t = 210 s
two-phase chip	7.5 %	15.0 %	28.0 %
three-phase chip	30.0 %	40.5 %	73.0 %

Table 1 Absolute yields obtained by using two- and three-phase chips at several flow velocities.

In the case of the three-phase chip, on the other hand, the absolute PyCN yield of 73 % was attained at u = 0.2 μl/min (t = 210 s) whereas the relevant value in the two-phase chip was 28 %, indicating that the yield in the three-phase chip is ~2.8 times higher than that in the two-phase chip. It is worth noting that the specific interface area (ratio of an interfacial area to a volume) between the water and PC phases in the three-phase chip is three times larger than that in the two-phase chip. Therefore, the higher reaction yield in the three-phase chip compared to that in the two-phase chip is concluded to be the larger specific interfacial area in the three-phase chip.

4. Conclusion

Photocyanation of PyH across an oil/water interface was shown to proceed very efficiently in polymer microchannel chips and, the absolute PyCN yield as high as 73 % was attained by the three-phase channel chip with the reaction time of 210 s. In the present system, the reaction yield can be controlled arbitrarily by the flow rate and the channel geometry, which is very unique characteristics of the microchip chemistry.

5. References

[1] W. Ehrfeld Eds., *Microreaction Technology: Industrial Prospects*, Springer Verlag, Berlin, 1999.

[2] W. Ehrfeld, V. Hessel and H. Löwe, Eds., *Microresctors: New Technology for Modern Chemistry*, WILEY-VCH Verlag, Weinheim, 2000.

[3] A. Manz and H. Becker, Eds., *Microsystem Technology in Chemistry and Life Sciences*, Springer Verlag, Berlin, 1999.

[4] H.-B. Kim, K. Ueno, M. Chiba, O. Kogi and N. Kitamura, *Anal. Sci.*, 16 (2000) 871-876.

[5] F. Kitagawa, M. Murase and N. Kitamura, *J. Org. Chem.*, 67 (2992) 2524-2531.

Microfluidic Protein Patterning Controlled By Hydrophobic Surface and Pneumatic Control

Chang-Soo Lee[1], Sang-Ho Lee[2], Yun-Gon Kim[1], Yong-Kweon Kim[2] and Byung-Gee Kim[1]

[1] School of Chemical Engineering and Institute of Molecular Biology and Genetics, Seoul National University, Korea

[2] School of Electrical Engineering and Computer Science, Seoul National University, Korea

Keywords : Protein patterning, Discrete flow control, Hydrophobic surface

Abstract

Protein patterning of nanoliter solution was performed in a microfluidic channels. Precise metering of nano-liquid was controlled by the external pneumatic pressure and modified surface engineering. As a model system, we demonstrated streptavidin/biotin reaction system.

1. Introduction

This study reports that protein patterning tactics using nanoliter fluidic control and surface engineering. It is based on microfluidic control technique and patterned hydrophobic surface rendering prevention of nonspecific binding [1,2].

2. Theory and basic design of variable metering in a microchannel

From the equation (1) (width; w, height; h, H_2O surface tension; γ), we designed the liquid flow direction in the microchannel networks. Positive pressure indicates capillary filling, but negative pressure indicates a repellent force at a solid-liquid-gas interface [3,4].

$$P = \gamma \cos\theta_{PDMS}\left(\frac{2h+w}{hw}\right) + \gamma\cos\theta_{SiO_2\ or\ FC}\left(\frac{w}{hw}\right) \quad (1)$$

When a contact angle is greater than 90° (hydrophobic) and a cross section of

Y. Baba et al. (eds.), Micro Total Analysis Systems 2002, Volume 2, 879–881.

microchannel is smaller, positive external pressure is necessary for push a fluid to the desired direction. Fig. 1 shows the basic design of microchannel network and hydrophobic surface for the variable metering.

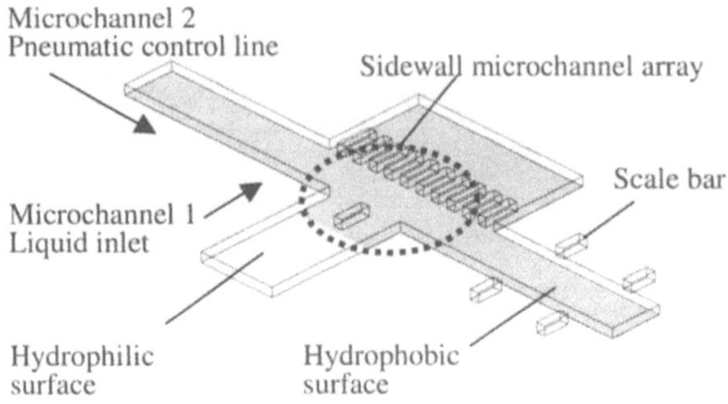

Fig. 1 Schematic design of microchannel network and hydrophobic surface for variable metering

3. Results and discussion

For the protein patterning on the desired area, photolithographic lift-off process was used to display two-dimensional patterns of spatially hydrophilic and hydrophobic region. Aldehyde modified surface was obtained from oxidation, aminosilanization and aldehyde binding procedure. In a microchannel, the injected protein solution halted in patterning reservoir and protein was covalently immobilized into only hydrophilic aldehyde patterns (Figure 2). Finally, functional activity of patterned streptavidin arrays was examined by various concentration of biotin-BSA-FITC (Figure 3).

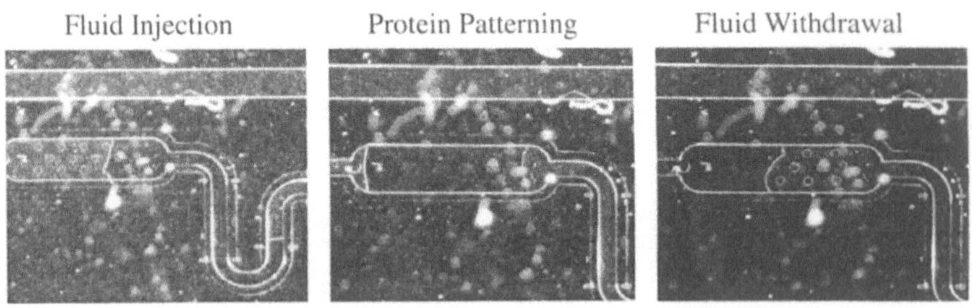

Fig. 2 Operating photograph for protein patterning and protein functional assay

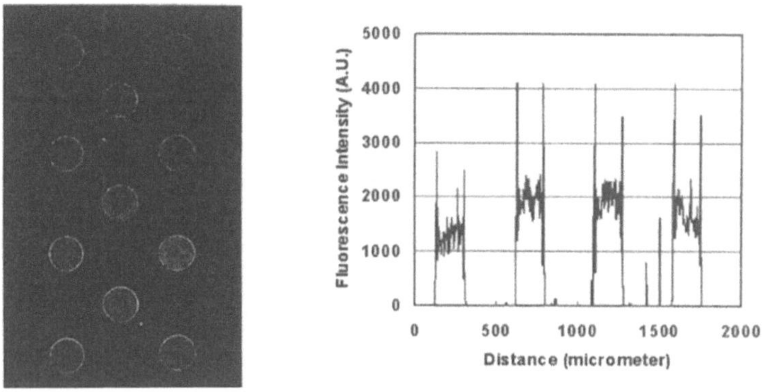

Fig. 3 Fluorescence image and fluorescence profile on protein pattern. In this experiment, streptavidin (125nl, 100 μg/ml) was firstly patterned and then FITC-BSA-biotin (125nl, 10 μg/ml) was reacted on the streptavidin array for 20min.

4. Conclusion

In this study, feasibility of a new protein patterning method was examined using modified surface and discrete fluidic control in nanoliter scale. From our proposed technique, protein patterning was successfully performed at nanoscale volume and patterned protein arrays are correctly reacted with binding partner.

5. Acknowledgements

This work was supported by the Nano Bioelectronics & Systems Research Center of Seoul National University.

References

[1]. S. H. Lee, Y. K. Kim, *Proc μTAS 2001*, Monterey, USA pp. 205-206, (2001).

[2]. C. S. Lee, S. H. Lee, H. S. Joo, Y. K. Kim, B. G. Kim, *NanoTech 2000*, Montreux, Switzerland, P29, (2000).

[3]. K. Hosokawa, T. Fuji and I. Endo, *Anal. Chem.* pp. 4781- 4785, (1999).

[4]. A. Puntambeker, H. J. Cho. C. C. Hong, J. Choi, C. H. Ahn, Proc. Transducers'01, pp. 1240-1243, (2000).

FABRICATION AND APPLICATIONS OF POLYMER-BASED MICROCHANNEL-HEATER CHIP AS MICROREACTOR

Yukiko Hosoda, Kosei Ueno, Shoji Ishizaka and Noboru Kitamura
Hokkaido University, Kita-ku, Sapporo, 060-0810 Japan
(e-mail:kitaumra@sci.hokudai.ac.jp)

Abstract

We fabricated polymer-based microchannel-microheater chips and applied the chip to a synthetic reaction to benzylidenemalononitrile (BMN), known as an anti-fouling agent. Under the optimum conditions, the BMN yield as high as 96% was attained with the reaction time of 84 s.

Keywords: polymer-based microchannel, microreactor, thermal-reaction

1. Introduction

Miniaturization of chemical systems has received current interests in various fields, since microfabricated structures can integrate with chemical reaction, separation, and/or detection systems on a single chip. Among various chip materials, polymer substrates are versatile and can be processed by a variety of techniques [1]. Furthermore, we recently reported that a metal thin film could be vacuum deposited directly onto a polymer substrate and, photolithography/etching of the metal layer gave a polymer microchannel chip integrated with microelectrodes [2]. On the basis of these results, we fabricated double Y-style polymer-based microchannel (100 μm width and 20 μm depth)-microheater (100, or 200 μm width) chips by photolithography and imprinting methods. In this study, we applied the chip to a synthetic reaction to benzylidenemalononitrile (BMN) known as an anti-fouling agent (Scheme 1).

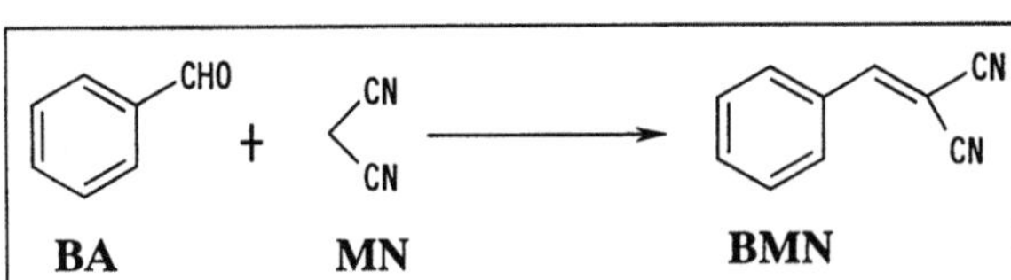

BA:benzealdehyde
MN:malononitrile
BMN: benzilydenemalononitrile

Scheme 1

Y. Baba et al. (eds.), Micro Total Analysis Systems 2002, Volume 2, 882–884.

2. Experimental

Microchannel substrates were fabricated on polystyrol plates (1.5 cm x 3.0 cm) by an imprinting method. As a microheater substrate, on the other hand, an Au thin-film (thickness 50 nm) was vacuum-deposited directly on an another polymer substrate (2.6 cm x 3.0 cm), and it was lithographed to fabricate a microheater. Finally, both substrates were pressed at 110 °C for 18 min to obtain a channel-heater chip (Fig.1). Sample solutions were introduced into the channel chip by using a programmable syringe pump. Product analysis was made by GC-MS.

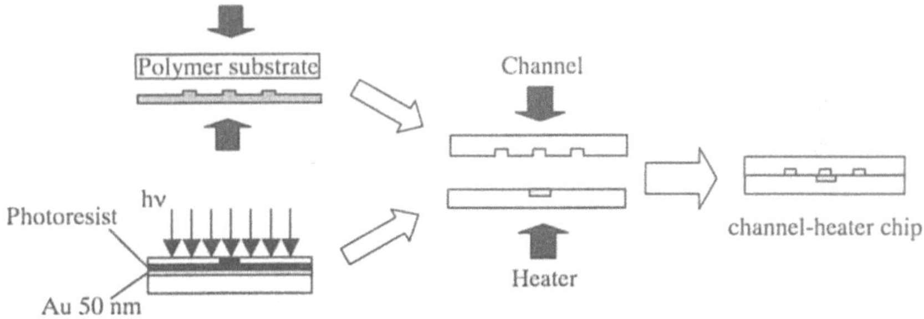

Figure 1. Fabrication of a microchannel-heater chip.

3. Results and discussion

A temperature profile around the microheater in the chip was estimated on the basis of a temperature-dependent fluorescence lifetime of Rhodamine B (RB). A methanol solution of RB was introduced to the chip and the fluorescence lifetime of RB were determined along the solution-flow direction at several applied voltages (V) to the heater. Figure 2 shows the fluorescence decay profiles which were determined by irradiating an excitation laser beam to the solution phase just above the microheater. The fluorescence decay was always fitted by a single exponential function and became faster with an increase in V, demonstrating that the fabricated electrode acted as a microheater. At V=0.7 V and the solution-flow rate of 0.1 μl min^{-1} as an example, the solution temperature above the heater was estimated to be 37 °C.

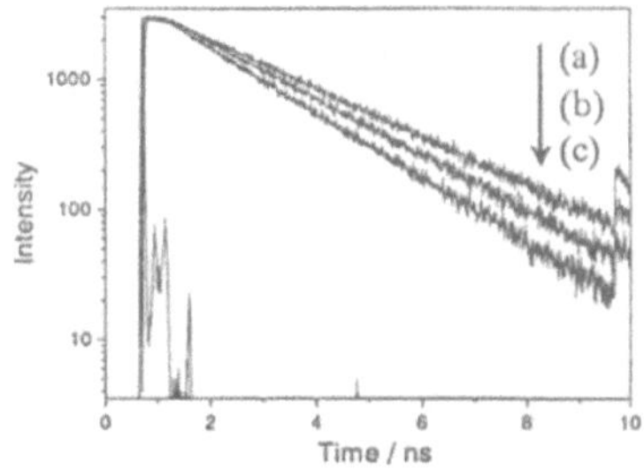

Figure 2. Fluorescence decay profiles of RB in the microchannel: (a) 21 °C, (b) 29 °C, and (c) 37 °C

Since the fabricated chip acted as a microchannel-microheater device, we explored a thermal reaction between benzaldehyde (BA) and malononitrile (MN) [3] in the chip. Methanol solutions of BA and MN were introduced separately to the microchannel with the same flow rate, and the thermal reaction was allowed by applying a voltage to the microheater (Fig.3). The gas chromatogram of the oil phase demonstrated formation of BMN (Fig.4), and the yield increased with an increase in V as well as with a decrease in the flow rate. The reaction yield was also dependent on the width and position of the heater in the microchannel. Under the optimum conditions, BMN was obtained in a 96 % yield with the reaction time (7.0 cm / 0.1 μl min^{-1}) of 84 s (Fig.4), while the yield in a bulk system (4 h reflux) was 70 %. The reaction proceeds efficiently because of efficient heat management in the microchip. One directional feed of the reactants to the microheater and subsequent flow of the product to the downstream side might be also the probable reason for the high reaction yield.

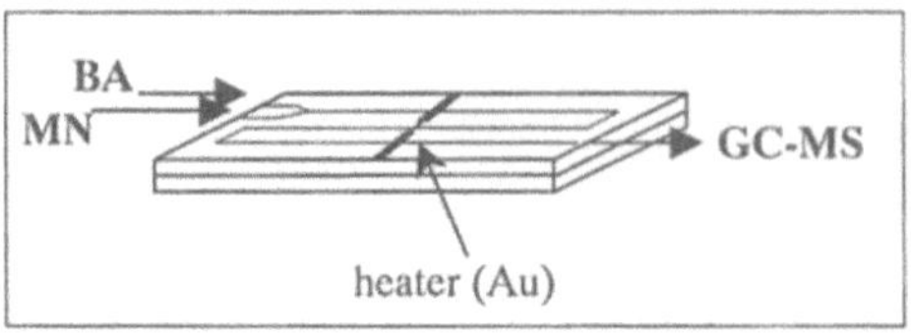

Figure 3. The structure of the microchip.

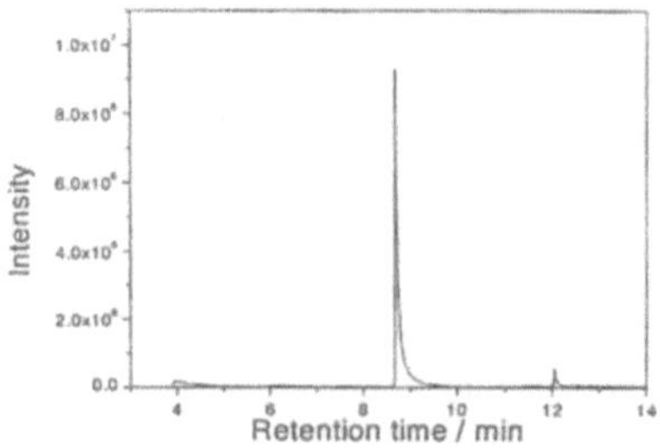

Figure 4. A gas chromatogram of the reaction mixture (channel).

4. Conclusions

We fabricated microchannel chips integrated with a microheater and applied the chip to a synthetic reaction of BMN. The reaction was shown to proceed more efficiently in the microchip as compared with that in a bulk system. The present polymer based microchannel-microheater chip was demonstrated to act as a microreactor.

References

[1] H.-B Kim et. al, *Anal. Sci.*, Vol. 16, No. 8, 871-876 (2000).

[2] K. Ueno et. al., *Chem. Lett.*, No. 8, 858-859 (2000).

[3] F. Bigi et. al., *Green Chemistry*, Vol. 2, No. 3, 101-103 (2000).

CdSe nanoparticles formation in a microspace and their properties.

Hiroyuki Nakamura, Masaya Miyazaki, and Hideaki Maeda
Micro-Space Chemistry Laboratolry, National Institute of Advanced Industrial Science and Technology,
807-1, Shuku, Tosu, Saga, 841-0052 Japan
(e-mail:maeda-h@aist.go.jp)

Asuka Tashiro, Kazuya Ogino, Takanori Watari
*Graduate School of Science and Engineering, Saga University, 1 Honjo, Saga, 840-8502 Japan

Abstract

A microreactor was utilized for a continuous and controlled CdSe nanocrystal preparation and effect of reaction conditions on optical properties of the nano-crystals were investigated. In the reaction temperature from 230 to 340℃, it was possible to prepare the nanocrystal of ~ 2 to 4.5nm in diameter was possible to prepare, and the fluorescence peak shifted from ~480-600 nm with particle size. Higher temperature and longer reaction time gave larger particle size thus longer fluorescence peak.

Keywords: Micro-reactor, quantum dots, CdSe, flow reactor, Photo-luminescence, particle size distribution

1.Introduction

As it is well known, Micro-reactor has a high heat exchange efficiency and is regarded as an effective reactor with an exact and quick temperature control[1]. Moreover, a parallel operation of multiple reactors can make it possible a industrial production without any intrinsic scaling-up of the reactor[1]. Therefore, a microreactor can be an industrial reactor for a high-precision chemical reaction. On the other hand, recently, CdSe nano-particles are attracting an increasing attention, as a fluorescence tag for biological molecules, a raw materials for tunable LED and Laser, and an optical memory and so on[2],[3]. CdSe based nano-particles can be found in market, however, as it requires a precise control of reaction conditions. Therefore, production process is basically a laboratory manner and will have a difficulties in industrial reproducible production to satisfy the increasing demand of CdSe. From this point of view, the authors had applied an micro-reactor for a continuous CdSe nanoparticle preparation process, and found a extremely high reproducibility and controllability of average particle size[4]. In this study, effect of the reaction conditions such

Y. Baba et al. (eds.), Micro Total Analysis Systems 2002, Volume 2, 885–887.

as, reaction temperature, reaction time, capillary materials on particle size distributions and fluorescence properties were explored.

2.Experimental

Experimental Section

CdSe nanocrystal preparation method reported by Qu et al.[2] was modified for this reaction system. Typically, Se powder was added to trioctyl phosphine (TOP) to prepare a TOP-Se stock solution. Meanwhile, $Cd(CH_3COO)_2$ was added to of stearic acid and heated at 130 °C for 15 min. Then of trioctyl phosphine oxide (TOPO) was added under a nitrogen flow. After the solution was cooled to below 100 °C, it was mixed with TOP-Se to make a raw material solution. The raw material solution was loaded into a glass syringe, which is connected to a glass capillary. The residence time of the reaction solution in the heating part of the reactor was determined by the flow rates and capillary length. A schematic representation of the reaction system is shown in Fig.1.

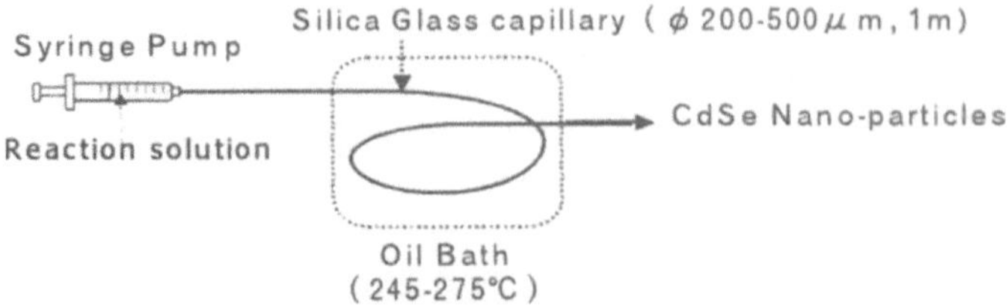

Fig.1 Schematic representation of flow reactor for CdSe nano-particles.

The product was collected in a test tube and part of it was diluted by 20times with chloroform for UV-VIS spectroscopy, fluorescence spectroscopy. The average particle diameter was estimated from UV-VIS absorption peak on the basis of the literature reported by Dushkin et.al.[5].

3.Results and discussion

By utilizing the microreactor, it was possible to produce nanocrystaline CdSe continuously, at temperature from 230 to 300°C and reaction time of 1 to 150s. Effect of these parameters on the estimated average particle size is shown in Fig.2. The estimated particle size was able to be varied from around 1.5nm to 4nm and the larger particle size was obtained at higher reaction temperature and longer reaction time. The particles growth curve fits the formula reported by Dushkin et. al. and the particle growth mechanism seemed to be the same with those reported for batch system[6]. The peak location of photoluminescence was able to be varied from ~480 to 600 nm with particle size (Fig.3). The full width of half maxima of fluorescence peak was around 40nm.

However, for particles smaller than ~2nm, the intensity of the fluorescence was relatively weak and had emission peak.

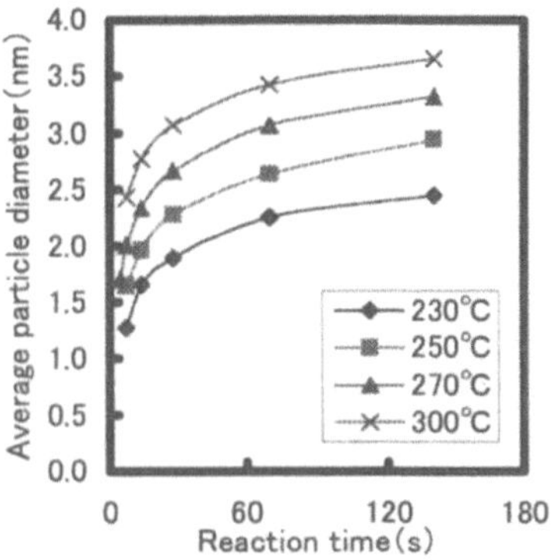

Fig.2 Effect of Reaction time and temperature on average particle size of CdSe nano-particles.

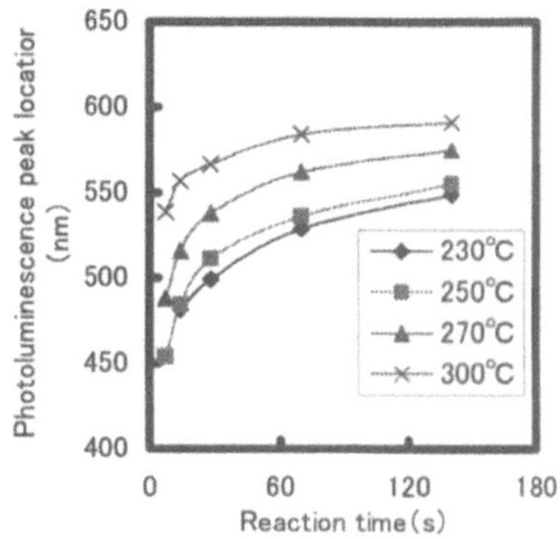

Fig.3 Effect of Reaction time and temperature on photoluminescence peak location.

Effect of operation conditions on particle size distribution was also checked. It was found that a 200 μm diameter channel gave narrower particle size distribution than 500 μm diameter channel. From this result, it is considered that homogeneity of heating rate of the raw material solution in the capillary would be effective on the particle size distribution. On the other hand, segmentation of the reaction solution by inert gas was effective for a narrow particle size distribution. This shows the flow rate distribution, which arose from a friction between capillary and reaction fluid, would increase the reaction time distribution, and was reduced by the segmentation.

Reference

[1] W.Ehrfeld, V.Hessel, H.Lowe, "Microreactors", WILEY-VCH, Germany(2000).

[2] L.H. Qu, Z.A. Peng, X.G.Peng, *Nano-Lettrers*, **1**, 333-37 (2001)

[3]W.C.W.Chan, D.J.Maxwell, X.H.Gao, R.E.Bailey, M.Y.Han, S.Nie, *Analytical Biotechnology*, **13**,40-46(2002)

[4]H.Nakamura, Y.Yamaguchi, M.Miyazaki, M.Uehara, H.Maeda, P.Mulvaney, *Angewandte Chemie.,* (Submitting).

[5] C.Albe, C. Jouanin, and D. Bertho, *Phys. Rev.*, **58**, 4713-20(1998).

[6] C.D. Dushkin, S.Saita, K.Yoshie, Y.Yamaguchi, *Adv. Colloid Interface Sci.*, **88**, 37-78 (2000).

TO PLACE CELLS AS AN ARRAY USING ASPIRATION TECHNIQUE

Agnès Tixier-Mita[1], Yoshio Mita[2], Katell Cozic[3], Marie Frénéa[3], Bruno Le Pioufle[3] and Hiroyuki Fujita[1]
[1]CIRMM/Institue of Industrial Sciences, The University of Tokyo, Japan,
[2]Department of Electric Engineering, the University of Tokyo, Japan,
[3]BIOMIS, ENS Cachan, Britany Branch, Campus Ker-Lann, France
e-mail: agnes@fujita3.iis.u-tokyo.ac.jp

Abstract

Thousands of cells are simultaneously aligned in array in grace of a silicon micro-strainer. The cells alignment was performed in few seconds with sub-micron precision.

Keywords: aspiration, micro-system, micro-holes, arrayed cells, KOH and TMAH etching

1. Introduction

One of the difficulties when working with biological materials (proteins, cells etc...) is to handle and/or to place precisely the material on specific zones, the active area of a micro-system for instance. In the case of cells, their manipulation is becoming very important in the sphere of biotechnology (cells sorting, cells fusion, gene transfection...) [1-2]. And being able to place them very precisely on specific areas, treating thousands of cells in parallel improves the efficiency of these studies. In this article, we propose to place thousands of cells as an array on a micro-systen, using aspiration through a micro-strainer made in silicon. The realisation of the micro-strainer uses an original process utilising only anisotropic etching to perform through-holes until size of one hundred nanometer in width.

2. Principle

The principle of the micro-strainer is as follows. It consists on a silicon wafer with micro-holes of diameters from one hundred of nanometer to few microns in diameter. A manual driven small seringue-pump is connected to the micro-system (Figure 1). When the solution of cells (usually 0.5ml; 100 000 cells/ml) is placed on it and the aspiration is performed by means of the seringue, the solution begins to flow through the holes. This flow creates a depression around each micro-hole and cells are dragged towards them. As the holes are smaller in diameter (maximum 6 micrometers) than the cells diameter (from 10 to 15 micrometers), the cells remain on the micro-holes.

3. Experimental

The first series of micro-systems were made using deep etching of silicon with

Y. Baba et al. (eds.), Micro Total Analysis Systems 2002, Volume 2, 888–890.

ICP-RIE to fabricate micro-holes (see Figure 2). First results with Jurkat cells [3], realised using an absorbant paper for the aspiration, showed that almost all the micro-holes were occupied by one cell after the aspiration process (see a close view of a cell placed on a micro-hole by aspiration on Figure 3). However, fatal drawbacks was that some of these very deformable cells disappeared inside the holes, because of too large holes (6 micrometers in diameter) and a too strong aspiration.

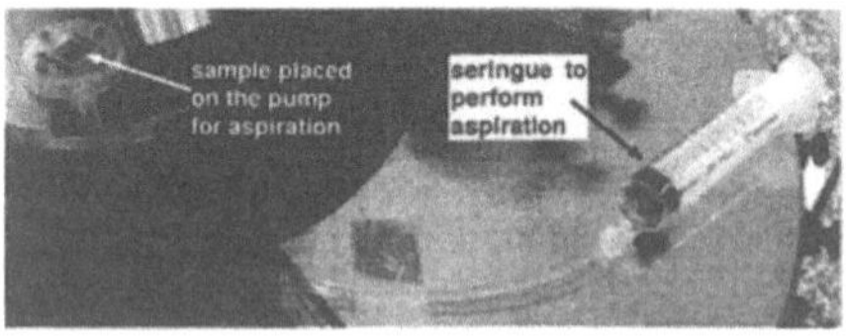

Figure 1: Micro-system for the experimental set-up.

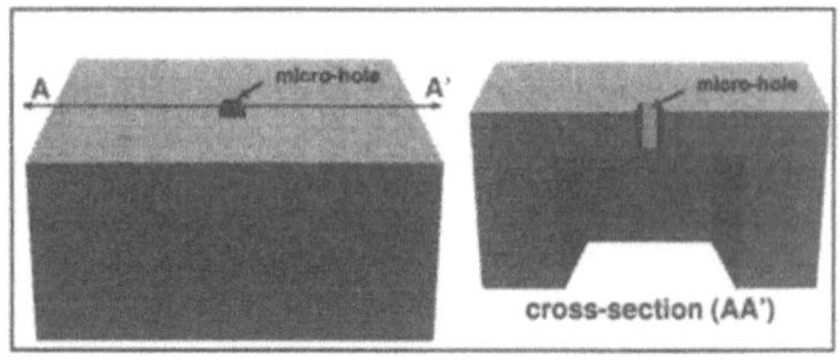

Figure 2: 3D scheme of the frontside microholes and backside big holes made by ICP-RIE.

Figure 3: Close view of a hole of the system of Figure 2 with a cell placed on it after aspiration.

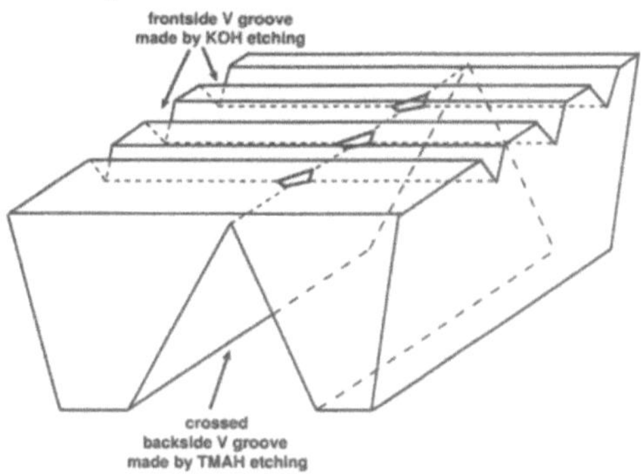

Figure 4: 3D scheme of the micro-holes performed by front-side KOH and backside TMAH etching of a silicon wafer.

4. Results and Discussion

The ICP-RIE process didn't allow to obtain holes smaller than 4 micrometers in diameter. So the process was modified in order to reduce the holes size. This process used only KOH and TMAH etching of silicon on both frontside and backside parts of the wafer. Small holes appeared at the intersection of the backside V grooves and the orthogonal frontside ones as schemed on Figure 4. The smallest aperture size obtained with this process was 90 nanometers on 180 nanometers. Larger holes were also obtained until dimensions of 4 micrometers in width. Square or rectangular holes can be obtained with different dimensions just modifying the design (mask width) and the backside etching time with TMAH.

Aspiration process was performed on samples having rectangular holes with width of 1 micrometers, with yeast cells (6 microns in diameter). Figure 7 a-b-c-d show photos of the system seen under the microscope (X40) before (a) and during (b-c) aspiration and after pushing back the seringue of the seringue pump(d).

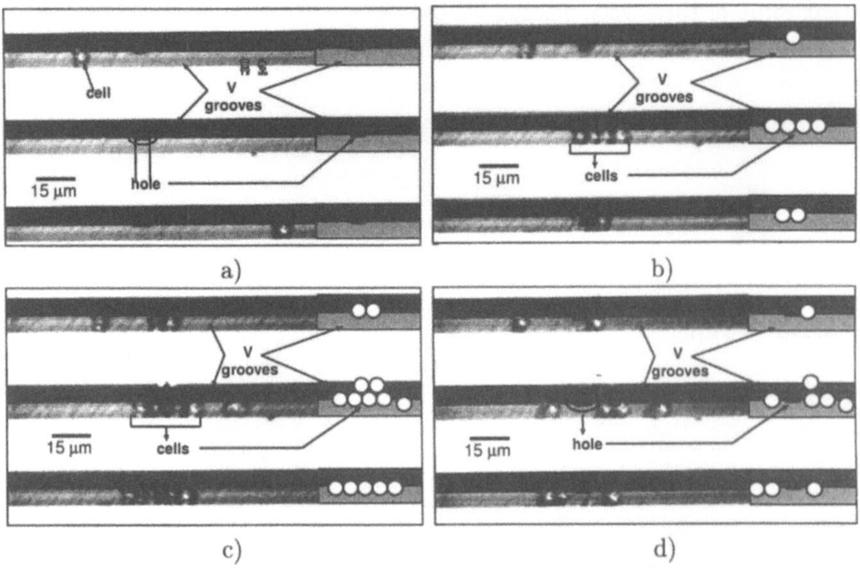

Figure 7: Micro-system seen under the microscope (X40): a) before, and b-c) during asiration, and d) after pushing back the seringue of the seringue-pump.

Thanks to this system it was possible to place small groups of cells in lines. We think that with an improved design giving square small holes it is possible to catch only one or two cells for each hole.

Acknowledgments.

Photo mask fabrication for this project is supported by VDEC (VLSI Design and Education center) the University of Tokyo.

References.

1. M. Washizu, T. Nanba, S. Masuda; *Handling biological cells Using a Fluid Integrated Circuit*; IEEE transactions on industry applications; Vol. 26, 2; pp. 352-357S; (1990).

2. N. G. Greena, H. Morgana, and Joel J. Milnerb; *Manipulation and trapping of sub-micron bio-particles using dielectrophoresis*; Journal of Biochemistry Biophysical Methods; Vol. 35, Issue 2; pp. 89-102S; (1997).

3. A. Tixier, L. Griscom, K. Cozic, H. Nagai, B. Le Pioufle, Y. Murakami, E. Tamiya, H. Fujita; *Catching and Attaching Cells using an Array of Microholes*; IEEE-EMBS 2000; Lyon, France, pp. 36, (2000).

AN INTEGRATED CHEMILUMINESCENCE DETECTOR FOR MEASURING ENZYMATICALLY GENERATED HYDROGEN PEROXIDE

Anders M. Jorgensen, Daria Petersen and Oliver Geschke
MIC – Mikroelektronik Centret, 2800 Lyngby, Denmark
(amj@mic.dtu.dk)

Abstract

We present an integrated detector for chemiluminescently generated light. The detector contains a fluidic network monolithically integrated with backside photodiodes. The detector is used together with a porous polymer enzyme reactor to examine the performance by measuring glucose concentrations in the range of 0.1 g/l to 1 g/l. In this range the system exhibits linear behavior.

Keywords: Backside photodiode, Chemiluminescence, Porous polymer, Enzyme

1. Introduction

We present an integrated chemiluminescence detector, combining microfluidic handling and optical detection on the same chip (Figure 1). The detector works by use of the chemiluminescent reaction of luminol with hydrogen peroxide in the presence of Fe^{3+}. Hydrogen peroxide results from the conversion of glucose to gluconolactone by the enzyme glucose oxidase (GOD). The chemiluminescent light is collected by integrated backside photodiodes, which convert the optical signal into an electrical current, which is then measured using off-chip electronics. The microsystem is characterized by fast response time, high sensitivity, automation and low consumption of sample and reagents.

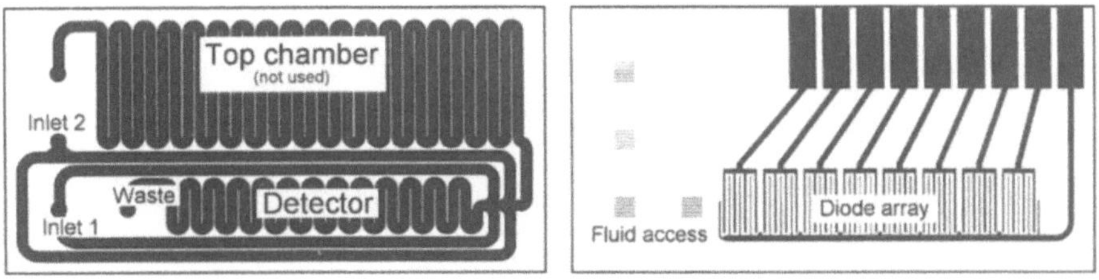

Figure 1 Pictures of the chip layout (chip footprint: 10 mm x 20 mm). The frontside of the chip (left) contains the fluidic network. The channels are 400 μm wide and 80 μm deep. The bottom meandering structure is where the chemiluminescent reaction takes place. The top chamber is designed to be an enzyme reactor, however, it is not used here. The backside of the chip (right) contains 4 fluid access ports and an array of 8 photodiodes at the bottom. The 9 terminal connector at the top is designed for interfacing with 1/20 inch pitch electronic connectors.

Y. Baba et al. (eds.), Micro Total Analysis Systems 2002, Volume 2, 891–893.

2. Experimental

The footprint of the chip is 10 mm x 20 mm and it was fabricated using MEMS technologies. The most important steps were: channel fabrication through the use of reactive ion etching (RIE), metal contacts defined by use of the lift-off technique and, finally, hermetical sealing of the channel network by anodic bonding of a borofloat glass lid.

Measurements using the system were carried out using a flow injection analysis approach with an external injector with a 20 µl sample loop. The chip was placed in a holder and fluid connections were made to the backside access ports (Figure 2). Only the detector region was used and therefore the inlet to the large meander was blocked. The waste outlet was connected to a piece of opaque PEEK tubing to prevent light generated after leaving the chip from influencing the measurement. Inlet 1 was connected to a teflon tube containing the luminol/catalyst mix. Inlet 2 was connected to the injector by tubing containing a 50 mm long porous polymer enzyme reactor. The porous polymer was created inside a teflon tube by UV-photo-polymerization of ethyleneglycoldimethacrylate and butylmethacrylate initiated by azobisisobutyronitrile in the presence of a propanol/butanediol porogen solution. On the surface of this porous polymer the GOD was immobilized by photoimmobilization. The enzyme reactor was then cut to the desired length and mounted using fluidic fittings. The whole system including injector was placed in a metal box to suppress interfering light and achieve electrical shielding. Three syringe pumps controlled the flow rate of the carrier liquid (phosphate buffer at pH 6.5), luminol solution (2.2 mM, pH 10.5) and Fe^{3+} catalyst (25 mM). The luminol and Fe^{3+} solutions were mixed prior to entering the chip. Samples of different glucose concentrations were prepared by dilution with phosphate buffer.

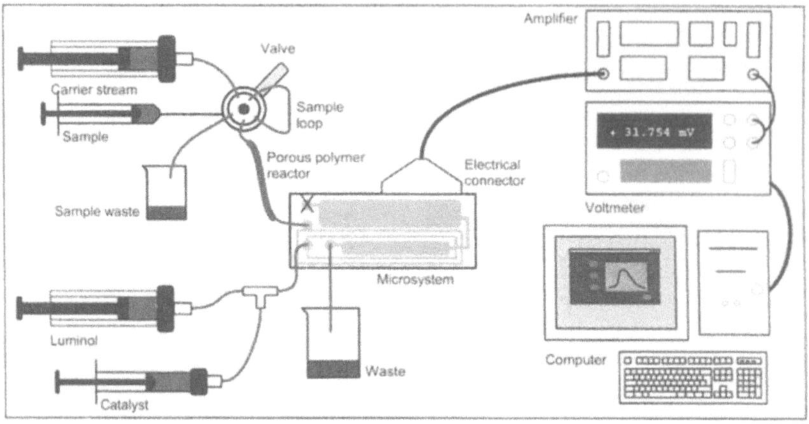

Figure 2 Sketch of the experimental setup, showing how the different fluids are connected to the system. The fluid connector in the upper left-hand corner of the microsystem is blocked. The luminol and catalyst streams are mixed at the T-section just before the chip.

3. Results and discussion

Electrical signals were fed to a transconductance amplifier connected to a programmable voltmeter. Data was then transferred to a computer. Any offset was subtracted and the peak of the signal was used for calculating the concentration (Figure 3). A series of measurements were done for glucose concentrations ranging from 0.1 g/l to 1.0 g/l. The results (Figure 4) indicate that all concentrations in the investigated range can be measured and that there is a linear relationship between the concentration and the signal.

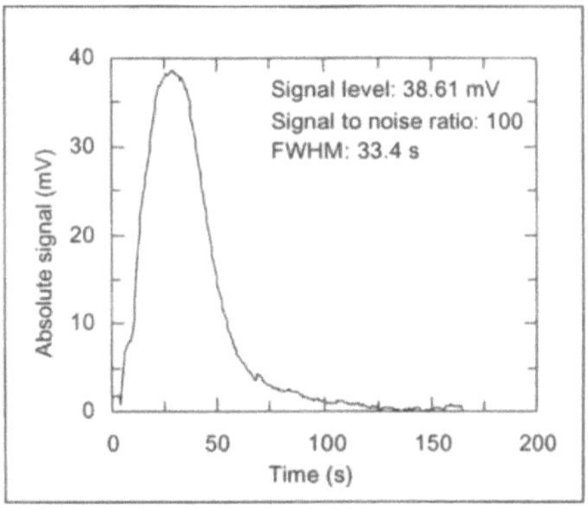

Figure 3 A measurement of generated light as a function of time. Sample size: 20 μl with a concentration of 0.7 g/l glucose. Flow of carrier, luminol and catalyst: 60 μl/min, 38 μl/min and 2 μl/min respectively. The noise was measured as the RMS signal of the baseline from 130 s to end of scan.

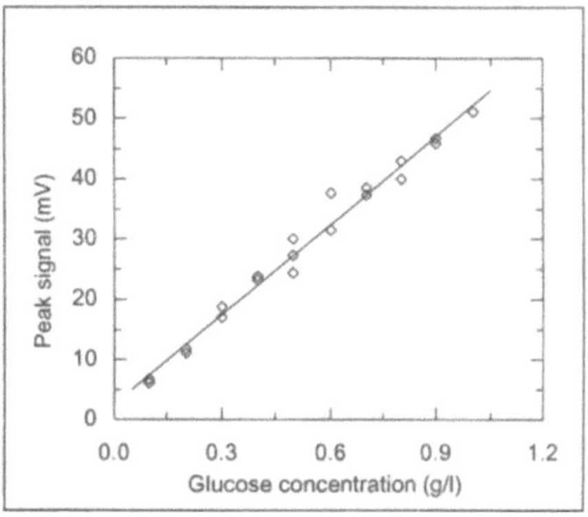

Figure 4 The maximum peak signal measured as a function of glucose concentration. The flow settings were the same as in Figure 3. The straight line is a best linear fit of the data.

4. Conclusion

We have shown that the chemiluminescent detection scheme implemented with the integrated chemiluminescence detector is able to measure glucose concentrations in a range that is biologically relevant. In the future even higher performance is expected by using an on-chip enzyme reactor. This will reduce dispersion and is expected to allow even smaller sample volumes and/or sample concentrations to be measured at higher sample rates.

Acknowledgements

AMJ wishes to thank The Technical University of Denmark for financial support.

AN ATMOSPHERIC-PRESSURE MICROPLASMA JET SOURCE FOR OPTICAL EMISSION SPECTROSCOPY OF LIQUID SAMPLE

T. Koidesawa[1], T. Ichiki[1] and Y. Horiike[2]
[1]Bio-Nano Electronics Research Center, Toyo University, Japan
[2]Department of Materials Science, School of Engineering, The University of Tokyo, Japan
e-mail address: ichiki@eng.toyo.ac.jp

Abstract

An atmospheric-pressure microplasma jet source has been developed as a core device for the miniaturized ICP-OES (inductively coupled plasma - optical emission spectroscopy) system. This paper presents the production of high density plasma jets using a compact VHF transmitter for automobile and the optical detection of emitted spectra from tiny amounts of sample solution injected into the microplasma jet.

Keywords: on-site chemical analysis, microplasma source, optical emission detection, liquid sample

1. Introduction

A portable chemical analysis system has been requested to reply various potential needs of on-site analysis such as detection of accidental mixing of toxic substance in a water purification plant, monitoring of factory effluent contamination, urgent analysis of food poisoning or chemical pollution and so forth. Peculiar issue of the miniaturization is, however, the relatively low detection sensitivity owing to the small sample volume. Recently, in order to solve this problem, on-chip plasma sources have been developed as a sensitive detector for miniaturized gas analysis systems [1-3]. These miniaturized plasma sources developed so far are designed for gas analysis and do not have sufficient plasma density and temperature for the analysis of liquid sample. In this paper, an atmospheric-pressure and high-density ICP source on a chip has been studied for the establishment of the miniaturized liquid analysis system that integrates microcapillary electrophoretic (μCE) separator and inductively coupled plasma optical emission spectroscopy (ICP-OES) as depicted in Fig. 1 Analytic procedure of the system is as follows. Ingredients of the liquid sample separated on a μCE separator are sprayed into the atmospheric-pressure high-density microplasma jet via the micronebulizer, and are detected by measuring the emission spectrum from the plasma.

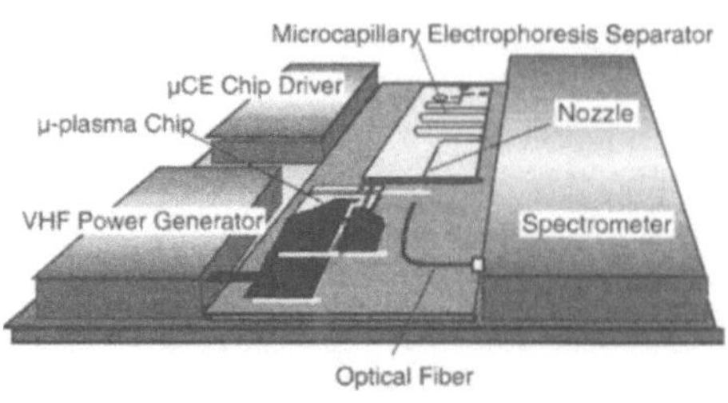

Figure 1: A schematic diagram of an integrated analytical system of microcapillary electrophoresis and microplasma emission spectroscopy.

Y. Baba et al. (eds.), Micro Total Analysis Systems 2002, Volume 2, 894–896.

2. Experimental

A VHF-driven miniaturized ICP jet source was composed of a microplasma chip and a compact power supply system as shown in Fig. 2. A microplasma chip has a simple structure, that is, a 100-μm-thick electroplated antenna with serpentine structure and a fine discharge tube with a cross section of 1 by 1 mm and a length of 30 mm that was engraved on a 15 by 30 mm dielectric plate. The antenna was located near the edge of the microplasma chip so as to produce jet-like plasmas. Since downsizing of peripheral equipments such as a power supply and a matching network is also inevitable to make the whole system portable, a compact power supply system composed of a commercially available VHF transmitter for automobiles (freq.; 144 MHz, max. power; 50 W) and a handmade matching network were employed in the present study. The matching network was tuned so as to locate the current antinode of the standing wave at the antenna of a microplasma chip.

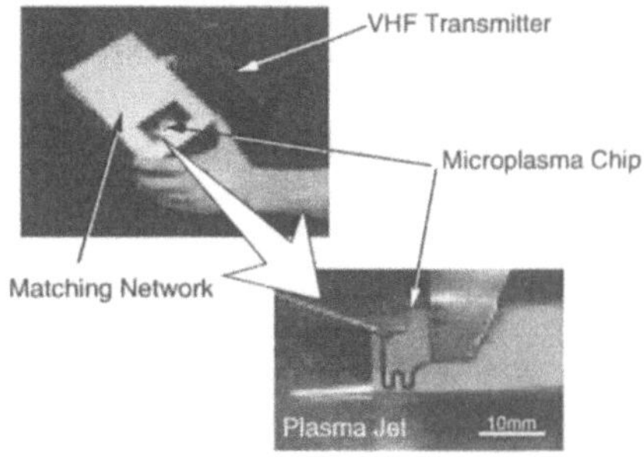

Figure 2: A microplasma chip connected with a compact VHF power supply system.

3. Results and Discussions

Figure 3 shows thermographs of microplasma chips made of quartz and alumina plates, which clarify the advantage of alumina plates in heat dissipation ability. In the case of the quartz-made chip, significant temperature rise was observed in the neighborhood of the antenna due to Joule heating effect, while gradual temperature rise on the whole chip surface was observed for the alumina-made one. Since the thermal conductivity of alumina is approximately 15 times higher than that of quartz, the increase in electrical resistance of antenna due to temperature rise is appreciably sup-

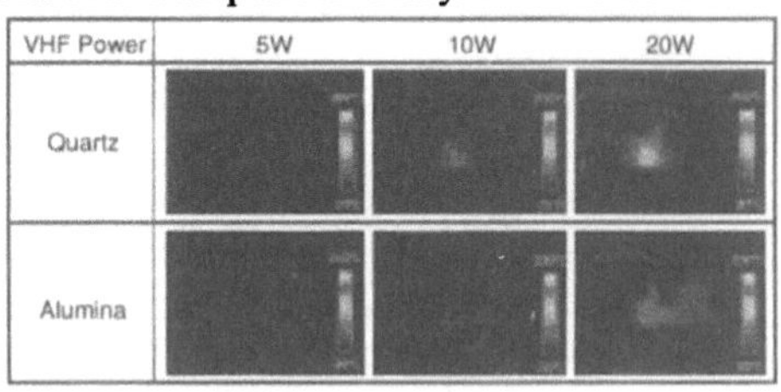

Figure 3: Comparison o temperature distribution of microplasma chips made of quartz and alumina.

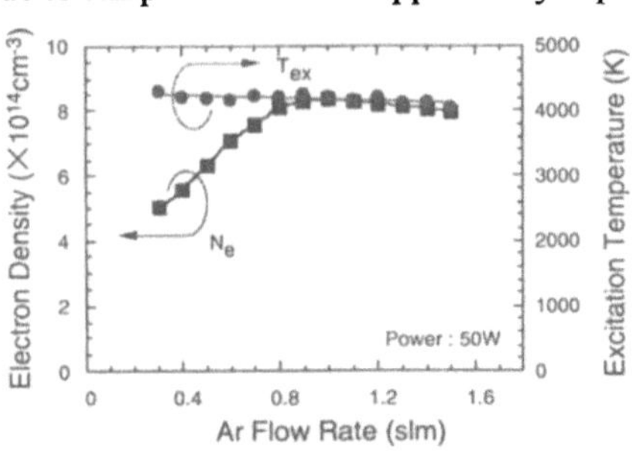

Figure 4: Effects of argon flow rate on electron density and electron excitation temperature of atmospheric-pressure microplasma jet.

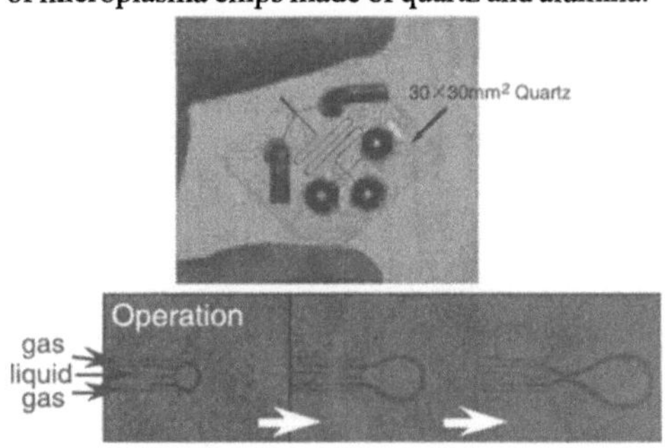

Figure 5: Photos of the microfluidic chip integrated with a microcapillary electrophoresis separator and a micronebulizer.

pressed. As a result, the power transfer efficiency to the plasma was significantly improved since the power absorbed by the ICP is given by the following equation based on the transformer circuit model.

$$P_{plasma}=(R_{plasma}/(R_{plasma}+R_{system}))(P_f-P_r) \tag{1}$$

where P_{plasma} is the power absorption, R_{plasma} is the plasma resistance, R_{system} is the system resistance, P_f is the forward power and P_r is the reflected power. Thus the alumina-made microplasma chip is preferable for the portable analysis system that cannot adopt water-cooling. Figure 4 shows the gas flow rate dependence of plasma characteristics measured at 2 mm away from the antenna edge. Though a rapid decrease in the plasma density occurred below 0.8 slm due to the thermal loss on the discharge tube wall, the high plasma density of 8×10^{14}cm^{-3} was attained above 0.8 slm. As for the injection of sample solution into the plasma, two types of interfaces between a microcapillary separator and a microplasma jet, i.e., a micronebulizer (see Fig. 5) and an electrospray (ES) nozzle, have been fabricated and examined, and better injection performance is attained for ES injection at present. Finally, detection of sodium in the standard solution (Conc. 100 ppm) was demonstrated using a multichannel photodetector (Otsuka Electronics Inc. MCPD-7000, Japan) and the ES injection method. As shown in Fig. 6, intense emission peak of NaI was observed at 589 nm when aqueous solution was injected into the microplasma jet. For the present, lower detection limit of 5 ppm has been attained using a monochromator (JASCO CT-50CS, Japan) with a photomultiplier (Hamamatsu Photonics Inc. R374, Japan), and will be further advanced by the improvement of the sample injection scheme and the optics configuration.

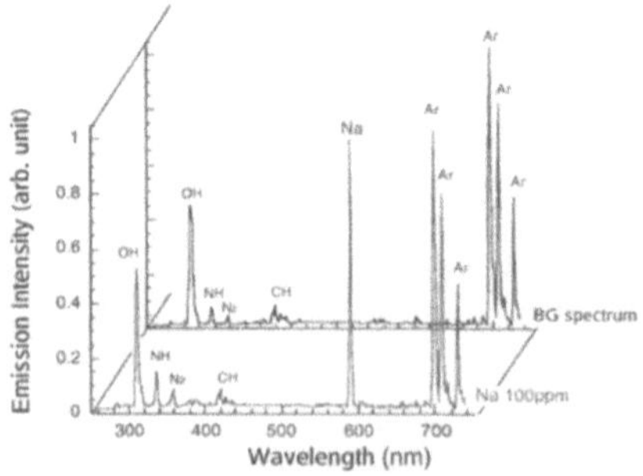

Figure 6: An optical emission spectrum from the standard solution for water quality analysis injected into the microplasma jet.

4. Conclusions

An atmospheric-pressure microplasma jet source has been developed for a portable analysis system of the liquid solution. The plasma density of 8×10^{14} cm^{-3} and the plasma temperature of 4000 K were attained at the VHF power of 50 W, and optical emission detection of sodium in the aqueous solution has been demonstrated.

Acknowledgment

This work was supported in part by a Grant-in-Aid for Scientific Research from the Ministry of Education, Sports and Culture of Japan (No. 13025242).

References

1. J. C. T. Eijkel, H. Stoeri and A. Manz, Anal. Chem. 71, pp. 2600-2606 (1999).
2. U. Engel, A. M.. Bilgi , O. Haase, E. Voges and J. A. C. Broekaert, Anal. Chem. 72, 193-197 (2000).
3. Y. Yin, J. Messier and J. Hopwood, IEEE Trans. Plasma Sci. 27, 1516-1524 (1999).

RAPID DETECTION OF MICROORGANSIMS USING MICROWAVES

Grant Hollis[1], Daniel B. Oerther[2], and Ian Papautsky[1]
[1]Department of Electrical and Computer Engineering and Computer Science,
[2]Department of Civil and Environmental Engineering,
University of Cincinnati, Cincinnati, Ohio, USA

Abstract
In this paper, we discuss the development of a novel method of bacteria detection using a microwave resonant cavity. The approach is rapid, on the order of minutes, and provides an alternative to the slower fluorescence-based assay currently used with culture-based biochips.

Keywords: microwave cavity, microorganisms, bacteria, culture-based biochips

1. Introduction

Recently, molecular biology-based bacterial identification technologies have begun to compete with conventional culture-based technologies as they offer tremendous promise for detecting many different microorganisms using a common experimental platform [1]. One of the primary advantages is the relatively rapid (several hours) analysis of samples due to enormous signal amplification from biochemical reactions such as PCR-based detection of nucleic acid. However, molecular biology-based technologies lack the power to distinguish between living and dead organisms. Thus, culture-based methods remain the "gold standard" for monitoring pathogens in environmental samples despite the lack of convenient turn-around time that hallmarks molecular biology-based techniques.

In recent papers [2,3], we reported on the development of a culture-based biochip to selectively culture nocardiaforms and mycobacteria from environmental samples. The organisms were isolated on biochips by taking advantage of their paraffinophilic nature [4] and were detected using fluorescence *in situ* hybridization (FISH) targeting 16S rRNA. Our results showed that target organisms could be selectively isolated from complex samples, such as sewage from wastewater treatment plant, after only 5 min of incubation. However, despite the rapid culturing with the biochip (5 min vs. up to 78 h for traditional cultures), the final detection step using FISH takes hours.

In this paper, we report development of a novel bacterial detection approach using a cylindrical microwave cavity. The approach is rapid, on the order of minutes, and is intended to replace the slower FISH assay. Further, the new approach is non-destructive, allowing for subsequent confirmation using molecular biology assays, if necessary.

2. Methods

A cylindrical microwave cavity has many resonant frequencies [5]. These frequencies are found by solving Maxwell's equations and confirmed by viewing the S_{11} parameter (reflection coefficient) of the cavity. At a given resonant frequency, the reflection coefficient is the ratio of the magnitude of the output to input voltage injected into the cavity and will drop to nearly zero when cavity is resonating, indicating complete

Y. Baba et al. (eds.), Micro Total Analysis Systems 2002, Volume 2, 897–899.

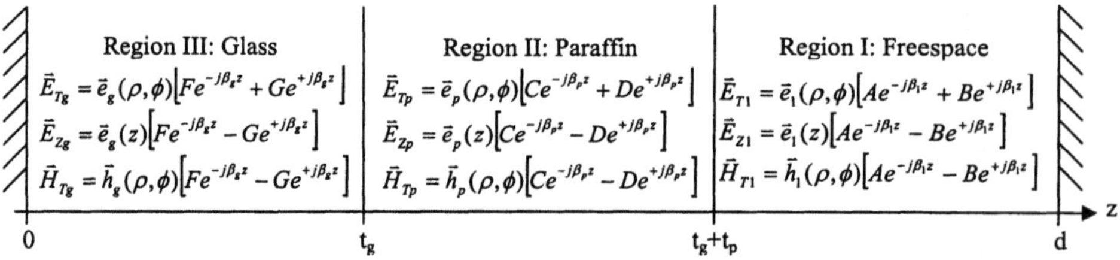

Figure 1. Mathematical model of a 3-material cavity problem.

absorption of energy. When material different from freespace (air) is present in the cavity, these resonant frequencies shift in amplitude and frequency due to changes in permeability and permittivity. If size, shape, and dielectric constant (ε_r) of the material are known, the new resonant frequencies can be predicted.

Our detection approach was demonstrated using the previously mentioned culture-based biochips. The biochips consisted of a 230 μm thick, 18 mm diameter glass wafers (ε_r = 4.62) coated with a 10 μm thick layer of paraffin (ε_r = 2.24). Resonant shifts of the empty cavity and bacteria-free biochips in the 10-26.5 GHz range were examined and mapped using HP8510C vector network analyzer. The biochips were then exposed to pure cultures of bacteria and organisms were isolated in a 5-min incubation step. The tested organisms, *Gordonia terrae, Rhodococcus rhodochrous*, and *Tsukamurella pauromycobacterium*, were selected due to their paraffinophilic nature and importance in wastewater treatment and public health. Following incubation in Czapek broth [2,3], the biochips were placed into the cavity, and the resulting resonance spectra were collected.

A mathematical model of a 3-material cylindrical cavity was constructed, as shown in Figure 1 for transverse electric mode in cylindrical coordinates, to predict the shifts due to a blank biochip. By applying the appropriate boundary conditions at the cavity walls and interfaces, the new resonant frequencies can be determined using the following equations:

$$\beta_{1,p,g} = \sqrt{(\omega_{TE}\sqrt{\mu_{1,p,g}\varepsilon_{1,p,g}})^2 - k_c^2}\,, \tag{1}$$

$$\beta_p \tan\left(\beta_1(d - t_g - t_p)\right)\left(\beta_p \tan(\beta_p t_p)\tan(\beta_g t_g) - \beta_g\right) + \beta_1\left(\beta_g \tan(\beta_p t_p) + \beta_p \tan(\beta_g t_g)\right) = 0\,, \tag{2}$$

where β is propagation constant for glass ($_g$), paraffin ($_p$), and freespace ($_1$); ω is frequency, t is thickness of the three materials, d is total cavity thickness, μ is permittivity, and ε is permeability.

3. Results and Discussion

When a bare biochip was placed in the cavity, shifts in the resonant frequencies were clearly observed. Output of an empty cylindrical cavity (10-17 GHz) used in this study is shown in Figure 3. The "valleys" in Figure 3 indicate frequencies at which the cavity is resonant. Since the S_{11} is a measure of reflected energy, these valleys are frequencies where little energy was reflected back to the network analyzer and thus all incident energy was adsorbed. When materials are placed inside the cavity, there are shifts in the resonant frequencies. All shifts were accounted for by the constructed mathematical model using Maxwell's Equations.

When inserted into an empty cavity, bacteria-free biochips produced a noticeable (78 MHz) shift in the resonant frequency at 13.535 GHz. Following incubation, further shifts beyond that of the bacteria-free biochip were observed in the third resonant frequency, as clearly shown in Figure 2. The shifts were 1.5 MHz for *G. terrae*, 3 MHz for *R. rhodochrous*, and 5 MHz for *T. pauromycobacteria*. Similar shifts were observable at other resonant frequencies from 10 GHz to 26.5 GHz.

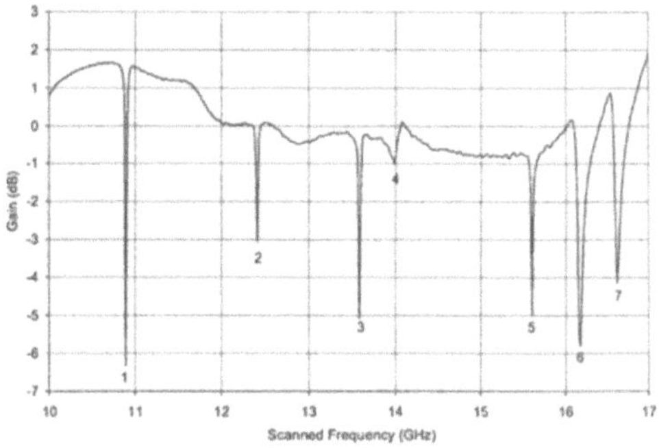

Figure 3. Sweep of the empty microwave cavity 29 mm in height and 18 mm in diameter.

4. Conclusions

The described approach of using a microwave resonant cavity for detecting bacteria was successfully demonstrated with culture-base biochips and three types of bacteria. This rapid method is intended to provide an alternative to the slower fluorescence-based assay currently in use. Although the described approach uses a bulky vector network analyzer, in the future the system could be miniaturized using a Gunn or IMPATT diode to generate the appropriate microwave frequencies [5].

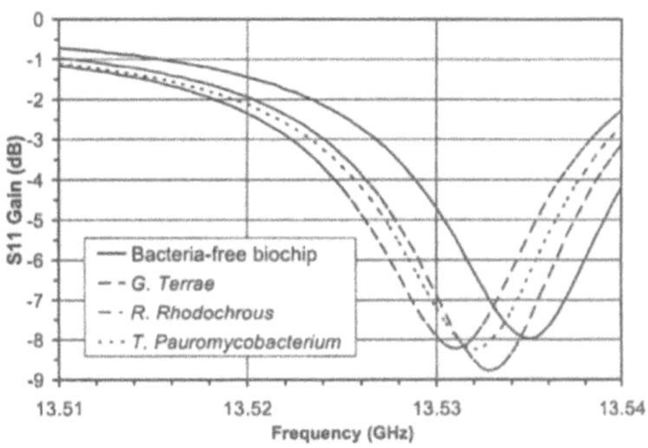

Figure 2. Frequency shifts at the third resonant frequency due to bacteria isolated on biochips.

Acknowledgments

The authors would like to acknowledge the support from the National Science Foundation (BES-0116912) and Emerging Concepts Inc.

References

[1] D. Ivnitski, I. Abdel-Hamid, P. Atansov, E. Wilkins, "Review: Biosensors for Detection of Pathenogenic Bacteria," *Biosensors & Bioelectronics*, vol. 14, 1999.
[2] H. Eluru, A. Polaczyk, R. Chhabra, B. Kinkle, D. B. Oerther, and I. Papautsky, "Culture-Based Biochips for Detecting Microorganisms in Wastewater Treatment," *IEEE-EMBS 2nd Conference on Microtechnologies in Medicine and Biology*, Madison, WI, May 2-4, 2002.
[3] A. Polaczyk, R. Chhabra, H. Eluru, I. Papautsky, B. Kinkle, and D. B. Oerther, "Culturing via BioMEMS Strikes Back in the Epic Saga Between Cultivation and Molecular Methods," *102nd Meeting of American Society for Microbiology*, Salt Lake City, UT, May 19-23, 2002.
[4] R. Ollar, *Molecular Mycobacteriology*, New York, NY: Marcel Dekker, Inc., 1999.
[5] M. Pozar, *Microwave Engineering*, New York, NY: John Wiley & Sons, Inc., 1998.

QUANTIFICATION OF MOLECULAR MIXING WITH EPIFLUORESCENCE IMAGING

Sneha Madhavan-Reese, Daniel Lim, Jyotirmoy Mazumder, E. F. Hasselbrink, Jr.
University of Michigan, Department of Mechanical Engineering,
2350 Hayward Street, Ann Arbor, MI 48109 USA
sneham@umich.edu

Abstract

Measurements of mixing in microdevices using epifluorescence imaging can be highly misleading due to resolution limitations. An "unmixedness" method using a chemical reaction is demonstrated for quantifying mixing in a laser-machined micromixer.

Keywords: Mixing, microscopy, laser-machining, microdevices

1. Introduction

Epifluorescence imaging of a passive fluorophore always overestimates the amount of mixing that has occurred between two streams, because the span of concentration gradients can be smaller than the resolution limit of the imaging system (including usually very poor axial resolution) [1, 2]. Here we demonstrate an approach using a fast chemical reaction to quantify the mixing.

2. The "Unmixedness" Method

The concentration of all elements and/or any nonreacting equidiffusive species (*i*) in two fluid streams (*1* and *2*) can be deduced from the mixture fraction $\xi = (c_i - c_{i2})/(c_{i1} - c_{i2})$. For two streams being mixed together, the goal of the mixing measurement is to obtain the rate (with time or with axial distance) at which the mixture fraction, initially 1 or 0 at the inlets, is homogenized asymptotically to a uniform value of ξ_m, where $\xi_m = Q_1/(Q_2+Q_1)$ is the ratio of the flowrate of stream 1 to the total flowrate. Since ξ profiles are difficult to resolve using an epifluorescence microscope (particularly in the direction of the optical axis), the next best option is to measure total signal from the product of a fast fluorogenic reaction. Unfortunately, most fluorogenic labelling reactions (e.g., labelling proteins with fluorescamine) are relatively slow (on the order of 1 sec) compared to microfluidic convective timescales. Fluorescamine also requires organic solvent and is therefore incompatible with polydimethylsiloxane (PDMS), a popular microfluidic substrate.

In this work we demonstrate a binary "unmixedness" method [2,3], which retains the discrimination of a chemical reaction, but employs a fast acid-base titration reaction and makes use of the pH dependence of fluorescein fluorescence efficiency to measure the volume of *un*mixed fluid. Stream 1 is a strong base, stream 2 a strong acid, and both are equimolar in fluorescein (~50μM). The pH vs. ξ is a typical titration curve (Fig. 1a), and since fluorescein fluorescence is a strong function of pH (Fig. 1b), fluorescence is nearly a binary function of mixture fraction

Y. Baba et al. (eds.), Micro Total Analysis Systems 2002, Volume 2, 900–902.

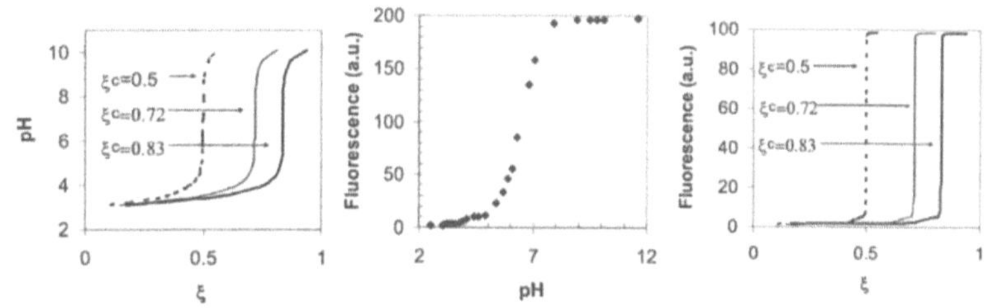

Figure 1: (a) Curves of pH versus mixture fraction for various acid/base strength ratios (which result in various "cutoff" mixture fractions, ξ_c). (b) Fluorescein fluorescence yield as a function of pH. (c) Signal as a function of mixture fraction, ξ, showing a nearly binary on-off behavior.

(Fig. 1c). This allows us to interpret any signal as an indication of the amount of fluid with $\xi > \xi_c$. By changing the pH of the basic stream, the local mixture fraction ξ_c at which the fluorescence is essentially completely quenched can be varied (Fig. 1c). By conducting experiments at different ξ_c, it is possible to determine how rapidly the two fluids are mixed at least to a particular ξ_c.

3. Experimental

We apply this method to the characterization of a "chaotic" micromixer similar to that presented by Stroock et al. [4], comprised of a channel with herringbone divots in the channel bottom; the present device (Fig. 2) is novel in that it is created by rapid prototyping via laser machining of silicon using an Nd:YAG laser at 1064nm. The two streams of HCl and fluorescein/KOH are brought together via a Y-junction at the inlet of the mixing channel. Fluorescence images were captured on a CCD camera and distance to fluorescence turn-off was measured.

4. Results and discussion

Experimental results are shown in Fig. 3. As ξ_c decreases towards 0.5 (ξ_m),

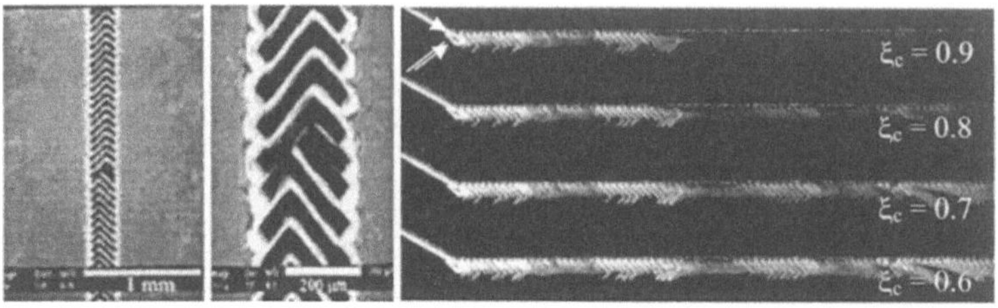

Fig. 2. (a, b) SEM images of the mixing channel showing herringbone ridges. (c) Successive images of mixing of two streams. Inlet for second stream (at left) is highlighted by an overlay drawing in the first image, and arrows show the inlet flow. As cutoff mixture fraction ξ_c approaches 0.5, the signal disappears more slowly, as expected. Total signal in any channel cross-section is interpreted as the total amount of fluid with $\xi > \xi c$.

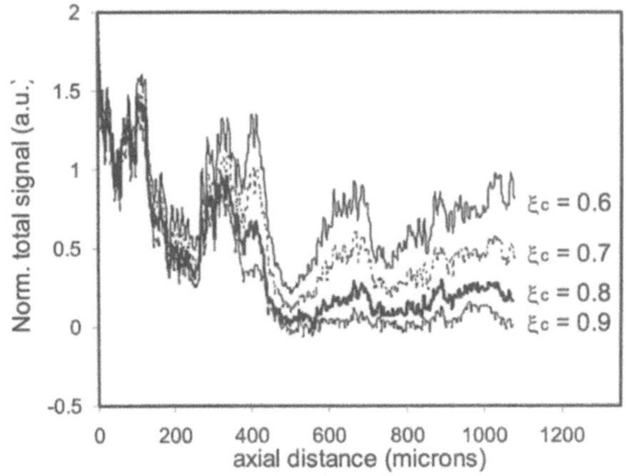

Fig. 3. Normalized total signal in channel cross-section versus axial distance along the channel length. As expected, higher ξc results in a more rapid decay of fluorescence signal due to mixing. Large fluctuations are observed, implying that unmixed fluid is fluctuating between low-speed and high-speed regions of the channel, since the flux of "unmixed" fluid changes more slowly.

the fluorescence signal decays more slowly, as expected. Large fluctuations in the signals result from the unmixed fluid moving through high-speed and low-speed regions of the channel. Straight channels (not shown) show much slower decay of signal, indicating much slower mixing.

5. Conclusions

A chemical reaction based method for quantification of mixing has been shown to easily determine when full mixing has been achieved in these microchannels. This method is a rapid, inexpensive, and accurate alternative to confocal scanning laser microscopy. Ongoing work is seeking to identify a fast fluorogenic reaction that occurs in aqueous solvent.

Acknowledgements

This material is based upon work supported under a National Science Foundation Graduate Research Fellowship.

References

1. M. Koochesfahani and P. Dimotakis, *J. Fluid Mech.*, vol. 170, pp. 83-112, (1986).
2. P.S. Karasso and M.G. Mungal, *J. Fluid Mech.*, vol. 323, pp. 23-63, (1996).
3. N. T. Clemens and P. H. Paul, *Phys. Fluids*, vol. 7(5), pp. 1071-1081, (1995).
4. A. Stroock, et al. *Science*, v. 295, pp. 647-651, (2002).

EVALUATION OF THE CAPTURE EFFICIENCY FOR DNA IN A FLOW THROUGH DEVICE

Olena Melnychuk and D. Jed Harrison
Dept. of Chemistry, University of Alberta, Edmonton, AB, T6G 2G2, CANADA

Abstract
In designing a microchip system for preparing a cDNA library on a chip the crucial initial step is obtaining high-yield, isolated mRNA; especially for the low abundance transcripts. We demonstrate, for the first time, a quantitative method of assessing the capture efficiency $(dT)_{25}$ coated magnetic beads trapped within a micro-fluidic device.

keywords: capture efficiency, DNA hybridization, magnetic beads, design

Introduction

Isolated mRNA molecules, and c-DNA libraries formed from them, can reveal the differences in expression levels between normal tissues, and ones affected by disease. In order to prepare cDNA libraries on-chip an effective strategy for isolating mRNA on-chip is required. Recently, it was shown chip-based mRNA capture and isolation could be successfully performed. The isolated mRNA yields suggested that a better fluid flow design would be required to improve the capture efficiency [1]. However, those studies used the reverse transcription polymerase chain reaction (RT-PCR), which is very difficult to use quantitatively, so that accurate determination of the efficiency was difficult. To further evaluate the effect of the flow design we have developed a quantitative assay, in which fluorescein labeled poly-A-tailed-ss DNA probes and oligo $(dT)_{25}$ paramagnetic Dynabeads were used to evaluate hybridization on a bed bead within a microfluidic device. Direct fluorescence measurement of the probe was then possible.

Results and Discussion

In the study of a magnetically trapped bead bed a simple Y-intersection microfluidic device was employed (Figure 1).

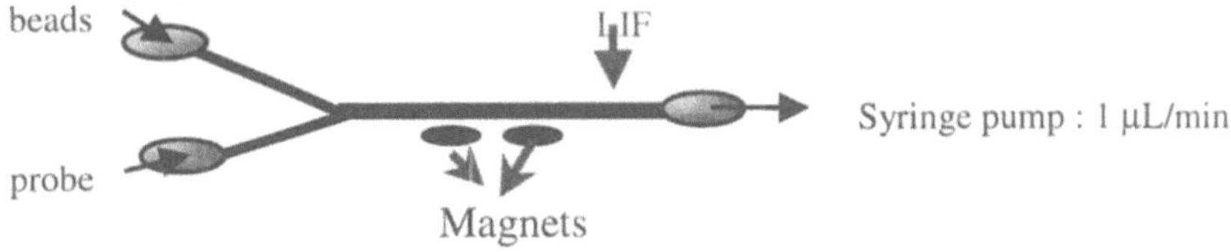

Figure 1. Y-intersection device for hybridization of probe and beads.

Y. Baba et al. (eds.), Micro Total Analysis Systems 2002, Volume 2, 903–905.

These beds are conveniently formed and then released, allowing for effective harvesting of the isolated DNA or mRNA. However, the loose packing that occurs in these beds causes channels to form, which may lead to poor mixing and poor capture efficiency. A probe : bead ratio of 1:10 [2] (0.005 μg probe/μL of beads) was used with a 0.5x binding buffer (10 mM Tris-HCl (pH 7.5), 0.5 M LiCl, 1 M EDTA)[3]. In Figure 2, two different approaches for measuring capture efficiency were employed, (a) a normalized probe strategy, and (b) a bed saturation method. Probe in one reservoir and beads in the other were drawn towards the Y-intersection and the magnet, using hydrodynamic pumping (flow rate ranging of 1 to 5 μL/min.). Further downstream, an LIF detector was set up to measure capture efficiency. For method "a", the ratio of signal downstream of the bed to that upstream of the bed gave the capture efficiency. For method "b", the ratio of bed capacity to volume needed to saturate the bed gave the capture efficiency.

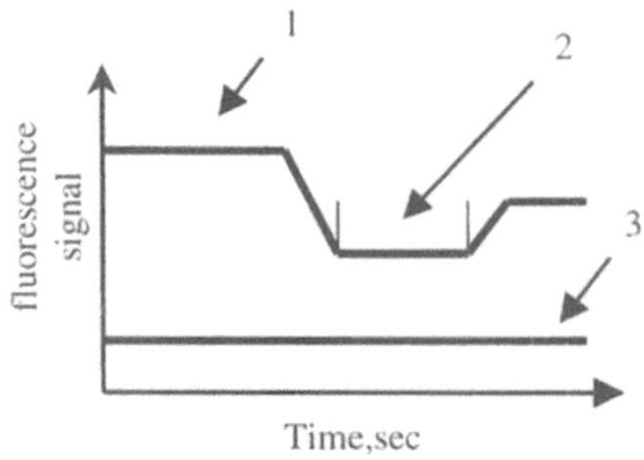

Figure 2. Cartoon illustrating quantitative measurement of the cap efficiency.
1-probe, 2-bed saturation, 3-buffer

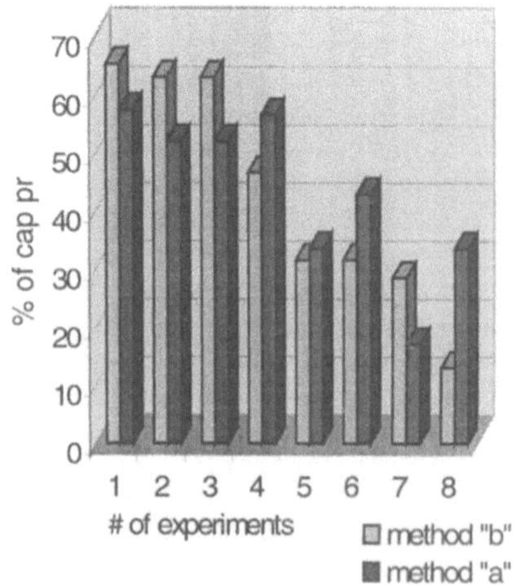

Figure 3. Capture efficiency obtained using magnetically trapped bead bed

The calculated capture efficiency is presented in Figure 3. The results show 30-60% of captured efficiency with considerable variability. This was attributed to the variable effects of channeling in the beds.

The effects of channel width on capture within the open channels of the Y-intersection device were also evaluated. Y-intersection devices with widths of 50, 100, 150 μm were tested. The capture efficiency was in the range of 18-33%, 30-65%, and 70-88%, respectively, when operated at a flow rate of 1 μl/min. The fact that wider channels give better capture efficiency is probably a result of the reduced linear flow rate in the larger channel, which allows more time for hybridization to occur, and reduces the pressure on the bead bed, resulting in less channeling of the bed.

In a search for increasing capture efficiency at higher flow rates a physically trapped bead bed was used. Figure 4 shows a projection view of the design. Beads were packed into a 200 μm long bed. The capture efficiency was determined using (a) the normalized probe strategy. The capture efficiency was 84% for these beds when used at a flow rate of 1 μl/min, indicating on improvement compared to the more open magnetically trapped beds.

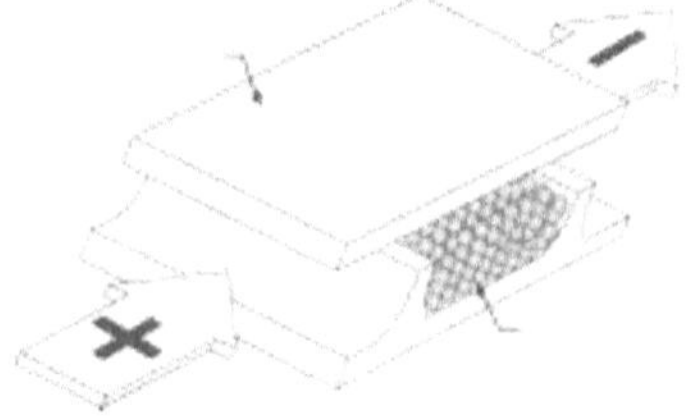

Figure 4. Projection view of weir design.

Conclusions

These results indicate that magnetically trapped beds give quite variable capture efficiencies. The variation can be reduced, and the efficiency improved by reducing the flow rates and by increasing the number of beads trapped in the bed. The physically trapped beds appear to be more homogeneous and routinely give high capture efficiency. However, these latter beds are less convenient in terms of subsequent release and harvesting of the beads. Thus the appropriate bed design depends upon the eventual application. Where the beads are to be captured the magnetic trapping method is preferable, and using longer beds can reduce the variation seen with this approach. When the sample is to be eluted off the bed again, then the physically trapped bed will be more efficient at concentrating sample.

References:

1. J.Guifeng, D.J. Harrison, Analyst, 2000, 125,pp.2176-2179.
2. D.V.Morrissey, M.Lombardo, J.K.Eldredge, K.R.Kearney, P.Groody, M.L.Collins,Anal. Biochem.,1989,181,pp.345-359
3. Protocol of Dynal beads mRNA purification, Dynal A.S.Oslo, Norway.

Practical Studies on Multi-channel Operation of a Novel Highly Sensitive Quartz Crystal Microbalance-based Planar Biosensor

Hiroyuki Sota[1], Hiroshi Yoshimine[1], Yasuro Shinohara[1], Lars Fägerstam[2], Patric Fricking[2], Mattias Vangbo[2] and Yoshio Okahata[3]
[1]Department of Research and Development, Amersham Biosciences K.K., Japan
[2]Protein & Separation Sciences Research, Amersham Biosciences, Sweden
[3]Department of Biomolecular Engineering, Tokyo Institute of Technology, Japan
(e-mail: hiroyuki.sota@jp.amershambiosciences.com)

Abstract

With the aim of realizing a multi-channel sensor system using our advanced quartz crystal microbalance (QCM)-based planar sensor chip design, practical studies were carried out on undesired coupling among sensor elements. Carefully designed shielding and grounding onto each multiplexed oscillation circuit were found to effectively suppress coupling attributable to remote electromagnetic transmission between the multiplexed sensors.

Keywords: Biosensor, Quartz Crystal Microbalance, QCM chip, Multi-channel Planar Sensor, Non-labeling Molecular Detection

1. Introduction

In the last symposium (µTAS 2001), we demonstrated a novel quartz crystal microbalance (QCM)-based micro-biosensor having the potential for integration into microfluidic devices aimed at state-of-the-art analysis of biological samples [1]. Our central innovation relies on a novel chip-based design (QCM chip, Figure 1) that raises practical sensitivity and reproducibility by lowering the noise usually encountered by piezoelectric devices operating in an aqueous environment. From the practical viewpoint of the user, our design enables reproducible measurements down into the sub-picogram range, and even more remarkably, this low noise level is flow-rate independent [2]. Here, we report our practical studies to build on these advantages by realizing a multi-channel sensor system.

2. Experimental

The key barrier to multiplexing piezoelectric sensor units within a miniaturized format is the undesired coupling phenomenon [3-5] that takes place between neighboring resonators. Closely spaced resonators oscillating at substantially identical frequencies are highly susceptible to this problem. Since piezoelectric resonation encompasses both acoustic and electromagnetic components, either or both components may mediate the undesired coupling. In order to dissect the contributions of these two

Y. Baba et al, (eds.), Micro Total Analysis Systems 2002, Volume 2, 906–908.

separate components to real signal behavior, we first configured a system, in which the electromagnetic interaction should predominate. Two QCM chips, each harboring a single resonator, were installed separately into two flowcells, each connected to a separate oscillation circuit (Figure 2). The details of the QCM chip and flowcell designs and basic instrumentational set-up were described elsewhere [2].

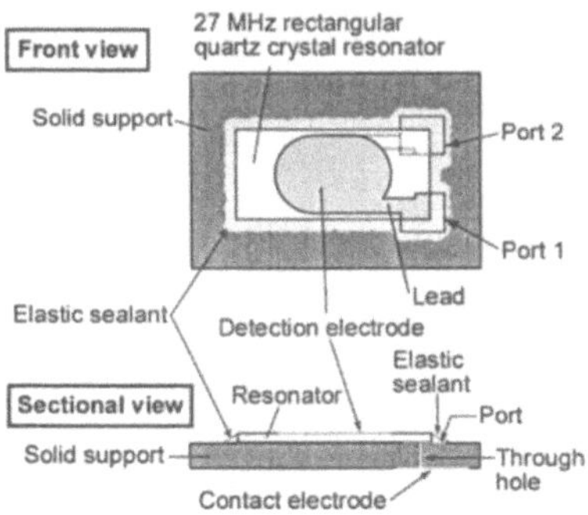

Figure 1. Schematic View of the QCM chip.

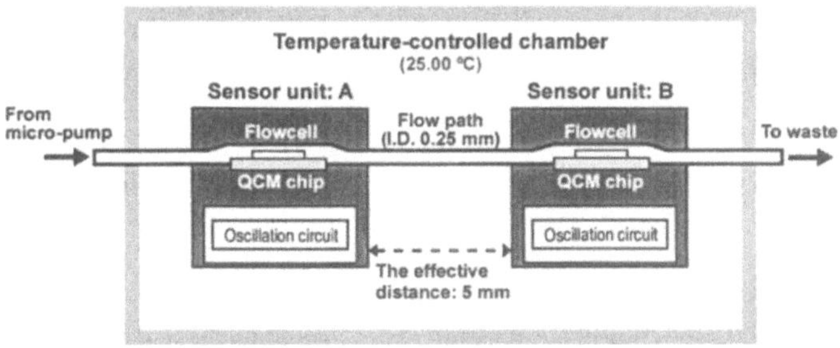

Figure 2. Experimental setup for studying the effect of electromagnetic coupling between the sensor units. Each unit consisting of QCM chip, flowcell and circuit is sequentially connected via an HPLC flow path of 5 mm.

3. Results and discussion

As shown in Figure 3 (a), the signals from the two independent sensors showed evidence of coupling; Sensor B exhibited pulse-shaped frequency changes when the driving current and hence resonation frequency of Sensor A changed. Such behavior was reproduced independent of whether or not the flow path was filled with an electrolyte. Therefore we infer that the coupling was mediated via remote electromagnetic transmission between the two sensors. In order to suppress such coupling, we installed carefully designed shielding and grounding onto each oscillation circuit. As shown in Figure 3 (b), Sensor B no longer exhibited synchronized signal behaviors in the modified design. These results indicate that the high-frequency (27 MHz) oscillation circuit itself is susceptible to electromagnetic wave transmission/reception that leads to coupling, and hence signal synchronization. Suppression of such electromagnetic transmissions is essential to diminish undesired coupling.

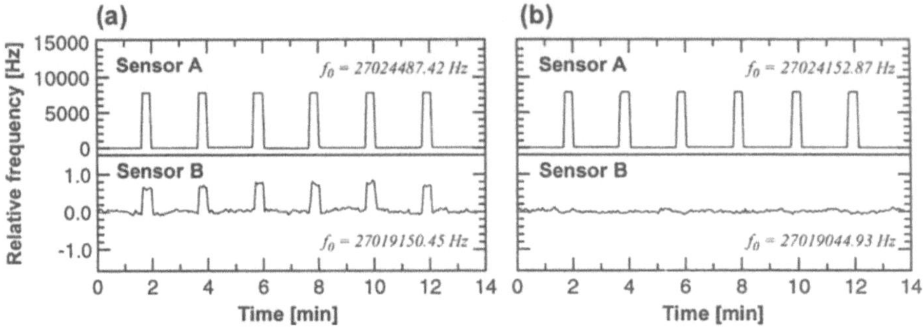

Figure 3. Signal synchronization tests for two QCM chip sensors (A, B) installed separately into different sensor units upon a repeated frequency shift operation on Sensor A. (a) An initial data showing pulse-shaped frequency change of Sensor B in response to the frequency shifts of Sensor A, indicating coupling between two sensors. (b) A reexamined data obtained after making some electromagnetic modifications on each oscillation circuit involving carefully designed shielding and grounding. f_0: initial absolute resonant frequency.

5. Conclusions

With the electromagnetic modification described above, the envisioned multi-channel sensor system will achieve coupling-free signal, allows us to realize the first practical miniature multi-channel piezoelectric liquid-phase sensor.

Acknowledgements

This work was partly supported by the Ministry of Agriculture, Forestry and Fishery of Japan (MAFF).

References

1. Sota, H. *et al.*, *Proc. μTAS 2001 Symposium*, Kluwer Academic Publisher, Dordrecht, The Netherlands, pp. 579-580 (2001).
2. Sota, H. *et al.*, *Anal. Chem.*, **74**, 3592-3598 (2002).
3. Berg, S. and Johannsmann, D., *Anal. Chem.*, **73**, 1140-1145 (2001).
4. Abe, T. and Esashi, M., *Sens. Actuators A*, **82**, 139-143 (2000).
5. Tatsuma, T. *et al.*, *Anal. Chem.*, **71**, 3632-3636 (1999).

SENSITIVITY ENHANCEMENT IN CHIP-BASED CAPILLARY ELECTROPHORESIS SYSTEMS BY A HADAMARD TRANSFORM APPROACH

Tao Zhang[1], Qun Fang[2] and Zhao-Lun Fang[1,2]

[1] Research Center for Analytical Sciences, Northeastern University, Shenyang, China

[2] Institute of Microanalytical Systems, Department of Chemistry, Zhejiang University, Hangzhou, China

Abstract

Hadamard transform was applied to a microfluidic chip-based capillary electrophoresis system to improve its signal-to-noise level. Multiplexed sample injection was effected by an electrokinetic gated injection approach, employing a 7-bit, 127-step pseudorandom binary sequence. A 5-fold improvement in S/N ratio was achieved, with only 27% increase in analysis time for Cy5 labeled amino acid samples.

Keywords: microfluidic chips, capillary electrophoresis, Hadamard transform, multiplexed injection, sensitivity enhancement

1. Introduction

Small detection volumes, often in the picoliter range, and short path lengths of the detector limit the sensitivity achievable on microfluidic chips, and various approaches were pursued to improve the detection power of such devices. In addition to measures involving preconcentration, and increase of detector path length, these also included the use of mathematical calculations based on a cross-correlation approach [1]. Hadamard transform, a powerful mathematical approach for enhancing the signal-to-noise ratio via multiplexed injections, has been applied successfully to IR spectroscopy [2], TOF-MS [3,4], and imaging techniques [5]. Recently, Kaneta et al. applied Hadamard transform to conventional CE, employing an optically gated sample injection approach [6-8], achieving very encouraging results in S/N enhancement. However, multiplexed sample injection was achieved by relatively sophisticated and expensive equipment, including a 150 mW high-power gating laser. The approach is also limited to fluorescence dyes that are readily bleached using the laser. Although these authors have postulated that a much more efficient and cheaper approach might be realized on chip-based CE system [8], to our best knowledge, hitherto such attempts have not been carried out. This paper reports the first demonstration of Hadamard transform capillary electrophoresis (CE) on a microfluidic chip using a simple electrokinetically gated injection approach.

2. Theory

High throughput signals for CE may be obtained using a multiplexing technique

Y. Baba et al. (eds.), Micro Total Analysis Systems 2002, Volume 2, 909–911.

involving alternative injections of sample and running buffer into the separation channel according to a pseudorandom binary sequence (PRBS) generated by a shift register method [6] (Fig. 1). The resulting Hadamard electropherogram obtained by the detector will be a convolution of the PRBS and single injection electropherograms. Conventional single injection electropherograms could be recovered with inverse Hadamard transform, with S/N ratios improved by a factor of $(n+1)/(2n^{1/2})$, where n is the length of the PRBS.

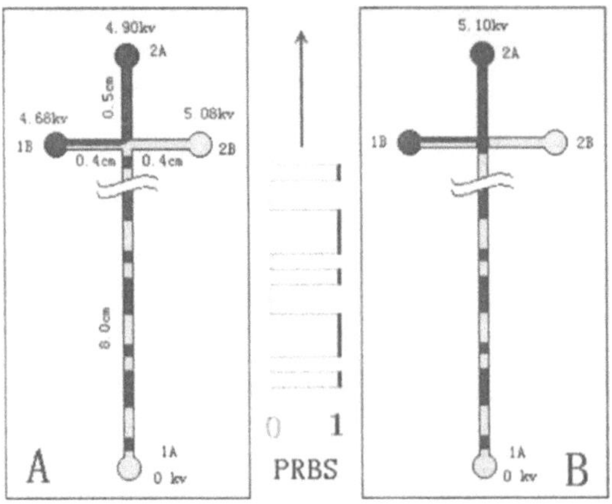

Figure 1. Gated injection scheme. 2A, sample; 2B, running buffer; and 1A and 1B; waste reservoirs. By sequentially controlling the voltages at 1A, 1B, 2A, 2B, sample and running buffer were injected into the separation channel alternatingly according to the PRBS scheme. 0, buffer injection; 1, sample injection.

3. Experimental results

A microfluidic tool kit from Micralyne (Alberta, Canada) equipped with a 635 nm diode laser and standard crossed-channel glass chips (45 mm effective separation length) was used for CE-LIF determinations, employing 0.1 nM Cy5 (Amersham Pharmcia Biotec) labeled amino acids as a test sample. A 7-bit, 127-step PRBS was used in obtaining the Hadamard electropherograms. Each step corresponded to 0.5 s in the Hadamard electropherograms, and the sample injection time was 0.5 s in the conventional single injection electropherogram. The sequence of gated injection was achieved by a preset program of the tool kit, controlling the voltages and timing of the electrodes in the chip reservoirs. In order to minimize fluctuations due to toggling between injection and separation states, the potential at the injection cross was maintained constant throughout the pseudorandom inject sequence, the field strength in the separation channel remained at ~600V/cm throughout the experiments. In this work, Hadamard and conventional electropherograms required 127 and 100 s, respectively, for recording. Although a little more time was consumed for the Hadamard transform mode, a 5-fold improvement (5.6 in theory) in S/N ratio was achieved, as compared with conventional CE (Fig. 2).

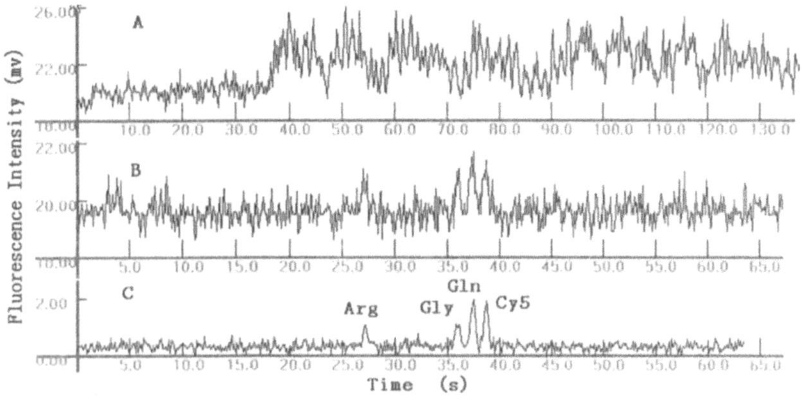

Figure 2. Electropherograms of 0.1 nM Cy5-labeled amino acids (a) Hadamard electropherogram obtained by a 127-step PRBS, (b) electropherogram obtained by conventional single injection, (c) electropherogram obtained by inverse Hadamard transformation of (a).

4. Conclusions

This preliminary study demonstrates the feasibility and potentials of employing Hadamard transform CE on microfluidic chips for improving concentration detection limit. In this study, the performance was limited by the number of programmable steps allowed by the instrument (200 steps), and further enhancement of sensitivity may be achieved by increasing the order of the matrix. The multiplexed gated-injection approach could be adapted readily to chip-based CE systems employing other detection systems for achieving Hadamard transform.

Acknowledgements

This work was funded by Natural Science Foundations of China (project 20299030).

References

1. J. C. Finster, III, S. C. Jacobson, and J. M. Ramsey, *Anal. Chem.*,71, 4460-4464 (1999).
2. J. A. Decker, Jr., *Anal. Chem.*, **44**, 127 A–134 A (1972).
3. A. Brock, N. Rodriguez, and R.N. Zare, *Anal. Chem.*, **70**, 3735–3741 (1998).
4. F. M. Fernandez, J. M. Vadillo, J. R. Kimmel, M. Wetterhall, K. Markides, N. Rodriguez, and R. N. Zare, *Anal. Chem.*, **74**, 1611-1617 (2002).
5. P. J. Treado, and M. D. Morris, *Anal. Chem.*, **61**, 723 A–734 A (1989).
6. T. Kaneta, Y. Yamaguchi, and T. Imasaka, *Anal. Chem.*, **71**, 5444–5446 (1999).
7. T. Kaneta, *Anal. Chem.*, **73**, 540A-547 A (2001).
8. T. Kaneta, K. Kosai, and T. Imasaka, *Anal. Chem.*, **74**, 2257-2260 (2002).

MICRO/NANO MACHINED DEVICES PROBING INTO NANO/BIO WORLD

Hiroyuki Fujita
CIRMM/IIS, The University of Tokyo
4-6-1 Komaba, Meguro-ku, Tokyo 153-8558 Japan
Also with CREST/JST

Abstract

This paper deals with miniaturized nano electro mechanical systems (NEMS) and applications to nano and bio technologies. Fabrication methods for NEMS are overviewed. Some examples on NEMS made by IC-compatible batch fabrication processes applied to the investigation of nano/bio field are described. Future prospects and implications of NEMS in terms of bio chemical application are also discussed.

Keywords: Nano machining, nano electro mechanical systems (NEMS), nano technology, bio technology, molecular handling

1. Introduction

Micromachining technology has created a great variety of micro electro mechanical systems (MEMS) through the integration of moving mechanisms, sensors and electronics in a chip-size system. Although dimensions of mechanical parts generally used in MEMS are limited in a range of micrometers or larger, rapidly expanding interest in the nanotechnology urges the development of mechanical parts in nanometer sizes for the creation of new class of nano electro mechanical systems (NEMS).

One of the promising applications of NEMS is the tool for the investigation of nano and bio worlds. As the size decreases comparable to micro/nano objects such as cells and macro molecules, direct and localized interaction is anticipated between NEMS and the object. NEMS serves as the interface between the nanoscopic world and our macroscopic world (Fig. 1). We can reveal material properties specific to the nanoscopic region by evaluating nano matters with NEMS tools. Scanning probe microscopes (SPM's) are the first precursor of such tools. Please note that the real nano structure in SPM is only the apex of the tip. NEMS will provide higher degree-of-freedom in observation and manipulation, multi-probe devices, wider functionality and parallel processing capability.

This paper gives the overview of nano fabrication methods, followed by current application examples and future prospects of NEMS in nano and bio technologies.

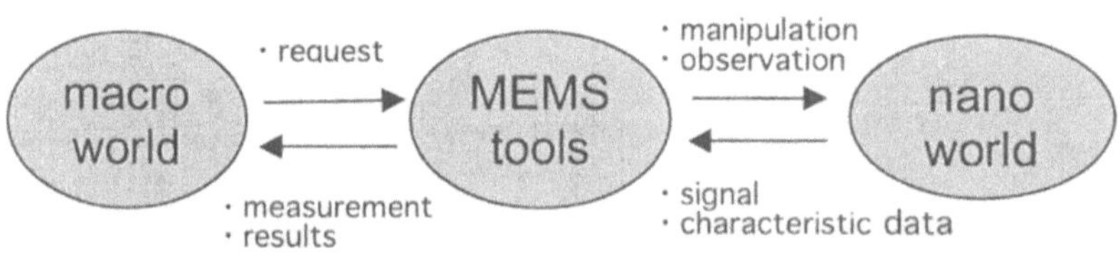

Fig.1 MEMS tools are useful for investigation of new properties and phenomena in the nano domain through providing means to handle and measure nano objects such as quantum dots and molecules.

Y. Baba et al. (eds.), Micro Total Analysis Systems 2002, Volume 2, 912–917.

2. Fabrication methods for NEMS

In order to miniaturize MEMS to nano scales, it is important to establish fabrication technologies for nano mechanical parts. There are two approaches; one is to push the semiconductor-related micro fabrication technology further down in scale, while another is to utilize alternative methods such as self-assembly technology. Major nano fabrication methods are listed in Table 1. Methods listed from the first to the fifth row, from epitaxy to X-ray lithography, are related to semiconductor processes. Ultra-thin films can be deposited single atomic layer by layer with well-established molecular beam epitaxy (MBE) process. Because photo lithography can not be applied due to the limitation of wavelength of light, electron beam, X-ray, or SPM probes are used to draw nano patterns. Precise etching follows such pattern definition process and transfer nano patterns to thin films or substrates.

In the sixth and seventh rows, beam-based machining methods are listed. Atomic or molecular beams can etch nano patterns or deposit nano structures directly at the desired position. Even with photons, we can take advantage of nonlinear effect, called two-photon absorption, to solidify a very small volume of photo-sensitive resin because the reaction occurs only in a minimal fraction of strongly focused light spot. Kawata, et al. developed nano stereo lithography by using the phenomena [1].

We can obtain the replica of a nano mold by filling some plastic material in the mold and solidifying it (nano molding method). If the replica is made of soft material such as PDMS (poly di-methyl siloxane), we can use the replica as a nano stamp. Usually, a self assembled monolayer (SAM) of certain molecules is deposited on the stamp. Only the parts on elevated patterns are transferred to the substrate when the stamp is pushed against it. Transfer of nanometric patterns has been verified [2]. Etching or oxidation of the substrate can be performed using SAM patterns as the mask. Direct patterning of less stable materials, e.g. bio molecules, is possible by nano imprinting.

Last four methods are in the category so-called bottom-up technology. Atoms are manipulated and assembled to form complicated nano structures through chemical reactions or crystalline formation. Sometimes, well-regulated self-assembly process may occur resulting in production of many regular nano structures.

3. NEMS applied to nano/bio technologies

This section deals with NEMS tools fabricated by IC-compatible micromachining. Static nano structures have been utilized widely. Membranes with minute holes are used as filters, dense arrays of nano columns in a DNA separation channel, and nano fluidic channels as spatially confined reactors. Selective and patterned surface modification has been successfully applied to microarrays and cell culture on a chip [3]. Three dimensional active mechanisms are about to be introduced.

3.1 Nano twin probes

The first type of nano twin probe has thin and sharp probes arranged orthogonally to each other. Typical dimensions are 50 ~ 100 nm in probe width and 4 ~ 20 nm in tip radius. The initial gap spacing between tips is in the range of 300 ~ 500 nm; the spacing can be adjusted by integrated microactuators. Figure 2 is the scanning electron micrographs of the device with different magnifications. Please note that the actuator is thousand times bigger than nano probes but still measures less than a millimeter. This

Table1 Nano fabrication technologies

Fabrication technologies	Machining methods	remarks
epitaxy	MBE, MOCVD	precise deposition of single atomic layer level; self-assembled quantum dots formation
etching	plasma, etchant	precise etching down to single atomic layer level; nano structure formation by crystallographic orientation dependent etching
electron-beam lithography	electron beam	direct exposure of resist layer by electron beam to draw patterns down to a few nm
micro probe lithography	scanning probe microscope	direct drawing of ~10nm patterns on substrate by localized oxidation at probe contact
X-ray lithography	SOR	transfer of mask patterns of ~10nm size to resist layer by X-ray exposure
molecular / atomic / ion beam machining	molecular /atomic / ion beam	localized deposition / etching / modification by various beams
nano stereo lithography	UV laser light	3-D nano structures made by solidification of resin through two-photon absorption
nano molding	nano mold	replica formation of nano machined mold
nano imprinting	nano stamp	pattern transfer by putting a nano stamp coated with SAM
supra molecule	chemical synthesis	molecular design and synthesis of supra molecule having specific functionality
protein synthesis	gene recombination	gene recombination and synthesis of protein having specific functionality
fullerene, nano tube	arc discharge, CVD, etc.	synthesis and growth of fullerene and nano tubes in electric arc or plasma
atomic manipulation	SPM	moving atoms on clean and flat surface to form patterns

device operates in an ultra-high vacuum environment of a transmission electron microscope. Possible measurement targets include tunneling phenomena between tips, cold electron emission under high electric field, and four probe measurement of electrical potential distribution along quantum nano wires.

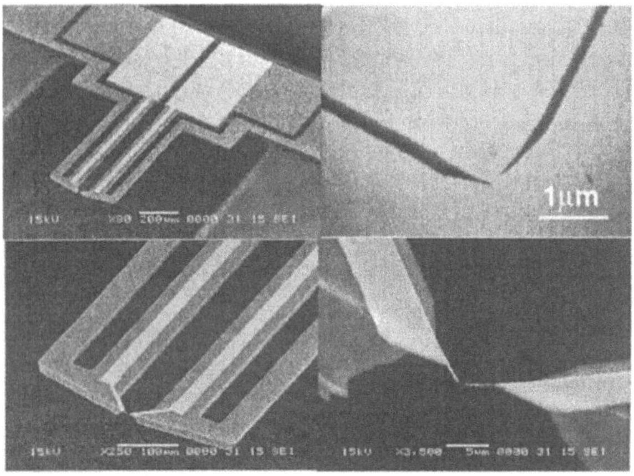

Fig.2 Twin nano probes with slender tips

The second type is more suitable for bio application and shown in Fig. 3. The tips are as sharp as the first type but mechanically more robust. We are planning to use the device to capture and retrieve DNA molecules from aqueous solution. Handling DNA in MEMS and with NEMS has been studied for ten years. Washizu, et al. demonstrated alignment and immobilization of DNA molecules between fixed electrodes [4]. Nakao, et al. fabricated a hook-like nano probe by electron beam deposition and extracted DNA filaments from a pre-treated rice cell nucleus [5]. Baba and Washizu, et al. also captured a single DNA molecule between an Al electrode and an Al coated probe tip of an atomic force microscope [6]. We are going to coat the twin nano probes with Al layers, to immerse it in a water droplet containing DNA and to apply AC voltage between probes. The electrical field that concentrates at and around tips should attract DNA molecules.

Fig.3 Twin nano probes with mechanically (robust) tips

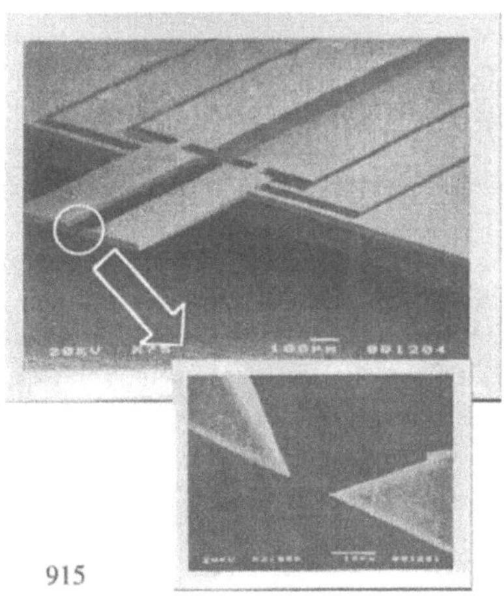

3.2 Micro/nano capillary array for DNA injection

Among many insertion methods of gene into cells, microinjection by sharp pipettes has the lowest toxicity. This method, however, is essentially a manual method and suffered from low productivity. We have been working on the parallelized DNA injection to living cells using a micromachined capillary array and a cell capture device [7] that holds cells in an array matching with the capillary spacing. Previous study demonstrated successful transfection of genes to tobacco cells [8]. However the diameter of the capillary was 5 micrometers; this is too large to be applied to smaller cells. Current investigation enabled us to fabricate a hollow capillary of 1 micrometer for outer diameter and 500 nm for inner diameter. Figure 4 shows the micrograph of the hollow capillary.

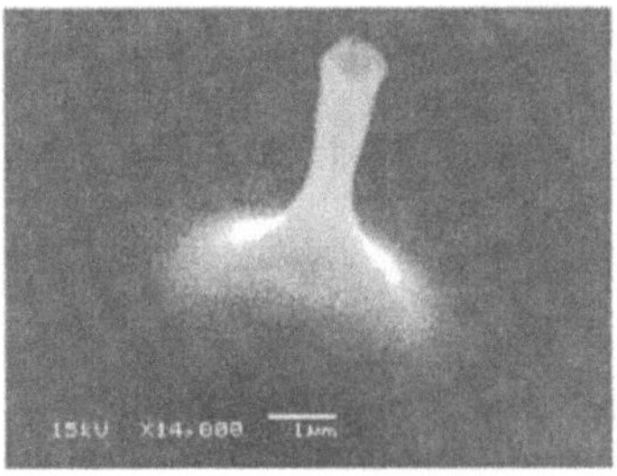

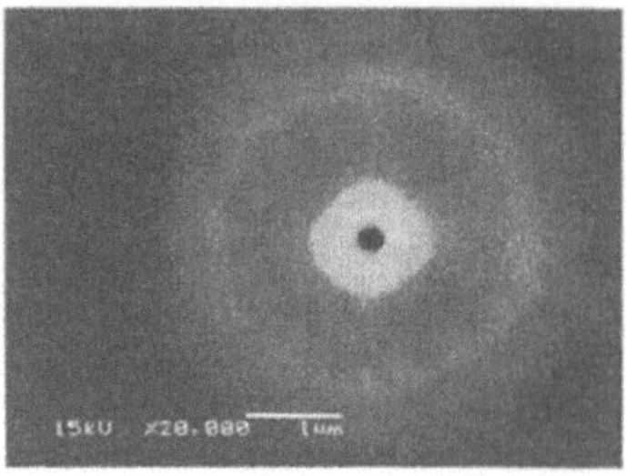

Fig.4 Hollow micro capillary of 1 micrometer in diameter

The first generation of the capillary array offer different possibilities. We have demonstrated the electric cone jet to eject liquid of very high viscosity from the capillaries [9]. The purpose was to draw fine lines directly. We have achieved the minimum line width of 800 nm with the ink of 5800 mPa s. This device has potential applications in nano liquid dispenser and/or parallel nozzles for mass spectrometers.

4. Future prospects

The introduction of micrometer scale structures in chemistry and bio technology has provided a lot of advantages through miniaturization of conventional chemistry. The governing principles are, however, very similar to those for macroscopic chemistry. When the scale goes down further into nanometers, what will happen? By the discussion of my colleagues working in chemistry and bio technology, I come to believe some new phenomena can happen. Also NEMS will enable observation of individual molecular reactions in detail.

With NEMS, we can expect to perform the following:

(1) To confine small amount of molecules produces by a cell or a enzyme molecule in a nano space for detection.
(2) To control the concentration of chemicals in ms or microseconds and to observe the fast transient phenomena in chemical reaction.
(3) To measure time course or distribution of single molecular activities in large quantity.
(4) To enhance the effect of walls of a reaction space and to observe the effect of

molecular clustering forces over chemical reaction.
(5) To apply strong but local electric or magnetic field to the reaction space and to control chemical reaction by forced molecular reorientation.
(6) To manipulate molecules directly and physically.
(7) To attach nano tags to a single molecule of protein and to study the relationship between the reaction and the structural change of the molecule.
(8) To construct hybrid nano systems by combining bio molecules in NEMS.

5. Conclusion

Scaling down bio chemical systems does not mean mere miniaturization of macroscopic counterparts but requires revolutionary change in fabrication and also brings revolutionary change in chemistry. NEMS will provide crucial functionalities and environment for investigating and utilizing bio chemistry in nano world. Even stronger collaboration between bio chemists and MEMS/NEMS engineers than now will be indispensable. In Japan, we have established a dedicated academic society to promote this growing field; Society for Chemistry and Micro-Nano Systems. Collaboration among various disciplines and countries should be promoted further.

Acknowledgement

The author wishes to acknowledge financial support from CREST/JST, MEXT, and NEDO for experimental works and fruitful discussion with my colleagues including Prof. Takehiko Kitamori, Prof. Shuichi Shoji, Prof. Yoshinobu Baba and Prof. Hiroyuki Noji.

References

1. S. Maruo, O. Nakamura, S. Kawata, Opt. Lett., **22**, 132 (1997)
2. Y. Xia, D. Win, G.M. Whitesides, Advn. Mat., **8**, 1015 (1996)
3. L. Griscom, P. Degenaar, B. LePioufle, E. Tamiya, H. Fujita, Jpn. J. Appl. Phys. **40**, 5485 (2001)
4. T. Yamamoto, O. Kurosawa, H. Kabata, N. Shimamoto, M. Washizu, IEEE Trans. On IA, **37**, 1625 (2001)
5. T. Ooi, K. Matsumoto, M. Nakao, M. Otsubo, S. Shirakashi, S. Tanaka, Y. Hatamura, IEEE Conf. on MEMS-2000, 580, Miyazaki, Japan (Jan. 2000)
6. M. Ueda, Y. Baba, H. Iwasaki, O. Kurosawa, M. Washizu, Jpn. J. Appl. Phys. **38**, 6568 (1999).
7. B. Le Pioufle, P. Surbled, H. Nagai, Y. Murakami, K. S. Chun, E. Tamiya and H. Fujita · Mater. Sci. Engin.: C, **12**, 77(2000)
8. K.S. Chun, G. Hashiguchi, H. Toshiyoshi, H. Fujita, Jpn. J. Appl. Phys. **38**, L279 (1999).
9. R. Ohigashi, K. Tsuchiya, Y. Mita, H. Fujita, IEEE Conf. on MEMS-2001, 389, Interlaken, Switzerland (Jan. 2001).

BIO-CHEMICAL ANALYSIS ON MICROFABRICATED POLYMER CHIPS

Koji Shimoide[1], Kazuma Mawatari[1], Shigemi Mukaiyama[2] and Hiroyuki Fukui[1]
Asahi KASEI Corporation
[1]Central Technology Laboratory, 2-1 Samejima, Fuji, Shizuoka 416-8501 Japan
[2]Fabrication Technologies Laboratory, 1-3-1 Yakoh, Kawasaki, Kawasaki, Kanagawa 210-0863 Japan
(shimoide.kb@om.asahi-kasei.co.jp)

Abstract

Polymer μfluidic chip was integrated with thermal lens spectroscopy detection system to analyze bio-chemical substances. Polymer chips were fabricated with both of injection molding and embossing technique. AMOTEC® technology was applied as the injection molding to accomplish fine and accurate pattern transferring. For detections in μcapillaries, the optically induced thermal lens spectroscopy was applied.

Keywords: polymer microfabrication, thermal lens spectroscopy, bio-chemical analysis, clinical diagnosis

1. Introduction

Facing the pressure on reducing the cost in clinical environment, and preparing for the aging society, portable and easy-to-use instruments are craved.

Especially for clinical diagnostic, it is desirable that the devices are disposable, and polymers are more suitable material than conventional Si and glasses. And all sample handling should be carried out in the chip to attain easy-to-use.

Detections in μcapillaries are another aspect to be solved. While a laser-induced fluorescence is a popular on-chip detection option, there still is a limit to be applied only to fluorescent materials.

2. Experiments

Integrated μanalysis system was constructed to detect bio-chemical substances. Figure 1 shows schematic image of the system. Sample was introduced to a polymer chip, and chemical reactions were carried out in the chip without any manual operations followed by thermal lens spectroscopy detection.

Polymer chips

μfluidic devices were fabricated with Polymethylmetacrylate, PMMA, by both of injection molding[1], and embossing technique. AMOTEC®[2][3] technology was applied as the injection molding to accomplish fine and accurate pattern transferring.

Y. Baba et al. (eds.), Micro Total Analysis Systems 2002, Volume 2, 918–921.

AMOTEC® is a new technology developed by AsahiKASEI. This technology uses CO_2 as a plasticizer of resin. CO_2 dissolves in molten resin to reduce the viscosity, and cavities are easily filled with the resin as a result.

Advantages to use hot embossing technique are significant in low costs for embossing tools, a simple replication process, and high structural accuracy.[4] From the point of throughput, however, an injection molding technique has advantages. Figure 2 shows fabricated μfluidic chip.

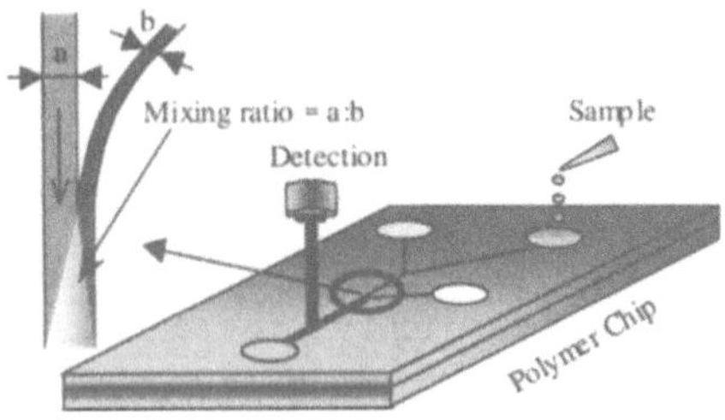

Figure 1. Schematic image of the system

(a) Injection molding chip. (b) Capillary(emboss)
Figure 2. Polymer chip

μFluidics design

Quantitative accuracy is crucial especially for the medical diagnostic system. A parenthesis in Figure 1 shows μfluidics principle adapted here. A sample and a reagent are flowing continuously, and two flows meet each other at a confluence with designed flux ratio. This ratio is garanteed by accurate on-chip pumping mechanisms, and certain channel length before detection point gives appropriate reaction time to the chemicals.

On-chip pumping

For accurate liquid handling on the chip, inexpensive on-chip pumping mechanism was developed. Figure 3 shows schematic image of the developed pump. Reservoirs for a sample and reagents were sealed by porous and hydrophobic membrane, e.g. PTFE membrane. In the condition that the reservoir was filled with liquid (Fig.3(a)), the membrane was pushed by accurately controlled plunger, then the liquid in reservoir was pushed into capillary according to the internal volume change of the reservoir (Fig.3(b)). This on-chip pumping mechanism make the system free from troubles like leaks caused by joints between chips and the outer pumps.

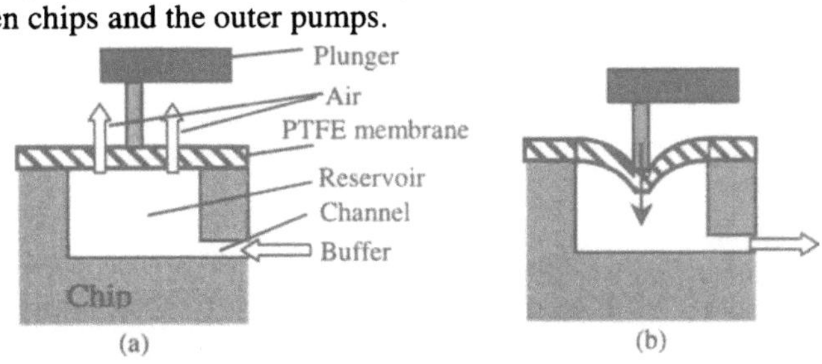

Figure 3. On-chip pumping mechanism

Thermal Lens Spectroscopy detection

Aiming at integration of analysis system on a chip for manifold applications, thermal lens spectroscopy detection technique [5] was applied. Thermal lens spectroscopy under a microscope, thermal lens microscopy, could be applied to even sub-micro sized capillaries [6], and no other detection methods can achieve the same. Figure 4 shows principle of the thermal lens microscopy detection. The detail of the mechanism is written somewhere in the reference.

Though this method was shown to have capabilities to detect almost one molecule in a channel [6], the optical parameters of the system here was fitted to obtain high sensitivity to the concentrations of substances in a channel to adapt the system to bio-chemical measurement.

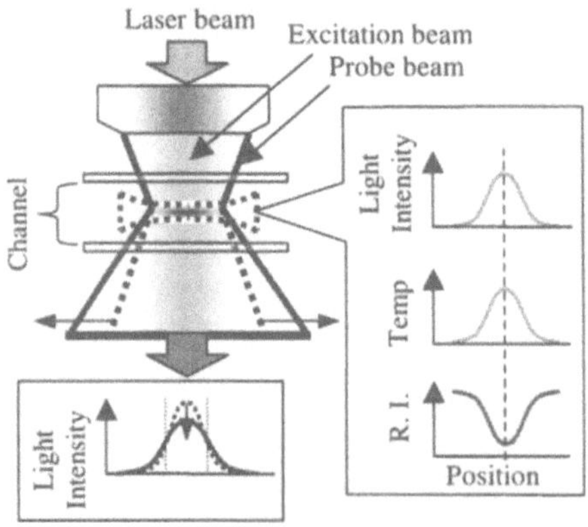

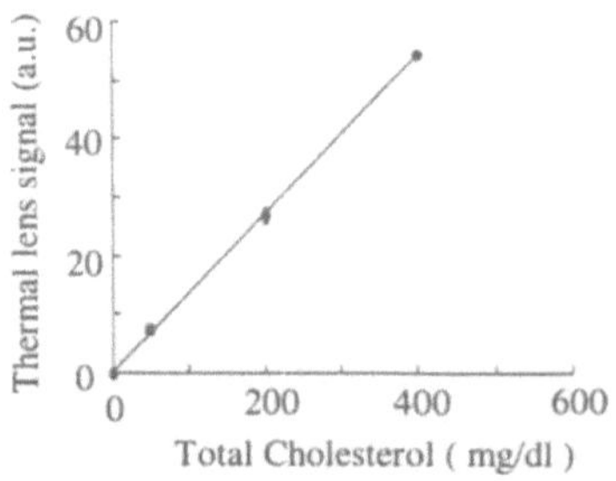

Figure 4. Principle of thermal lens spectroscopy   Figure 5. Results of detecting total cholesterol

3. Results

Concentrations of cholesterol in serum were measured by the system. Layout of channels in the chip and protocol of reagents to detect cholesterol were designed to fit the system. Diffusion and chemical reaction time between sample and reagents should be considered.

Figure 5 shows the results of detecting 400mg/dl, 200mg/dl, and 50mg/dl cholesterol in serum.

4. Conclusion

A polymer chip with chemical operation units, and the thermal lens detection units were integrated, and successfully detected bio-chemical substances.

Acknowledgements

The authors wish to thank Prof. T. Kitamori (University of Tokyo) for his useful discussions and advice on thermal lens microscopy.

References

1. T. D. Boone, H. H. Hooper, Proc. of μTAS'98 Workshop, Canada, Kluwer Academic Publishers, Dordrecht, The Netherlands, pp.257-260 (1998).
2. H. Yamaki, Plastics age(Japanese) **47**(6) pp.136-139 (2001).
3. Modern Plastics, pp62, March(2001).
4. H. Becker, W. Dietz, P. Dannberg, Proc. of μTAS'98 Workshop, Canada, Kluwer Academic Publishers, Dordrecht, The Netherlands, pp253-256, (1998).
5. M. Harada, K. Iwamoto, T. Kitamori, T. Sawada, Anal. Chem. **65** pp. 2938-2940 (1993).
6. T. Kitamori, M. Fujinami, T. Odake, M. Tokeshi, T. Sawada, Proc. of μTAS'98 Workshop, Canada, Kluwer Academic Publishers, Dordrecht, The Netherlands, pp.295-298 (1998).

BIOCHEMICAL SIGNAL TRANSFER USING LIPOSOMES IN MICROFLUIDIC CHANNELS

J. A. Gimm, A. E. Ruoho* and D. J. Beebe
Department of Biomedical Engineering and *Department of Pharmacology,
University of Wisconsin-Madison, Madison, WI. 53706, USA
(jagimm@facstaff.wisc.edu)

Abstract

We demonstrate signal transfer using liposomes loaded with a "secondary messenger" and chemically sensitive hydrogels. The device consists of emulsion filter, chemically responsive hydrogel posts and filter-captured liposome complexes which contain the secondary chemical messenger. In response to external stimuli (in this case, a detergent or protein), the liposome complexes unload the encapsulated chemical messenger. This release in turn triggers dissolution of hydrogel structures downstream, thereby transferring the original signal.

Keywords: signal transfer, liposome, responsive hydrogel

1. Introduction

One of our goals is to relay and amplify an initial biochemical signal downstream for visual detection. One such of secondary messenger signal transfer can be achieved through utilization of lipid membranes.

Lipid membranes provide a natural environment for immobilization of bioactive molecules such as enzymes, antibodies, or antigens [1]. In addition, vesicle forms of lipid membrane or liposomes, allow encapsulation of molecules inside for drug delivery or other types of load delivery [2]. Utilization of liposomes as both sensor and delivery vehicle adds new potential to specificity and sensitivity of future microfluidic biological detection systems by mimicking nature. For example, there are numerous transmembrane proteins which function both as sensors and as signal transducers across the membrane into the cytosol. Similarly, functionalized liposome surfaces (where variety of proteins or molecules could be embedded) can be activated by specific biochemical reactions triggered by external stimuli. This causes lysis of vesicles and release of encapsulated "drugs" downstream.

Here we demonstrate encapsulation of a specific chemical in liposomes which upon release causes a visible secondary reaction downstream. The encapsulated chemical, once released, reduces disulfide (S-S) bonds. The reduction of disulfide crosslinkers in hydrogel posts downstream causes their dissolution which can be visually detected.

Y. Baba et al. (eds.), Micro Total Analysis Systems 2002, Volume 2, 922–924.

2. Experimental

The layout of the glass device used is shown in Figure 1. It is composed of an emulsion filter [3] and polyacrylamide posts, some of which contain dissolvable disulfide crosslinkers. The dissolvable polyacrylamide posts were polymerized with a cleavable crosslinker N,N'- cystaminebisacrylamide.

Liposomes were prepared following a standard protocol for small unilamellar vesicles [4] using phospholipid mixtures with traces of fluorescent and biotinylated lipids for functionalization and for visualization, respectively. Liposomes were prepared in presence of TCEP-HCl (tris-(2-carboxyethyl) phosphine hydrochloride) to encapsulate the reducing agent. Purified TCEP-HCl-encapsulated liposomes were incubated with avidin.

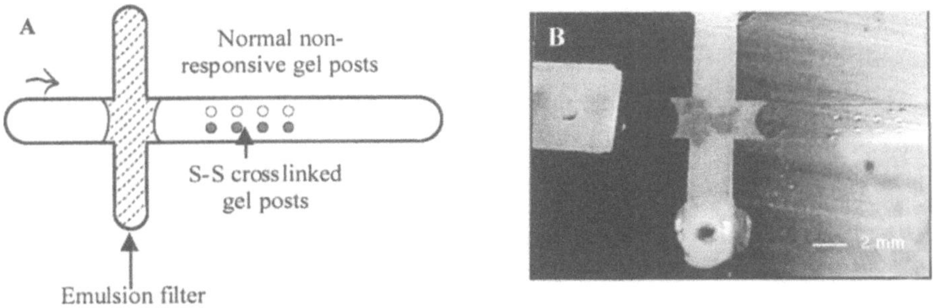

Figure 1. Schematic (A) and actual image (B) of the microfluidic device. The shaded portion denotes the location of the polymerized emulsion filter where liposome complexes are captured. Clear circles downstream indicate normal polyacrylamide gel posts, while gray circles denote disulfide crosslinked posts which dissolve in presence of a reducing agent. The direction of flow is left to right. The diameter of the posts is 350 μm and the width of the horizontal channel is 2.5 mm.

3. Results and discussion

The liposome quality (i.e. encapsulation efficiency, leakage) was determined by measuring the reducing potential of the liposome solutions before and after lysis (results not shown). Liposome aggregation due to avidin/biotin interaction is visible in the emulsion filter (Figure 2a). Avidin, by binding biotin on liposome surface, caused aggregation of liposomes, thereby effectively increasing the liposome size. Avidin binding to liposome had two significant effects. First, it demonstrates functionalization of liposome surface where biotinylated lipids were used to bind avidin. Furthermore, larger liposome size allowed more efficient capture of liposome complexes by the emulsion filter.

Addition of a detergent (10% triton-X) or the bee venom peptide melittin (1-10 mM) solubilizes liposome complexes, thereby unloading the encapsulated TCEP-HCl downstream (Figure 2). Complete dissolution of S-S crosslinked posts takes place in minutes, leaving only the normal polyacrylamide posts in the channel (Figure 3).

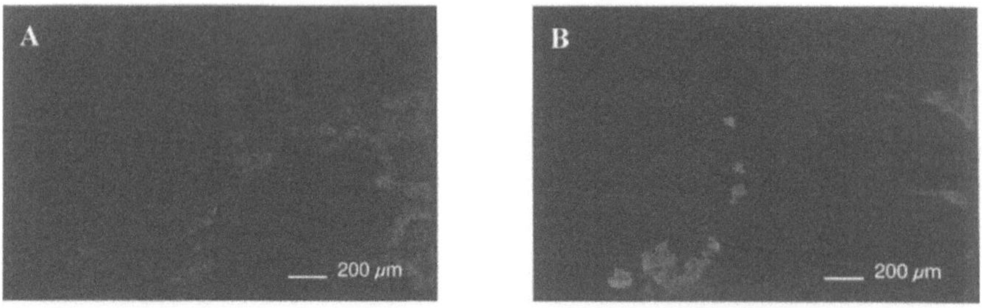

Figure 2. Fluorescence images of emulsion filter with TCEP-HCl encapsulated liposomes before and after lysis. Large clumps of vesicles are visible within the filter and on the upstream surface (A). With the addition of a detergent (10% triton-X) or bee venom peptide melittin (1-10 mM), vesicles are solubilized (B) and unload the encapsulated TCEP-HCl. This relays the signal downstream to disulfide-crosslinked gel posts. Direction of the flow is left to right.

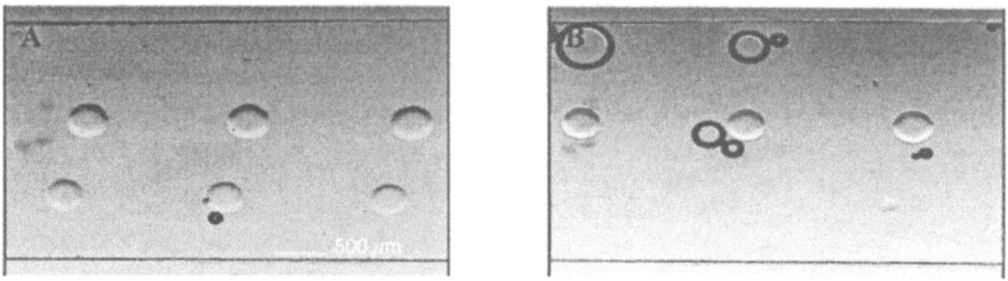

Figure 3. Effect of liposome released TCEP-HCl on normal and disulfide-crosslinked polyacrylamide posts in a microfluidic channel. Normal and disulfide-crosslinked posts are located on the top and bottom rows, respectively (A). After lysis of filter-sequestered liposome complexes upstream, disulfide-crosslinked posts dissolve away in 14 minutes as seen in (B), leaving only the normal polyacrylamide posts. The dissolution of these posts is visually detectable without instrumentation.

4. Conclusions

We have demonstrated first steps in realizing an integrated biological microfluidic system where upstream biochemical stimuli is relayed and amplified for visual detection downstream.

Acknowledgements

This research was supported by a grant from DARPA (#F30602-00-2-0570).

References

1. Sackmann E. and M. Tanaka. *Trends in Biotechnology* 2000, **18**: 58-64.
2. Lasic, D. *Trends in Biotechnology* 1998, **16**: 307-321.
3. Moorthy, J. and D. J. Beebe. *2nd Annual International IEEE-EMBS Special Topic Conference on Microtechnologies in Medicine and Biology*, Madison 2002: 514-517.
4. New, R. R. C., Editor. *Liposomes : a practical approach.* Oxford University Press, Oxford; New York; 1990.

A Novel Biomolecule Sorter Using Thermosensitve Hydrogel in Micro Flow System

Yoshitaka Shirasaki[1], Hiroshi Makazu[1], Koichi Tashiro[2], Shingo Ikeda[2], Tetsushi Sekiguchi[2], Shuichi Shoji[2], Shoichiro Tsukita[3], and Takashi Funatsu[1]

[1]Department of Physics, and [2]Department of Electronics, Information and Communication Engineering, School of Science and Technology, Waseda University, 3-4-1 Ohkubo, Shinjuku, 169-8555 Tokyo Japan

[3]Department of Cell Biology, Faculty of Medicine, Kyoto University, Sakyo-ku, Kyoto 606-8501, Japan. (E-mail: funatsu@waseda.jp)

Abstract

A novel biomolecule sorter using thermoreversible hydrogel was developed. Thermosensitive block copolymer changes its state from sol to gel when it is heated. The solution containing fluorescent biomolecules and thermosensitive polymer was introduced to a Y-shaped microchannel. The sol-gel transition was locally induced by site directed IR (infrared) laser irradiation in order to plug the microchannel and sort out the biomolecules. The sorting time of 120 ms was achieved. As an application of this sorter, fluorescently labeled λ-phage DNA molecules were separated and collected.

Keywords: Biomolecule sorting, Micro flow system, Thermal gelation, Infrared Laser, Automatic sorting system

1. Introduction

There has been no machine that can sort out biomolecules or supermolecules such as organelles, while their purification is indispensable to study their functions. In order to realize the biomolecule sorter, we proposed a micro flow switch using sol-gel transition of methylcellulose [1]. The sorting time of 2 s was achieved by switching the flow by manually deflecting IR laser. To achieve faster sorting, thermosensitive block copolymer composed of poly (*N*-isopropylacrylamid) and poly (oxyethylene) was used instead of methylcellulose [2]. A sensitive fluorescence detector was installed to obtain fluorescence signals from individual biomolecules, and to control scanner mirrors that switch the illumination area of IR laser. These improvements realized the sorting of individual λ-phage DNA molecules with the sorting time of 120 ms.

2. Materials and methods

λDNA (48502 bp) molecules were labeled with either Cy3 or Cy5 using Label IT kit (Mirus) with dye to DNA molar ratio of 900 : 1. The labelled λDNA molecules were mixed with a carrier solution containing thermosensitive block copolymer which was composed of poly (*N*-isopropylacrylamid) and poly (oxyethylene). They were introduced into a Y-shaped micro channel as shown in Fig.1a. λDNA molecules labeled

Y. Baba et al. (eds.), Micro Total Analysis Systems 2002, Volume 2, 925–927.

with Cy3 were detected with excitation wavelength of 553nm and emission wavelength of 570nm, respectively.

3. Principle of the sorting system

Fig.1 shows the principle of our sorting system, and Fig.2 shows schematics of the apparatus. By irradiating the focused IR laser at the right branch near the junction of Y-shaped micro channel, the local temperature exceeds 37 °C. Gel is formed to block the flow, and wastes flow into the left branch (Fig.1a). Two scanner mirrors are used for switching and scanning (Fig.2). Scanner mirror 1 directs the irradiation of IR laser whereas scanner mirror 2 scans the laser as it crosses the channel and blocks the flow completely (see Fig.3a). The fluorescence signals are detected at 60 μm upstream from the junction. The fluorescence signals are transformed to electric signals using a photomultiplier tube (PMT). When the labelled DNA flows into detection area, the electric signal level exceeds the threshold voltage and the focused IR laser is switched to the left channel (Fig.1b, Fig.4). A gelled layer is formed at the left channel, and the target DNA is sorted into the right branch (Fig.1c).

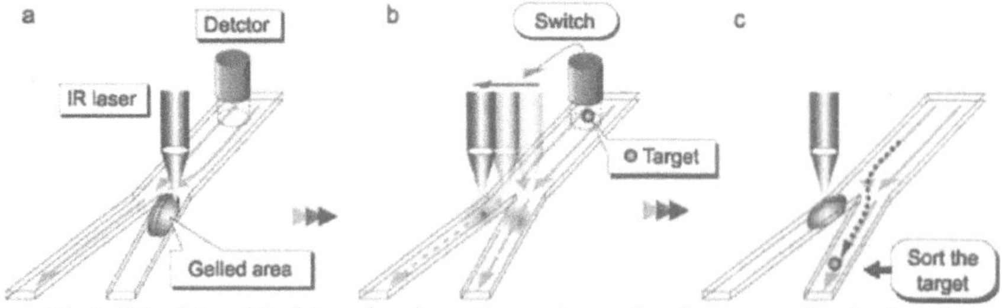

Fig.1 Principle of the biomolecule sorting system using thermosensitve hydrogels

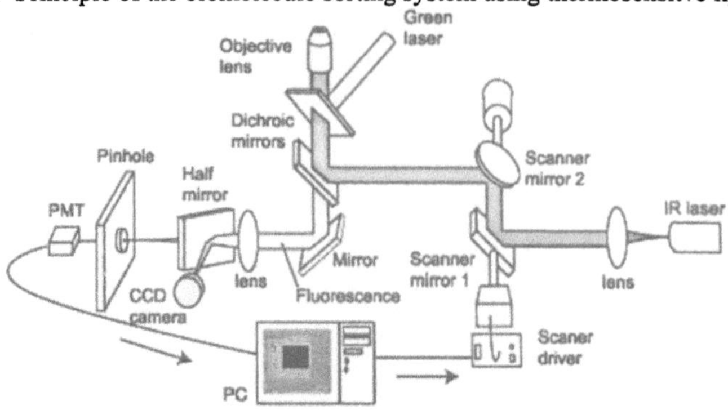

Fig.2 Schematics of the apparatus.

4. Results and discussion

λDNA molecules labeled with Cy3 were visualized in a micro flow system (Fig.3b). Fig.4 shows the time course of fluorescence signals from PMT. The signals from Cy3-labeled λDNA were distinguishable from a background noise. When the signal exceeded the threshold (Fig.4), an illumination area of infrared laser was switched to the other branch of Y-shape channel to sort λDNA molecules. These procedures continued over 6 h and Cy3-λDNA molecules were collected.

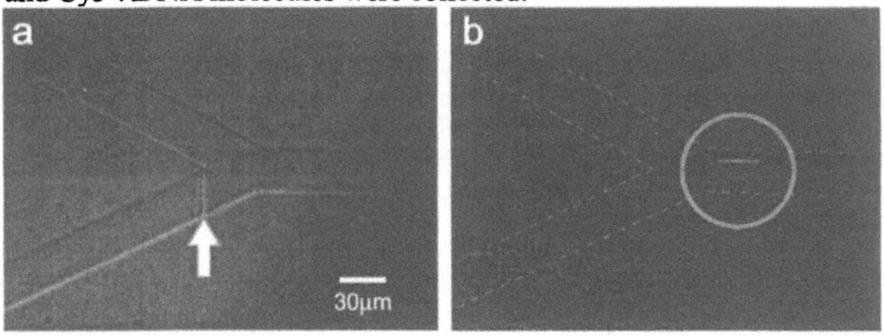

Fig.3 Photographs of microchannel. a, a phase contrast micrograph of microchannel. The arrow indicates the area where gels were formed to block the flow. b, fluorescence micrograph of a λDNA molecule passing through the detection area shown by circle.

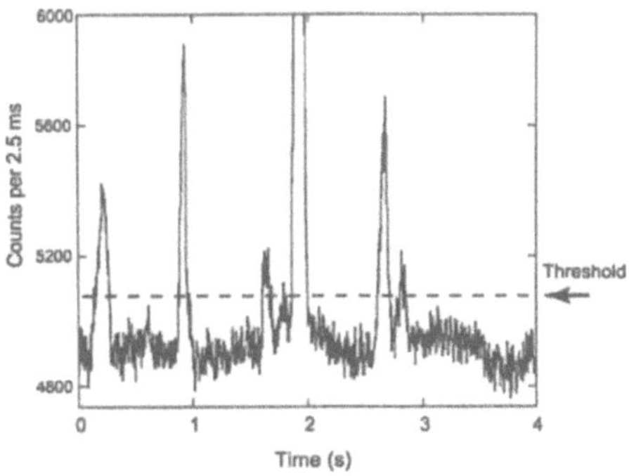

Fig.4 Time course of the fluorescence intensity from the detection area

5. Conclusions

Fluorescent λDNA molecules could be sorted with sorting time of 120 ms using thermoreversible hydrogel. This method is applicable to sorting various kinds of biological specimen.

Acknowledgements

This research was partly supported by Grants-in-Aid for COE Research of Waseda University, Scientific Research Priority Area (B) No.13124209, Scientific research (B) No.12450167, and 13558088 from the Ministry of Education, Culture, Sports, Science and Technology of Japan

References

1. K. Tashiro, S. Ikeda, T. Sekiguchi, S. Shoji, H. Makazu, T. Funatsu, and S. Tsukita. "A Particles and Biomolecules Sorting Micro Flow System Using Thermal Gelati0on of Methyl Cellulose Solution" Micro Total Analysis System 2001, pp471-473, (2001).
2. Hiroshi Yoshioka, "Molecular Design and Phase Transition Mechanism on Thermogelling Thermoreversible Hydrogel" Doctoral Thesis, Waseda University (1995)

FIXATION AND ISOLATION OF MICROORGANISMS BY LOCAL VISCOSITY CONTROL OF METHYL CELLULOSE SOLUTION

Fumihito Arai[1,2], Akihiko Ichikawa[1], Toshio Fukuda[1] and Tohoru Katsuragi[3]
[1]Nagoya University, 1 Furo-cho Chikusa-ku Nagoya-City 464-8603 Japan
[2]PRESTO, JST
[3]Nara Institute of Science and Technology, 8916-5 Takayama-cho Ikoma-city Nara 630-0101 Japan

·Abstract

We developed a novel screening system for random separation of target microorganisms from the randomly suspended samples in the microchamber by using a local viscosity control of methyl cellulose. The methyl cellulose is mixed with the sample liquid and we controlled the state from sol to gel and gel to sol reversibly based on the temperature change controlled by heating of the microelectrode with the electric current and laser irradiation.

Keywords: Microbe, Methyl cellulose, Viscosity control, Thermal gelation

1.Introduction

Recently, there has been great interest in the high throughput screening of microorganisms, for example, for finding of the novel microbes. It is estimated less than 10% within the whole in which the microorganisms known at present. For remarkable advancement of biology and bioengineering, it is important to develop a new technology, with which we can manipulate and separate randomly suspended microorganisms with high speed and high purity. The purpose of our research work is to separate target microorganisms randomly suspended in the liquid. By the effect of turbulence and adhesion, it's too difficult to pick up a microbe from solution by the mechanical manipulation. So we have studied on the selective separation of the microbe suspended in the liquid. In the previous works [1] [2], we integrated the laser manipulator for local position control of the target and the dielectrophoresis for exclusion of the other objects around the target to isolate the target. Then, we succeeded in the high-speed separation and transportation by the hydrodynamic force in the microchannel. However, the method based of the laser trapping of the target has a problem of direct irradiation of high power laser. Here we propose a novel separation method based on the fixation or immobilization of the target using a local control of the viscosity around it.

2.Principle of system

Here we used the thermal sol-gel transformation of methyl cellulose solution. Figure 1 shows the schematic diagram of the separation method. There are transparent microelectrodes of ITO (indium tin oxide) for heating on the glass surface of the microchamber. The microorganisms are guided there in the mixed flow of the methyl cellulose. The microelectrodes can be heated widely by the applied voltage and locally by the irradiation of the Nd:YVO_4 laser(maximum output 4.8 W, wave length 1064 nm, TEM_{00}, M^2<1.1). The methyl cellulose 6 wt% solution gels from 40°C. It completely gels over 48°C while it becomes opaque and the viscosity rapidly increases. On the other hand, the viscosity gradually decreased, when it is cooled down after gelation. At first, the ITO electrode is preheated beneath the gelation level by the applied voltage (1.1 V), then the operator selects the target through the microscope and fix it by heating

Y. Baba et al. (eds.), Micro Total Analysis Systems 2002, Volume 2, 928–930.

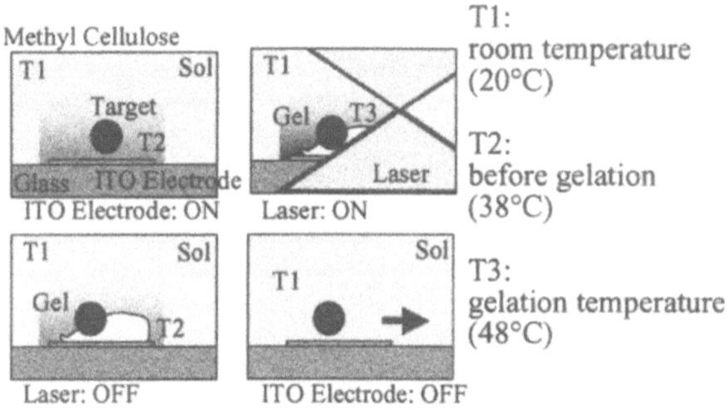

Figure 1. Process of the targets retaining

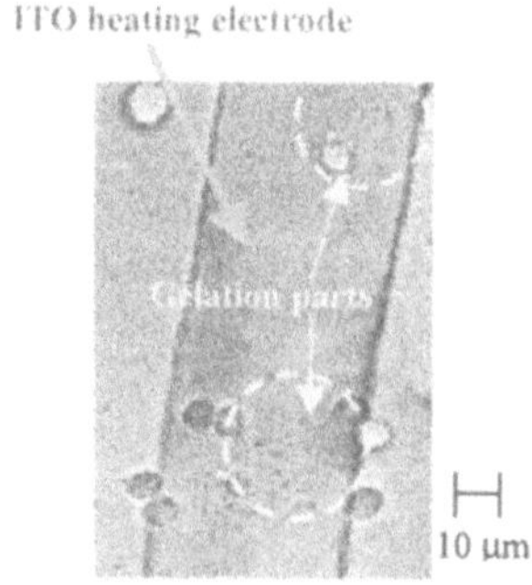

Figure 2. Multiple gelation

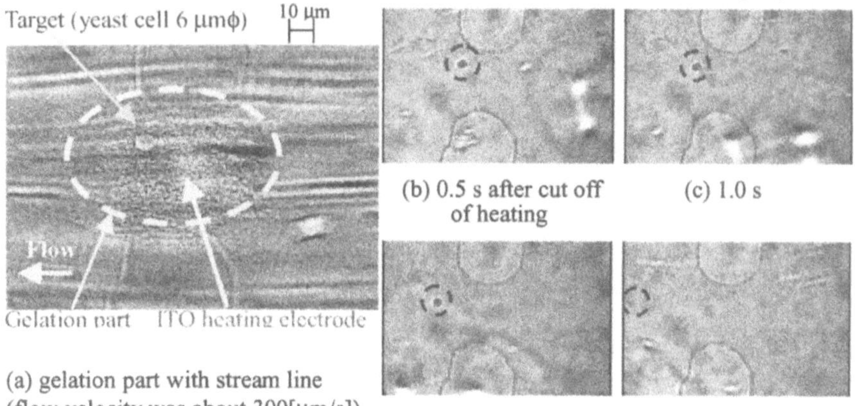

(a) gelation part with stream line (flow velocity was about 300[μm/s])

(b) 0.5 s after cut off of heating

(c) 1.0 s

(d) 1.5 s

(e) 2.0 s

Figure 3. Photo of immobilization and release of target microbe

the electrode around it by the laser. The laser must be cut off immediately since the gelation is fast. By repeating this process, we can retain the multiple target objects on the electrode since the geration process has hysteresis and the geration state is kept as long as the electrode is heated by the electric current (Figure 2). By cutting off the voltage to the microelectrode, the targets are released from the glass basal plane, since the gel changes to sol immediately (Figure 3).

3.Experimental

Figure 4 shows the schematic diagram of separation system and the experiment of isolation and extraction of target (yeast, 5 μmϕ). Retaining, cleaning and extraction were possible in about 1 min. The minimum resolution of local gelation by the laser is about 10 μmϕ. The separation as well as cltivation will be performed in an integrated chip. The force of methyl cellulose gel adhesion to the basal plane is stronger than the liquid cross-link force, so it is also possible to take out the object in the air by the removal of water (Figure 5). We can easily access to the selected target by the mechanical manipulator too.

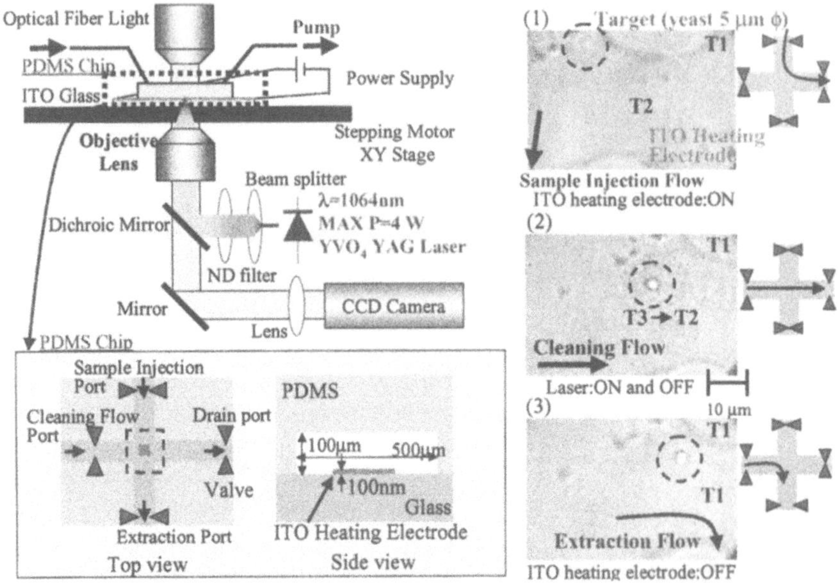

Figure 4 Screening system and separation experiment result

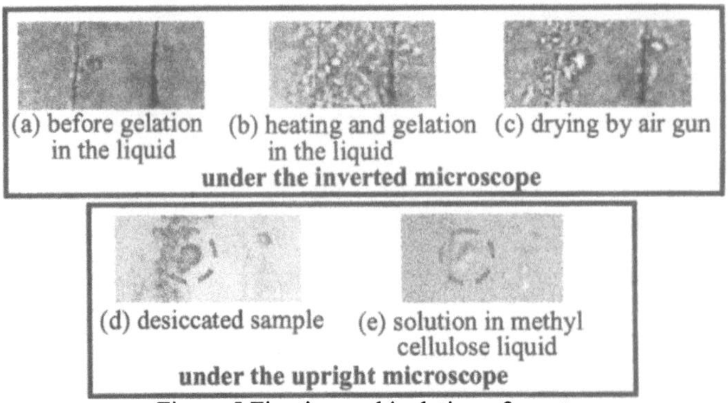

Figure 5 Fixation and isolation of yeast

4.Conclusion

An arbitrary selected single yeast was extracted even though there are a large number of yeasts in solution. We confirmed the effectiveness of the proposed method based on the strategy of controlling viscosity of methyl cellulose. Our method is extremely useful in the pure cultivation of the microbe and will be a promising method for biologists in screening useful microbes

References

[1] F. Arai et al., Electrophoresis 2001, 22 No.2 , p. 283-288
[2] F. Arai et al., μ-TAS2001, p. 474-476

CHEMICO-FUNCTIONAL MEMBRANE FOR INTEGRATED CHEMICAL PROCESSES ON A MICROCHIP

Hideaki Hisamoto[1], Yuki Shimizu[1], Kenji Uchiyama[2], Manabu Tokeshi[2], Yoshikuni Kikutani[2], Akihide Hibara[1], and Takehiko Kitamori[1, 2]
[1] Department of Applied Chemistry, Graduate School of Engineering, The University of Tokyo, 7-3-1 Hongo, Bunkyo-ku, Tokyo 113-8656, Japan
[2] Integrated Chemistry Project, Kanagawa Academy of Science and Technology, 3-2-1 Sakado, Takatsu-ku, Kawasaki, Kanagawa 213-0012, Japan
(e-mail: hisamoto@appchem.t.u-tokyo.ac.jp)

Abstract

A chemically-functional polymer membrane was synthesized inside a microchannel by an interfacial polycondensation reaction between the monomer multilayer flows and membrane surface modification. The polymer membrane had an important role in ensuring stable contact of different phases such as gas/liquid or liquid/liquid, and the permeation of chemical species through the membrane. Furthermore, membrane surface modification chemistry allowed chemical transformation of permeated chemical species.

By using the membrane, dissolved ammonia gas permeation through inner channel membrane was investigated and separation of neutral and ionic species was demonstrated. Furthermore, horseradish peroxidase was immobilized on one side of the membrane surface, and substrate permeation through the membrane and subsequent chemical transformation at the membrane surface were realized.

Keywords: enzymatic reaction, gas permeation, interfacial polycondensation, nylon membrane, surface modification

1. Introduction

We have been focusing on the integration of various kinds of micro unit operations (MUOs), and we successfully developed continuous flow chemical processing (CFCP) [1]. During its development, we observed that the formation of a stable organic/aqueous interface along the longitudinal direction of the microchannel played an important role for integration of various kinds of MUOs. By using aqueous / organic multi layer flow, we successfully demonstrated highly efficient molecular transport between two phases for analytical and synthetic applications, which exploited the characteristics of a liquid microspace such as short diffusion distance and large specific interfacial area [1-2].

Previously, we briefly reported on the preparation of a polyamide membrane at the organic / aqueous interface along the longitudinal direction which was accomplished by a well-known interfacial polymerization reaction and linker group modification of the inner microchannel surface [3].

In our opinion, active addition of chemical functions to the inner channel membrane is one promising way to open up new directions for microchip technologies in chemistry. In this paper, we present the membrane preparation, and integration of chemical

Y. Baba et al. (eds.), Micro Total Analysis Systems 2002, Volume 2, 931–933.

functions of the membrane such as dissolved gas-permeation and chemical transformation of permeated species by membrane surface modification.

2. Experimental

Microchannel was fabricated on a pyrex glass plate (30 mm x 70 mm) by the photolithographic wet etching method (240 μm width and 115 μm depth, or 210 μm width and 40 μm depth). Inner surface of the microchannel was chemically modified by 3-aminopropyl triethoxysilane as a linker layer. For interfacial polycondensation, adipoyl chloride in 1,2-dichloroethane (0.01M) and hexamethylene diamine in 0.1M NaOH solution (0.01 M) were used as organic and aqueous solutions, and membranes were prepared under multilayer flow. Chemical modification of membrane surface by an horseradish peroxidase was performed by conventional glutaraldehyde method.

3. Results and discussion

Figures.1 (a) and (b) show the channel patterns, and top and cross-sectional views of the nylon membranes prepared inside the microchannels. After introducing organic and aqueous phases containing membrane monomers, membrane-type 3D-structures were successfully formed at the liquid / liquid interface position.

In order to confirm the molecular transport function of the membrane, we carried out the simple dissolved gas permeation experiment (Figure.2(a)) using ammonium chloride and sodium hydroxide solutions. Figure 2(b) shows the time course of the thermal lens microscope (TLM) signal generated at the phenolphthalein (PP) phase. In order to confirm leaking of sodium hydroxide solution to PP phase through the membrane, introduction of sodium hydroxide solution without ammonium chloride was also investigated. As can be seen, the TLM signal at the PP phase did not change when sodium hydroxide solution was introduced without ammonium chloride, on the other hand, the signal rapidly increased when ammonia gas was generated at the other side of the membrane. These results clearly suggested that the nylon membrane prepared inside the microchannel successfully worked as a gas permeation membrane.

Substrate permeation and subsequent enzyme reaction at the membrane surface was also investigated. Figure 3 (a) illustrates the experiment. Horseradish peroxidase was immobilized on one side of the membrane surface, and hydrogen peroxide was supplied from the other side. The enzyme-immobilized side of the channel was filled with TOOS and 4-AAP of dye forming substrates. When the flow stopped, the TLM signal increased rapidly and reached a

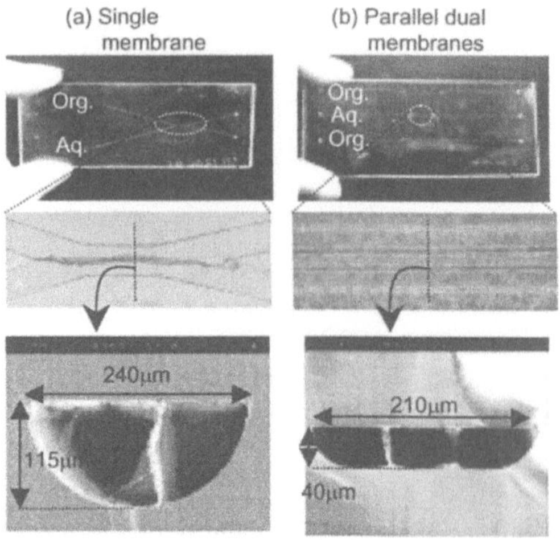

Figure.1 Channel pattern, top and cross sectional views of the nylon membranes.

plateau at approximately 250 seconds (Figure 2 (b)). The result indicated the hydrogen peroxide permeation and subsequent enzyme reaction at the membrane surface were successfully completed. These results suggested that surface modification chemistry was successfully implemented, and complicated and sophisticated chemical processes involving substrate permeation and subsequent chemical reaction could be realized.

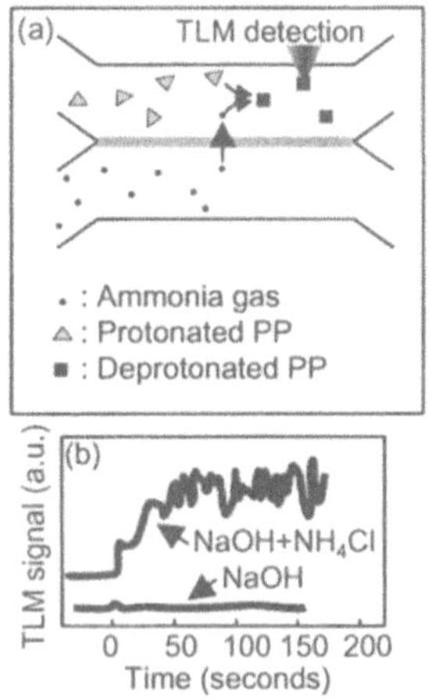

Figure 2 Gas permeation experiment.
(a) Experimental system.
(b) Time course of the TLM signal.

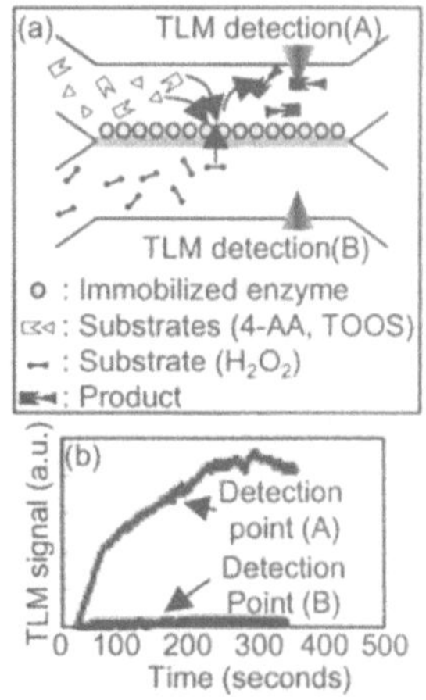

Figure 3 Substrate permeation and enzyme reaction experiment.
(a) Experiment system.
(b) Time course of the TLM

4. Conclusions

We have demonstrated the preparation of nylon membrane structures by an interfacial polycondensation reaction and multilayer flow inside a microchannel. The prepared membrane served as a gas permeation membrane inside the microchannel.

Surface modification by an enzyme allowed active addition of a molecular transform function to the inner channel membrane. By using the enzyme-modified membrane, substrate permeation and subsequent molecular transformation were realized.

We expect the chemico-functional membrane proposed here will lead to a novel 3D-structure construction methodology in future microchip technology.

Acknowledgements

This work was partially supported by Grants for Scientific Research from the Ministry of Education, Science, and Culture, Japan.

References

1. M. Tokeshi, T. Minagawa, K. Uchiyama, A. Hibara, K. Sato, H. Hisamoto, T. Kitamori, *Continuous Flow Chemical Processing on a Microchip by Combining Micro Unit Operations and a Multiphase Flow Network*, Anal. Chem., **74**, 1565-1571, (2002).
2. H. Hisamoto, T. Saito, M. Tokeshi, A. Hibara, T. Kitamori, *Fast and High Conversion Phase-Transfer Synthesis Exploiting Liquid/Liquid Interface Formed in Microchannel Chip*, Chem. Commun., 2662-2663. (2001).
3. Y. Shimizu, H. Hisamoto, A. Hibara, M. Tokeshi, T. Kitamori, *Inner-Channel Preparation of Polymer Membrane by Interfacial Polycondensation*, Digest of Papers, 2001 International Microprocesses and Nanotechnology Conference, Japan Society of Applied Physics: Tokyo, Japan, 2001; pp20-21.

IN SITU FABRICATED MICROCHANNELS USING POROUS POLYMER AND XENON DIFLUORIDE ETCHANT

George C. Lopez[1], Alan J. Rosenbloom[2], Victor W. Weedn[2] and Kaigham J. Gabriel[1,3]
[1]The Robotics Institute, [2]Science and Technology Center, [3]Electrical and Computering Engineering, Carnegie Mellon University
Pittsburgh, Pennsylvania, 15213, U.S.A Email:glopez@cs.cmu.edu

Abstract

A novel microfluidic fabrication technique which allows in situ formed silicon microchannels and does not necessitate substrate bonding is presented. A gas-permeable polymer membrane acts as the top surface of the microchannel, allowing xenon difluoride (XeF_2) etchant to pass unobstructed to the silicon substrate. Exposed silicon areas on the wafer are then etched isotropically.

Keywords: Microchannels, Xenon Difluoride, Porous Polymer

1. Introduction

A microchannel is a fundamental component of most microfluidic systems, which is typically used for reactant delivery, biochemical reactions, particle separations as well as for heat exchanging. Application areas where microchannels have been used include miniaturized systems of electrophoresis, gas and liquid chromatography and chemical analysis systems [1]. Standard microchannel fabrication techniques are usually divided into two categories: 1) Discrete fabrication such as substrate bulk etching followed by various bonding techniques in silicon, glass and plastic, and 2) Integrated fabrication which involves using thin films to create structural layers of a microchannel; these techniques include substrate bulk etching with subsequent sealing by film deposition (such as LPCVD), as well as surface micromachined channels created by the removal of a sacrificial material.

A major limitation of most fabrication methods is the final sealing step, microchannels are usually open-roofed after being created. Substrate bonding usually requires surfaces to be relatively clean and flat in order to produce intimate contact between substrates. Other problems encountered are complication in packaging, process compatibility and fabrication techniques that can only provide a certain microchannel geometry, shape or arraying capability.

2. Fabrication

The fabrication process begins with a lithography step that defines the location where the microchannel will be created using XeF_2 (see Figure 1). Silicon dioxide is used as the masking material, defining the exposed silicon regions and protecting the areas that

Y. Baba et al. (eds.), Micro Total Analysis Systems 2002, Volume 2, 934–936.

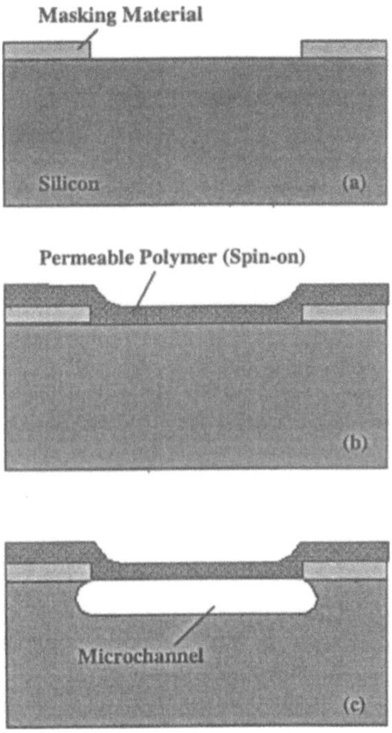

Figure 1. The fabrication process starts with (a) a patterned masking material, (b) deposition of a gas-permeable polymer over the substrate by spin-casting, (c) XeF_2 etch through the polymer to create microchannels.

will not be etched. Other masking materials, such as metal or photoresist, can be used instead of silicon dioxide. A gas-permeable polymer is then deposited over the entire substrate. This polymer can be deposited by various methods such as spin-casting, plasma-deposition or vapor-deposition. Spin-casting was used due to its simplicity, speed and cost. Finally, XeF_2, a small molecule that passes through the permeable polymer, is used to etch exposed silicon regions. Chemical species that are by-products of the XeF_2 and silicon reaction must be able to pass through the membrane as well. The major by-product during the etching process is silicon tetrafluoride.

Xenon difluoride is a gas-phase, room-temperature, plasmaless, isoptropic silicon etchant that can be used for bulk or thin film etching [2]. Since XeF_2 does not require external energy or ion bombardment for silicon etching, it easily penetrates the polymer.

3. Experimental

Poly(dimethylsiloxane) (PDMS), a readily available polymer commonly used in microfluidic applications, was used during the experimentation process. Polysiloxanes have a much higher permeability to gases than most polymers, yet due to their hydrophobic nature, fluids are excluded. The PDMS prepolymer was spun at speeds up to 6000 RPM to create a 5-10 micron thick film. To create thinner films, the mixing of thinning agents can be used to lower the viscosity of the polymer. After curing the polymer film, the sample was then placed into a XeF_2 etching system where it was pulsed at 60 second intervals using 3 Torr of XeF_2 for a total of 30 minutes (see Figure 2).

4. Results and Discussion

At various locations across the sample, the PDMS demonstrated some neglible reaction to the XeF_2 etch. Since small concentrations of hydrofluoric acid are produced

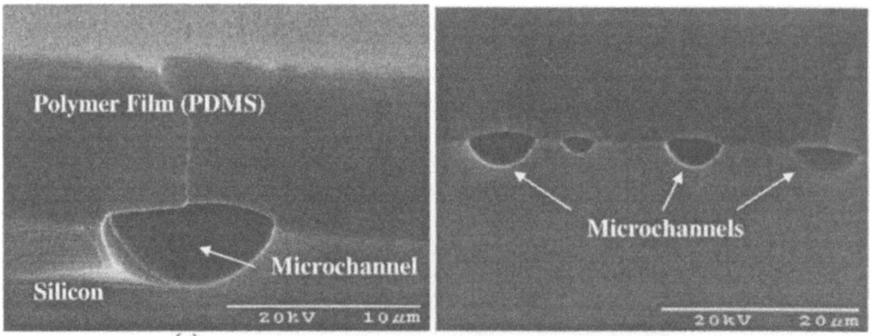

(a) (b)

Figure 2. Micrographs showing: (a) cross-section of an 8 μm wide microchannel with a 10 μm PDMS film above, (b) cross-section with multiple microchannels.

in the etching chamber, due to the combination of fluorine radicals and moisture, this could explain the phenomena observed. Polymer permeability values for xenon difluoride are being researched to determine the optimal polymer. Polymers deposited on the silicon substrate must be sufficiently thin to allow easy transmembrane flux of reactants and products. This technology is particularly suited for developing separation membranes over microchannels. By depositing a liquid-permeable polymer over the substrate, it is possible to create suspended membranes that can be used for analyte concentration or isolation, such as with microdialysis applications.

5. Conclusion

A new microchannel fabrication technique is presented. The favorable integration of silicon's micromachining properties along with the flexibility of polymers, allows a monolithic microfluidic device.

Acknowledgements

The authors would like to thank the W.M. Keck Foundation for providing support. As well as *XACTIX* for providing the Xetch Xenon Difluoride System to Carnegie Mellon University.

References

1. N. Chiem, C. Colyer, D.J. Harrison, *Microfluidic systems for clinical diagnostics*, Proc. Int. Conf. On Solid-State Sensors and Actuators (Transducers '97), pp. 183-186, (1997).
2. P.B. Chu, J.T. Chen, R. Yeh, G. Lin, J.C.P. Huang, B.A. Warneke, S.J. Pister, *Controlled Pulse-Etching with Xenon Difluoride*, Proc. Int. Conf. On Solid-State Sensors and Actuators (Transducers '97), Chicago, pp. 665-668, (1997).

DIRECT NANOMANIPULATION TOOLS FOR BIOLOGICAL SAMPLES

Shoji Maruo, Koji Ikuta and Hayato Korogi
Nagoya University, Furo-cho, Chikusa-ku, Nagoya 464-8603, Japan
(e-mail: maruo@mech.nagoya-u.ac.jp, ikuta@mech.nagoya-u.ac.jp)

Abstract

Optically driven micromanipulators have been developed by using an assembly-free process based on two-photon microstereolithography. Since these manipulators are driven and controlled by optical trapping, they are suitable for biotechnologies required aqueous solution work. The femtonewton order force-control of the manipulator arm was also demonstrated by adjusting the focal position of the trapped laser beam. In addition, nanotweezers with probe tips of diameter 250 nm were developed and driven in a liquid. Such remote-controlled micro/nano manipulation tools provide a unique and effective handling method of biological samples such as living cell, protein and DNA.

Keywords: Nanomanipulators, Optically driven micromachine, Two-photon microstereolithography

1. Introduction

Manipulation techniques at micro and nanometer scale are crucial for progress in biotechnology as well as nanoscale science and technology. Recently several types of nanomanipulators such as carbon nanotube nanotweezers and micromachined nanotweezers have been developed for grabbing or probing nanoscale objects. However, since these tweezers are based on electrostatic force, they are not ideal for aqueous solution work such as manipulation of cells, microbes and single molecules.

In this paper, we report promising types of micromanipulators suitable for aqueous solution work. Our manipulators were fabricated by using an assembly-free process based on two-photon microstereolithography that was previously developed by us [1, 2]. These manipulators are driven by optical trapping based on radiation pressure from a tightly focused laser beam. Optical trapping enables remote manipulation of microobjects in biological fluid environments, so that our nanomanipulators are especially suited for the application in bionanotechnology unlike electrostatic types.

2. Optically driven micro/ nano manipulators

To make optically driven micromanipulators, we utilize two-photon micro stereolithography, in which any 3D microstructure are fabricated by scanning a focused near-infrared pulsed laser beam inside photocurable resin [3]. In the two-photon microstereolithography, direct laser writing allows us to fabricate various types of optically driven micromanipulation tools.

As an example of micromanipulation tools, we designed a pair of micromanipulators driven by a focused laser beam (Fig. 1). The arms are trapped and swung by scanning a laser beam. Fig. 2 shows an SEM image of prototypes of micromanipulators. By optimizing the scanning pattern of the laser beam, the arms (length: 8.7μm, width: 1.7μm) were precisely fabricated without any supporting columns or sacrificial layers. To drive the manipulators, a laser beam was focused on one of the micromanipulator arms floating in a liquid, and

Y. Baba et al. (eds.), Micro Total Analysis Systems 2002, Volume 2, 937–939.

then the arm was turned by scanning the focus in an arc as the other arm floated. Fig. 3 shows sequential optical microscope images of the micromanipulators under the control of the focused laser beam. The trapped arm was successfully pivoted through an arc of 40 degrees. In our experiments, a fabrication system with Ti: sapphire laser (λ: 763 nm, pulse width: 130 fs, repetition: 82 MHz) was used for both fabrication and driving of micromanipulators.

In addition, we succeeded to control the torque of the manipulators by adjusting the position trapped by the laser beam along the length of the arm. Fig. 4 shows the change in the maximum torque (maximum rotation speed) of the arm as the focal position of the laser beam is shifted under a laser power of 120 mW. This result indicates that the torque of the micromanipulators is proportional to the turning radius of the laser beam. As the focal position moves toward the tip of the arm, the torque of the manipulators increases without any change in the laser power. The torque of the arm were estimated at 0.4 – 1.5 $\times 10^{-19}$ Nm. Such precisely controlling of generative torque cannot be performed with electrostatic-type micromanipulators.

Nanotweezers with submicron probe tips was also fabricated by using an optimal laser scanning. Fig. 5 shows SEM images of nanotweezers. Each probe tip is 250 nm in diameter. Fig. 6 shows optical microscope images of nanotweezers opened and shut. The probe tips were precisely closed without any gap between them.

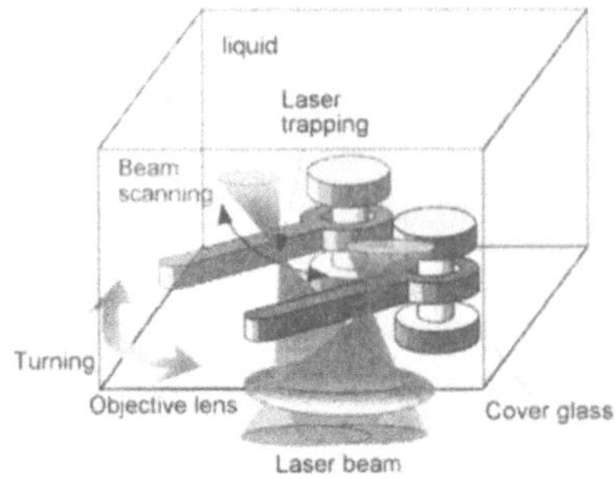

Fig. 1 Micromanipulators driven and controlled by focused laser beams

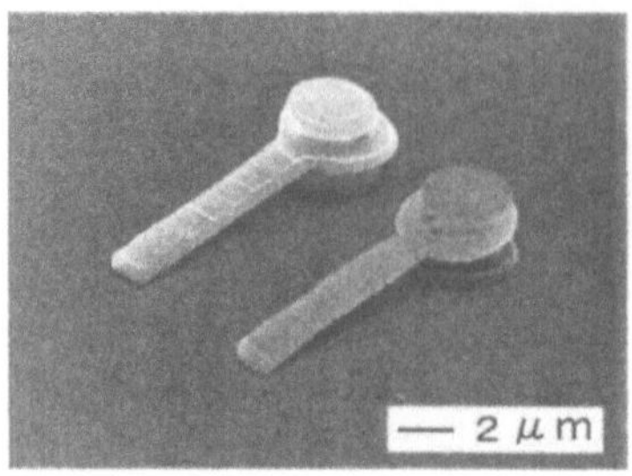

Fig. 2 SEM image of micromanipulators made by the two-photon microstereolithography

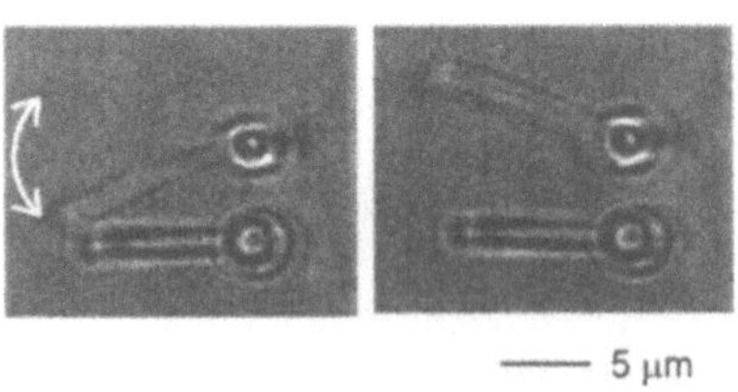

Fig. 3 Optical drive of micromanipulators in a liquid. A manipulator arm was opened and closed by the scanning of a focused laser beam in an arc.

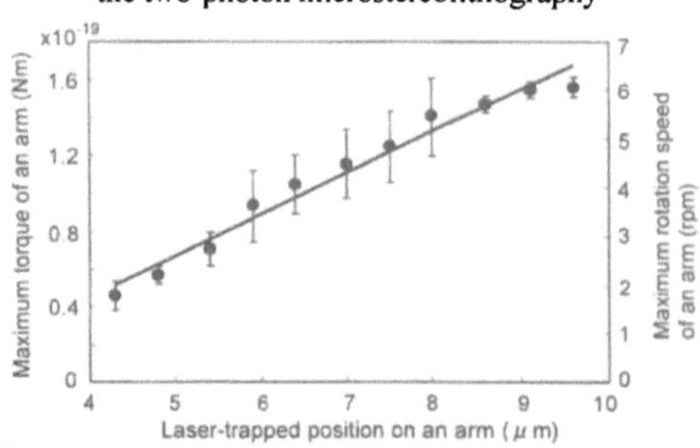

Fig. 4 Force control of a manipulator arm by adjusting the position trapped by a laser beam. Femto-newton force control was demonstrated.

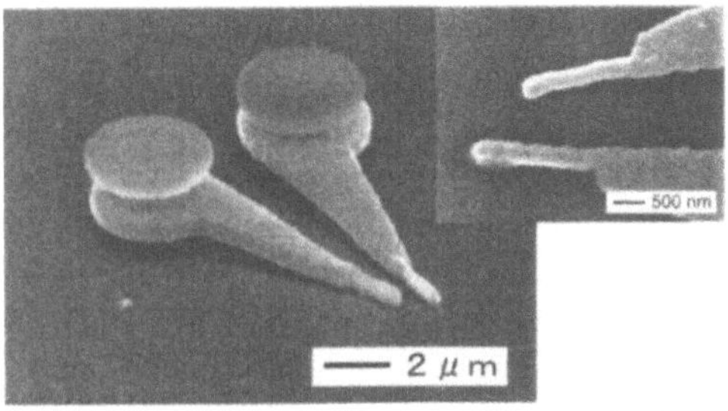

Fig. 5 Optically driven nanotweezers made by the two-photon microstereolithography (probe diameter: 250 nm)

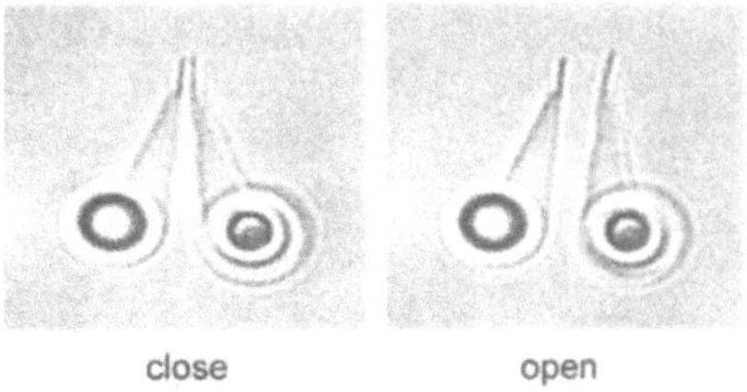

Fig. 6 Optical drive of nanotweezers in a liquid. Two probes were open and shut precisely.

3. Advantages of optically driven micro/ nano manipulators

The main advantages of these optically driven manipulators are as follows: remote drive in an aqueous solution, force control on the order of femto-Newton, and elimination of the need to use ultrahigh precision 3D stages to move the probe tips. In addition, these manipulators are very small, whereas other types of manipulator reported earlier were considerably larger and more complicated owing to scanning probe systems such as AFM or a long pipette to attach the probe tips. Furthermore, since the size and shape of our manipulators can be quickly modified using 3D CAD, they can be easily optimized for particular biological samples. The flexible, rapid fabrication using the two-photon microstereolithography technique enables custom-built nanotools to be created for individual applications. The resolution of our current fabrication system has attained 140 nm. This resolution will exceed 100 nm through the further optimization of the fabrication system and use of higher-order nonlinear effects such as three-photon absorption.

4. Conclusions

We have developed novel micro/nano manipulators driven by light through the use of an assembly-free, rapid fabrication method in two-photon microstereolithography. Such optically driven manipulation tools can be widely applied for not only biotechnology but also μ-TAS. In the next stage, the optically driven manipulation tools will be integrated into 3D polymeric microfluidic circuits by using microstereolithography techniques developed by our group [4, 5]. The optically controlled microfluidic devices including various types of manipulation tool are useful for bionanotechnology applications such as nanosurgery systems for living cells and nanoanalysis systems for single molecules.

Acknowledgments

This work was partially supported by Grant-in-Aid for Scientific Research from the Ministry of Education, Culture, Science and Technology and a reseach grant from Foundation Advanced Technology Institute

References

1. S. Maruo and K. Ikuta, Proc. of SPIE **3937**, 106 (2000).
2. S. Maruo, K. Ikuta and H. Korogi, Proc. of MEMS 2001, p. 594.
3. S. Maruo, O. Nakamura, and S. Kawata, *Opt. Lett.* **22**, 132 (1997).
4. K. Ikuta, S. Maruo, T. Fujisawa and A. Yamada, Proc. of MEMS 99, p. 376 (1999).
5. S. Maruo and K. Ikuta, *Appl. Phys. Lett.* **76**, 2656 (2000).

Microfluidic Circuit for Integrated DNA Sequencing Product Purification and Analysis

**Brian M. Paegel, Stephanie H. I. Yeung,
James R. Scherer and Richard A. Mathies**
Department of Chemistry, University of California, Berkeley, CA 94720 USA

Abstract

A microfluidic circuit for DNA sequencing sample desalting, template removal, and preconcentration has been integrated with on-chip electrophoretic analysis. A 60-nL chamber containing a DNA affinity copolymer is used to capture and preconcentrate sequencing extension products in only 120 s. Thermal denaturation of the affinity hybrid is followed by high-speed microchip electrophoresis resulting in 560 bases with an accuracy $\geq 99\%$ in only 30 min.

Keywords: DNA sequencing, integrated sample preparation

1. Introduction

The DNA sequencing process is based on a microliter-scale paradigm that is incompatible with the nanoliter sample handling capabilities of μTAS devices. While a variety of DNA sequencing microdevices have been presented in which separation times and reagent consumption are dramatically minimized [1], most of the upstream sample handling is on a microliter scale and requires hours for completion. In particular, the crucial sample purification step preceding capillary electrophoresis must be improved. We have thus devised an electrokinetically driven microfluidic circuit for desalting, template removal, and preconcentration of DNA sequencing samples in a fraction of the time and volume required by conventional techniques and integrated this process with state-of-the-art microchip-based separation.

2. Experimental

A microfluidic circuit for integrated high-speed, nanoliter-scale purification of DNA sequencing products followed by capillary electrophoresis is presented in Figure 1. The separation system is based on previous studies [1, 2], but we have added a 1-mm-wide capture chamber connected to input and output wells via two arms of width 100 μm. The inlet arm is also connected to the injection cross via a 75-μm-wide coupling arm. This 60-nL chamber is filled with oligonucleotide purification matrix complementary to the 20-nt sequence just 3' of the M13 -40 universal forward sequencing priming site, thus capturing only enzymatic extension products. To operate the device, sequencing sample taken directly from the thermal cycler is placed in well (2) and the sample is driven through the matrix at 50° C and 250 V for 90 s. Fresh buffer is placed in (2), and the product is washed for 30 s under indentical conditions. The bottom left inset to Figure 1 shows a fluorescence image after a 2 μL sample has been applied and concentrated into a 10-nL space within the chamber. The device temperature is then elevated to 67° C, whereupon the product-purification matrix duplex is denatured, product is electrophoretically loaded into the injection cross, and high-speed microchip electrophoresis is performed.

Y. Baba et al. (eds.), Micro Total Analysis Systems 2002, Volume 2, 940–942.

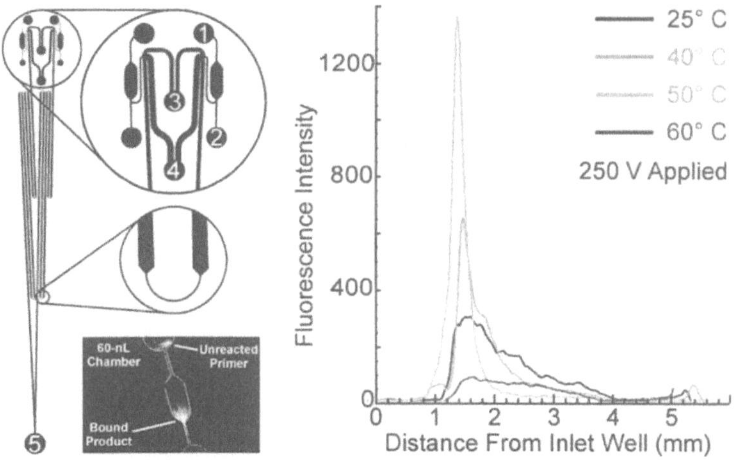

Figure 1. Left: microfluidic circuit for integrated DNA sample purification followed by CE analysis. The chamber has a capture outlet (1) and capture inlet (2) for sample introduction, and common cathode (3), waste (4), and anode (5) reservoirs for electrophoresis. The fluorescence image shows the sequencing products bound to the oligonucleotide capture matrix in the chamber. Right: fluorescence profiles of product bound in the capture chamber at the indicated binding temperatures. Optimal binding occurred at 50° C. A similar optimization of voltage at constant temperature yielded optimal binding at 250 V across the capture chamber.

3. Results and Discussion

High-speed, nanoliter-scale purification has been accomplished using a methacrylated oligonucleotide copolymer of acrylamide. Binding optimization experiments indicated sample was most effectively concentrated and localized when binding was performed at 50° C and 250 V driving potential. Product is expected to localize approximately 1 mm from the inlet well because the chamber flares out at this point and the local electric field drops by a factor of 10, thus encouraging product hybridization.

Four-color sequencing of dye-primer cycle sequencing reactions is shown in Figure 2. For clarity, only the C terminations are presented. Integrated purification of the product (trace A) yields peaks 85% the intensity of ethanol-precipitated samples (trace B) with 560 bases of $\geq$ 99% accuracy called in 30 min (Figure 2, inset plot). Samples injected without purification give $<$ 10% intensity peaks, yielding no sequencing information.

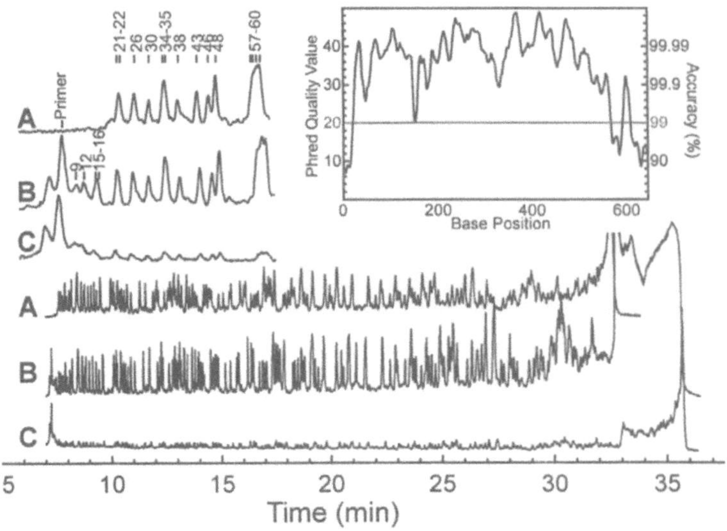

Figure 2. CE analysis of M13mp18 4-color fluorescence DNA sequencing products. The C-terminations are shown for (A) on-chip purified product, (B) ethanol-precipitated product, and (C) unpurified product. The left expansion shows that on-chip purification removed excess primer and all terminations in the first 20 bases containing the capture sequence. The right inset presents a phred analysis of the capture-purified 4-color sequencing run.

Magnification of the early regions (Figure 2, top left) of each trace shows successful specific removal of primer and terminations that fall within or before the capture sequence on the vector when oligo capture is employed [3].

4. Conclusions

Integrated, high-speed, nanoliter-scale purification of DNA sequencing products has been successfully demonstrated. Compared to conventional precipitation-based techniques, this approach represents a 10-fold decrease in time, a 100-fold decrease in process volume, and paves the way for further integration of DNA sequencing processes. Generalization to other affinity capture-based applications using antibodies or aptamers, for example, should also be possible.

1. Paegel, B. M., Emrich, C. A., Wedemayer, G. J., Scherer, J. R., and Mathies, R. A. *Proc. Natl. Acad. Sci. U. S. A.* **2002,** *99,* 574-579.
2. Paegel, B. M., Hutt, L. D., Simpson, P. C., and Mathies, R. A. *Anal. Chem.* **2000,** *72,* 3030-3037.
3. Paegel, B. M., Yeung, S. H. I., Scherer, J. R., and Mathies, R. A. *Anal. Chem.* **2002,** (in press).

FULLY INTEGRATED MCM MICROFLUIDIC SYSTEM FOR DNA ANALYSIS

Robin H. Liu, Justin Bonanno, Jianing Yang, Dale Ganser, Ralf Lenigk, Dave Rhine, Thomas Smekal, Randy Stevens, and Piotr Grodzinski
Microfluidics Laboratory, PSRL, Motorola Inc., Tempe, AZ, USA

Abstract

A fully integrated and self-contained microfluidic multi-chip modular (MCM) device having sample-to-answer capability was developed. The device performs complete genetic analysis, including sample preparation (i.e., target cell capture, cell purification/concentration, cell lysis, DNA amplification) and microarray detection.

Keywords: MCM, sample preparation, DNA analysis

1. Introduction

Integration of sample preparation with back-end detection on the chip level is a significant challenge for two reasons: 1) sample preparation from crude bio-samples represents an extremely complex procedure that requires numerous functional and robust microfluidic components (e.g., pump, mixer, valves); 2) integration of individual microfluidic components into a system has proven to be non-trivial. A universal integration protocol is not currently available. Moreover, many issues such as bio-compatibility and flow instability induced by surface tension or air bubbles must be addressed.

2. Device Designs and Experiments

1) System Design

A novel MCM integration approach (Fig.1), analogous to standard electronic IC, was developed to provide a module and assembly standard for integrated microfluidic systems. The MCM device for complete DNA analysis consists of seven modules (Fig. 1c). Paraffin films are sandwiched between the modules as fluidic interfaces, while circuit boards supply electrical interfaces. This integration approach allows easy connection, excellent chip/assay reconfiguration, and use of many disparate materials and fabrication processes. The MCM device incorporates micropumps, mixers, microvalves, a target cell capture/PCR channel, and a DNA array chip (Fig. 1b).

2) Acoustic Micromixer

A bubble-induced acoustic mixer (Fig. 2a-d) was developed to enhance the mixing between the bio-sample and magnetic capture beads for target cell capture, and to accelerate hybridization kinetics of the DNA microarray. This mixing technique is based on bubble-induced acoustic microstreaming principle [1]. The mixer consists of a piezoelectric disk that is attached to a reaction chamber, which is designed in such a way that a set of air bubbles with desirable size is trapped in the solution. Fluidic experiments showed that air bubbles resting on a solid surface and set into vibration by the sound field generated steady circulatory flows, resulting in global convection flows and thus rapid mixing. The time to fully mix a 50-μL chamber is significantly reduced from hours (a

Y. Baba et al. (eds.), Micro Total Analysis Systems 2002, Volume 2, 943–945.

pure diffusion-based mixing) to 6 seconds. Immuno-magnetic cell capture experiments showed acoustic microstreaming provided efficient mixing of bacterial cell (*Esherichia coli* K12) matrix suspended in blood with magnetic capture beads, resulting highly effective immuno-magnetic cell capture (Fig.2e). Hybridization kinetics study in the electro-chemical detection-based chips showed that acoustic microstreaming results in up to 5 folds kinetics acceleration (Fig.2f). Acoustic microstreaming has many advantages over most existing chamber micromixing techniques, including simple apparatus, ease of implementation, low power consumption (2 mW), and low cost.

3) Electrochemical Micropump

An electrochemical pump that relies on an electrochemical reaction (electrolysis) between two electrodes to produce a gas that is used to push liquids through microchannels was developed (Fig. 3). The pump is very effective in transporting large volume of liquids within a reasonable time (e.g., 1 mL liquid within 10 min). Since this micropump does not require check valves and mechanical membranes, it is simple and easy to be integrated in the microfluidic systems, compared to conventional micropumps.

4) Paraffin Microvalves

Single-use paraffin microvalves were developed to regulate flow (Fig. 4). Paraffin (wax) is used as the valve actuator material. A small plug of paraffin can be used to close a microchannel at any temperature below the melting point of the paraffin (Fig. 1a), and open the channel when subjected to heat. Using this principle, a variety of microvalves, including “close-open”, “open-close-open”, “T”, and “toggle” valves, were developed. PCR experiments demonstrated that the paraffin is bio-compatible.

5) Assays and Results

Large volume (~ 1 mL) of bio-sample mixed with magnetic capture beads was run through the capture/PCR channel. Rare target cells were captured and concentrated in the channel by a magnet, followed by thermal lysis and PCR target amplification. The amplicon was then mixed with hybridization buffer and flowed onto the DNA array chip for detection. The capture/PCR device is ultra-thin so that rapid PCR (~ 30 min) can be achieved and PCR sensitivity is high (10 E.coli K-12 cells can be detected) (Fig. 4). Electrochemical detection of DNA was performed using the Motorola eSensorTM devices (Fig. 5). Detection of 1000 E.coli K-12 cells from a 1 mL bio-fluid was successfully demonstrated using the MCM device.

3. Conclusions

We have developed a microfluidic multi-chip modular (MCM) device that integrates bubble-induced acoustic micromixers, electrochemical micropumps, paraffin microvalves, etc. to perform complete genetic analysis, including sample preparation (i.e., target cell capture, cell purification/concentration, cell lysis, DNA amplification) and microarray detection.

Acknowledgements

This work has been sponsored by DARPA BioFlip contract #MDA972-01-3-0001.

References

1. Elder, S.A., "Cavitation Microstreaming," *J. Acoust. Soc. Am.*, 1959. **31**: pp. 54-62.

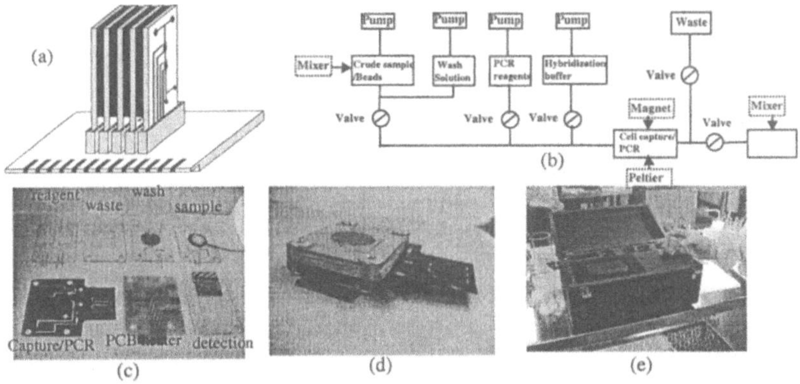

Fig.1: (a) Schematic of a MCM system; (b) design diagram; (c) (d) (e) Photographs of the MCM device.

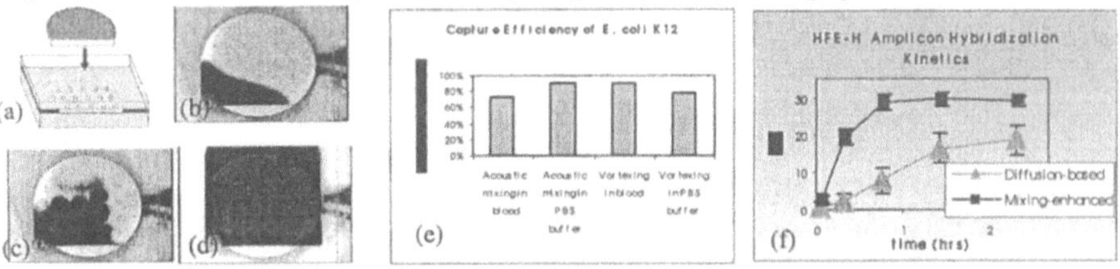

Fig.2: (a) Schematic of an acoustic mixer; Dye mixing in a 16 x 16 x 0.2 mm chamber at (b) time 0; (c) 2 sec; (d) 6 sec. (e) Cell capture efficiency using acoustic mixing vs. vortex. (f) Hybridization kinetics enhancement.

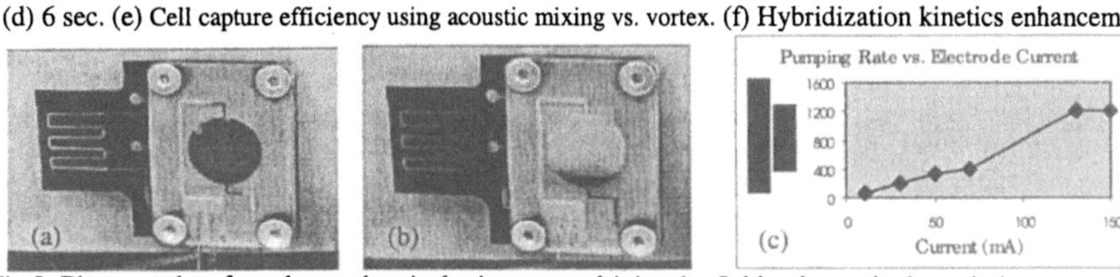

Fig.3: Photographs of an electrochemical micropump driving 1 mL blood sample through the capture channel at (a) time 0; (b) 5 min. (c) pumping rates vs. electrode current.

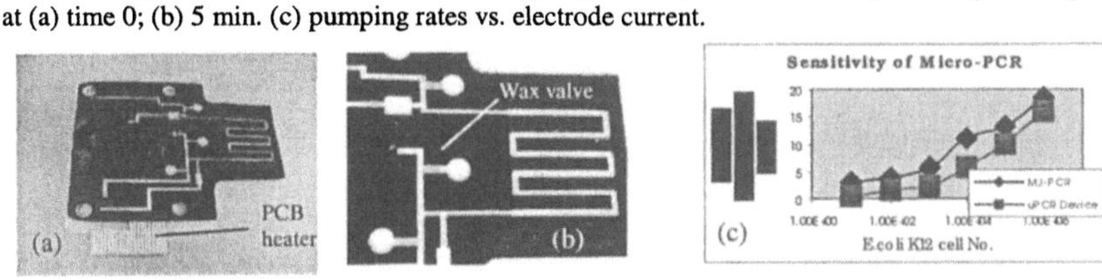

Fig.4: (a) (b) Photographs of the cell capture/PCR device. (c) PCR sensitivity (~ 10 E.coli K-12 cells).

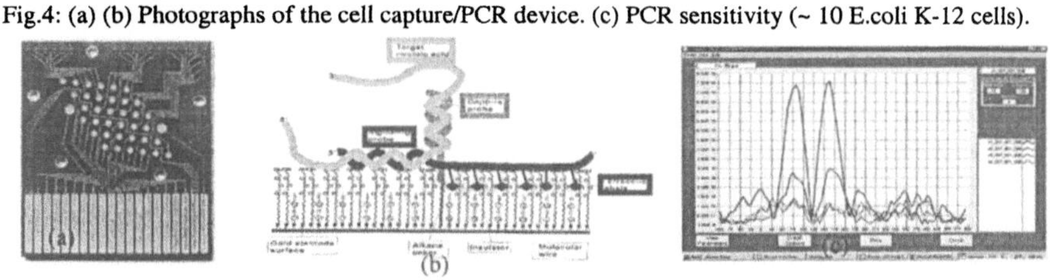

Fig.5: (a) Photograph of Motorola eSensor™ array chip; (b) Diagram of e-detection; (c) Final detection signal of 1000 E.coli cells captured from 1 mL bio-fluid using the MCM device.

AUTOMATED DNA FRACTION COLLECTION ON GLASS MICROCHIPS

Abdel Minalla, Li Ding, Luc Bousse, and Robert Dubrow
Caliper Technologies Corp., 605 Fairchild Drive, Mountain View, CA 94043

Abstract

We have developed a glass microchip capable of automation and integration of all steps required for DNA fraction collection, and which also integrates the downstream mixing of reagents needed for cloning the collected DNA. Out of five different samples, the device allows us to collect five different fragments, which can be used for further investigation. A fraction can be selectively collected from a mixture of samples without any detectable impurities in less than 70 seconds. Also, we describe multiple methods to evaluate the concentrations and purity levels of collected fractions.

Keywords: Fraction collection, Cloning, DNA separation

1. Introduction

One of the goals of Lab-on-a-Chip technology is to integrate and automate assay steps that can currently only be performed manually. An example of such a step is fraction collection after electrophoretic separation, which is now done by manually cutting out bands from gels with a knife. Another approach based on CE fraction collection was first reported by Guttman *et al.* [1]. Both techniques have limitations in terms of automation, speed, and output concentration. On-chip fraction collection of a single DNA fragment has been reported by Effenhauser *et al.* [2] and Hong *et al.* [3].

2. Experimental work

2.1. Fraction collection chip:

Figure1 shows the layout of the glass microchip used to automate and integrate all steps involved in fraction collection. It consists of 16 wells: five sample wells (S_1 to S_5); five output wells (O_1 to O_5); all other wells (G_1 to G_6) are used to control the separation and fraction collection steps. This novel configuration has the advantage that the user can select the destination of any peak of any sample. For instance, five fractions from one sample can be sent to different output wells, or the same fraction from five sample wells can be sent to one output well.

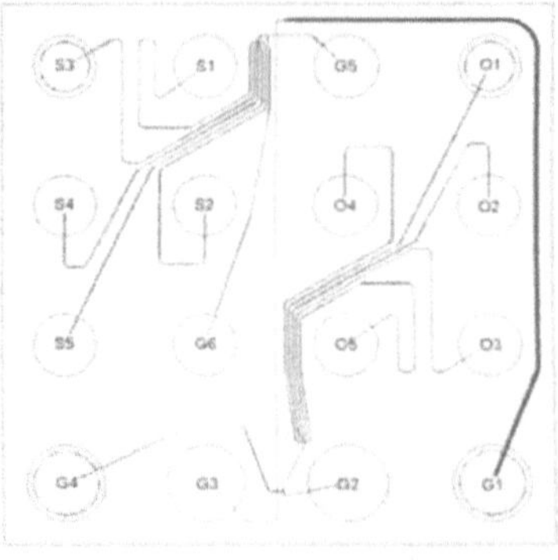

Fig.1: Fraction collection chip.17.5X17.5 mm
Channels are 36 µm wide and 13 µm deep

2.2. Chip preparation:

All channels and wells G_1 to G_6 were filled with a polyacrylamide sieving matrix. The output wells were filled with 5 µl DNA marker in 2 mM buffer. All sample wells were filled with 5 µl of sample mix. A typical sample is a mixture of four dsDNA fragments of sizes 150, 300, 500, and 850 bp. The final concentration of each fragment is 5 ng/µl in 20 mM buffer. The fraction collection chip was run in an Agilent 2100 Bioanalyzer [4]. For

Y. Baba et al. (eds.), Micro Total Analysis Systems 2002, Volume 2, 946–948.

video images, we have run chips on a fluorescent inverted microscope [5] connected to a CCD camera and the power supply of a 2100 Bioanalyzer.

2.4. Procedure:

An initial electrophoretic separation of the sample is needed to provide the timing of the arrival of each fragment. Subsequently, the fraction collection runs consist of:

1. Loading the sample to the injection intersection and starting the separation as usual.
2. During the correct time interval, diversion of the selected fraction while the separation current remains constant.
3. Continuing the separation without diversion, while simultaneously moving the collected fraction toward the selected output well.

For the purpose of DNA cloning, on-chip mixing of topoisomerase is needed, and it can be done simultaneously during the last two steps inside the output channels.

3. Result and discussion

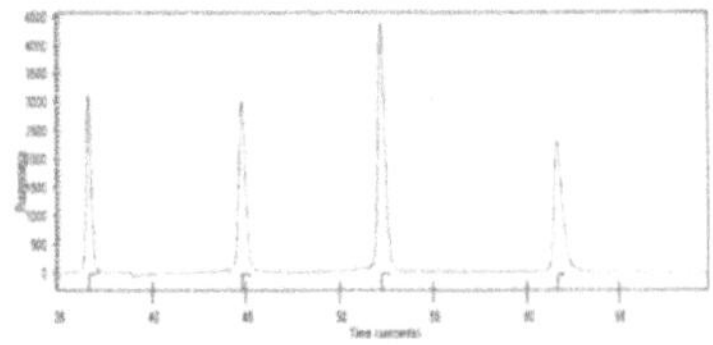

Fig. 2: Electropherogram of sample

As shown in Figure2, separation of dsDNA mixture described above was achieved in less than 65 seconds. About five more seconds is required to divert a selected fraction from detection point into the collection channel. Well-resolved peaks and very consistent time of arrival are among important parameters for this assay. Elsewhere [4], Mueller *et al.* shows that microfluidic systems are capable of providing reproducible, fast, high-resolution electrophoretic separation, which leads to a feasible, automated fraction collection assay. The chip has five sample wells and five fraction collection wells. The choice of which fractions from which samples are sent to which wells is entirely flexible, and all permutations can be chosen by controlling the script of voltages and currents as a function of time. In-house built software was used to monitor and control all steps.

To minimize the impurities in the output wells, we have to identify exactly a specified fragment to be collected, the start of the peak collection, and the duration of collection period. To increase the amount of collectable material: multiple runs were performed; an inverse pinch (which expands the sample plug rather than confines it) was applied; dilute buffer in the output wells was limited to only 5 µl; and, injected sample was increased by running the chip under stacking conditions.

In contrast with previous work [2,3], multiple methods were used to evaluate the fraction collection runs: 1) Figure3 shows direct video observation of collected (3a) and uncollected (3b) prelabeled fragments moving into collecting channel and past the collection intersection to waste well, respectively.

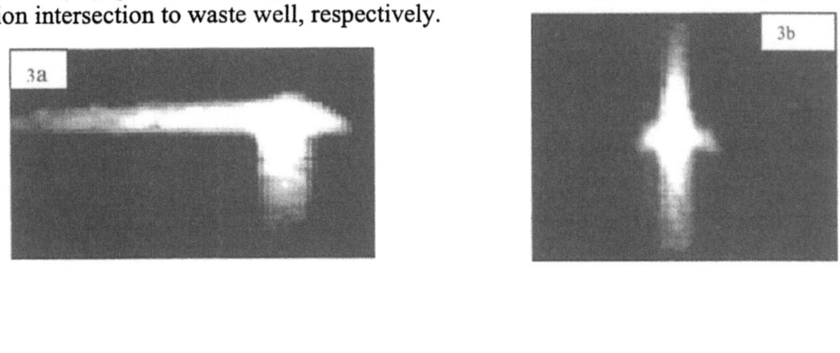

2) The contents of the output wells were reintroduced in a DNA separation chip [4] and analyzed with an ultra-sensitive version of the separation assay. Figure 4 shows that the expected peaks were recovered. The concentration of the collected fraction is measured to be around 40 ng/ml.

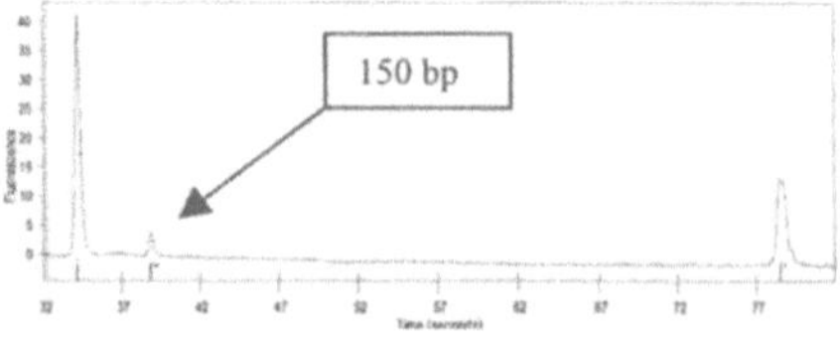

Figure 4a: 150 bp collected from S1 into O1

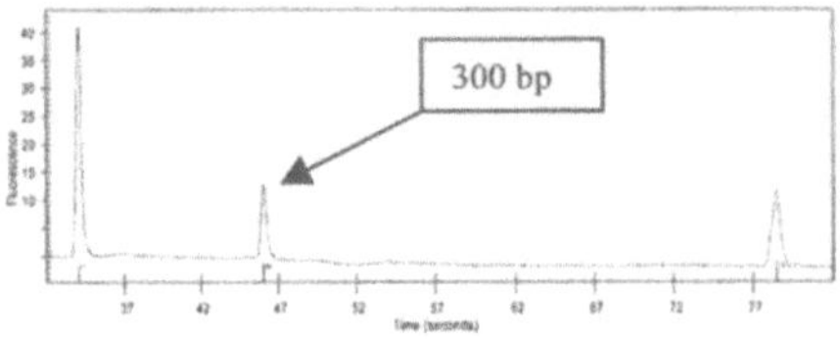

Figure 4b: 300 bp collected from S2 into O2

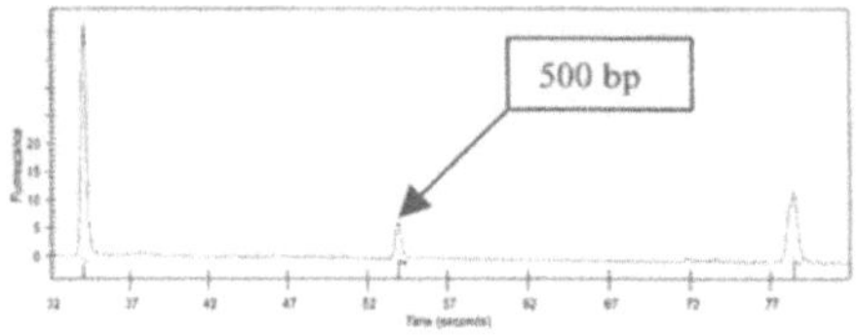

Figure4c: 500 bp collected from S3 into O3

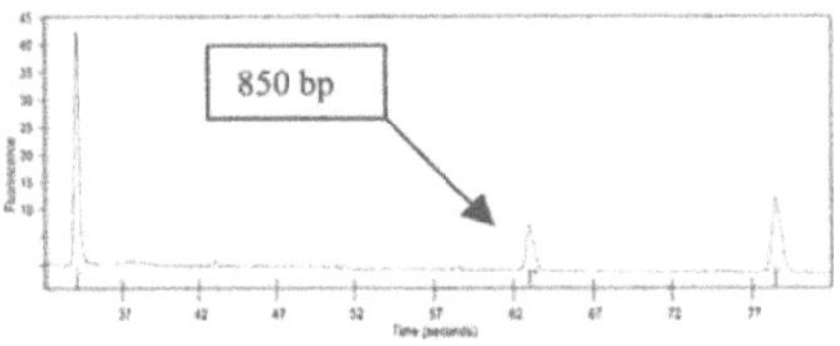

Figure 4d: 850 bp collected from S4 into O4

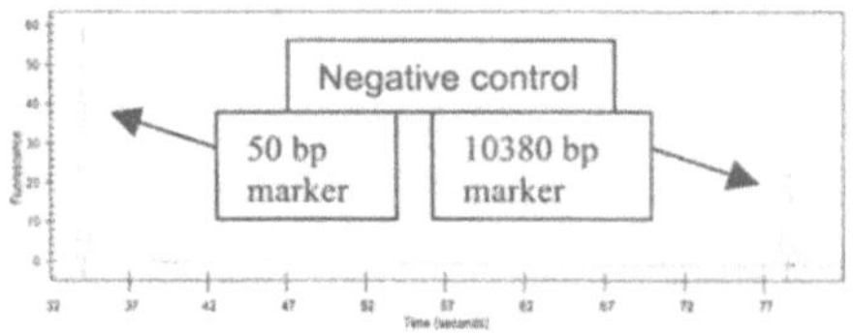

Figure 4e: Blank collected from S5 into O5

3) When PCR products are fractionated, the contents of the output wells were successfully cloned via topoisomeraseI mediated TA cloning. Typically 10 to 100 recombinants with the expected DNA insert are obtained from an output well sample containing a PCR fragment. In comparison, the cloning of the negative control output sample only yields up to 3 spontaneous recombinants with no DNA inserts. Well G2 is designed to allow on-chip topoisomeraseI mediated ligation reaction. We believe this is the first demonstration of cloning of DNA fractions collected on a microchip.

4. Conclusion

A key advantage of this chip is the speed and accuracy it achieves: it takes only 70 seconds to collect one fragment from the sample mix without any detectable impurities. It is clear that fraction collection from samples other than DNA can be done in the same way; for instance, mixtures of peptides or proteins could be separated, and the collected fractions analyzed with mass spectrometry.

Acknowledgements
The authors thank Mac McReynolds, Luke Liang, and Kader Mokrane for their assistance.

Reference:

[1] Guttman, A.S. *et al.*, Anal. Chem., 1990, Vol. 62, pp. 137-141.
[2] Effenhauser, C.S. *et al.*, Anal. Chem., 1995, Vol. 67, pp. 2284-2287.
[3] Hong, J.W., *et al.*, MicroTAS, 01, pp. 113-114.
[4] Mueller, O., *et al.*, Electrophoresis, 2000, Vol. 21, 128-143.
[5] Minalla, A., *et al.*, Proceedings of SPIE Conference, 2000,Vol. 4177, pp. 134-140.

A DISPOSABLE ON-CHIP AIR DETONATOR FOR DRIVING FLUIDS ON POINT-OF-CARE SYSTEMS

Chien-Chong Hong, Jin-Woo Choi, and Chong H. Ahn
MicroSystems and BioMEMS Lab
University of Cincinnati, P.O. Box 210030, Cincinnati, OH 45221-0030, USA

Abstract

This paper presents a disposable on-chip air detonator for driving fluids on points-of-care systems. Pressurized air pre-stored in a chamber works as a source of pressure for driving fluids upon breaking membrane that seals the pressurized air. By applying an electrical pulse through the localized microheater, the thermoplastic membrane can be easily broken and consequently the pressurized air can be detonated to drive liquids through microchannels. In this paper, we report the development of new device fully microfabricated using plastic substrate and plastic membrane, which is fabricated as one body to reduce fabrication steps for low-cost disposable alternative on-chip pressure sources. Dynamic response of pressure in microfluidic channel has also been fully characterized.

Keywords: On-chip air detonator, disposable on-chip pressure source, point-of-care system, disposable biochip

1. Introduction

Micropumps are the most common devices to produce and control pressure for microfluidic systems [1]. For disposable microfluidic biochips or systems, the active micropumps should be integrated with disposable batteries, which increases its cost and involves many technical difficultes. So, an alternative new power source to the battery is desirable for the disposable biochips or lab-on-a-chips, combining passive-type microfluidic components. For this reason, we have reported a novel concept for an alternative on-chip power source using SU-8 membrane on a silicon-based device, which made use of thermal cracking technique to break the membrane in our previous work [2]. Although we have proved the basic concept of a disposable on-chip pressure source, the silicon-based device needs complicated processing steps and is cost-ineffective as a part of disposable biochips or point-of-care systems. In this work, we present a new device that is fully micro-fabricated using plastic substrate and plastic membrane using plastic injection molding to reduce fabrication steps for low-cost disposable alternative on-chip pressure source as illustrated in Figure 1.

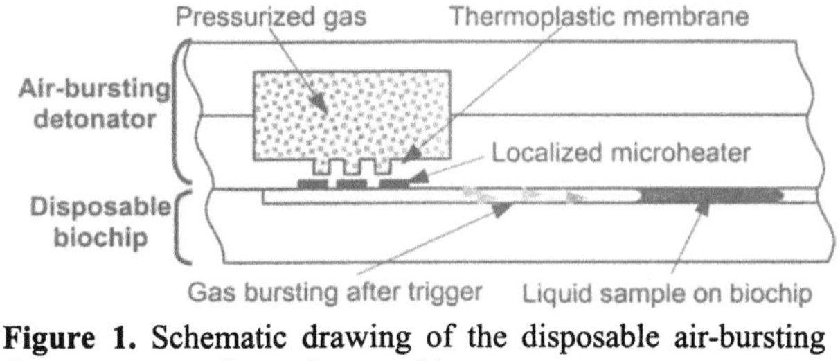

Figure 1. Schematic drawing of the disposable air-bursting detonator as an alternative on-chip power source.

Y. Baba et al. (eds.), Micro Total Analysis Systems 2002, Volume 2, 949–951.

For a conventional micropump, it needs a continue power input to convert electrical energy to drive fluids. However, as described in Figure 2, the novel air-bursting detonator needs small consumed energy comparing to a conventional micropump since only a short electric pulse breaks the membrane and then the pressurized air drives fluids in microchannel without further power consumption.

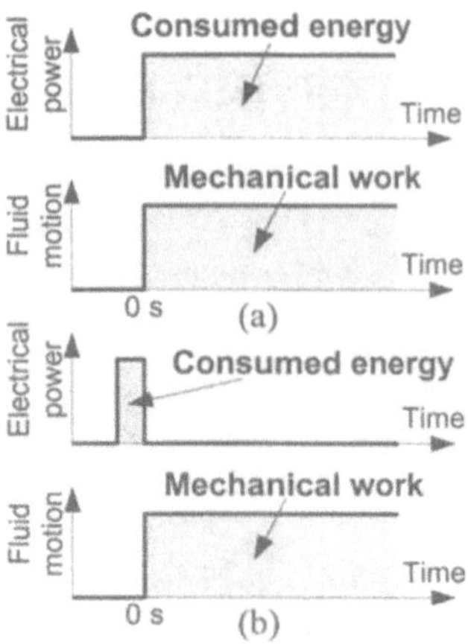

Figure 2. Consumed electrical power in driving fluids for: (a) conventional micropump and (b) air-bursting detonator.

2. Design and Simulation

Dynamic thermo-mechanical simulation was performed by CFD-ACE+ to investigate the temperature profile and thermal stress on the heater and membrane as shown in Figure 3. Obviously, lower power input has slower response due to heat capacity of the materials and heat dissipation to air by convection. Pulsed electric current starts to trigger the detonator around 350 ms for the detonator with plastic membrane (the glass transition temperature is around 140 °C) for 100 mW of the power input. Dynamic flow simulation was also performed to investigate the flow response when the air-bursting detonator integrated with meander-type long microchannel as shown in Figure 4.

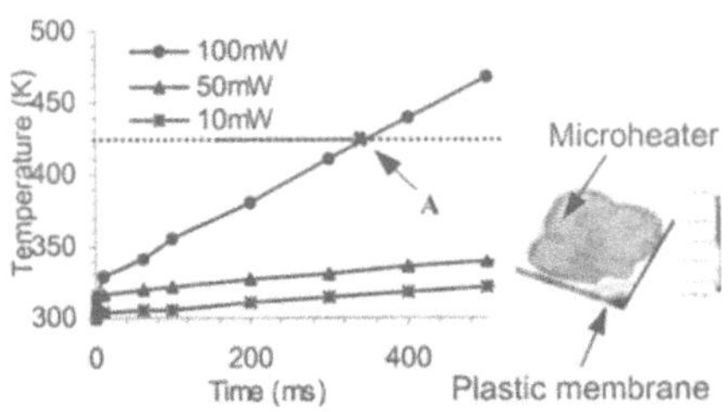

Figure 3. Dynamic thermmo-mechanical simulation of the heater. Trigger point A is the glass temperature for plastic membrane.

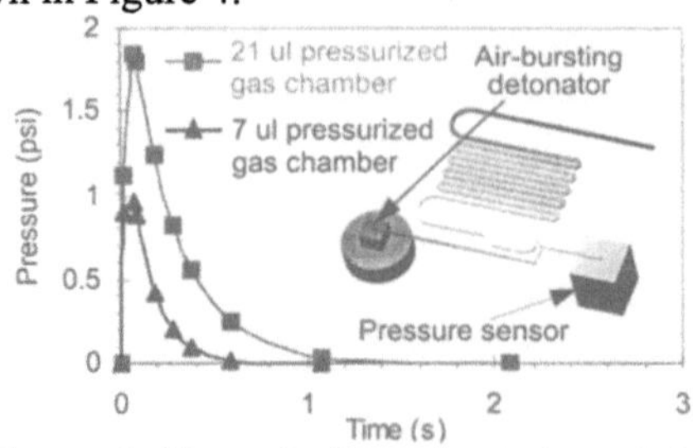

Figure 4. Dynamic flow simulation of the air-bursting detonator with meander-type long microchannel.

3. Experimental Results

To realize the air-bursting detonator, plastic substrate (olefin copolymers) with plastic membrane has been micro- fabricated using a plastic hot embossing technique. Nickel microheater was electroplated on the top of the plastic membrane. Figure 5 shows the fabricated device

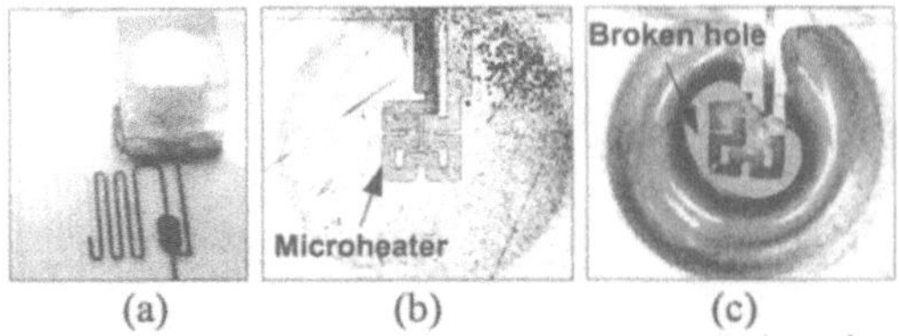

Figure 5. Fabricated device: (a) air-bursting detonator with microchannel; (b) detonator before bursting; and (c) detonator after bursting.

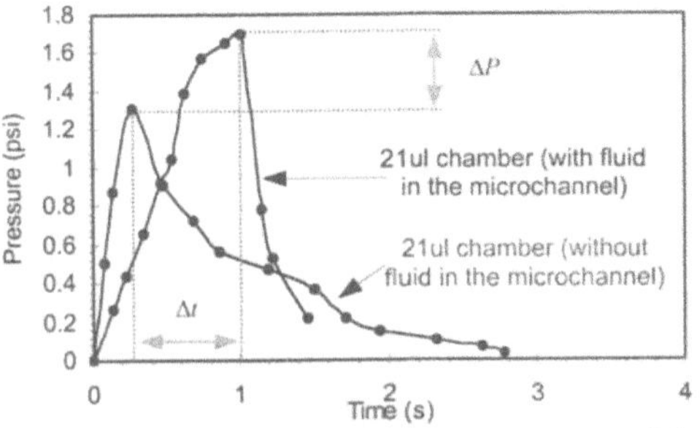

Figure 6. Dynamic pressure measurement of the air-bursting detonator with serpentine micro-channel. With fluid in the microchannel, dynamic pressure drop makes pressure difference (ΔP) and time delay (Δt).

that was integrated on a plastic microfluidic system and micro photographs of plastic membrane structure before and after bursting action. Figure 6 shows measurement results of dynamic pressure response after bursting of membrane structure. For demonstration of the device, dyed water was injected into a microreservoir and then the liquid inlet was sealed. After the trigger of the detonator, pressurized air in the chamber pushes the liquid into a serpentine microfluidic channel as clearly demonstrated in Figure 7.

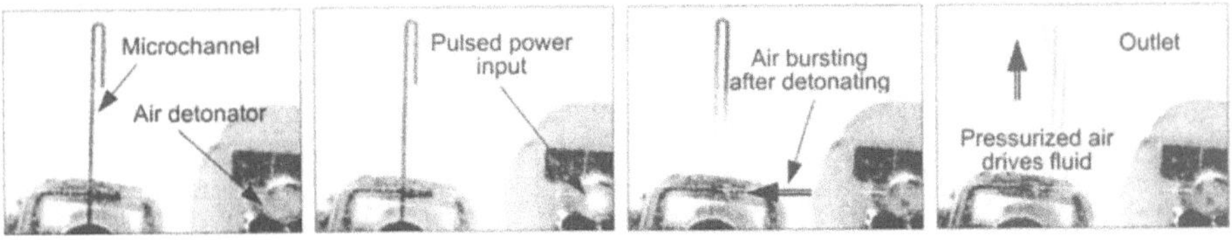

Figure 7. Recorded video clips of the air-bursting detonator integrated on a disposable plastic biochip. After breaking the membrane with a pulsed electric power input, the pressurized air pushes the fluid into serpentine microchannel without additional electric power input.

4. Conclusions

The disposable air-bursting detonator realized in this work clearly showed promising performances in driving liquid samples through the microchannel with low power consumption. Compact and simple structure makes it easy to integrate on microfluidic systems for applications in disposable biochips and point-of-care systems that require fast response and reliable pressure source with low power consumption.

Acknowledgements

This research was fully supported by a DARPA grant under contract AF F30602-00-1-0569 from the *BioFlips* program, DoD, USA.

References

1. D. Maillefer, S. Gamper, B. Frehner, and P. Balmer, "A High-Performance Silicon Micropump for Disposable Drug Delivery Systems," Proceedings of the 14th IEEE MEMS Workshop (MEMS '01), 2001, pp. 413-417.
2. C.-C. Hong, J.-W. Choi, and C. H. Ahn, "Disposable Air-Bursting Detonator As an Alternative On-Chip Power Source," Proceedings of the 15th IEEE MEMS Workshop (MEMS '02), 2002, pp. 240-243.
3. E.V. Mukerjee, A.P. Wallace, K. Y. Yan, D. W. Howard, R. L. Smith, and S. D. Collins, "Vaporizing Liquid Microthruster," Sensors and Actuators, Vol. 83, pp. 231-236, 2000.

MICROFLUIDIC CAPILLARY SYSTEMS FOR THE AUTONOMOUS TRANSPORT OF BIO/CHEMICALS

D. Juncker [1,2], H. Schmid[1], U. Drechsler[1], H. Wolf[1], B. Michel[1], N. de Rooij[2], and E. Delamarche[1]

[1]IBM Research, Zurich Research Laboratory, 8803 Rüschlikon, Switzerland
[2]University of Neuchâtel, Institut de Microtechnique, 2000 Neuchâtel, Switzerland
(jun@zurich.ibm.com)

Abstract

The transport of minute amounts of reagents is the key to the miniaturization of bio/chemical processes but represents a considerable technological challenge. The microfluidic systems demonstrated hitherto can be very small, but require large and cumbersome peripherals for actuation and control, as well as electrical and fluidic connections from the chip to the world, all of which limit further scaling and integration. Here we present a microfluidic capillary system (CS) that solely relies on capillary phenomena to pump, valve, and synchronize the transport of multiple liquids delivered in sequence to its service port, autonomously. The "capillary functions" are integrated into the system, that is to say "coded" into the structure of the flow path. The CS is self-contained (free of outside connections), and therefore amenable to arraying. Moreover, it is simple to use, robust, free of dead volumes, free of moving parts, and straightforward to fabricate.

Keywords: **Microfluidic, capillary system, self-contained, autonomous**

1. Introduction

Capillary action is widely used for filling microfluidic devices, but until now only one solution could be flowed into the microchannel using this pumping scheme because in practice loading a microchannel with even a small volume will fill it entirely [1–3], which prevents the placing of a second liquid. Here we solved this limitation (i) by adding a *capillary pump* at the distal end of the microchannel that creates both a large volume and a high capillary pressure, and (ii) by implementing a *capillary retention valve* (CRV) that stops the flow of liquid when the service port is drained and protects the reaction chamber from adventitious drying, Fig. 1. Thus, the task of the operator is reduced to delivering the reagents to the service port using a pipette, in the correct sequence.

2. Fabrication

CSs were fabricated using deep reactive-ion etching in a procedure requiring four photolithographic steps [4], Fig. 2. The reaction chambers were designed to be between 50 and 5000 µm long, and the depth was determined by the last etching step and was varied between 2 and 50 µm. A fabricated array of four CSs is shown in Fig. 3. The Au layer

Y. Baba et al. (eds.), Micro Total Analysis Systems 2002, Volume 2, 952–954.

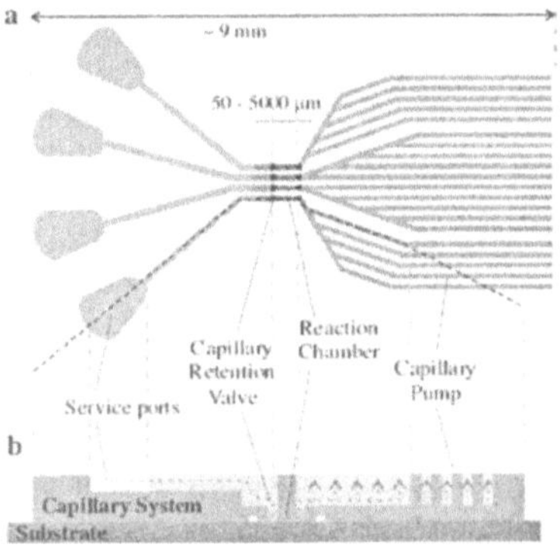

Figure 1: (a) Top and (b) cross-sectional view of four similar microfluidic CSs. The devices are characterized by large service ports and capillary pumps, which generate low and high capillary pressure, respectively. The arrows in (b) indicate the flow of liquids loaded into the service ports.

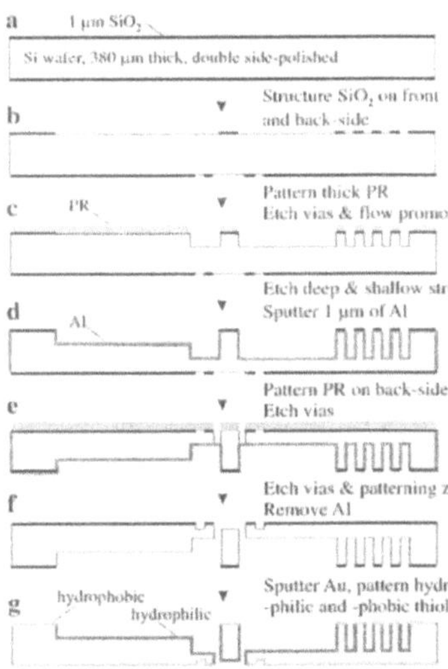

Figure 2: Fabrication procedure of a CS.

allows using microcontact printing to pattern hydrophobic thiols (eicosanthiol) on the outer surfaces of the CS, and subsequently hydrophilic poly(ethyleneglycol)thiols on the inner surfaces of the channels by simple dip coating. The close up view in Fig. 3b reveals the detail of the CS adjacent to the reaction chamber.

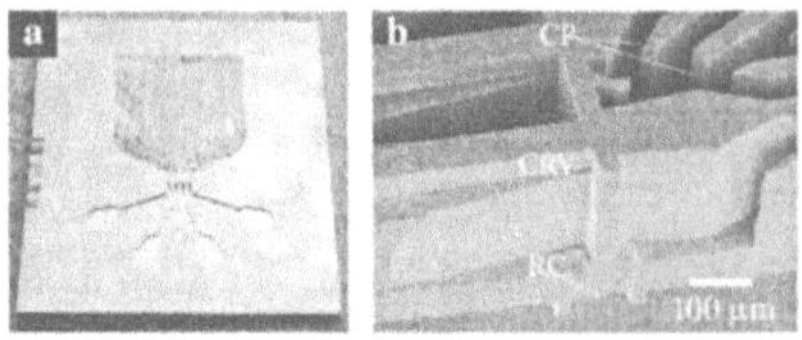

Figure 3: Micrographs of a CS. (a) An Au-covered CS that was diced to size. (b) SEM image showing a close-up of the capillary retention valve (CRV), the reaction chamber (CR), and the fan-out structure of the capillary pump (CP).

3. Autonomous Capillary Flow

The capillary pressure P_c of a liquid-air meniscus in a rectangular microchannel is

$$P_c = -\gamma\left(\frac{\cos\alpha_b + \cos\alpha_t}{d} + \frac{\cos\alpha_l + \cos\alpha_r}{w}\right), \qquad (1)$$

where γ is the surface tension of the liquid, $\alpha_{b,t,l,r}$ are the contact angles of the liquid on the bottom, top, left, and right wall, respectively, and d and w are the depth and width of the channel, respectively. If the cross section varies, however, smaller cross sections will produce a correspondingly higher differential capillary pressure. Thus, from Eq. (1), it is possible to calculate precisely how to vary the cross section of a channel to drain a finite volume of liquid from one region to another (taking into account the difference in receding and advancing contact angles of the solution on the channel walls), and further to predict the flow rate between these regions. Here, the liquid is drained from the service port to the capillary pump using this strategy. To prevent the reaction chamber from being drained too, a CRV is necessary. The CRV has the highest capillary pressure of the CS and therefore pins the dewetting liquid-air interface, and halts the flow of liquid exactly when the service port has been drained. A subsequent aliquot delivered to the service port flows and joins the pinned liquid-air interface at the CRV, and suppresses it; the filling-draining-pinning process repeats itself autonomously upon delivery of a sequence of aliquots to the service port. The arborescent structure of the capillary pump allows large quantities of liquid to be displaced unidirectionally whilst not (considerably) increasing the overall flow resistance of the CS.

4. Results & Discussion

The functionality of the CSs is illustrated by sequentially loading ~ 200 nl aliquots of colored aqueous solutions into the service port, and monitoring the successive filling and flushing of the reaction chamber, Fig. 4. The flow in this 50-μm-deep chamber was about 50 mm s^{-1}, and the displacement of a solution by another one was done in less than 1/15 s. The flow speed can be tailored by varying the dimensions and the wettability of the different sections of the CS. For example, by reducing the depth of the reaction chamber, the flow resistance increases and the flow rate decreases. Such a CS is ideally suited to perform processes that require flushing many solutions in sequence. A PDMS substrate that can reversibly close the reaction chamber (see Fig. 1) has been used for the local adsorption of fluorescently labelled antibodies (IgGs) with high homogeneity on the substrate surface, as defined by the 50 × 50 × 6 μm^3 (15 pl) reaction chamber (results not shown here).

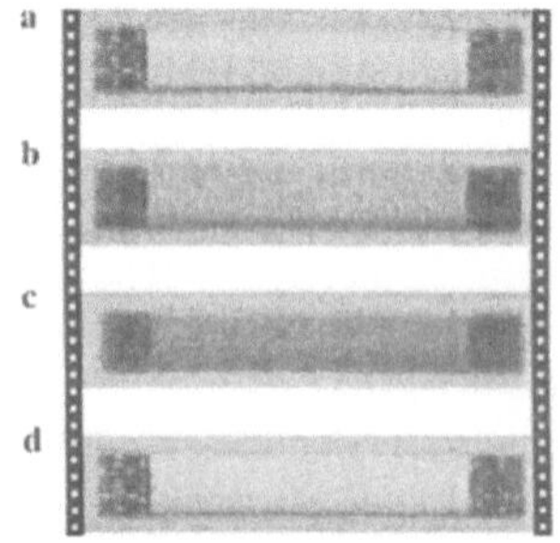

Figure 4: Four images (extracted from a video) of a 500-μm-long reaction chamber viewed through a PDMS substrate using an inverted microscope. They show a reaction chamber containing a liquid that was exchanged sequentially by loading the service port with colored water (a-d).

5. Conclusion

Such a CS is ideal for miniaturized multi-step bio/chemical processes because (i) it is self-contained (no need for peripherals or connections), (ii) it can robustly and autonomously pump different solutions delivered manually or by a pipetting robot to its service port, and (iii) the flow rate can be tailored for a specific bio/chemical reaction.

References

[1] Juncker, D. *et al.*, Proc. Micro Total Analysis Systems 2001, pp. 429- 431.
[2] Delamarche, E. *et al.*, *Science* **276**, 1997, pp. 779-781.
[3] Bernard, A. *et al.*, *Anal. Chem.* **73**, 2001, pp. 8-12.
[4] Juncker, D. *et al*, *J. Micromech. & Microeng.* **11,** 2001, pp. 532-541.

ELECTRIC DAMAGE FREE SEPARATION CHIP OF LYMPHOCYTES SUB-UNIT EMPLOYING HIGH PRESSURE ELECTROOSMOSIS PUMP

Hiroyuki Onoda, Yuzuru Takamura, and Yasuhiro Horiike
Dept. of Materials Engineering, The University of Tokyo, 7-3-1, Hongo, Bunkyo-ku, Tokyo, Japan
(onoda@micro.mm.t.u-tokyo.ac.jp)

Abstract

A separation chip of lymphocytes to each sub-set was developed by combination of a pressure flow and an electroosmosis flow (EOF) pump consisting of both shallow and deep channels. Since the inlet/outlet of the EOF pump is electrically grounded and channels are separated from electrodes through salt bridges, the cell sorter chip demonstrated electrically damage-free and no pH change characteristics, and further low pulsating within 5% of total flow.

Keywords: lymphocytes, electroosmosis pump, non-pulsating flow, cell-sorter, MPC polymer

1. Introduction

Lymphocytes play a main role of an immune system. The previous work reported an electrophoresis separation of T and B cells in a microcapillary chip [1]. However lymphocytes consist of huge number of kinds. The separation of individual lymphocytes to each sub-unit is required before their cultivations and monitoring to study the immune system and therapy. Disposable cell separation chips are necessary to diagnose infection diseases such as HIV. To separate a number of cells to an individually specific cell on a chip, not only serially connectable and electric damage-less characteristics, but reduced pulsating flow are required for the cell sorter. This paper reports that an electric field and pulsating flow free cell-sorter whose function is driven by hydraulic pressures generated by an integrated electroosmosis flow (EOF) pump without leak of electric fields and disturbance of electric potentials.

2. Experimental

Figure 1 shows a photograph of the cell sorter chip fabricated with a quartz plate employing a process of electron-beam lithography, dry etching, and HF-bonding techniques [2]. Lymphocytes were extracted from musenteriums of a rat of five weeks. To avoid sticking of lymphocyte cell on an inner wall of a capillary, the wall surfaces were coated by a biocompatible MPC (2-methacryloyloxyethy-phosphorylcholine)

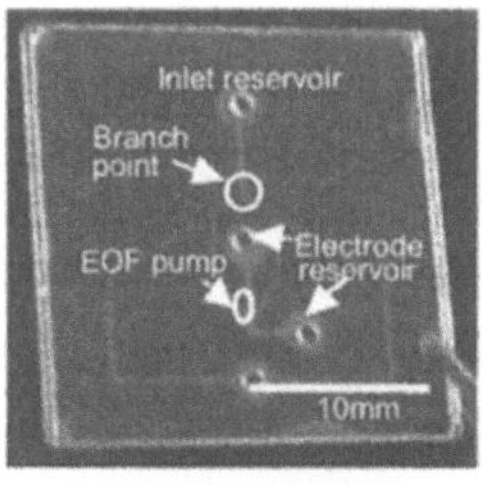

Fig.1 Cell sorter chip made by quartz plate.

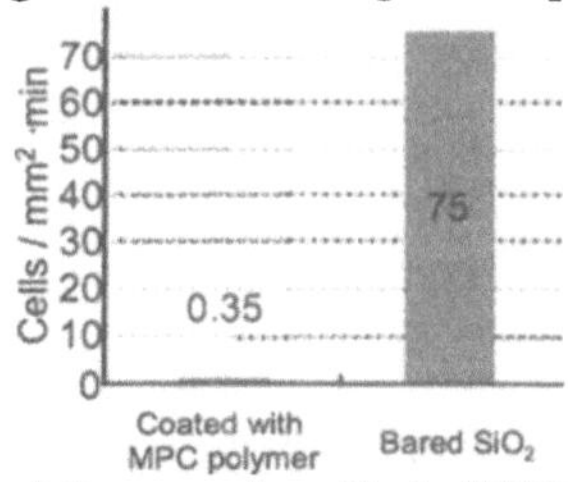

Fig.2 Suppression effect of MPC polymer coating on quartz surface for sticking of cells

Y. Baba et al. (eds.), Micro Total Analysis Systems 2002, Volume 2, 955–958.

polymer [3]. Indeed, MPC polymer coating was so effective that the number of adsorption cells in the coated capillary was far less (<1/200) than that one in a bared quartz surface as shown in Fig. 2. Figure 3 shows a detail structure of the EOF pump. Channels of a shallow gap with a 120nm depth and a deep gap with a 25um depth are parallel and sequentially connected through a flow pass channel. Edges of both channels are connected with reservoir electrodes supplying a ground potential and voltages through gel electrodes patterned by a photo-polymerizing technique [4]. Our previous report demonstrated that a high pressure pumping action was dominated by the shallow gap channel, controlling the flow beyond the pressure generated by the deep gap channel [4]. Therefore, positive and negative voltages added an electrode enables us to change the flow direction. Noteworthy, the same ground potentials at the inlet/outlet in this configuration does not affect any leak of the voltage and overflow of low/high pH solution [5], even if 200 volts resulting in high ξ potential due to coating of the MPC polymer on the wall surface was added to electrodes through the gel electrodes.

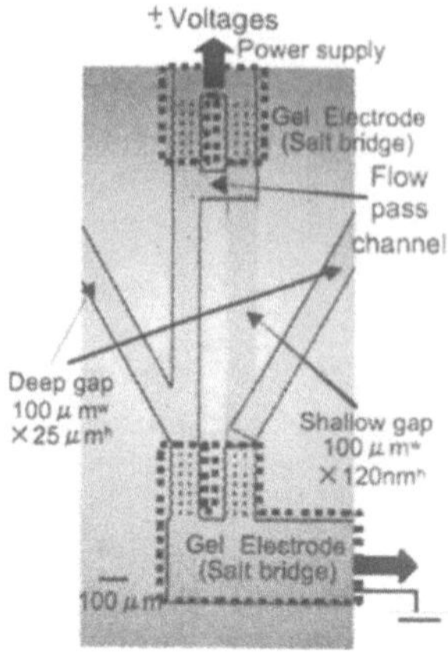

Fig.3 High pressure EOF pump.

3. Results and discussion

A principle of the cells sorting is shown in Fig. 4. It consists of an inlet capillary (i), a left blanch (ii), a right blanch (iii), a left outlet (iv), a right outlet (v) and an EOF pump. At the initial state of Fig. 4(a), the EOF pump is off, and steady flows are made from the inlet to the outlets by a syringe pump, where the flow ratio of (i), (ii), (iii) is 2:1:1. When a cell arrives at the blanch point, the EOF pump is turned on at the flow rate 1, thus sending it to (ii) as shown in Fig. 4(b). At this moment, the flow ratio of (i), (ii), (iii) is 2:2:0. After the cell is sent to (ii), the pump is turned on to a reverse direction for the same duration in order to prevent pH changes [5] at the pump electrodes (Fig. 4(c)), where the flow ratio of (i), (ii), (iii) is 2:0:2. In each case, the flow ratio of (i), (iv), (v) is always kept at 2:1:1.

Figure 5 (a, b, c and d) is a series of movie shots during actual cell sorting. Fig. 5(a) and (b) correspond to photographs before and after sorting a cell at the Fig. 4(b) state, respectively. A cell in the inlet capillary is sent to (ii), while a cell in (iii) does not move and the cell in the right capillary moves. Then the sorter returns to a steady state (Fig. 4(a)) and cells move with flow ratio of 2:1:1 in (i), (ii), (iii). Figure 6 shows the flow rate in (i), (ii), (iii) corresponding to each state. The pulsating flow in (i) was limited to 5% of the total flow.

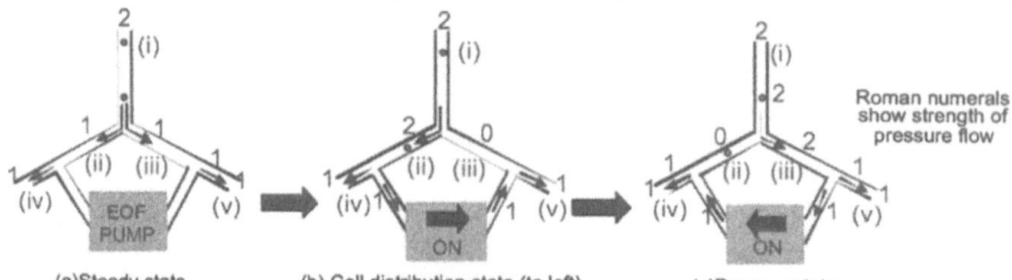

Fig.4 Cell sorting principle.

The time from a pumping signal to a steady cell motion changed quickly less than 50 ms. This cell sorter was operated stably at least more than 30 min.

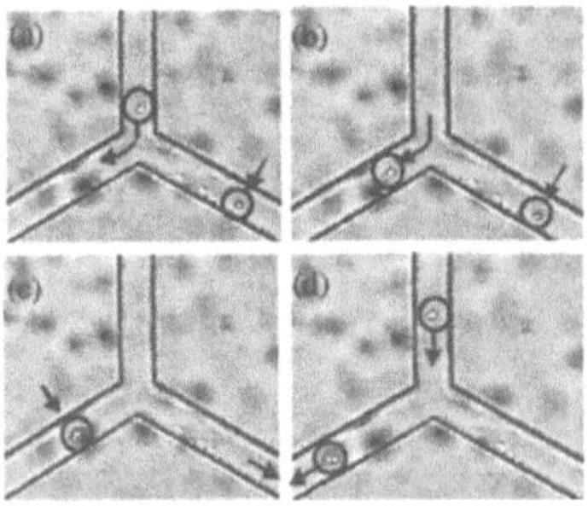

Fig.5 Movie shots during lymphocytes separation. (a) and (b) show the distribution state, (c) shows the recovery state,and (d) shows the steady state.

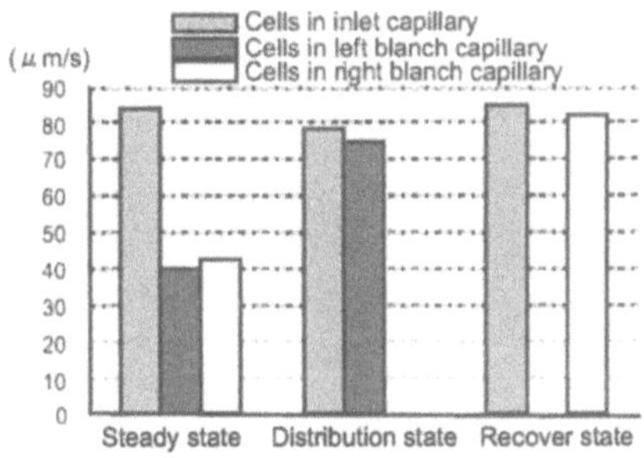

Fig.6 Velocity of lymphocytes in each capillary at each state under constant velocity of inlet capillary.

The cell sorter was applied to actual separation of T cells for fluorescence labeled B cells. Figure 7 shows a photograph during sorting. Since it was difficult to distinguish luminous and non-luminous cells, a detection level was set as shown in Fig. 8. As a result, living T cells without labeling were obtained successfully, by eliminating luminous B cells.

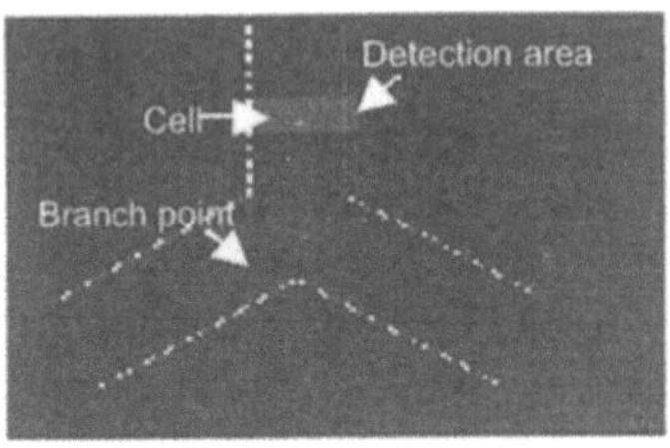

Fig.7 Photograph of Lymphocytes separation.

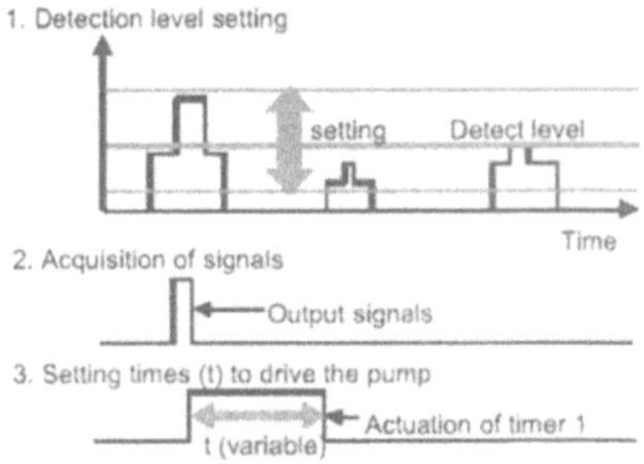

Fig.8 Detection pulse sequence.

4. Conclusions

We have developed the cell sorter which is free from the electric field and a pulsating flow based on the hydraulic pressure force by the special arrangement of the EOF pump. The cell-sorter chip is expected to connectable with following micro-fluidic devices without electric potential differences at both inlets and outlets, pulsating flow outwards the devices and damages due to the electric field.

References

1. A. Oki, et al., Micro Total Analysis Systems 2001, Kluwer Academic Publisher, pp 505-506, (2001).
2. A. Oki, et al., Electrophoresis, pp 341-347, Vol. 22, (2001).
3. K. Ishihara, et al., J.Biomedical Material Research, pp1543, Vol. 26, (1992).
4. Y. Takamura, et al., Micro Total Analysis Systems 2001, Kluwer Academic Publisher, pp 230-232, (2001).
5. A. Oki, et al., Electrophoresis, to be printed in August issue, 23, (2002).

AUTHORS INDEX

L

SUBJECT INDEX

N

Z

MIX
Papier aus verantwortungsvollen Quellen
Paper from responsible sources
FSC® C105338

If you have any concerns about our products,
you can contact us on
ProductSafety@springernature.com

In case Publisher is established outside the EU,
the EU authorized representative is:
Springer Nature Customer Service Center GmbH
Europaplatz 3, 69115 Heidelberg, Germany

Printed by Libri Plureos GmbH
in Hamburg, Germany